IV International Conference on Nuclear Structure and Dynamics (NSD2019)

EPJ Web of Conferences Volume 223 (2019)

Venice, Italy
13 – 17 May 2019

Editors:

Giacomo de Angelis
Lorenzo Corradi

ISBN: 978-1-5108-9932-2

Printed from e-media with permission by:

Curran Associates, Inc.
57 Morehouse Lane
Red Hook, NY 12571

Some format issues inherent in the e-media version may also appear in this print version.

For additional information, please contact EDP Sciences – Web of Conferences
at the address below.

EDP Sciences – Web of Conferences
17, Avenue du Hoggar
Parc d'Activité de Courtabœuf
BP 112
F-91944 Les Ulis Cedex A
France

Phone: +33 (0) 1 69 18 75 75
Fax: +33 (0) 1 69 28 84 91

contact-edps@webofconferences.org

Additional copies of this publication are available from:

Curran Associates, Inc.
57 Morehouse Lane
Red Hook, NY 12571 USA
Phone: 845-758-0400
Fax: 845-758-2633
Email: curran@proceedings.com
Web: www.proceedings.com

TABLE OF CONTENTS

Preface: Nuclear Structure and Dynamics IV - NSD2019

Giacomo de Angelis[1,*] and *Lorenzo* Corradi[1,*]

[1]INFN - Laboratori Nazionali di Legnaro, Italy

The fourth International Conference on Nuclear Structure and Dynamics NSD2019 was held on May 13-17, 2019 at "Centro Culturale Don Orione Artigianelli" in Venice within the exciting context of the 58th International Art Exhibition. The Center is a large complex, built as a monastery and later used as an orphanage and vocational school. The entire building develops around three cloisters of the 18th century and one cloister of the 15th century near the Church of Santa Maria della Visitazione. It provided a beautiful and peaceful atmosphere for stimulating discussions. The conference, organized by the INFN Laboratori Nazionali di Legnaro and the Physics and Astronomy Department of the Padova University, was a follow-up of the three previous conferences held first in 2009 in Dubrovnik (Croatia), continued in 2012 in Opatija (Croatia) and last in 2015 in Portoroz (Slovenia), and belongs to a series of conferences devoted to the most recent experimental and theoretical advances in the field of nuclear structure and reactions. NSD2019 maintained this tradition, promoting exchange of ideas and collaboration among researchers, post-docs and graduate students with experimental, theoretical and phenomenological background. According to the tradition of this series of conferences, the focus was on topics covering most of the research areas in low energy nuclear physics:

- Nuclear structure and reactions far from stability
- Collective phenomena and symmetries
- Dynamics and thermodynamics of light and heavy nuclei
- Sub and near barrier reactions
- Fusion and Fission dynamics
- Ab initio calculations, cluster models and shell model
- Nuclear energy density functionals
- Nuclear astrophysics
- Fundamental interactions.

About 150 physicists from all over the world attended the conference. The opening ceremony took place in the morning of May 13th. After a short introduction of the conference chairs (Giacomo de Angelis and Lorenzo Corradi), the participants were welcomed by the director of the Legnaro National Laboratory (Diego Bettoni). 30 invited speakers and 60 regular talks, some of which scheduled in parallel sessions, covered a wide range of topics. More than 30 posters were presented at the poster session. The three best voted ones were discussed as short oral presentations. We paid tribute to our colleague Graziano Fortuna, former director of the Legnaro National Laboratory, who very recently passed away.

Important new results were outlined in the various subfields, spanning the whole range from neutron-deficient to neutron-rich nuclei and from light to heavy ions. The talks, covering state-of-art experimental methods and theoretical developments, had as main reference the tight interconnection between structure and dynamics, enhancing attendees to an active participation and prompting lively discussions.

We sincerely appreciated the cooperative contribution of all members of the Organizing Committee: Enrico Fioretto, Tommaso Marchi, Marco Mazzocco, Daniele Mengoni, José Javier Valiente Dobón. Special thanks to our secretaries Anna D'Este and Adriana Schiavon for their skill and effort in following all aspects of the conference and making the event so successful.

We wish to warmly thank the members of the International Advisory Committee for helping us in the selection of speakers and precious suggestions: Rob Bark, Piotr Bednarczyck, Klaus Blaum, Francesco Cappuzzello, Michael Patten Carpenter, Giovanni Casini, Wilton Catford, Bo Cederwall, José Luis Egido, Angela Gargano, Paul Greenless, Alessandra Guglielmetti, Joseph Hamilton, Denis Lacroix, Chengjian Lin, Nicolae Marginean, Matko Milin, Petr Navratil, Gerda Neyens, Katsuhisa Nishio, Robert Page, Norbert Pietralla, Riccardo Raabe, Berta Rubio, Hendrik Shatz, Sergey Sidorchuck, Cedric Simenel, Ram Prakash Singh, Olivier Sorlin, Siem Sunniva, Tomohiro Uesaka, Andrea Vitturi, Dario Vretenar, Yu Hu Zhang.

We are finally very thankful to Dario Vretenar for having accepted the challenging task to held the Summary talk. His presentation underlined some of the major achievements of modern nuclear physics research presented during the conference.

* Corresponding author: deangelis@lnl.infn.it, corradi@lnl.infn.it

edp sciences

EPJ Web of Conferences

Statement of Peer review

In submitting conference proceedings to EPJ Web of Conferences, I certify to the Publisher that I adhere to the **Policy on Publishing Integrity** of the journal in order to safeguard good scientific practice in publishing.

1. All articles have been subjected to peer review administered by the proceedings editors.
2. Reviews have been conducted by expert referees, who have been requested to provide unbiased and constructive comments aimed, whenever possible, at improving the work.
3. Proceedings editors have taken all reasonable steps to ensure the quality of the materials they publish and their decision to accept or reject a paper for publication has been based only on the merits of the work and the relevance to the journal.

IV International Conference on Nuclear Structure and Dynamics - NSD2019, 13-17 May 2019, Venice

Giacomo de Angelis, Lorenzo Corradi

Date and editor's signature

29 Oct. 2019

EPJ Web of Conferences **223**, 01001 (2019) https://doi.org/10.1051/epjconf/201922301001
NSD2019

Challenging nuclear structure of the heaviest – opportunities at S^3

Dieter Ackermann[1,*] on behalf of the S^3 collaboration

[1]Grand Accélérateur National d'Ions Lourds – GANIL, CEA/DRF-CNRS/IN2P3, Bd. Becquerel, BP 55027, F-14076 Caen, France

Abstract. After more than half a century of research addressing the synthesis and nuclear structure of superheavy nuclei (SHN) a boost for its progress is expected from the advent of new instrumentation. An order of magnitude in beam intensity increase is envisaged to be provided by new powerful accelerators like the new DC280 cyclotron at the SHE factory of FLNR/JINR or the superconducting linac at SPIRAL2 of GANIL. In addition new ion-optical installations like the separator-spectrometer set-up S^3 with two complementary detection systems SIRIUS and LEB will provide a substantial sensitivity increase for classically pursued routes like decay spectroscopy after separation (DSAS), and alternative and complementary methods like high precision mass measurements and laser spectroscopy. Decay spectroscopy has proven in the past to be a powerful tool to study the low lying nuclear structure of heavy and superheavy nuclei. Single particle levels and other structure features like *K* isomerism, being important in the fermium-nobelium region as well as for tracing deformation towards spherical shell stabilised SHN, have been investigated almost up to the limit posed by the sensitivity of present-day instrumentation. Precision mass measurements and laser spectroscopy will offer the possibility to study alternative features like atomic and nuclear binding energies, nuclear charge radii and quadrupole moments.

1 Achievements in SHN research and challenges

Nuclei beyond fermium-rutherfordium owe their existence solely to quantum mechanics. In that region of the chart of nuclei the fission barrier derived from a macroscopic liquid drop approach would vanish [1], which gives rise to the nuclear physicists definition of a superheavy nucleus (SHN). These quantum mechanics features, often referred to as shell effects and taken into account as shell corrections in so called microscopic-macroscopic models (see e.g. [2]), lead to a modification of the nuclear potential which is the basis for the extension of the chart of Segré to high Z and A. This makes these nuclei, in turn, an ideal laboratory to study the strong nuclear interaction by in-beam methods as well as decay spectroscopy after separation (DSAS) [3]. The nuclear structure investigation discussion in this paper will be restricted to DSAS of these deformed nuclei in the region Z=100-112 and N=152-162. These studies have the potential to provide links to the next heavier spherical closed shell nuclei, by investigating single particle levels [4] close to the fermi energy which according to some models play a major role in defining the spherical shell gaps in the region of the so-called *island of stability*. For in-beam spectroscopy which gives access to nuclear structure at higher spins like e.g. rotational bands, we refer to the recent review by Theisen et al. and references therein [5].

Particularly interesting features like K isomers, can be used to trace the spherical superheavy nuclei (SHN) and to locate the island of stability [7]. A number of cases have been found in the region of fermium isotopes and beyond mainly for isotopes of all even-Z elements up to darmstadtium (Z=110) apart from seaborgium (Z=106), but also for some even-odd and odd-odd nuclei. High intensity accelerators, targets withstanding those high beam intensities and, in particular, actinide targets [8], efficient in-flight separators and spectrometers, and highly efficient detectors with fast electronics are the essential ingredients for the success of the field. The new SPIRAL2 facility with the separator-spectrometer set-up S3 [6] presently under construction at the accelerator laboratory GANIL in Caen, France, will offer great perspectives for the field [7]. At this facility a number of ground breaking experiments are envisaged, including DSAS examining features like the single particle structure, deformation and the search for K-isomers for those exotic nuclear species.

The experimental tools employed in the field, composed of highly efficient and selective separators and spectrometers combined with highly sensitive detection arrays for particles and photons, are recently complemented by alternative approaches with precision mass measurements, using devices like Penning traps [9] and Multi-Reflection Time of Flight Mass Spectrometers (MR ToF MS), and laser spectroscopy installations [10], which give access to basic properties of those heavy nuclei like binding energies, charge radii and quadrupole moments. The existing facilities like e.g. SHIP with its Penning trap set-up SHIPTRAP and the gas-filled separator TASCA at GSI in Darmstadt, Germany, RITU and the MARA set-up at the cyclotron laboratory of the University of Jyväskylä,

*e-mail: Dieter.Ackermann@ganil.fr

https://doi.org/10.1051/epjconf/201922301001

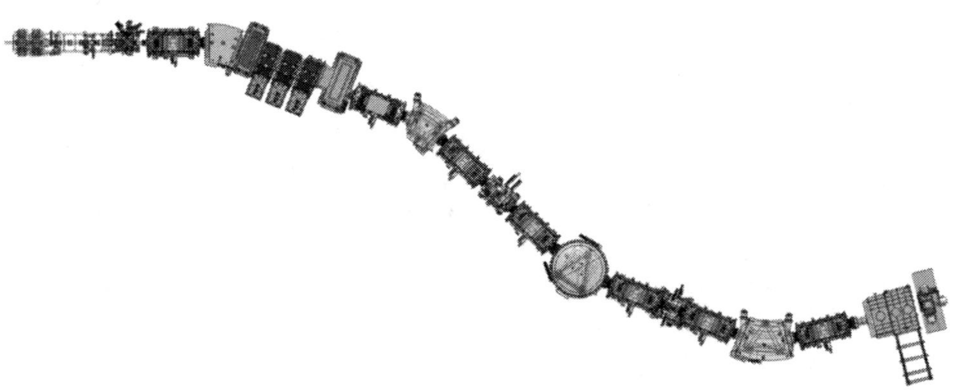

Figure 1. Technical drawing of the arrangements of the ion optical components of S^3, consisting of three magnetic dipoles, one electric dipole, a room-temperature quadrupole triplet and seven super-conducting multiplets (SMT), in a QQQ-MD-QQQ-QQQ-MD-QQQ-QQQ-ED-QQQ-QQQ-MD-QQQ configuration [6].

Finland, SHELS (former VASSILISSA) at FLNR/JINR in Dubna, Russia or BGS at LBNL, will soon be joined by the Superconducting Separator-Spectrometer set-up S^3 of SPIRAL2 at GANIL in Caen, France. This new facility will profit from high beam intensities similar to the ones expected from the new DC280 cyclotron of the SHE-factory, in operation since recently at FLNR/JINR, and the highly sensitive decay spectroscopy set-up SIRIUS, and low energy branch (LEB) instrumentation.

In a special issue of Nuclear Physics A in 2015, volume 944, all major topics concerning the field of superheavy element (SHE) research have recently been reviewed. Asai and collaborators [4] summarised the recent achievements of DSAS for isotonic chains in the vicinity of the closed neutron subshell N=152 in a region ranging from fermium to meitnerium. A general problem to investigate single particle states is the admixture of other components to the initial and final states for which transitions are observed. In regions of low level density, however, the single-particle character is enhanced. Moreover, a spin and parity assignment is supported by the selectivity provided by transition selection rules in stepwise construction of level schemes, built on states for which quantum numbers are known. A more realistic picture can be achieved by employing models which take into account correlations beyond the single particle picture by coupling to the relevant configurations which involve e.g. vibrational states. In particular, the N=151 and 153 isotones with one hole in and one particle more than the sub-shell closure, respectively, are suited to trace single particle levels which are important for the predicted spherical shell gaps.

Deformation, as present in the region of deformed shell stabilised nuclei around Z = 100 to 108 and N = 152 to 162, gives rise to a class of metastable states at high values of K. The K quantum number is defined by the projection of the total spin, i.e. the sum of the nuclear spin and the orbital angular momenta of quasiparticle states excited in the nucleus, on the symmetry axis of a deformed nucleus. In cases of high K numbers, isomeric states can be caused by a large difference in angular momentum and possibly opposite parity of the isomeric state and the next available state into which the isomer can decay. Such isomeric states have been predicted for the whole region of deformed heavy and superheavy nuclei [11].

The heaviest nucleus for which such an isomeric state has been observed is ^{270}Ds [12]. Recently such a state has been reported also for its α decay daughter ^{266}Hs [13]. For both nuclei the isomeric state is longer lived than the g.s., a feature which is rarely occurring in atomic nuclei. As the necessary precondition for the formation of K isomers is nuclear deformation, their development towards increasing Z can be used to trace deformation towards the predicted spherical shell stabilised SHN for which it should vanish. Together with the nuclear structure information gained from the respective implications of the involved quasiparticle states which form those metastable states, this can be used to develop and adjust theoretical model predictions to eventually localise the long sought for *island of stability* of SHN. Table 2 in ref. [7] lists as an update of the table from [14] of known K-isomers in even-even nuclei in heavy and SHN in the region from curium to darmstadtium. An outlook into the expectations for higher Z and the next deformed neutron shell gap is given by V. Prassa et al. [15]. Employing Energy Density Functional (EDF) calculations, they predict a particular pattern for high K states in N=162 isotopes with respect to their neighbours for isotopic chains from rutherfordium to darmstadtium. For those the low lying neutron single-particle states predicted at excitation energies below 1 MeV for the neighbours are elevated to higher energies. The resulting low energy gap is a feature which should observable experimentally.

2 Opportunities at S^3

As the heart of the new SIRAL2 facility at GANIL in Caen, France, a high intensity super conducting linear accelerator (SC LINAC) is presently under construction with specifications which respond to the need of highest intensities [16]. The planned intensities for the two envisaged construction phases are listed in ref. [3, 16]. The superconducting separator spectrometer (S^3) set-up [6] combined with the decay spectroscopy detection array SIRIUS (Spectroscopy and Identification of Rare Isotopes Using S^3) in combination with the high intensity beams from SC LINAC will be one of the most powerful facilities for SHN/SHE research worldwide together with the SHE Factory presently under construction at JINR/FLNR.

Figure 2. Technical drawing of the mechanical arrangement for the DSAS detection array SIRIUS, consisting of a Double-sided Si Strip Detector (DSSD), surrounded by silicon detectors with an adopted pad structure for the detection of α particles escaping from the the implantation detector, fission fragments and conversion electrons (CE), surrounded by large-volume germanium detector for γ ray detection

The community of scientists interested in exploiting the capabilities of the SC LINAC-S^3 facility for SHN/SHE research has over the years developed their envisaged research topics in various letters of intent. The possible features for the various heavy and superheavy nuclei range from evaporation residue (ER) cross section measurements to detailed spectroscopy topics like *K* isomerism, α fine structure, or γ, conversion electron (CE) and X-ray spectroscopy. The latter has the potential of settling the still open question of Z identification for the heaviest nuclei produced in ^{48}Ca induced reactions for the first time at the FLNR (see e.g. reference [17, 18]), due to the precise prediction capabilities for atomic transitions by theory. A summary of the collected ideas together with a description of the facility and the state of the art of the development at that time can be found in [19].

The detection instrumentation of two complementary set-ups in the focal plane of S^3 consists of the particle and photon detection array for DSAS, SIRIUS and the S^3 Low Energy Branch (LEB) for precision mass measurements and laser spectroscopy. SIRIUS is composed of a combination of a box-like arrangement of silicon detectors in a thin aluminium housing which is transparent for γ radiation and can be surrounded by an array of large volume Ge detectors. Tracker detectors in beam direction before the Si array serve for veto and time-of-flight purposes. Fig. 2 shows an overview of the set-up. Apart from allowing for the investigation of *K*-isomers, isotopic and isotonic trends of low lying nuclear excitations by exploiting γ-electron-α/fission and X-ray coincidences, SIRIUS is also an ideal tool to study delayed processes like isomeric states and β-delayed fission. The S^3 LEB consists of a combination of a gas-stopping cell, various ion guides, a laser spectroscopy installation and a Multi-Reflection Time-o-Flight Mass Spectrometer (MRTof MS), which can be combined with additional particle and photon detectors. It will be used to study nuclei in the N=Z region as well as the heaviest nuclear species.The major components of the LEB mounted as an almost complete set-up in a test laboratory at LPC Caen, is shown in Fig. 3.

In a farer future the synthesis and investigation of, also so far unknown, highest-*Z* systems is envisaged, for which the earlier experiments will establish the fundament. A necessary prerequisite for his type of research is a new injector for the SC LINAC with the capability to accelerate ions with an A/Q ratio of 7 which will allow for the treatment of ions up to uranium. The present A/Q = 3 injector causes efficiency losses for ion masses beyond 40 to 50.

In a recent meeting of the S^3 community in June 2018 the day one experimental campaign at S^3 has been discussed in further detail. In preparation of this meeting a call for so-called pre-proposals, providing a detailed scientific and technical discussion of realistically feasible experiments, was launched. It resulted in the submission of eleven projects for both focal plane instrumentation arrangements. For an appealing S^3 "day 1" program the five most promising submissions, concerning scientific excellence, feasibility and uniqueness, were chosen. The resulting projected campaigns comprise for the LEB, apart from mass measurements and laser spectroscopy in lighter nuclei in the A = 80 to 100 region, high-resolution in-gas laser spectroscopy and deformation studies for various actinides, and the investigation of fundamental properties of nuclei in the vicinity of the deformed shell gaps at Z = 100 and N = 152. For SIRIUS the exploration of the limits of stability in the vicinity of the proton drip line in the fermium-nobelium region, the search for new heavy isotopes, the more detailed investigation of known and the search for new isomeric states, and the investigation of odd-Z nuclei are the building blocks of this envisaged first DSAS campaign at S^3.

The laboratories primarily involved in this work are: Grand Accélérateur National d'Ions Lourds (GANIL), Institut de Recherche sur les Lois Fondamentales de l'Univers (CEA, IRFU), Institut de Physique Nucléaire d'Orsay (IPNO), Institut Pluridisciplinaire Hubert Curien (IPHC), Centre de Sciences Nucléaires et de Sciences de la Matière (CSNSM), Argonne National Laboratory (ANL), KU Leuven, Laboratoire de Physique Corpusculaire de Caen (LPC), and the Institut des NanoSciences de Paris (INSP).

S3 has been funded by the French Research Ministry, National Research Agency (ANR), through the S3 EQUIPEX (EQUIPment of EXcellence) reference ANR-

Figure 3. Photograph of the S^3 LEB set-up as presently mounted in a test laboratory at LPC Caen. The MRToF MS set-up in the left foreground is completed by the laser spectroscopy arrangement in the background of the picture.

10EQPX-46, the FEDER (Fonds Européen de Développement Economique et Régional), the CPER (Contrat Plan Etat-Région), and partially supported by the U.S. Department of Energy, Office of Nuclear Physics, under contract No. DE-AC02-06CH11357, the KU Leuven by the European Commission FP7-INFRASTRUCTURES 2007, SPIRAL2 Preparatory Phase Grant agreement No. 212692. SIRIUS has been funded by the CPIER (Contrat de Plan Interrégional Etat-Région).

References

[1] A. Baran, M. Kowal, P.G. Reinhard, L. Robledo, A. Staszczak, M. Warda, Nuclear Physics A **944**, 442 (2015), special Issue on Superheavy Elements

[2] A. Sobiczewski, K. Pomorski, Prog. Part. Nucl. Phys. **58**, 292 (2007)

[3] D. Ackermann, C. Theisen, Phys. Scr. **92**, 083002 (2017)

[4] M. Asai, F.P. Heßberger, A. Lopez-Martens, Nucl. Phys. A **944**, 308 (2015), special Issue on Superheavy Elements

[5] C. Theisen, P. Greenlees, T.L. Khoo, P. Chowdhury, T. Ishii, Nucl. Phys. A **944**, 333 (2015), special Issue on Superheavy Elements

[6] F. Déchery, A. Drouart, H. Savajols, J. Nolen, M. Authier, A.M. Amthor, D. Boutin, O. Delferriére, B. Gall, A. Hue et al., Eur. Phys. J A **51**, 1 (2015)

[7] D. Ackermann, Nucl. Phys. A **944**, 376 (2015)

[8] C. Stodel, J. Libin, C. Marry, F. Lutton, M. Saint-Laurent, B. Bastin, J. Piot, E. Clement, S. Moal, V. Morel et al., J. Radioanal. Nucl. Ch. **305**, 761 (2015)

[9] M. Block, D. Ackermann, K. Blaum, C. Droese, M. Dworschak, S. Eliseev, T. Fleckenstein, E. Haettner, F. Herfurth, F.P. Heßberger et al., Nature **463**, 785 (2010)

[10] M. Laatiaoui, W. Lauth, H. Backe, M. Block, D. Ackermann, B. Cheal, P. Chhetri, C.E. Düllmann, P. van Duppen, J. Even et al., Nature **538**, 495 EP (2016)

[11] F. Xu, E. Zhao, P.W. R. Wyss, Phys. Rev. Lett. **92**, 252501 (2004)

[12] S. Hofmann, F.P. Heßberger, D. Ackermann, S. Antalic, P. Cagarda, S. Ćwiok, B. Kindler, J. Kojouharova, B. Lommel, R. Mann et al., Eur. Phys. J. A **10**, 5 (2001)

[13] D. Ackermann, F.P. Heßberger, S. Antalic, M. Block, H.G. Burkhard, V.F. Comas, P.T. Greenlees, S. Heinz, S. Hofmann, S. Ketelhut et al., GSI Sci. Rep. 2011 p. 208 (2012)

[14] R.D. Herzberg, P. Greenlees, Prog. Part. Nucl. Phys. **61**, 674 (2008)

[15] V. Prassa, B.N. Lu, T. Nikšić, D. Ackermann, D. Vretenar, Phys. Rev. C **91**, 034324 (2015)

[16] E. Petit, *STATUS REPORT ON THE SPIRAL2 FACILITY AT GANIL*, in *Proceedings of NA-PAC2016, Chicago, IL, USA* (2016), TUA1IO02

[17] Y. Oganessian, V. Utyonkov, Nucl. Phys. A **944**, 62 (2015)

[18] Y.T. Oganessian, A. Sobiczewski, G.M. Ter-Akopian, Phys. Scr. **92**, 023003 (2017)

[19] C. Theisen, EPJ Web Conf. **163**, 00059 (2017)

EPJ Web of Conferences **223**, 01002 (2019)
NSD2019
https://doi.org/10.1051/epjconf/201922301002

Fission modes in fermium isotopes with Brownian shape-motion model

Martin Albertsson[1,*], *B. Gillis* Carlsson[1], *Thomas* Døssing[2], *Peter* Möller[1,3], *Jørgen* Randrup[4], and *Sven* Åberg[1]

[1]Mathematical Physics, Lund University, 221 00 Lund, Sweden
[2]Niels Bohr Institute, DK-2100 Copenhagen Ø, Denmark
[3]P. Moller Scientific Computing and Graphics, Inc. P.O. Box 1440, Los Alamos, New Mexico 87544, USA
[4]Nuclear Science Division, Lawrence Berkeley National Laboratory, Berkeley, California 94720, USA

Abstract. Fission-fragment mass and total-kinetic-energy distributions following fission of the fermium isotopes ^{256,258,260}Fm at low excitation energy have been calculated using the Brownian shape-motion model. A transition from asymmetric fission in ^{256}Fm to symmetric fission in ^{258}Fm is obtained. The total-kinetic-energy distributions for the three isotopes show radically different behaviour due to varying contributions from different fission modes, with a double-humped distribution for ^{258}Fm.

1 Introduction

The co-existence of two modes in fission leading to very different values of the total kinetic energy (TKE) of the fragments, known as bimodal fission, has been observed in the fermium region [1]. This phenomenon is argued to arise due to two fission paths in the potential-energy landscape; one path is described by the liquid-drop model leading to elongated scission configurations with low TKE, while the other path is influenced by both fragments being close to the doubly magic nucleus ^{132}Sn leading to compact scission with high TKE.

In our endeavour to understand fission modes of nuclei in the superheavy region [2], we first consider fission in neutron-rich fermium isotopes, which is often used as a benchmark for theoretical studies. We calculate fission-fragment mass and TKE distributions for the fermium isotopes ^{256,258,260}Fm using the Brownian shape-motion (BSM) model [3]. The calculational details are explained in Sect. 2. In Sect. 3 the calculated fission-fragment mass and TKE distributions are presented with a short discussion about the two fission modes.

2 Calculational details

Fission-fragment mass distributions are calculated using the BSM method on the five-dimensional potential-energy surfaces as specified in Refs. [3, 4]. The potential energy of the fissioning nucleus is calculated within the macroscopic-microscopic model [5] in a grid of more than 6 million shapes, where the macroscopic part is calculated within the finite-range liquid-drop model and the microscopic part is calculated with the folded-Yukawa single-particle model. In the Strutinsky shell-correction procedure we use a larger smoothing range [2] compared to Refs. [3, 4], since this is particularly important for nuclei in the vicinity of fermium [6, 7]. Since most of the CPU time is used to calculate the single-particle levels, we use the same set of levels of ^{258}Fm to calculate the shell corrections also for ^{256}Fm and ^{260}Fm.

The starting point for the dynamical simulations is in the isomer minimum at low compound excitation energies, just above the outer barrier. This corresponds to excitation energies $E^* = 3.88$, 2.77 and 1.71 MeV for $^{256-260}$Fm, respectively. The shape evolution of the fissioning nucleus is performed using the effective level density of Ref. [4] with a bias potential of 60 MeV in order to decrease the computational time[1]. The scission point is defined as when the neck radius reaches the critical value $c_0 = 1.5$ fm. If the critical neck radius is not reached within 2×10^6 steps or if the walk reaches the maximum value of the elongation coordinate, then the walk is discarded. A total of 10000 walks are performed for each nucleus, with less than 1% of the walks being discarded.

The calculation of the TKE is done as specified in Ref. [2]. The total available energy in the fission process, given by the sum of the compound nucleus excitation energy E^* and the Q-value, is shared between the TKE and the total excitation energy (TXE) of the two fragments. The TXE is composed of two parts; intrinsic excitation energy in the compound nuclues at scission and distortion energy released in the fragments when they relax their respective scission shapes (taken as the cap deformations in the 5D shape parametrization) to their ground-state forms.

The shape-dependent masses entering in the calculation of the Q-values, as well as distortion energies are calculated in the same macroscopic-microscopic model that was used to obtain the potential-energy surfaces [8]. We only consider $\varepsilon_2, \varepsilon_4, \varepsilon_6$ for the ground-state deformations.

*e-mail: martin.albertsson@matfys.lth.se

[1]Using a bias potential of 30 MeV slightly increases the full width at half maximum in the mass yield for ^{258}Fm by about 8 nucleons while the computational time increases by roughly a factor of 20.

The number of protons Z and neutrons N in a fragment are determined by requiring the same Z/N ratio as for the fissioning nucleus. In the present study only fragments with even Z and N are considered.

3 Results and discussion

The calculated fission-fragment mass distributions for ^{256,258,260}Fm are shown in Fig. 1. A transition from asymmetric fission in ^{256}Fm to symmetric fission in ^{258}Fm is obtained. This transition is in agreement with experimental data for spontaneous fission [1, 9], and has been ascribed to the strong shell effects emerging from fragments approaching the doubly magic nucleus ^{132}Sn.

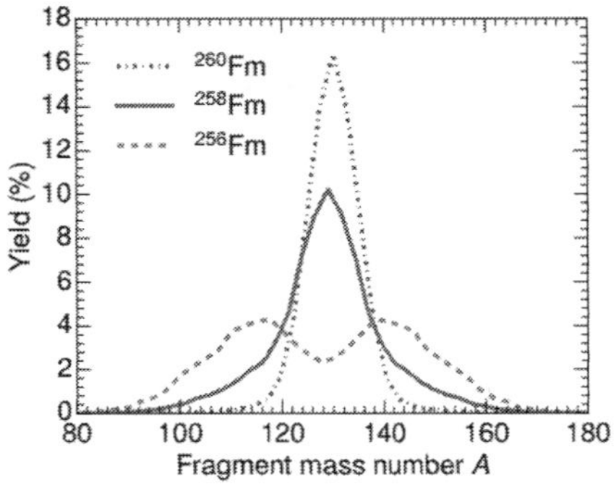

Figure 1. Calculated fission-fragment mass yields for the fermium isotopes ^{256,258,260}Fm.

Figure 2 shows the typical scission shapes obtained in the calculations for the three isotopes. The most common scission configuration for ^{256}Fm corresponds to an asymmetric mass division with a prolate light fragment and a slightly oblate heavy fragment. This mode is identified as the standard fission mode introduced by Brosa [10]. Fission of ^{258}Fm mainly leads to symmetric mass divisions with a very compact scission configuration with two spherical fragments. This mode is identified as the super-short fission mode. However, the standard fission mode seen in ^{256}Fm is also present in ^{258}Fm resulting in bimodal fission and relatively broad mass distribution in this nucleus. In ^{260}Fm it is mainly the super-short fission mode that is appearing and the mass yield is therefore very narrow.

The calculated TKE distributions are shown in Fig. 3. Since the TKE is largely determined by the strong Coulomb repulsion between the fragments, the TKE value mainly depends on the elongation of the scission configuration. The distribution for ^{256}Fm is well reproduced by a single Gaussian with an average TKE close to the measured value of 197.9 MeV in spontaneous fission [11]. This TKE value is also close to the value expected by the Viola systematics for liquid-drop fission [12]. The distribution for ^{258}Fm on the other hand contains a mixture of the two modes, leading to a double-humped distribution. Peaks near 200 and 235 MeV in the TKE distribution are also seen in measurements of spontaneous fission of ^{258}Fm [1]. The distribution for ^{260}Fm is peaked around 235 MeV due to the super-short fission mode, though a minor contribution from the standard mode is also visible.

Since the super-short fission mode is related to strong shell-effects, it has the most influence at low energies. The high-TKE component in ^{258}Fm is therefore slightly less pronounced in the calculations, due to the excited nuclei, compared to experimental data for spontaneous fission [1]. When the excitation energy is increased, the shell-effects gradually disappear and the mass yields broaden, approaching what would be obtained based on liquid-drop model potential energies. The calculated TKE for both ^{258}Fm and ^{260}Fm then approaches the TKE value described by the standard asymmetric fission mode.

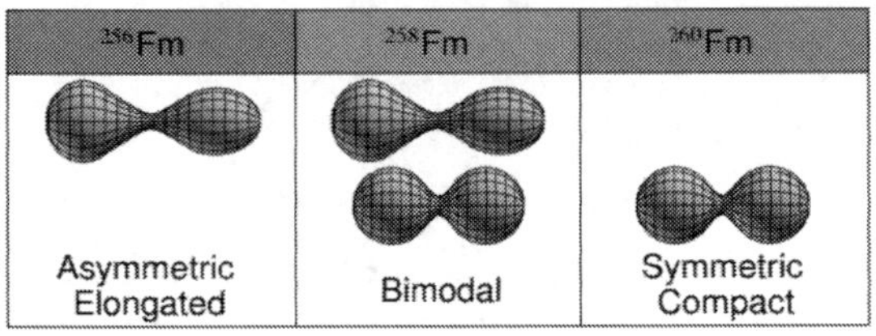

Figure 2. Typical scission shapes for ^{256,258,260}Fm obtained in the calculations.

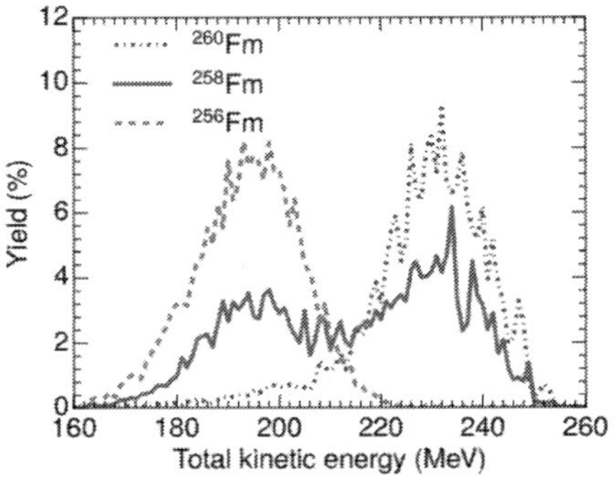

Figure 3. Calculated TKE distributions for ^{256,258,260}Fm.

This work was supported by the Swedish Natural Science Research Council (S.Å.) and the Knut and Alice Wallenberg Foundation Grant No. KAW 2015.0021 (M.A., B.G.C. and S.Å.); J.R. was supported in part by the NNSA DNN R&D of the U.S. Department of Energy.

References

[1] E.K. Hulet *et al.*, *Phys. Rev. Lett.* **56**, 313 (1986)
[2] M. Albertsson *et al.*, manuscript in preparation
[3] J. Randrup *et al.*, *Phys. Rev. Lett.* **106**, 132503 (2011)
[4] J. Randrup, P. Möller, *Phys. Rev. C* **88**, 064606 (2013)
[5] P. Möller *et al.*, *Phys. Rev. C* **79**, 064304 (2009)
[6] P. Möller *et al.*, *Nucl. Phys.* **A492**, 349 (1989)
[7] P. Möller, J.R. Nix, *J. Phys. G* **20**, 1681 (1994)
[8] P. Möller *et al.*, *Phys. Rev. Lett.* **92**, 072501 (2004)
[9] K.F. Flynn *et al.*, *Phys. Rev. C* **5**, 1725 (1972)
[10] U. Brosa *et al.*, *Phys. Rep.* **197**, 167 (1990)
[11] D.C. Hoffman *et al.*, *Phys. Rev. C* **21**, 972 (1980)
[12] V.E. Viola *et al.*, *Phys. Rev. C* **31**, 1550 (1985)

EPJ Web of Conferences **223**, 01003 (2019) https://doi.org/10.1051/epjconf/201922301003
NSD2019

Measurement of proton-distribution radii of neutron-rich nitrogen isotopes

Soumya Bagchi[1,2,3,*], *Rituparna* Kanungo[1,4], *Wataru* Horiuchi[5], *Gaute* Hagen[6,7], *Titus D.* Morris[6,7], *Steven Ragnar* Stroberg[4], *Toshio* Suzuki[8], *Frederic* Ameil[2], *Joel* Atkinson[1], *Yassid* Ayyad[9], *Dolores* Cortina-Gil[9], *Iris* Dillmann[2], *Alfredo* Estradé[1,2], *Alexey* Evdokimov[2], *Fabio* Farinon[2], *Hans* Geissel[2,3], *Giulia* Guastalla[2], *Rudolf* Janik[10], *Satbir* Kaur[1,11], *Ronja* Knöbel[2], *Jan* Kurcewicz[2], *Yury* Litvinov[2], *Michele* Marta[2], *Magdalena* Mostazo[9], *Ivan* Mukha[2], *Chiara* Nociforo[2], *Hooi Jin* Ong[12], *Stephane* Pietri[2], *Andrej* Prochazka[2], *Christoph* Scheidenberger[2,3], *Branislav* Sitar[10], *Peter* Strmen[10], *Maya* Takechi[2], *Junki* Tanaka[12], *Yoshiki* Tanaka[1,2,3], *Isao* Tanihata[12,13], *Satoru* Terashima[13], *Jossitt* Vargas[9], *Helmut* Weick[2], and *John Stuart* Winfield[2]

[1]Astronomy and Physics Department, Saint Mary's University, Halifax, NS B3H 3C3, Canada
[2]GSI Helmholtzzentrum für Schwerionenforschung GmbH, D-64291 Darmstadt, Germany
[3]Justus-Liebig University, 35392 Giessen, Germany
[4]TRIUMF, Vancouver, BC V6T 4A3, Canada
[5]Department of Physics, Hokkaido University, Sapporo 060-0810, Japan
[6]Physics Division, Oak Ridge National Laboratory, Oak Ridge, TN 37831, USA
[7]Department of Physics and Astronomy, University of Tennessee, Knoxville, TN 37996, USA
[8]Department of Physics, Nihon University, Setagaya-ku, Tokyo 156-8550, Japan
[9]Universidad de Santiago de Compostela, E-15706 Santiago de Compostela, Spain
[10]Faculty of Mathematics and Physics, Comenius University, 84215 Bratislava, Slovakia
[11]Department of Physics and Atmospheric Science, Dalhousie University, Halifax, NS B3H 4R2, Canada
[12]RCNP, Osaka University, Mihogaoka, Ibaraki, Osaka 567 0047, Japan
[13]School of Physics and Nuclear Energy Engineering and IRCNPC, Beihang University, Beijing 100191, China

Abstract. Measurement of root-mean-square radii of proton distributions of $^{17-22}$N from charge-changing cross section shows the emergence of thick neutron skin towards the neutron-drip line. Signature of $N = 14$ shell gap has been found in nitrogen isotopes along with the emergence of neutron halo in ^{22}N. The measured radii are in good agreement with the shell model calculations.

1 Introduction

With large neutron-to-proton ratios far from the valley of stability, nuclei develop exotic structures such as neutron skin, nuclear halo, modifications of conventional shells and apperance of new shell gaps. Evidence for breaking down of $N = 8$ shell gap has been found in ^{11}Li and ^{11}Be [1–3] while a new shell gap at the neutron number $N = 16$ has been found in carbon, nitrogen and oxygen isotopes at the drip line [4]. Another sub-shell gap at $N = 14$ has been found in oxygen and is predicted to disappear in carbon [5]. Since nitrogen is in the transitional region between oxygen and carbon, it is interesting to investigate the presence of the $N = 14$ shell gap in nitrogen. Systematic studies of root-mean-square radii of proton-distribution along an isotopic chain can shed light on nuclear models as well as on the shell evolution [6].

2 Experimental setup

In this work, the determination of root-mean-square radii of proton-distribution (protons treated as point-like particles and therefore referred further as point-proton radii) from the measured charge-changing cross sections (σ_{CC}) of neutron-rich isotopes $^{17-22}$N and stable isotopes ^{14,15}N is presented. The experiment was performed at the fragment separator FRS at GSI. Beams of ^{14,15}N and $^{17-22}$N were produced by fragmentation of ^{22}Ne and ^{40}Ar, respectively, at 1 A GeV, impinging on a 6.3 g/cm^2 thick Be target. The isotopes of interest were separated in flight and identified using their magnetic rigidity ($B\rho$), time-of-flight (ToF) between the focal planes (F2-F4) of the FRS, and the energy-loss (ΔE) measured in a multi-sampling ionization chamber (MUSIC) placed at the final focal plane F4. The σ_{CC} was measured with a 4.010 g/cm^2 thick carbon target placed at F4. The experimental setup is discussed in more details in Ref. [7].

To obtain σ_{CC}, the number (N_{in}) of incident nuclei of interest ^{A}Z is determined from event-by-event counting. After the reaction target, the number of nuclei with the same charge Z as the incident nuclei (N_{SameZ}) are identified using the MUSIC detector. The σ_{CC} is then obtained from the relation $\sigma_{CC} = t^{-1}\ln(R_{T_{out}}/R_{T_{in}})$. Here $R_{T_{in}}$ and $R_{T_{out}}$ are the ratios of N_{SameZ} / N_{in} with and without the reaction target, respectively, and t is the target thickness. The term $R_{T_{out}}$ attributes for losses due to interactions with non-target materials. Phase-space restriction of the inci-

*e-mail: s.bagchi@gsi.de

dent beam at the target is required to eliminate events having large incident angles and positions far from the center. Details of the measurement procedure can be found in Ref. [7].

3 Results and discussion

To obtain the proton-distribution radii from the measured cross section σ_{CC}, finite-range Glauber model framework is used [8]. For stable isotopes (e.g. ^{10}B, $^{12-14}$C, ^{14}N), the obtained proton-distribution radii are in good agreement with the electron-scattering data [7, 9, 10]. The extracted point-proton-distribution radii (R_p) from σ_{CC} are shown in Fig. 1.

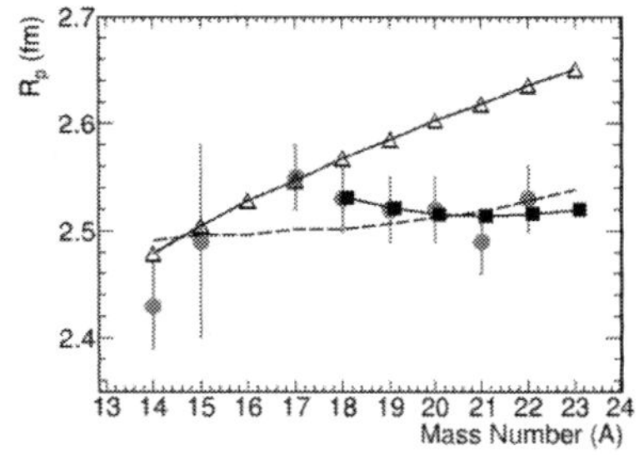

Figure 1. The experimental point-proton radii are shown with red filled circles. Shell model predictions using the YSOX interaction and WS (HO) potentials are shown with black filled squares (black open triangles). RMF theory predictions are given with the black dashed curve from Ref. [11].

The decrease of R_p from ^{17}N to ^{21}N within the uncertainties suggests the reduction in deformation and the transition towards sphericity at the $N = 14$ shell closure. An increase of R_p has been observed within uncertainties from ^{21}N to ^{22}N. In ^{22}N, the valence neutron is in the $2s_{1/2}$ orbital with a closed-shell core of ^{21}N. As a consequence, the center-of-mass (c.m.) of ^{22}N is different from that of the ^{21}N core which leads to the c.m. motion and hence it smears the core density, resulting in a larger core size [7]. Similar phenomenon has been observed in the two neutron halo ^{11}Li [12] and ^{6}He [13] isotopes.

Shell model calculations are compared with the experimental data and are shown in Fig. 1. Both harmonic oscillator (HO) and Woods-Saxon (WS) wave functions are used in the YSOX hamiltonian. The combination of both HO and WS wavefunctions depicts that the proton radii of the nitrogen isotopes increase from ^{13}N to ^{17}N and then a decrease is seen with a shallow minimum at the neutron number $N = 14$. In Fig. 1, the relativistic mean field (RMF) predictions [11] are shown which are consistent with the experimental data. Ab initio, in-medium similarity re-normalization group, calculations with a state-of-the-art chiral nucleon–nucleon and three-nucleon interaction reproduce well the data towards the neutron dripline isotopes rather than the whole isotopic trend of the radii [7].

The matter-distribution radii can be obtained from the measured interaction cross sections [14] using the finite-range Glauber model framework. Together with the information on the matter-distribution and proton-distribution radii, thick neutron skins for $^{19-21}$N, have been found, while for ^{22}N a neutron halo-like structure develops [7].

For nitrogen isotopes, the proton–neutron tensor interaction is attractive for p($1p_{1/2}$)–n($1d_{5/2}$) orbitals, thereby reducing the gap between proton $1p_{1/2}$ and $1p_{3/2}$ orbitals when more neutrons are added to the $1d_{5/2}$ orbital, resulting in a small point proton radius in ^{21}N. This attractive interaction also lowers the neutron $1d_{5/2}$ orbital leading to the $N = 14$ shell gap [7] in neutron-rich nitrogen isotopes.

References

[1] A. Navin, *et al.*, Phys. Rev. Lett. **85**, 266 (2009).
[2] H. Iwasaki, *et al.*, Phys. Lett. B **491**, 8-14 (2000).
[3] H. Simon, *et al.*, Phys. Rev. Lett. **83**, 496 (1999).
[4] Z. Dlouhý, J. Mrázek and D. Baiborodin, AIP Conf. Proc. **610**, 736 (2002).
[5] M.J. Strongman, *et al.*, Phys. Rev. C **80**, 021302(R) (2009).
[6] I. Angeli, K.P. Marinova, At. Data Nucl. Data Tables **99**, 69 (2013).
[7] S. Bagchi, *et al.*, Phys. Lett. B **790**, 251-256 (2019).
[8] Y. Suzuki, *et al.*, Phys. Rev. C **94**, 011602(R) (2016).
[9] A. Estradé, *et al.*, Phys. Rev. Lett. **113**, 132501 (2014).
[10] R. Kanungo, *et al.*, Phys. Rev. Lett. **117**, 102501 (2016).
[11] J. Meng, *et al.*, Phys. Lett. B **532**, 209-214 (2002).
[12] R. Sánchez, *et al.*, Phys. Rev. Lett. **96**, 033002 (2006).
[13] P. Mueller, *et al.*, Phys. Rev. Lett. **99**, 252501 (2007).
[14] A. Ozawa, *et al.*, Nucl. Phys. A **693**, 32-62 (2001).

A 21st Century View of Nuclear Structure

Robert B. Baker[1,*], *Jerry P.* Draayer[1,**], *Grigor H.* Sargsyan[1], *Alison C.* Dreyfuss[1], *Tomáš* Dytrych[1,2], *David* Kekejian[1], *Kristina D.* Launey[1], and *Alexis* Mercenne[1]

[1]Department of Physics and Astronomy, Louisiana State University, Baton Rouge, LA 70803-4001, USA
[2]Nuclear Physics Institute, 250 68 Řež, Czech Republic

Abstract. Exploiting exact and special symmetries to unmask simplicity within complexity, which remains the "holy grail" of nuclear physics, will be considered within its historical context and as evolving through 21st century *ab initio* methods, including emerging results linked to the internal structure of nucleons. Some exemplar results for very light to medium mass nuclei will be presented, and what these may portend for heavier systems, including species beyond known lines of stability, will be proffered.

1 Introduction

This contribution is organized as follows: after this introduction (Sec. 1), we discuss the development of various nuclear models which could capture collective and cluster features in atomic nuclei in Sec. 2. This includes an introduction to the symmetry-adapted no-core shell model (SA-NCSM). We then apply the SA-NCSM to study a number of different nuclei in Sec. 3, including intermediate and medium mass nuclei from an *ab initio* perspective. Afterwards, we discuss and summarize key findings in our conclusions (Sec. 4).

2 Historical Overview

Mayer and Jensen's independent particle model (IPM) [1, 2] and its ability to reproduce the so-called 'magic' numbers lead to the first microscopic theory of nuclear structure. Many modern nuclear structure techniques can trace their initial underpinning to more sophisticated versions of the IPM. These models, when combined with the breath of high-performance computing resources available today, have allowed for good descriptions of the structure of many low-lying states in light nuclei, including descriptions starting from an *ab initio* perspective. Principle among these is the no-core shell model (NCSM) [3, 4] and its extensions, e.g. Refs. [5–7].

The NCSM takes all of the nucleons in a nucleus and uses them to begin filling shells in a harmonic oscillator (HO) potential. Notably, it treats all nucleons as active and allows any of them to be excited to higher shells by an amount $N_{\max}$, where $N_{\max}$ is the total number of HO quanta available above the ground state configuration. Each of these HO shells is separated in energy by a value $\hbar\Omega$ and in the full, infinite space, i.e. a real nucleus, independence must be shown for both $\hbar\Omega$ and $N_{\max}$ since both are external parameters.

Further, and of particular relevance here, was the development of Elliott's SU(3) model [8, 9]. Starting from a microscopic perspective, it was able to realize the ability of the SU(3) symmetry to describe deformation in nuclei. The multi-shell generalization of this approach, the symplectic shell model [10, 11], uses the Sp(3,$\mathbb{R}$) symmetry (of which SU(3) is a subgroup) to study the one-particle-one-hole/$2\hbar\Omega$ excitations of the underlying equilibrium deformation, thereby allowing us to study a variety of collective phenomena.

The *ab initio* symmetry-adapted no-core shell model (SA-NCSM) builds on the advantages of the NCSM and these symmetry-informed shell models to study this underlying collectivity (as well as clustering) starting from first principles. It generally employs a Hamiltonian of the form

$$H = T_{\rm rel} + V_{\rm NN} + V_{\rm Coul}, \qquad (1)$$

where $T_{\rm rel}$ is the relative kinetic energy of the nucleons, $V_{\rm NN}$ is the nucleon-nucleon interaction, and $V_{\rm Coul}$ is the Coulomb interaction between the protons. This $V_{\rm NN}$ includes the use of realistic nuclear interactions which start from the underlying quark-gluon dynamics. To control the growth of the model space for heavier systems, e.g. intermediate and medium mass nuclei, the SA-NCSM selects only the most physically-relevant basis states and thus can construct a model space whose dimension is significantly smaller than the complete NCSM model space for the same number of total HO excitations ($N_{\max}$). The ability to naturally truncate the model space to its most physically-relevant components arises from the emergent fact that realistic nuclear interactions have a preference for a smaller subset of all possible SA-NCSM basis states available. Previous *ab initio* studies of light nuclei have illustrated this fact, regardless of the specific nucleus chosen

*e-mail: rbake25@lsu.edu
**e-mail: draayer@lsu.edu

(e.g. ^{6}Li, ^{6}He, ^{8}Be, ^{12}C) and the choice of realistic interaction, including interactions based on chiral effective field theory. This emergent feature again points to the importance of these symmetries to describing nuclear structure.

3 Exemplary Results

We have performed *ab initio* SA-NCSM calculations for intermediate and medium mass nuclei. Further details on the structure results discussed here can be found in Refs. [12, 13, 15]. For the intermediate mass region, we present results for odd-*A* nuclei with negative parity states (e.g. Fig. 1a) ^{19}Ne in 12 HO major shells [12]) and nuclei near the dripline (e.g. Fig. 1b) ^{32}Ne in 7 HO major shells [13]). Medium mass nuclei (e.g. ^{48}Ti in 8 HO major shells [13]) are also within the reach of the SA-NCSM. These results use realistic chiral interactions and can include contributions from higher HO major shells than previously achievable, thus allowing for the development of the most important configurations in each nucleus. They show good agreement with experiment, particularly with regards to collectivity, which is a traditionally challenging feature for *ab initio* models to reproduce. For example, the quadrupole moment for the first 2^+ state in ^{48}Ti from experiment is known to be -17.7 e fm^2 [14] and SA-NCSM calculations, without the use of effective charges, show a value of -19.3 e fm^2 [15] based on the chiral NNLO$_{\rm opt}$ NN interaction [16] in 8 HO major shells. This indicates that the symmetry-adapted basis is capable of allowing the necessary collectivity to develop while also controlling the combinatorial growth associated with standard NCSM model spaces.

To further study collectivity, especially with respect to the isospin symmetry breaking effects in mirror nuclei, we also present results for B(E2) values in mirror nuclei. Isospin symmetry breaking has traditionally been studied by comparing the energy levels in mirror nuclei or their masses, but we can also use a variety of spectroscopic data, including B(E2) values. In Fig. 2, we present *ab initio* SA-NCSM calculation results for the mirror nuclei ^{21}Mg and ^{21}F. The B(E2) strengths were calculated using various symmetry-based selections of the SA-NCSM model space and various $\hbar\Omega$ values, and only the results with the most optimal parameters that show the fastest convergence trend are depicted in Fig. 2. Variations in the size of the model space and the HO parameter $\hbar\Omega$ are accounted for the extrapolated values and their uncertainties. These results compare to the recent experimental values [17] reasonably well, demonstrating the SA-NCSM capability to describe collectivity in these challenging mirror nuclei. See Refs. [15, 17] for further discussions.

In addition to being able to describe the structure of intermediate and medium mass nuclei, recent work with the SA-NCSM has illustrated its ability to capture other important characteristics of nuclei, including dynamic properties. For example, starting from the symmetry-adapted wave functions, we can calculate response functions via the Lanczos response function method [18, 19], which tell us about how a nucleus responds to an external probe. If a probe is able to excite the ground state, the final states it

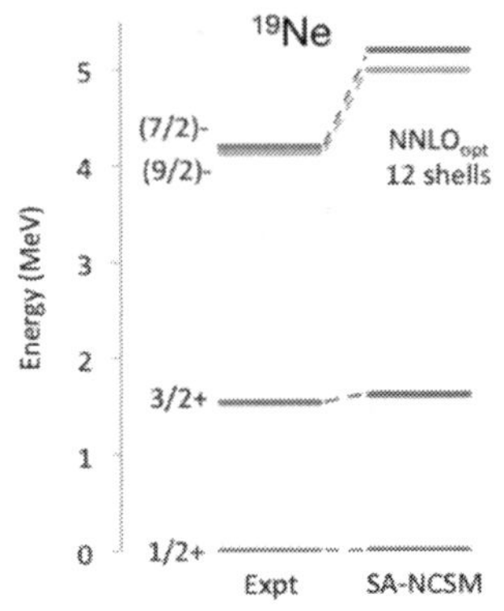

(a)

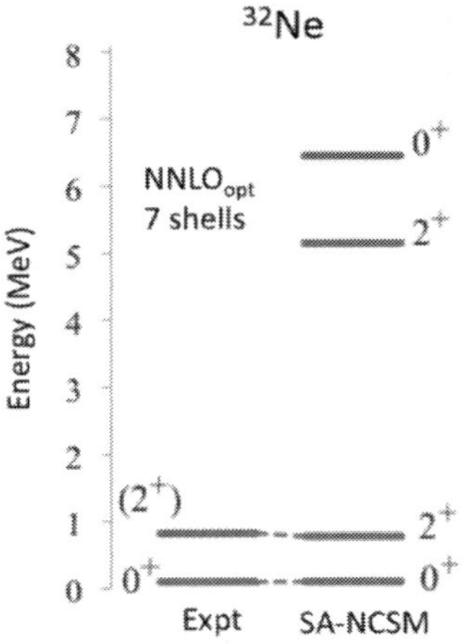

(b)

Figure 1. *Ab initio* SA-NCSM calculations using the chiral NNLO$_{\rm opt}$ *NN* interaction [16] for excitation spectra in (a) ^{19}Ne with 12 HO major shells and (b) ^{32}Ne with 7 HO major shells. Simulations were performed on the Blue Waters system.

then connects to and the strength of those transitions can also tell us something about the similarity of those two states.

More specifically, a typical response function can be expressed as

$$R(\omega) = \sum\!\!\!\!\!\!\int_f \left|\langle\psi_f|\hat{O}|\psi_0\rangle\right|^2 \delta(E_f - E_0 - E_x), \qquad (2)$$

where E_0 (E_f) and $|\psi_0\rangle$ ($|\psi_f\rangle$) are the energy and wave function for the ground (final) state, E_x is the energy transfered between the nucleus and the external probe, and $\sum\!\!\!\!\!\!\int_f$ indicates a sum over all possible discrete and continuum final states. The operator $\hat{O}$ describes the external probe – here, we will use the monopole operator, though any operator could be employed. In the Lanczos response function method, this response function $R(\omega)$ is folded with a Lorentzian of width Γ and the wave function information

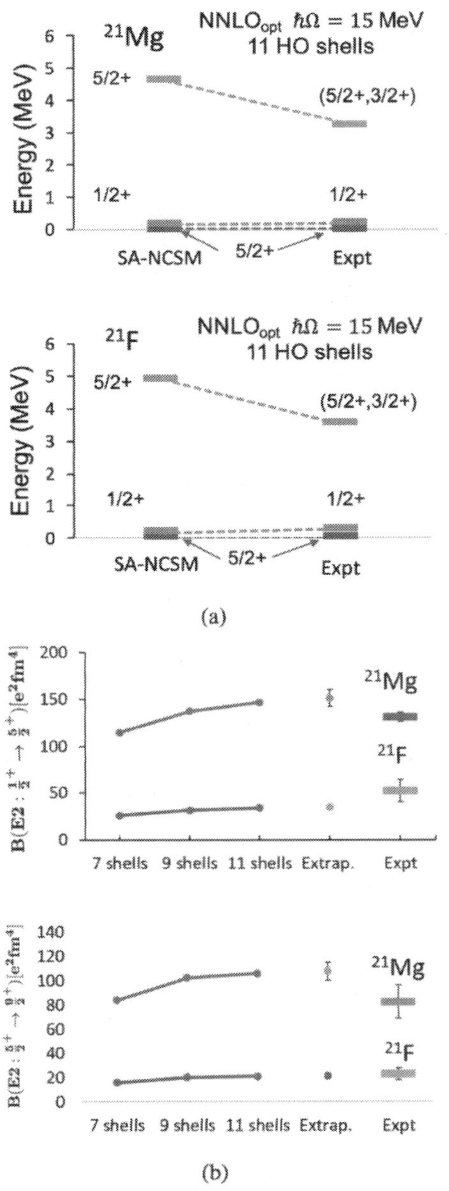

Figure 2. *Ab initio* SA-NCSM calculations using the chiral NNLO$_{opt}$ *NN* interaction in ultra-large model spaces ($\hbar\Omega$ = 15 MeV). (a) Energy spectrum of ^{21}Mg and ^{21}F in 11 HO major shells, and (b) convergence of the B(E2: $\frac{1}{2}^+ \to \frac{5}{2}^+$) (top) and B(E2 $\frac{5}{2}^+ \to \frac{9}{2}^+$) (bottom) strengths with increasing model space, and the extrapolated values with uncertainties from model space and $\hbar\Omega$ variance. Simulations were performed on the Blue Waters system. Experimental values for the B(E2) are from Ref. [17].

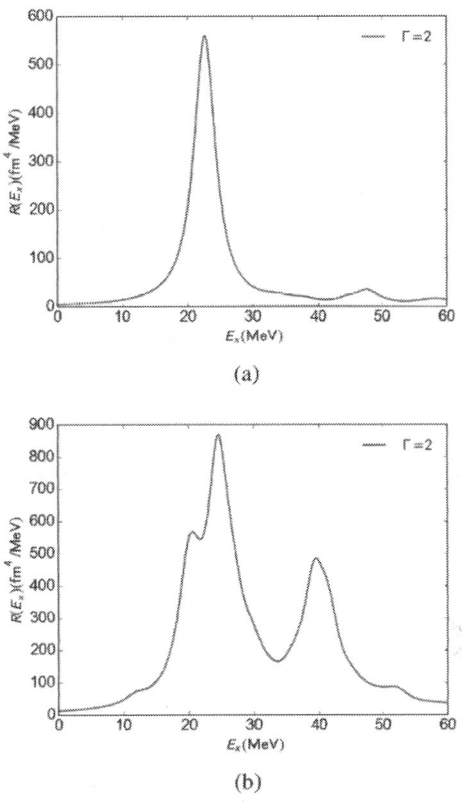

Figure 3. *Ab initio* SA-NCSM calculations using the chiral NNLO$_{opt}$ *NN* interaction for monopole response functions in (a) ^{16}O at $\hbar\Omega$ = 16 MeV in 12 HO shells and (b) ^{20}Ne at $\hbar\Omega$ = 15 MeV in 11 HO major shells. The response are folded with a Lorentzian of width Γ = 2 MeV. Simulations were performed on NERSC systems.

is included from solving the many-body Hamiltonian via the Lanczos algorithm.

Given the SA-NCSM's connection to the symplectic symmetry (and subsequently giant resonances [11]), we can use this approach to study the structure of giant resonances from first principles. For example, in Fig. 3a), we can see a single, strong peak in the monopole response function of ^{16}O, as calculated with the NNLO$_{opt}$ chiral interaction. This peak is dominated (> 60%) by a (2 0) state of correlated $2\hbar\Omega$ one-particle excitations, which is inline with expectations we would have for a giant monopole resonance based on previous work with the symplectic shell model [20]. Similarly, in Fig. 3b), we can see the monopole response function for ^{20}Ne, as calculated with NNLO$_{opt}$. Unlike the ^{16}O case, here the giant resonance peak is fragmented and signatures of it are found in various amounts in a few different peaks in this response function. See Ref. [21] for further analysis of these results.

4 Conclusions

We presented a short historical overview of the development of (no-core) shell model techniques capable of properly describing the observed collectivity in nuclei in Sec. 2. We focused on the development of the *ab initio* SA-NCSM, which can use realistic nuclear interactions and study nuclei from a first principles perspective. Through the use of emergent symmetries and high performance computing, the SA-NCSM has been able to control the growth of the model space and study up to, and including, intermediate and medium mass nuclei. In Sec. 3, we discussed results for the structure of neon isotopes, including the odd-A nucleus ^{19}Ne and a nucleus near the dripline, ^{32}Ne. We also examined collectivity in ^{21}F and ^{21}Mg through their associated B(E2) values. Lastly, we also discussed recent work with the SA-NCSM to examine giant monopole resonances in ^{16}O and ^{20}Ne. These results, through calculations of the response function, point to the SA-NCSM's unique applicability to study giant resonances (and their fragmentation) from first principles.

Acknowledgement: Support from the U.S. National Science Foundation (OIA-1738287 and ACI-1713690) and the U.S. Department of Energy (DE-SC0005248), as well as those from Louisiana State University, especially computational resources under the Louisiana Optical Network Initiative, the Blue Waters sustained-petascale computing project, the National Energy Research Scientific Computing Center (NERSC), and the Southeastern Universities Research Association are all gratefully acknowledged.

References

[1] M. G. Mayer, Phys. Rev. C **75**, 1969 (1949).
[2] O. Haxel, J. H. D. Jensen and H. E. Suess, Phys. Rev. **75**, 1766 (1949).
[3] P. Navrátil, J. P. Vary and B. R. Barrett, Phys. Rev. Lett. **84**, 5728 (2000).
[4] B. Barrett, P. Navrátil and J. Vary, Prog. Part. Nucl. Phys. **69**, 131 (2013).
[5] P. Navrátil, S. Quaglioni, I. Stetcu and B. R. Barrett, J. Phys. G: Nucl. Part. **36**, 083101 (2009).
[6] R. Roth and P. Navrátil, Phys. Rev. Lett. **99**, 092501 (2007).
[7] T. Abe, P. Maris, T. Otsuka, N. Shimizu, Y. Utsuno, and J. Vary, Phys. Rev. C **86**, 054301 (2012).
[8] J. P. Elliott, Proc. Roy. Soc. A **245**, 128 (1958).
[9] J. P. Elliott and M. Harvey, Proc. Roy. Soc. A **272**, 557 (1962).
[10] G. Rosensteel and D. J. Rowe Phys. Rev. Lett. **38**, 10 (1977).
[11] D. J. Rowe, Reports on Progr. in Phys. **48**, 1419 (1985).
[12] J. P. Draayer, K. D. Launey, T. Dytrych, A. C. Dreyfuss, G. H. Sargsyan, R. B. Baker, 'Untangling Simple Patterns in Intricate Atomic Nuclei,' in Walter Greiner Memorial Volume, edited by Peter O. Hess and Horst Stöcker (World Scientific Publishing Co., 2018), pp. 97-121.
[13] K. D. Launey, A. Mercenne, G. H. Sargsyan, H. Shows, R. B. Baker, M. E. Miora, T. Dytrych, and J. P. Draayer, Proceedings of the 4th International Workshop on ?State of the Art in Nuclear Cluster Physics? (SOTANCP4), May 2018, Galveston, Texas, Volume 2038. AIP Conf. Proc., 2018.
[14] J. W. Lightbody Jr., Phys. Lett. B **38**, 475 (1972).
[15] J. P. Draayer, K. D. Launey, T. Dytrych, R. B. Baker, A. C. Dreyfuss, D. Kekejian, G. H. Sargsyan, Proceedings of the International Conference 'Nuclear Theory in the Supercomputing Era – 2018' (NTSE-2018), Daejeon, South Korea, October 29 – November 2, 2018, eds. A. M. Shirokov and A. I. Mazur. Pacific National University, Khabarovsk, Russia, 2019.
[16] A. Ekström, G. Baardsen, C. Forssen, G. Hagen, M. Hjorth-Jensen, G. R. Jansen, R. Machleidt, W. Nazarewicz, T. Papenbrock, J. Sarich, and S. M. Wild, Phys. Rev. Lett. **110**, 192502 (2013).
[17] P. Ruotsalainen *et al.*, Phys. Rev. C **99**, 051301(R) (2019); arXiv:1811.00774 (2018).
[18] V. D. Efros, W. Leidemann, G. Orlandini, N. Barnea, J. Phys. G: Nucl. Part. Phys. **34**, R459 (2007).
[19] R. B. Baker, K. D. Launey, N. Nevo Dinur, S. Bacca, J. P. Draayer, and T. Dytrych, Proceedings of the 4th International Workshop on "State of the Art in Nuclear Cluster Physics" (SOTANCP4), May 2018, Galveston, Texas, Volume 2038. AIP Conf. Proc., 2018.
[20] D. J. Rowe, G. Thiamova, J. L. Wood, Phys. Rev. Lett. **97**, 202501 (2006).
[21] R. B. Baker, Ph.D. dissertation, Louisiana State University, 2019.

Structure and reactions of N=7 isotones: parity inversion and transfer processes

Francisco Barranco[1,*], *Ricardo A.* Broglia[2,3,**], *Gregory* Potel[4,***], and *Enrico* Vigezzi[5,****]

[1]Departamento de Física Aplicada III, Escuela Superior de Ingenieros, Universidad de Sevilla, Sevilla, Spain
[2]The Niels Bohr Institute, University of Copenhagen, Copenhagen, Denmark
[3]Universitá degli Studi di Milano, Milano, Italy
[4]National Superconducting Cyclotron Laboratory, Michigan State University, East Lansing, USA
[5]Istituto di Fisica Nucleare, Sezione di Milano, Milano, Italy

Abstract. The interplay of particle and vibrations in N=7 isotones is considered according to nuclear field theory, focusing on the main many-body effects which renormalise the energy spectrum of the halo nucleus ^{11}Be, leading to parity inversion and to renormalization of the form factors determining the cross sections associated with one-nucleon transfer reactions.

1 Introduction

Many features of the N=7 halo nuclei ^{10}Li and ^{11}Be can be understood taking into account the strong many-body renormalization effects arising from the coupling of particles with the quadrupole collective vibrations of the core. In particular, this coupling leads to the inversion of the sequence of the $\frac{1}{2}^+$ and $\frac{1}{2}^-$ levels, as compared to the usual shell model order. In this contribution we outline the treatment of these many-body effects and the results obtained, also for other N=7 isotones, within nuclear field theory (NFT) [1–4], which has been shown to lead to a rather accurate description of the structure and reactions measured in these nuclei, in particular reproducing the absolute cross sections of the ^{10}Be(d,p)^{11}Be and ^{9}Li(d,p)^{10}Li transfer reactions [5, 6].

2 The model Hamiltonian

The starting Hamiltonian is $\mathcal{H} = H'_{sp} + H_{tb}$, where H'_{sp} is a one-body term and H_{tb} is the two-body (four-point vertex) interaction. The free (bare) fermion fields are the Hartree-Fock solution (H_{sp}) of $\mathcal{H}$. In the resulting (particle-hole) basis the collective (quadrupole) states are to be calculated by diagonalising H_{tb} in the random phase approximation (RPA). The corresponding solutions provide, aside from the collective Hamiltonian H_{coll}, the particle-vibration coupling vertices (H_{PVC}). These vertices and the two-body interaction should be allowed to act in all orders of perturbation theory to generate the variety of Feynman diagrams leading to the renormalisation of the single-particle and collective (free fields) modes and their interweaving, with the proviso of eliminating the contribution of virtual unperturbed loops (bubbles). Expressing it differently, one is confronted with the task of diagonalising the second term of the Hamiltonian $H = (H_{sp} + H_{coll}) + (H_{PVC} + H_{tb})$ in a basis consisting of single-particles and collective states, with the proviso of eliminating from the initial and final states particle (and hole) configurations which can be replaced by collective modes.

In what follows, we implement this program in connection with a series of $N = 7$ isotones, in terms of the program of renormalized NFT [3]. The one which parallels that of QED in the case of single-particle states: the bare energies are adjusted so that the dressed (renormalized) single-particle states best reproduce the experimental findings [5]. Empirical renormalization (see e.g. [3]) concerning collective motion is adopted. In particular we will calculate the quadrupole vibrations by diagonalizing in the RPA and in an empirical single-particle basis (Woods-Saxon parametrization as found in Eq. (2-181) of ref. [7]) a separable interaction tuned to reproduce the properties of the low lying modes. Within this approximation, the particle-vibration coupling form factor in the case of quadrupole vibrations is $\delta V_2(r)Y_{2\mu}(\hat{r})$, where $\delta V_2(r) = -\beta_2 R_0 dV/dr$. It is of notice that we will not consider those diagrams which renormalise the phonon states, in keeping with the fact that such diagrams lead to results which reproduce the experimental findings [8]. Furthermore, we will neglect the four point (fermion) vertices, as it is known that their contribution is small.

Although it is possible to iterate the desired processes to infinite order of perturbation theory, a useful parameter to assess the order of a graphic contribution is $1/\Omega$, Ω being the effective degeneracy of the model space (number of Kramers degenerate levels available to the particles lying

*e-mail: barranco@us.es
**e-mail: broglia@mi.infn.it
***e-mail: potel@nscl.msu.edu
****e-mail: vigezzi@mi.infn.it

https://doi.org/10.1051/epjconf/201922301005

close to the Fermi energy). The order of a given diagram is given by $n_\Omega = n_\Lambda/2 - n_f + n_{tb}$, where n_f, n_Λ, n_{tb} are the number of fermion loops over which summation can be carried out, the number of particle-vibration coupling vertices and the number of four-point vertices respectively.

The configurations including one or more phonons lie outside the single-particle (basic) space, and enter the calculations as intermediate, virtual states, their energies appearing in the energy denominators of the Feynman diagrams. This procedure leads to an energy-dependent self-energy function, producing a much greater number of solutions than the dimension of the explicit single-particle basis. The square amplitude of the single-particle component associated with each solution is in general smaller than one, reflecting the fact that the renormalised states contain admixtures with collective phonons. The 'hidden' configurations containing one or more phonons may be explicitly reconstructed following the Bloch-Horowitz formalism (see App. B of [1]).

The parameters of the mean field H_{sp} are chosen so as to give an optimal global fit to the experimental energy spectra in the $N = 7$ isotones ^{10}Li, ^{11}Be, ^{12}B and ^{13}C, after many-body renormalisation effects are taken into account. This bare mean field is parametrised as a Saxon-Woods potential well, characterised by the well depth V_{WS}, the diffuseness a_{WS}, the radius R_{WS} and by the spin-orbit strength parameter V_{LS}, as well as by a $r-$dependent effective mass. We obtain the values $V_{WS} = -82 + 54(N - Z)/A$ MeV, $a = 0.75$ fm, $R_{WS} = 0.99\,A^{1/3}$ fm and $V_{LS} = 0.3\,V_{WS}$. We will be particularly interested in the renormalization of $s_{1/2}$, $p_{1/2}$ and $d_{5/2}$ orbitals lying above the Fermi energy E_F. The renormalized radial wavefunction associated with these states will be denoted by $\tilde{R}_{ai}(r)$, where $a \equiv \{l j\}$, while i labels the various solutions. It obeys the equation

$$H_{sp}\tilde{R}_{ai}(r) + \int dr'\Sigma_a(r, r', \tilde{E}_{ai})\tilde{R}_{ai}(r') = \tilde{E}_{ai}\tilde{R}_{ai}(r), \quad (1)$$

where $\Sigma_a(r, r', \tilde{E})$ denotes the self-energy. This is a complex, non local, energy-dependent quantity, which represents the interaction between the odd neutron (halo neutron in the case of ^{10}Li and ^{11}Be) and the core, and is an essential element to compute the single-particle transfer cross sections together with the optical potentials in the entrance and exit channels. Within this context Σ is the structure input needed to calculate the polarisation contribution to the optical potential.

3 Many-body renormalization effects

The self-energy in the first order of perturbation in $1/\Omega$, i.e. including only intermediate configurations containing one phonon, can be written explicitly as a sum over the unperturbed wave functions and energies R^{WS}_{ai} and E_{ai} calculated in the (bare) Saxon-Woods potential well,

$$\Sigma_a(r, r', \tilde{E}) = \sum_{bi} \frac{\Xi^2_{a,b\lambda} R^{WS}_{bi}(r')R^{WS}_{bi}(r)\delta V(r')\delta V(r)}{\tilde{E}-(E_{bi}+\hbar\omega-i\eta)} + \sum_{ci} \frac{\Xi^2_{a,c\lambda} R^{WS}_{ci}(r')R^{WS}_{ci}(r)\delta V(r')\delta V(r)}{\tilde{E}+(E_{ci}+\hbar\omega-i\eta)}, \quad (2)$$

where ($E_{bi} > E_F; E_{ci} < E_F$), while $\Xi_{a,b\lambda}$ denote angular momentum recoupling factors. We have included continuum states up to 50 MeV, calculated in a box of radius equal to 50 fm, checking that this leads to converged results, within 100 keV, for the energy of the renormalised states. Note that expressions analogous to that of Eq. (2), i.e. of second order in β_2, appear in a number of many-body approaches in the calculation of the self-energy. The first term accounts for correlations due to H_{PVC} and acts mostly on the bound $2s_{1/2}$ state and on the $d_{5/2}$ levels lying in the continuum, while the second one accounts for the blocking of core ground state correlations and strongly renormalizes the $1p_{1/2}$ state, due to the quadrupole coupling to the $1p_{3/2}$ hole state. The first-order results are schematically shown in Fig. 1 (2nd column), where it can be seen that the inversion between $1/2^+$ and $1/2^-$ levels is reproduced. The different renormalized states ($i = 1, 2, ...$) of angular momentum and parity j^π will be denoted by $\widetilde{j^\pi_i}$, their energy by $\widetilde{E}_{j^\pi_i}$, and their amplitude over a single-particle state k by $(\widetilde{a}_{j^\pi})_{ik}$. The sum of the amplitudes squared $\widetilde{A^2}_{j^\pi_i} \equiv \sum_k(\widetilde{a}_{j^\pi})^2_{ik}$ for the states $\widetilde{j^\pi_i} = \widetilde{1/2}^+_1, \widetilde{1/2}^-_1$ is equal to 0.93 and 0.85 respectively, reflecting the sizeable admixture of the 2^+ phonon in the wavefunctions. The $5/2^+$ phase shifts show a resonance lying at an energy several MeV lower than its unperturbed value calculated with the bare potential.

In many applications, the processes described by Eq. (2) are sufficient to obtain a reasonable description of experimental data, in particular in heavy spherical nuclei ($\Omega \approx 6-8$ for $A \approx 120$ and $\Omega \approx 8-10$ for $A \approx 200$). In the case of the N=7 isotones under study, for which $\Omega \approx 3-5$, it turns out that further orders in the perturbation series have to be included in order not to miss important renormalization effects on specific configurations, as it will be shown below.

Substituting the unperturbed WS energies and wavefunctions E_{bi}, R^{WS}_{bi} appearing in Eq. (2) with perturbed ones to be determined, leads to a more complex expression for the self-energy, involving the unknown energies for states of different angular momentum, and the resulting equations should be solved simultaneously and consistently for all of them. This amounts to the summation of the so called rainbow series to infinite order (see Fig. 2), including configurations with an arbitrarily large number of phonons, subject only to the restriction of no crossing among them. An important property of this series is the systematic coherence of all the terms. Contributions involving configurations containing crossing phonons (vertex corrections) imply recoupling factors of varying sign, which in general cause strong cancellations, leading to smaller contributions, which we neglect.

The rainbow series can be summed to infinite order making use of the *empirical renormalization* method for the intermediate states in Eq. (2). This should represent a reasonable approximation, if the renormalized energies and wavefunctions resulting from Eq. (1) turn out to be indeed close to the experimental data, as has been extensively tested in ref. [8]. In practice, we implement it by calculating Eq. (2) inserting energies and wave functions

https://doi.org/10.1051/epjconf/201922301005

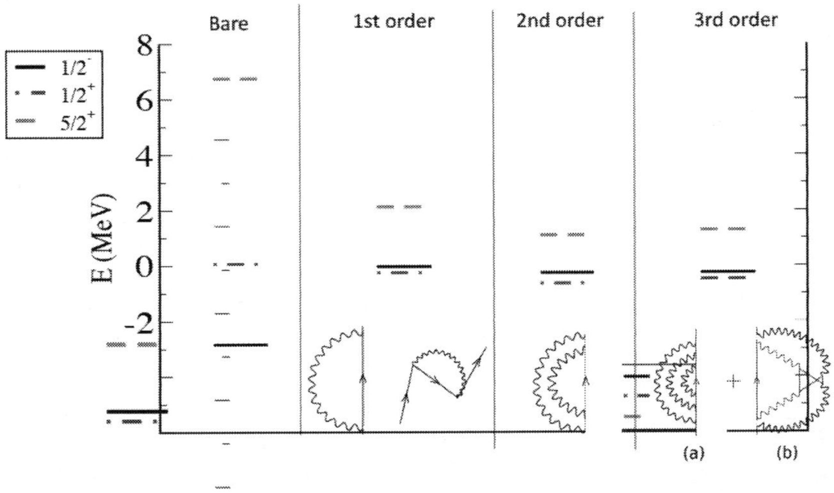

Figure 1. (color online) Theoretical spectrum of low-lying states of ^{11}Be, calculated in different perturbation orders in NFT. The lowest $1/2^-$,$1/2^+$ and $5/2^+$ levels are shown by solid, dot-dashed and dashed lines respectively. In the case of $5/2^+$, the energy of the resonant state is indicated. Representative diagrams included at each order are drawn at the bottom of the figure.

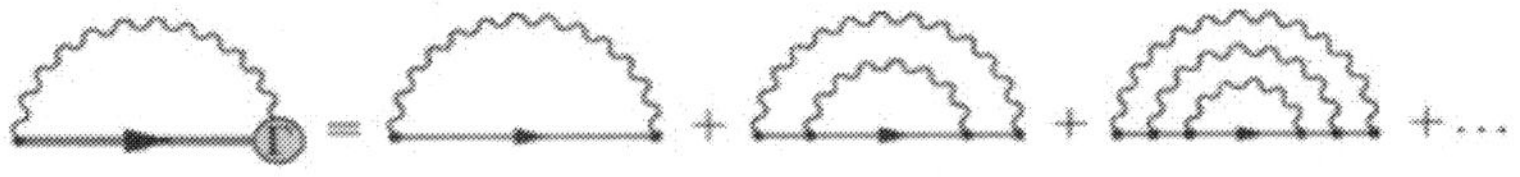

Figure 2. The rainbow series.

obtained from the empirical Saxon-Woods potentials mentioned above [7].

We then obtain that the summation of the rainbow series leads to a rather dramatic effect for the $\widetilde{5/2_1^+}$ state, which has a resonant, many-body character, displaying an admixture with the $s_{1/2} \otimes 2^+$ and $d_{5/2} \otimes 2^+$ configurations as large as 50%. This state becomes almost bound ($\widetilde{E}_{5/2_1^+}$ = 0.6 MeV), in contradiction with experimental data (1.3 MeV). This strong effect is mostly due to the fact that the virtual $2s_{1/2}$ state entering the rainbow series is bound (empirical renormalization, $E_{1/2^+}$= -0.5 MeV), leading to a enhanced coupling of the $\widetilde{5/2_1^+}$ state with the $s_{1/2} \otimes 2^+$ configuration, as compared to the first order, in which the $s_{1/2}$ state is calculated in the bare potential and lies in the continuum. The coupling of the $\widetilde{1/2_1^+}$ state with the $d_{5/2} \otimes 2^+$ configuration is also enhanced, and the value of $\widetilde{A^2}_{1/2_1^+}$ goes from 0.93 to 0.76.

Physically, the reason for these strong effects is the overcounting of the correlations appearing when two or more quadrupole phonons are present, as in rainbow-like processes. In other words, RPA phonons are quasi-bosons with an underlying fermionic structure, which eventually lead to Pauli principle corrections when two or more of them are present in a given configuration. The relevance of this interference increases as the p-h Hilbert configuration space becomes smaller, which is the case for light nuclei, where the density of bound/resonant levels is rather low. In the present case such problems manifest themselves first in the third order of perturbation, in keeping with the sensible results obtained for the $\widetilde{5/2_1^+}$ state in second order, obtained iterating the first order results ($\widetilde{E}_{5/2_1^+}$= 1.1 MeV, see Fig. 1, third column). In fact, Pauli blocking effect appears in the self-energy at the third order in $1/\Omega$, through the so called butterfly diagrams, shown in the bottom part of Fig. 1 (4th column, (b)). In ^{11}Be, these

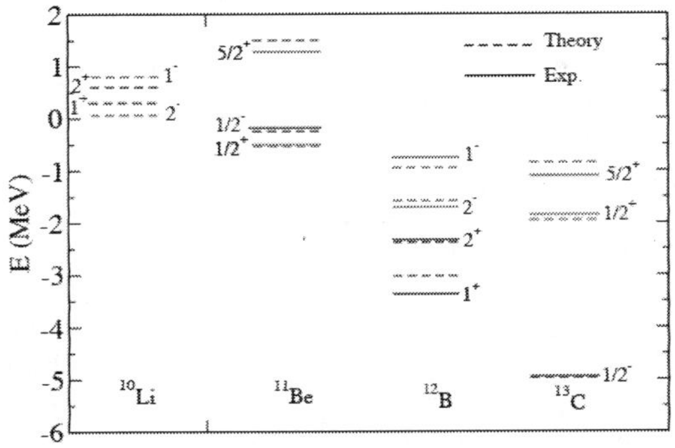

Figure 3. (color online) The experimental energies of the low-lying states in the N = 7 isotones ^{11}Be, ^{12}B and ^{13}C are displayed by solid lines. The corresponding theoretical energies are shown by dashed lines. Also shown are the predictions in ^{10}Li. States based on $1/2^-$, $1/2^+$ and $5/2^+$ neutron configurations are shown by blue, red and green lines respectively.

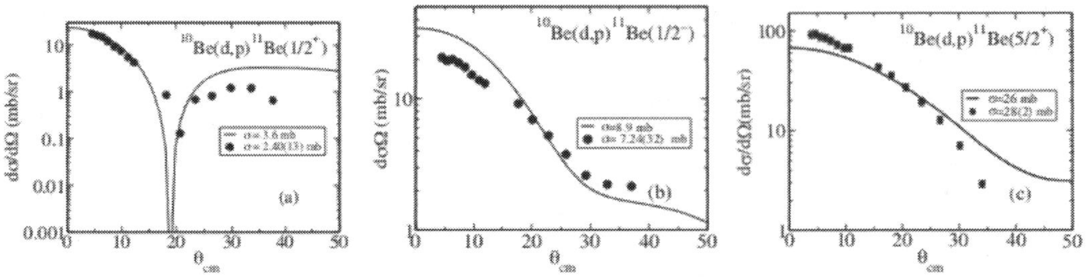

Figure 4. From [5]. Absolute differential and (insets) integrated cross sections associated with the reactions d(^{10}Be,^{11}Be)p at E = 107 MeV, populating the $1/2^+$, $1/2^-$ and $5/2^+$ low-lying states. The experimental data [9] are displayed in terms of solid dots.

diagrams increase the energy of the $2_1^+ \otimes 2_1^+$ two-phonon configuration by about 2.5 MeV, and decrease the value of the quadrupole deformation parameter β_2 by about 25%. Depending on the extent that the two phonon configuration appears in the rainbow series, the effect of the anharmonicity differs considerably from one partial wave to the other. Its contribution is particularly relevant for the $\widetilde{5/2}_1^+$ state, increasing its energy by 0.7 MeV and bringing the empirical renormalization result to $\tilde{E}_{5/2_1^+}$= 1.3 MeV, in agreement with the experimental data. This result coincides with energy calculated in perturbation, which increases by 0.2 MeV going from the second order to the third order, as a result of an attractive contribution of -0.5 MeV arising from diagram (a), and a repulsive one of +0.7 MeV due to diagram (b) (see Fig. 1, fourth column). Third order effects on the $\widetilde{1/2}_1^+$ state are modest and those on the $\widetilde{1/2}_1^-$ state are even weaker. Finally, one obtains $\widetilde{A^2}_{1/2_1^+}$ = 0.86 and $\widetilde{A^2}_{1/2_1^-}$ = 0.83. Overall, the results displayed in Fig. 1 testify to the convergence of the NFT results already at third order of perturbation theory.

It is of notice that the sensitivity of the $\widetilde{5/2}_1^+$ state to corrections beyond the first order is in keeping with the fact that, in order to reproduce the energy of this state in other theoretical schemes, important changes in the spherical mean field or the introduction of a deformed mean field are required. We furthermore remark that the effects discussed above are associated with many-body phenomena (core excitation), and their occurrence in microscopic models should be rather independent of the N-N interaction used.

The renormalized energies of ^{11}Be as well as of ^{13}C are compared to experimental data in Fig. 3. The calculated wave functions and associated self-energies have been used to calculate one-particle cross sections [5], resulting in an overall agreement with the data. In Fig. 4 the calculated cross sections associated with the reactions d(^{10}Be,^{11}Be)p at E=107 MeV performed in inverse kinematics, populating the $1/2^+$, $1/2^-$ and $5/2^+$ states are shown and compared with experimental data [9].

4 The case of odd-odd nuclei

In the case of the N=7 isotopes $^{12}_{5}$B and $^{10}_{3}$Li, which have an odd number of protons, one needs to take into account the interaction of this proton with the renormalized neutron, before being able to compare the NFT results with experiment. In fact the coupling of the $1p_{3/2}$ proton with the neutron $\widetilde{1/2}_1^+$ and $\widetilde{1/2}_1^-$ states leads in either case to a doublet, namely 1^- and 2^-, and 1^+ and 2^+ respectively, while its coupling with the $\widetilde{5/2}_1^+$ neutron state leads to a $1^-, 2^-, 3^-$ and 4^- quadruplet. These multiplets are degenerate in the mean field approximation, while they are splitted according to experiment. An accurate theoretical description of such splittings requires an accurate consideration of the n-p interaction. We have limited ourselves to the use of a rather simple, but extensively employed version of a zero range force [10], containing a direct and a spin dependent term. The resulting levels and splittings are shown in Fig. 3. In all these result the empirical renormalization and the anharmonicity corrections described above have been included. The comparison with data is rather satisfactory. The calculation of transfer cross sections using the calculated renormalized wavefunctions (form factors) and associated self-energies are in progress [6].

Acknowledgment

F.B and E.V. acknowledge funding from the European Union Horizon 2020 research and innovation program under Grant Agreement No. 654002. F.B. acknowledges funding from the Spanish Ministerio de Economía under Grant Agreement No. FIS2017-88410-P.

References

[1] D.R. Bès et al., Nucl. Phys. **A260** 1 (1976)
[2] P.F. Bortignon et al., Phys. Rep. **30C** 305 (1977)
[3] R.A. Broglia et al., Phys. Scr. **91** 063012 (2016)
[4] D.R. Bès, Phys. Scr. **91** 03010 (2016)
[5] F. Barranco et al., Phys. Rev. Lett. **119**, 082501 (2017)
[6] F. Barranco et al., arXiv:1812.01761
[7] A. Bohr and B.R. Mottelson, *Nuclear Structure*, Vol. I, Benjamin, New York (1969)
[8] A. Idini et al., Phys. Rev. C **92**, 031304(R) (2015)
[9] K.T. Schmitt et al, Phys. Rev. C **88**, 064612 (2013).
[10] K. Heyde, *The nuclear shell model*, Springer-Verlag (1990)

Phase transition dynamics in hot nuclei and N/Z influence

Bernard Borderie[1,*], *Nicolas* Le Neindre[2], and *Pierre* Désesquelles[3]for the INDRA Collaboration

[1]Institut de Physique Nucléaire, CNRS/IN2P3, Univ. Paris-Sud, Université Paris-Saclay, Orsay, France
[2]Normandie Univ., ENSICAEN, UNICAEN, CNRS/IN2P3, LPC Caen, Caen, France
[3]CSNSM, CNRS/IN2P3, Univ. Paris-Sud, Université Paris-Saclay, Orsay, France

Abstract. An abnormal production of events with almost equal-sized fragments was theoretically proposed as a signature of spinodal instabilities responsible for nuclear multifragmentation in the Fermi energy domain. On the other hand finite size effects are predicted to strongly reduce this extra production. High statistics quasifusion hot nuclei produced in central collisions between Xe and Sn isotopes at 32 and 45 MeV per nucleon incident energies have been used to definitively establish, through the experimental measurement of charge correlations, the presence of spinodal instabilities. N/Z influence was also studied. The nature of the phase transition dynamics i.e. the fragment formation was the last missing piece of the puzzle concerning the liquid-gas transition in nuclei.

1 Introduction

An important challenge of heavy-ion collisions at intermediate energies was the identification and characterization of the nuclear liquid-gas phase transition in hot nuclei, which has been theoretically predicted for nuclear matter. At present one can say that huge progress has been made even if some points can be deeper investigated [1, 2]. Statistical mechanics for finite systems appeared as a key issue to progress, revealing specific first-order phase transition signatures related to the consequences of the local convexity of the entropy (see [2] and references therein). By considering the microcanonical ensemble with energy as extensive variable, the convex intruder implies a backbending in the temperature (first derivative of entropy) at constant pressure [3] and correlatively a negative branch for the heat capacity (second derivative of entropy). Experimentally, these two converging signatures have been observed in hot nuclei from different analyses [4–7] of homogeneous event samples. It is important to recall here that signals of phase transition for finite systems are only meaningful at the level of statistical ensembles constructed from the outcome of carefully selected collisions. Another consequence of the entropy curvature anomaly manifests itself when systems are treated in the canonical ensemble. In this case a direct phase transition signature is the presence of a bimodal distribution of an order parameter [8] such as the charge (size) of the largest fragment (Z_{max}) of multifragmentation partitions [9].

The knowledge of the nature of the dynamics involved during phase transition in hot nuclei, i.e. fragment formation, is certainly the most delicate point. Two mechanisms have been proposed. On one side, stochastic mean field approaches predict the transition dynamics to follow the spinodal fragmentation scenario proposed very early on, triggered by phase-space fluctuations amplified in an unstable medium [10, 11] and, on the other side, molecular dynamics models such as AMD in which many-body correlations are sufficient to produce fragments at early times even in absence of unstable conditions [12, 13]. Both theoretical and experimental studies have evidenced for a radial extra energy boost (radial expansion energy) associated to multifragmentation products [14, 15]. It can be attributed either to a dominant compression-expansion phase in central nucleus-nucleus collisions or to thermal pressure for more gentle collisions: hadron-nucleus or semi-peripheral nucleus-nucleus collisions. The system might then reach densities and temperatures that correspond to the unstable spinodal region where exponential amplification of density fluctuations leads to clusterization: an inhomogeneous mixture of fragments (normal density region), nucleons and light fragments (low density region). This can be seen as an analogue for phase separation in a finite system, for which spinodal decomposition would be the microscopic mechanism.

2 Theoretical predictions

In infinite nuclear matter the signature of spinodal instabilities is the formation of equal-sized fragments due to density fluctuations which grow exponentially with time. The most unstable modes correspond to wavelengths lying around $\lambda \approx 10$ fm and the associated characteristic times are equal to around 30-50 fm/c, depending on density ($\rho_0/2$ - $\rho_0/8$) and temperature (0-9 MeV) [10, 11]. A direct consequence of the dispersion relation is the production of "primitive" fragments with size $\lambda/2 \approx 5$ fm which correspond to $Z \approx 8$ - 10 . However, this simple picture is expected to be largely blurred by several effects. The

*e-mail: borderie@ipno.in2p3.fr

beating of different modes occurs. Coalescence effects due to the nuclear interaction between fragments before the complete disassembly are also expected. For finite systems the situation is even more complicated. The presence of a surface introduces an explicit breaking of the translational symmetry with the important result that the growth rates are nearly the same for different multipolarities, which indicates that the unstable finite system breaks into different channels depending on multipolarity L [16]. Equal-sized "primitive" fragments are then expected to be produced with sizes in the range $A_F/2$ - A_F/L_{max}; A_F being the part of the system leading to fragments during the spinodal fragmentation. Moreover the finite system produced during the nucleus-nucleus collision has to stay or live long enough in the spinodal region (~ 3 characteristic time - 100-150 fm/c - for rather symmetric hot nuclei and - 200-300 fm/c - for nuclei with large N/Z [17]) to allow an important amplification of the initial density fluctuations. And finally, we experimentally detect fragments after secondary decay, which introduces a broadening of the fragment size distribution. Taking into account the accumulation of all these effects on the final extra production of equal-sized fragments, it is clear that any signature that spinodal fragmentation is responsible for the phase transition dynamics can only be what we have called a fossil signature. A full simulation of the spinodal decomposition of quasifused sources using Brownian One-Body dynamics calculations [18] already testified to this fact, with less than 1% of events with equal-sized fragments. It is the reason why the signature is difficult to observe experimentally.

3 Experimental results

The experiment was performed at GANIL (Grand Accélérateur National d'Ions Lourds) and two reactions were used $^{124,136}Xe + ^{112,124}Sn$ at two bombarding energies, 32 and 45 MeV per nucleon, The beam, impinging on thin targets (530$\mu g\ cm^{-2}$), had an intensity of about 3-5 10^7 ions per second to avoid event pile-up. Experimental data were collected with the 4π multidetector INDRA which is described in detail in [19, 20]. Accurate particle and fragment identifications were achieved and the energy of the detected products was measured with an accuracy of 4%. Further details can be found in [21–23].

The data used in the analysis were obtained with an on-line multiplicity trigger of 4 or more detected reaction products. The number of such recorded events was between 60 and 80 million events for each colliding system. Quasi-complete events are selected by requiring that at least 80% of the total charge of the system is measured. We then isolate quasifusion events by using a shape event sorting (see [24] for details). Measured cross-sections corresponding to selected events are ≈ 40 mb at 32 MeV per nucleon and 25 mb at 45 MeV per nucleon. Further details can be found in [24]. Total cross-sections for quasifusion events, taking into account detection efficiency and selection biases, are estimated to be ~ 250 mb at 32 MeV per nucleon, and ~ 180 mb at 45 MeV per nucleon.

Twenty years ago an intra-event correlation function called higher order charge correlations [25] was proposed to enlighten any extra production of events with specific fragment partitions. The high sensitivity of the method makes it particularly appropriate to look for small numbers of events as those expected to have kept a memory of spinodal fragmentation properties.

All fragments of one event with fragment multiplicity $M = \sum_Z n_Z$, where n_Z is the number of fragments with charge Z in the partition, are taken into account. By means of the normalized first order:

$$\langle Z\rangle = \frac{1}{M}\sum_Z n_Z Z \tag{1}$$

and second order:

$$\sigma_Z^2 = \frac{1}{M}\sum_Z n_Z(Z-\langle Z\rangle)^2 \tag{2}$$

moments of the fragment charge distribution in the event, one may define the correlation function (CF):

$$1 + R(\sigma_Z, \langle Z\rangle) = \left.\frac{Y(\sigma_Z,\langle Z\rangle)}{Y'(\sigma_Z,\langle Z\rangle)}\right|_M \tag{3}$$

Here, the numerator $Y(\sigma_Z, \langle Z\rangle)$ is the yield of events with given $\langle Z\rangle$ and σ_Z values. Because the measurement of the charge belonging to a given event is not subject to statistical fluctuations, we use here expression (2) rather than the "nonbiased estimator" of the variance, $\frac{1}{M-1}\sum_Z n_Z(Z-\langle Z\rangle)^2$, as proposed in [25]. The denominator $Y'(\sigma_Z, \langle Z\rangle)$ represents the uncorrelated yield. Its retained evaluation is based on the "intrinsic probability", $^{intr}P_Z$, of emission of a given charge. A unique set of intrinsic probabilities exists only if the chosen event sample is extracted from single-source events. The denominator partition probabilities are built to take into account in a combinatorial way the trivial correlations due to charge conservation. Thus, this denominator replicates all features of the partitions of the numerator, except the correlations due to other reasons than charge conservation. The probability to observe a given partition ($\mathbf{n} : (n_1, \ldots, n_{Z_{max}})$), at a given total multiplicity, $m = \sum_Z n_Z$, is obtained by the multinomial formula. If the total charge is fixed ($Z_{tot} = \sum_Z Z n_Z$), the partition probabilities are given by:

$$P(\mathbf{n}|m) = \alpha\ m! \prod_Z \frac{^{intr}P_Z^{n_Z}}{n_Z!}\ \delta_{Z_{tot},\sum_Z Z n_Z}\ , \tag{4}$$

where α is the normalization constant (so that $\sum_n P(\mathbf{n}|m) = 1$) and δ is the Kronecker symbol. All the details can be found in Refs. [18, 26]. The $^{intr}P_Z$ values are obtained by means of a recursive procedure of minimization which stops when the decreasing between two steps is lower than 10^{-12}. The minimization criterion is the normalized χ^2 between the experimental partition probabilities and the calculated partition probabilities for a considered set of data. The calculated χ^2/N_{dof} were always in the range 1.0-0.7.

The considered event samples for the study were those with fragment (Z ≧ 5) multiplicities, M, from 3 to 6 which correspond to higher statistics. Figure 1 of [24] shows one

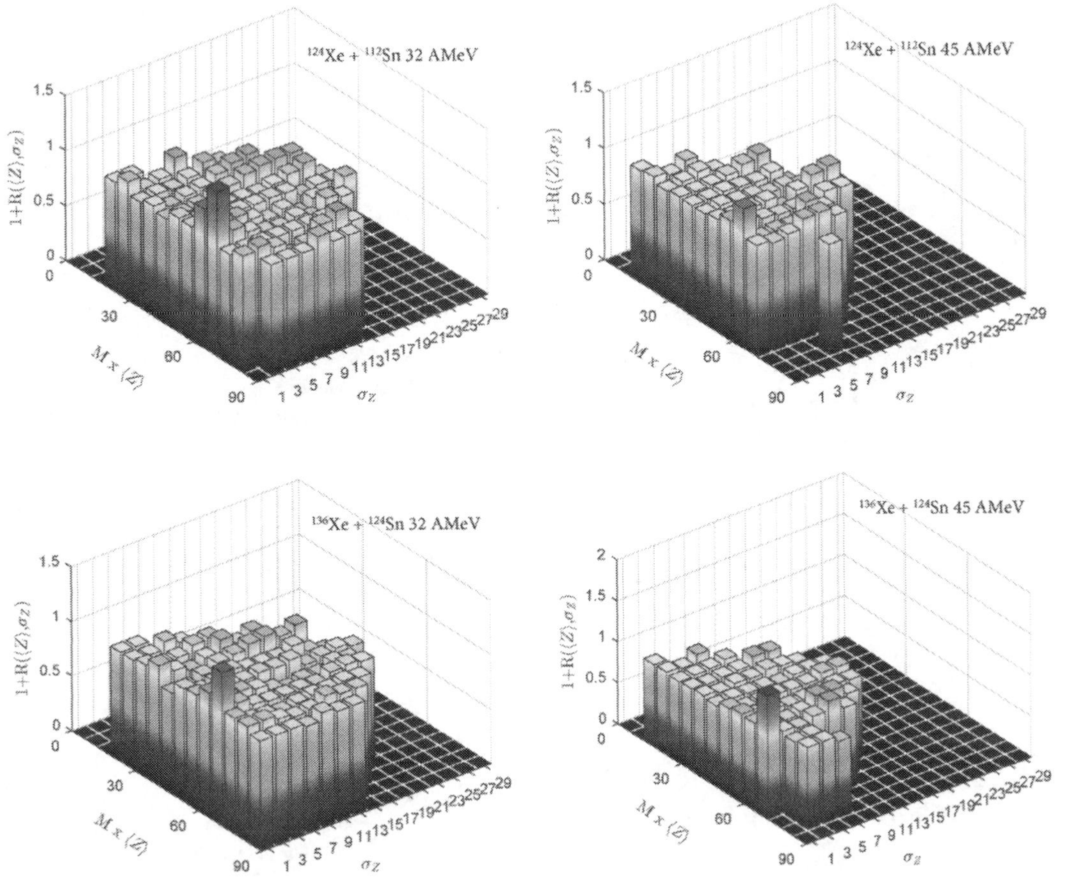

Figure 1. Experimental correlation functions for selected quasifusion events formed in central ^{124,136}Xe + ^{112,124}Sn collisions. Events with fragment multiplicities 3 to 6 are mixed. Correlation functions are calculated for a σ_Z bin equal to 2 and a $M\times <Z>$ bin equal to 6. The color scale for CFs has a maximum value for 1.6 which corresponds to dark red. From [24].

example on how the experimental fragment charge distributions are faithfully described by using the intrinsic probabilities, $^{intr}P_Z$, which have been calculated independently for each incident energy, for each reaction and for the different fragment multiplicities. CF values greater than one were observed at very low σ_Z (< 1) but also for σ_Z (1 - 2). This observation was used to fix the upper limit at 2 for the σ_Z of events with nearly equal-sized fragments. The complementary contribution (σ_Z 1 - 2) comes from the broadening of the fragment Z distribution introduced by the de-excitation of primitive fragments (see [24] for details). For the first time, the limited ranges of $\langle Z\rangle$ contributing to CF peaks were also clearly observed, which verifies what is theoretically expected for finite systems i.e. $M \times \langle Z\rangle \sim$ constant. To better visualize global results, CFs were built for all events of a reaction at a given beam energy, whatever their multiplicity, by summing the correlated yields of all M and by replacing the variable $\langle Z\rangle$ by $M\times\langle Z\rangle$. Uncorrelated yields are then constructed and weighted in proportion to real events of each multiplicity. Fig. 1 summarizes the results. For the four systems CF peaks are observed at low σ_Z. At 32 MeV per nucleon incident energy the neutron poor system exhibits two peaks with confidence levels greater around 6 - 7 σ and the neutron rich one peak above 6 σ and one around 3 σ, which definitively establishes the presence of spinodal fluctuations. At the higher incident energy the two systems have one peak with confidence level above 2 σ. Covered $M \times \langle Z\rangle$ domains are the same (60 - 72) for both reactions at 32 MeV per nucleon incident energy whereas at higher incident energy the neutron rich system covers a range a little bit higher (66 - 72) than the neutron poor one (54 - 66). Finally the percentages of events ($Y(\sigma_Z, \langle Z\rangle)$/total number of events) and extra events (($Y(\sigma_Z, \langle Z\rangle)$ - $Y'(\sigma_Z, \langle Z\rangle)$) /total number of events) are reported in Table 1. Within error bars, extra event percentages are similar for both systems at the lower incident energy. At higher incident energy we observe a strong reduction of percentages. The large reduction of the signal observed for the neutron rich system, a factor ten between 32 and 45 MeV per nucleon, can be un-

Table 1. Percentages of events and extra events with $\sigma_Z < 2$ for the different incident energies and reactions. Calculated errors are statistical. From [24].

E (AMeV)	reaction	evts	extra evts
32	^{124}Xe+^{112}Sn	0.27	0.068 ± 0.004
32	^{136}Xe+^{124}Sn	0.32	0.064 ± 0.004
45	^{124}Xe+^{112}Sn	0.34	0.025 ± 0.003
45	^{136}Xe+^{124}Sn	0.030	0.0065 ± 0.0017

derstood in terms of the expected N/Z influence. Indeed, if spinodal instabilities are at the origin of the dynamics of multifragmentation, as said before, theoretical calculations predict a reduction of instabilities for asymmetric systems in relation with an increase of the instability growth time. More precisely, it is shown in [17] that, for Sn isotopes, the most unstable modes associated to shorter instability growth times ($\simeq$ 50 fm/c) disappear when N/Z changes from 1.40 to 1.64. If we consider that quasifusion systems produced by the collisions, with N/Z varying from 1.27 to 1.50, have to stay long enough in the spinodal region ($\sim$ 3 characteristic times) to allow important amplification of the initial fluctuations, one can qualitatively understand the large extra reduction of the correlation signal for the neutron rich system at high incident energy as coming from insufficient reaction time. Such a situation also favours coalescence of primary fragments. Finally, note that the set of reaction trajectories in the density - temperature plane, close to the border of the spinodal region at 45 MeV per nucleon can be slightly different for the two reactions.

4 Conclusions

To summarize on these experimental results one can say that, using charge correlations, the fossil signature of spinodal instabilities i.e. the abnormal presence of nearly equal-sized fragments, even if very low as expected, was definitively established at a confidence level of around 6-7 σ. Associated to this weak signal, it is important to underline again the dominating role of chaotic/stochastic dynamics driven by spinodal instabilities for fragment formation, especially for finite systems. It has to do with beating of modes, breaking of translational symmetry and coalescence during the final step of fragment formation.

From universal fluctuations (Δ-scaling) the size of the largest fragment of each partition was determined as an order parameter [2, 27–29]. The determination of such an order parameter corresponds to an aggregation scenario which agrees with the process of spinodal fragmentation. Indeed, as far as phase transition dynamics is concerned the aggregation scenario recalls microscopic approaches in which fragments result from spinodal fluctuations occurring in the hot, expanding nuclear matter formed in collisions.

References

[1] P. Chomaz et al. (eds.) Eur. Phys. J. A **30**, (2006)
[2] B. Borderie, J.D. Frankland, Prog. Part. Nucl. Phys. **105**, 82 (2019)
[3] P. Chomaz et al., Phys. Rev. Lett. **85**, 3587 (2000)
[4] M. D'Agostino et al., Phys. Lett. B **473**, 219 (2000)
[5] B. Borderie, J. Phys. G: Nucl. Part. Phys. **28**, R217 (2002)
[6] N. Le Neindre et al. (INDRA Collaboration), Nucl. Phys. A **795**, 47 (2007)
[7] B. Borderie et al. (INDRA Collaboration), Phys. Lett. B **723**, 140 (2013)
[8] P. Chomaz et al., Phys. Rev. E **64**, 046114 (2001)
[9] E. Bonnet et al. (INDRA Collaboration), Phys. Rev. Lett. **103**, 072701 (2009)
[10] M. Colonna, P. Chomaz and A. Guarnera., Nucl. Phys. A **613**, 165 (1997)
[11] P. Napolitani and M. Colonna, Phys. Rev. C **96**, 054609 (2017)
[12] A. Ono, Phys. Rev. C **59**, 853 (1999)
[13] A. Ono et al., Phys. Rev. C **66**, 014603 (2002)
[14] E. Bonnet et al., Phys. Rev. C **89**, 034608 (2014)
[15] B. Borderie, M.F. Rivet, Prog. Part. Nucl. Phys. **61**, 551 (2008) and references therein.
[16] B. Jacquot et al., Phys. Lett. B **383**, 247 (1996)
[17] M. Colonna, P. Chomaz and S. Ayik, Phys. Rev. Lett. **88**, 122701 (2002)
[18] G. Tăbăcaru et al., Eur. Phys. J. A **18**, 103 (2003)
[19] J. Pouthas et al., Nucl. Instr. and Meth. in Phys. Res. A **357**, 418 (1995)
[20] J. Pouthas et al., Nucl. Instr. and Meth. in Phys. Res. A **369**, 222 (1996)
[21] G. Tăbăcaru et al., INDRA Collaboration, Nucl. Instr. and Meth. in Phys. Res. A **428**, 379 (1999)
[22] M. Pârlog et al., INDRA Collaboration, Nucl. Instr. and Meth. in Phys. Res. A **482**, 674 (2002)
[23] M. Pârlog et al., INDRA Collaboration, Nucl. Instr. and Meth. in Phys. Res. A **482**, 693 (2002)
[24] B. Borderie et al. (INDRA Collaboration), Phys. Lett. B **782**, 291 (2018)
[25] L. G. Moretto et al., Phys. Rev. Lett. **77**, 2634 (1996)
[26] P. Désesquelles, Phys. Rev. C **65**, 034604 (2002)
[27] R. Botet et al., Phys. Rev. E **62**, 1825 (2000)
[28] R. Botet et al., Phys. Rev. Lett. **86**, 3514 (2001)
[29] J.D. Frankland et al. (INDRA Collaboration), Phys. Rev. C **71**, 034607 (2005)

Coulomb excitation of pear-shaped nuclei

Peter Butler[1*], *Liam* Gaffney[1,2], *Pietro* Spagnoletti[3], *Joonas* Konki[2], *Marcus* Scheck[3], *John* Smith[3], *Kenzo* Abrahams[4], *Michael* Bowry[5], *Joakim* Cederkäll[6], *Timothy* Chupp[7], *Giacomo* de Angelis[8], *Hilde* De Witte[9], *Paul* Garrett[10], *Alina* Goldkuhle[11], *Corinna* Henrich[12], *Andres* Illana[8], *Karl* Johnston[2], *David* Joss[1], *James* Keatings[3], *Nicola* Kelly[3], *Michalina* Komorowska[13], *Thorsten* Kröll[12], *Miguel* Lozano[2], *Bondili* Nara Singh[3], *David* O'Donnell[3], *Joonas* Ojala[14,15], *Robert* Page[1], *Line* Pedersen[16], *Christopher* Raison[17], *Peter* Reiter[11], *Jose Alberto* Rodriguez[2], *Dawid* Rosiak[11], *Sebastian* Rothe[2], *Timur* Shneidman[18], *Burkhard* Siebeck[11], *Michael* Seidlitz[11], *Jacqueline* Sinclair[3], *Marek* Stryjczyk[9], *Piet* Van Duppen[9], *Silvia* Vinals[19], *Ville* Virtanen[14,15], *Nigel* Warr[11], *Kasia* Wrzosek-Lipska[13] and *Magda* Zielinska[20]

[1]Oliver Lodge Laboratory, University of Liverpool, Liverpool L69 7ZE, UK
[2]CERN, CH-1211 Geneva 23, Switzerland
[3]School of Computing, Engineering and Physical Sciences, University of the West of Scotland, Glasgow, PA1 2BE UK
[4]Department of Physics & Astronomy, University of the Western Cape, Private Bag X17, Bellville, 7535 South Africa.
[5]TRIUMF, Vancouver, V6T 2A3 BC, Canada
[6]Physics Department, Lund University, Box 118, Lund, SE-221 00 Sweden
[7]Department of Physics, University of Michigan, Ann Arbor, 48104 MI, USA
[8]INFN Laboratori Nazionali di Legnaro, Legnaro, 35020 PD, Italy
[9]Instituut voor Kern- en Stralingsfysica, KU Leuven, Leuven, B-3001 Belgium
[10]Department of Physics, University of Guelph, Guelph, N1G 2W1 Ontario, Canada
[11]Institute for Nuclear Physics, University of Cologne, Cologne, 50937 Germany.
[12]Institut für Kernphysik, Technische Universität Darmstadt, Darmstadt, 64289 Germany
[13]Heavy Ion Laboratory, University of Warsaw, Warsaw, PL-02-093 Poland.
[14]Department of Physics, University of Jyvaskyla, P.O. Box 35, Jyvaskyla, FIN-40014 Finland
[15]Helsinki Institute of Physics, P.O. Box 64, Helsinki, FIN-00014 Finland
[16]Department of Physics, University of Oslo, P.O.Box 1048, Oslo, N-0316 Norway.
[17]Department of Physics, University of York, York, YO10 5DD UK
[18]JINR Dubna, Dubna, 141980 Moscow Region, Russia.
[19]Consejo Superior De Investigaciones Científicas, Madrid, S 28040 Spain
[20]IRFU CEA, Université Paris-Saclay, Gif-sur-Yvette, F-91191 France.

Abstract. There is a large body of evidence that atomic nuclei can undergo octupole distortion and assume the shape of a pear. This phenomenon is important for measurements of electric-dipole moments of atoms, which would indicate CP violation and hence probe physics beyond the Standard Model of particle physics. Isotopes of both radon and radium have been identified as candidates for such measurements. Here, we have observed the low-lying quantum states in ^{224}Rn and ^{226}Rn by accelerating beams of these radioactive nuclei. We show that radon isotopes undergo octupole vibrations but do not possess static pear-shapes in their ground states. We conclude that radon atoms provide less favourable conditions for the enhancement of a measurable atomic electric-dipole moment.

1 Introduction

It is well established by the observation of rotational bands that atomic nuclei can assume quadrupole deformation with axial and reflection symmetry, usually with the shape of a rugby ball. The distortion arises from long-range correlations between valence nucleons which becomes favourable when the proton and/or neutron shells are partially filled. For certain values of proton and neutron number it is expected that additional correlations will cause the nucleus to also assume an octupole shape ('pear-shape') where it loses reflection symmetry in the intrinsic frame [1]. The fact that some nuclei can have pear shapes has influenced the choice of atoms having odd-*A* nuclei employed to search for permanent electric-dipole moments (EDMs). Any measurable moment will be amplified if the nucleus has octupole collectivity and further enhanced by static-octupole deformation. At present, experimental limits on EDMs, that would indicate charge-parity (CP) violation in fundamental processes where flavour is unchanged, have placed severe constraints on many extensions of the Standard Model. Recently, new candidate atomic species, such as radon and radium, have been proposed for EDM searches. For certain isotopes octupole effects are expected to enhance, by a factor 100-1000, the nuclear Schiff moment (the electric-dipole distribution weighted by radius squared) that induces the atomic EDM [2, 3, 4], thus improving the

* Corresponding author: peter.butler@liverpool.ac.uk

sensitivity of the measurement. There are two factors that contribute to the greater electrical polarizability that causes the enhancement: (i) the odd-A nucleus assumes an octupole shape; (ii) an excited state lies close in energy to the ground state with the same angular momentum and intrinsic structure but opposite parity. Such parity doublets arise naturally if the deformation is static (permanent octupole deformation).

The observation of low-lying quantum states in many nuclei with even Z, N having total angular momentum and parity of $I^{\pi} = 3^{-}$ is indicative of their undergoing octupole vibrations about a reflection-symmetric shape. Further evidence is provided by the sizeable value of the E3 moment for the transition to the ground state, indicating collective behaviour of the nucleons. However the number of observed cases where the correlations are strong enough to induce a static pear-shape is much smaller. Strong evidence for this type of deformation comes from the observation of a particular behaviour of the energy levels for the rotating quantum system and from an enhancement in the E3 moment [5]. So far there are only two cases, ^{224}Ra [6] and ^{226}Ra [7] for which both experimental signatures have been observed. The presence of a parity doublet of 55 keV at the ground state of ^{225}Ra makes this nucleus therefore a good choice for EDM searches [8]. In contrast to the radium isotopes, much less is known about the behaviour of radon (Rn) nuclei proposed as candidates for atomic EDM searches on account of possible enhancement of their Schiff moments [9-17]. For this reason, different isotopes of radon have been listed in the literature, for example ^{221,223,225}Rn [14], each having comparable half-lives and ground state properties. The most commonly chosen isotope for theoretical calculations [9,10] and the planning of experiments [11-13] is ^{223}Rn.

In this work [18] we present new data on the energy levels of heavy even-even Rn isotopes to determine whether parity doublets are likely to exist near the ground state of neighbouring odd-mass Rn nuclei. Direct observation of low-lying states in odd-A Rn nuclei (for example following Coulomb excitation or β-decay from the astatine parent) is presently not possible, as it will require significant advances in the technology used to produce radioactive ions. We observe that ^{224,226}Rn behave as octupole vibrators in which the octupole phonon is aligned to the rotational axis. We conclude that there are no isotopes of radon that have static octupole deformation, so that any parity doublets in the odd-mass neighbours will not be closely spaced in energy. This means that radon atoms will provide less favourable conditions for the enhancement of a measurable atomic electric-dipole moment.

2 Results

2.1 Measurement of the quantum structure of heavy radon isotopes

In the experiments described here, ^{224}Rn (Z = 86, N = 138) and ^{226}Rn (Z = 86, N = 140) ions were produced by spallation in a thick thorium carbide target bombarded by ~10^{13} protons s^{-1} at 1.4 GeV from the CERN PS Booster. The ions were accelerated in HIE-ISOLDE to an energy of 5.08 MeV per nucleon and bombarded secondary targets of ^{120}Sn. In order to verify the identification technique, another isotope of radon, ^{222}Rn, was accelerated to 4.23 MeV/u. The γ-rays emitted following the excitation of the target and projectile nuclei were detected in Miniball [19], an array of 24 high-purity germanium detectors, each with six-fold segmentation and arranged in eight triple-clusters. The scattered projectiles and target recoils were detected in a highly segmented silicon detector [20].

Prior to the present work, nothing was known about the energies and spins of excited states in ^{224,226}Rn, while de-exciting γ-rays from states in ^{222}Rn had been observed [21] with certainty up to $I^{\pi} = 13^{-}$. The chosen bombarding energies for ^{224,226}Rn were about 3% below the nominal Coulomb barrier energy at which the beam and target nuclei come close enough in head-on collisions for nuclear forces to significantly influence the reaction mechanism. For such close collisions the population of high-spin states will be enhanced, allowing the rotational behaviour of the nucleus to be elucidated. This is the method has been coined "unsafe Coulomb excitation" [22] as the interactions between the high-Z reaction partners is predominantly electromagnetic. It is not possible to precisely determine electromagnetic matrix elements because of the small nuclear contribution. The most intense excited states expected to be observed belong to the positive-parity rotational band, built upon the ground state. These states are connected by fast E2 transitions. In nuclei that are unstable to pear-shaped distortion, the other favoured excitation paths are to members of the octupole band, negative-parity states connected to the ground-state band by strong E3 transitions.

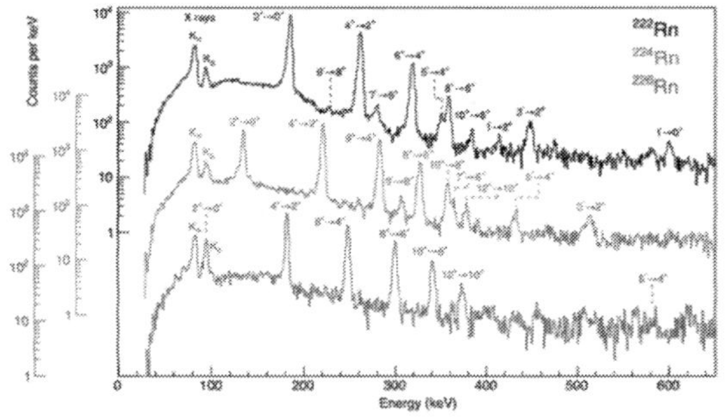

Figure 1. Spectra of γ -rays. The γ -rays were emitted following the bombardment of ^{120}Sn targets by ^{222}Rn (black), ^{224}Rn (blue), and ^{226}Rn (red). The γ-rays were corrected for Doppler shift assuming that they are emitted from the scattered projectile. Random coincidences between Miniball and CD detectors have been subtracted. The transitions which give rise to the observed full-energy peaks are labelled by the spin and parity of the initial and final quantum states. The assignments of the transitions from the negative parity states in ^{224,226}Rn are tentative (see text).

The spectra of γ-rays time-correlated with scattered beam and target recoils are shown in figure 1. The E2 γ-ray transitions within the ground-state positive-parity band can be clearly identified, as these de-excite via a regular sequence of strongly-excited states having spin and parity $0^+, 2^+, 4^+, \ldots$ with energies $\frac{\hbar^2}{2\Im} I(I+1)$. In this expression the moment-of-inertia $\Im$ systematically increases with increasing I (reducing pairing) and with number of valence nucleons (increasing quadrupole deformation). As expected from multi-step Coulomb excitation the intensities of the transitions systematically decrease with increasing I, after correcting for internal conversion and the γ-ray detection efficiency of the Miniball array.

The other relatively intense γ-rays observed in these spectra with energies < 600 keV are assumed to have E1 multipolarity, and to depopulate the odd-spin negative-parity members of the octupole band. In order to determine which states are connected by these transitions, pairs of coincident γ-rays were examined. In this analysis, the energy spectrum of γ-rays coincident with one particular transition is generated by requiring that the energy of this 'gating' transition lies in a specific range. Typical spectra obtained this way are shown in figure 2. Each spectrum corresponds to a particular gating transition, background subtracted, so that the peaks observed in the spectrum arise from γ-ray transitions in coincidence with that transition.

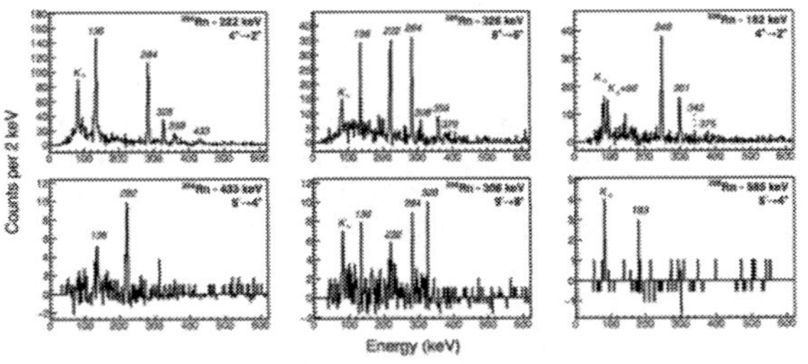

Figure 2. Coincidence γ-ray spectra. The representative background-subtracted γ-ray spectra are in time-coincidence with different gating transitions. Here the observed peaks are labelled by the energy (in keV) of the transition. The gating transition is additionally labelled by the proposed spin and parity of the initial and final states.

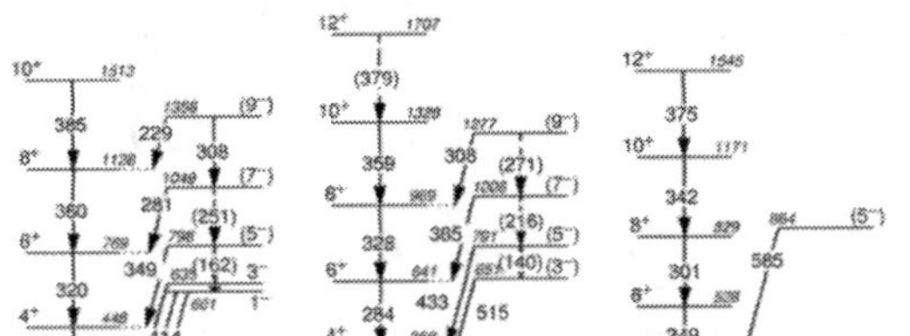

Figure 3. Level schemes. These partial level-schemes for ^{222,224,226}Rn show the excited states of interest. Arrows indicate γ-ray transitions. All energies are in keV. Firm placements of transitions in the scheme are from previous work [20] or have been made using γ - γ coincidence relations; otherwise in brackets.

The level schemes for ^{224,226}Rn constructed from the coincidence spectra, together with the known [20] scheme for ^{222}Rn, are shown in figure 3. For ^{226}Rn the energy of the strongly-converted $2^+ \rightarrow 0^+$ transition overlaps with those of the K_β X-rays, but its value can be determined assuming that the relative intensity of K_β, K_α X-rays is the same as for ^{222,224}Rn. The E2 transitions connecting the states in the octupole band are not observed because they cannot compete with faster, higher-energy E1 decays.

The only other plausible description for this band is that it has $K^\pi = 0^+$or 2^+, implying that the $K^\pi = 0^-$octupole band is not observed. This is unlikely as the bandhead would have to lie significantly lower in energy than has been observed in ^{222,224,226}Ra, and inter-band transitions from states with $I' > 4$ to states with I and I-2 in the ground-state band and in-band transitions to I'-2 would all be visible in the spectra. The spin and parity assignments for the positive-parity band that is strongly populated by Coulomb excitation can be regarded as firm, whereas the negative-parity state assignments are made in accord with the systematic behaviour of nuclei in this mass region.

2.2 Characterisation of octupole instability

From the level schemes and from the systematics for all the radon isotopes (figure 4) it is clear that the bandhead of the octupole band reaches a minimum around *N*=136. The character of the octupole bands can be explored [23] by examining the difference in aligned angular momentum, $\Delta i_x = i_x^- - i_x^+$, at the same rotational frequency ω, as a function of ω. Here i_x is approximately I for *K*=0 bands and $\hbar\omega$ is approximately $(E_I - E_{I-2})/2$.

For nuclei with permanent octupole deformation Δi_x is expected to approach zero, as observed for several isotopes of Ra, Th, and U [5]. For octupole vibrational nuclei in which the negative-parity states arise from coupling an octupole phonon to the positive-parity states, it is expected that $\Delta i_x \sim 3\hbar$ as the phonon prefers to align with the rotational axis. This is the case for the isotopes ^{218,220,222,224}Rn at values of $\hbar\omega$ (< 0.2 MeV) where particle-hole excitations do not play a role, see figure 4.

Thus we have clearly delineated the lower boundary at Z> 86 as to where permanent octupole deformation occurs in nature.

3. Discussion

The observation of octupole-vibrational bands in the even-even radon isotopes is consistent with several theoretical calculations [24-26], which predict that only nuclei with $Z > 86$ have stable octupole deformation. Other calculations suggest that radon isotopes with A~222 will have non-zero values of the octupole deformation parameter β_3[27-28]. For such nuclei which have a minimum in the nuclear potential energy at non-zero values of β_3, the positive- and negative-parity states are projected from intrinsic configurations having $K^\pi = 0^+, 0^-$ which are degenerate in energy. In the odd-*A* neighbours parity doublets arise by coupling the odd particle to these configurations.

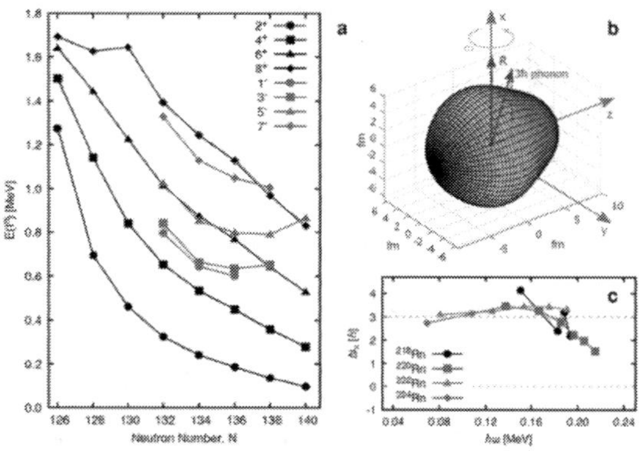

Figure 4. Systematic behaviour of radon isotopes. (a) Systematics of the energies for different spins of low-lying positive-parity (black) and negative-parity states (red) in radon isotopes; (b) cartoon illustrating how the octupole phonon vector aligns with the rotation (R) vector (which is orthogonal to the rotating body's symmetry axis) so that I=R+3$\hbar$ and $\Delta i_x = 3\hbar$; (c) difference in aligned spin for negative- and positive-parity states in $^{218\text{-}224}$Rn (re-analysed for $^{218\text{-}222}$Rn that have been presented earlier [23]). The dashed line at $\Delta i_x = 0$ is the expected value for static octupole deformation.

This is not the case for reflection-symmetric nuclei that undergo octupole vibrations around $\beta_3 = 0$. Bands of opposite parity with differing single-particle configurations can lie close to each other fortuitously [29-30] but in general those arising from coupling the odd nucleon to the ground state and octupole phonon will be well separated. The separation will be determined by the spacing of the bands in the even-even core, ~500 keV in the case of $^{222\text{-}226}$Rn (see figure 4), and will be in general much larger than that the value (~50 keV) observed for parity doublets in radium isotopes [1]. Quantitative estimates of Schiff moments for octupole-vibrational systems have yet to be made [31-32]. Nevertheless, it can be concluded that, if measurable CP-violating effects occur in nuclei, the enhancement of nuclear Schiff moments arising from octupole effects in odd-*A* radon nuclei is likely to be much smaller than for heavier octupole-deformed systems.

4. Outlook

We have also carried out measurements of γ-ray yields following the Coulomb excitation of ^{222}Rn, ^{222,228}Ra, at bombarding energies sufficiently below the Coulomb barrier to allow the extraction of multipole matrix elements using the GOSIA analysis technique [33]. Typical spectra are shown in figure 5. The analysis of these data is in progress; we expect to obtain E1, E2 and E3 matrix elements for transitions between states in the low-lying collective band in these nuclei.

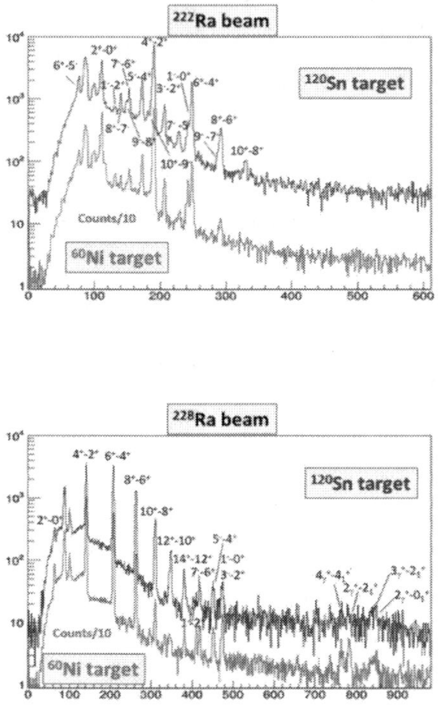

Figure 5. Spectra of γ-ray transitions in ^{222}Ra (upper) and ^{228}Ra (lower) following bombardment of ^{120}Sn (blue) and ^{60}Ni (red) targets. The γ-rays are corrected for Doppler shift assuming that they are emitted from the scattered projectile.

We are grateful to Niels Bidault, Eleftherios Fadakis, Erwin Siesling, and Fredrick Wenander who assisted with the preparation of the radioactive beams, and we thank Jacek Dobaczewski for useful discussions. The support of the ISOLDE Collaboration and technical teams is acknowledged. This work was supported by the following Research Councils and Grants: Science and Technology Facilities Council (STFC; UK) grants ST/P004598/1, ST/L005808/1; Federal Ministry of Education and Research (BMBF; Germany) grant 05P18RDCIA, 05P15PKCIA and 05P18PKCIA and the "Verbundprojekt 05P2018"; National Science Centre (Poland) grant 2015/18/M/ST2/00523; European Union's Horizon 2020 Framework research and innovation programme 654002 (ENSAR2); Marie Skłodowska-Curie COFUND grant (EU-CERN) 665779; Research Foundation Flanders (FWO, Belgium), by GOA/2015/010 (BOF KU Leuven) and the Interuniversity Attraction Poles Programme initiated by the Belgian Science Policy Office (BriX network P7/12); RFBR(Russia) grant 17-52-12015.

References

1. P.A. Butler, W. Nazarewicz, Rev. Mod. Phys. **68**, 349–421 (1996)

2. V. Spevak, N. Auerbach, V.V. Flambaum, Phys. Rev. C **56**, 1357–1369 (1997)
3. J. Dobaczewski, J. Engel, Phys. Rev. Lett. **94**, 232502 (2005)
4. J. Ellis, J. Lee, A. Pilaftsis, J. High Energy Phys. **2011**, 045 (2011)
5. P.A. Butler, J. Phys. G: Nucl. Part. Phys. **43,** 073002 (2016)
6. L.P. Gaffney et al., Nature **497**, 199-204 (2013)
7. H.J. Wollersheim et al., Nucl. Phys. A **556**, 261–280 (1993)
8. M. Bishop et al., Phys. Rev. C **94**, 025501 (2016)
9. V.A. Dzuba, V.V. Flambaum, J.S.M. Ginges, M.G. Kozlov, Phys. Rev. A **66**, 012111 (2002)
10. J. Dobaczewski et al., Phys. Rev. Lett. **121**, 232501 (2018)
11. E.R. Tardiff, et al., Phys. Rev. C **77**, 052501(R) (2008)
12. E.T. Rand et al., Journal of Physics: Conference Series **312**, 102013 (2011)
13. E.R. Tardiff et al., Hyperfine Interact **225**, 197–206 (2014)
14. J.A. Behr, G. Gwinner, J. Phys. G: Nucl. Part. Phys. **36,** 033101 (2009)
15. J. Engel, M.J. Ramsey-Musolf, U. van Kolck, Prog. Part. Nucl. Phys. **71**, 21-74 (2013)
16. N. Yamanaka, Eur. Phys. J. A **53**, 54 (2017)
17. T.E. Chupp, P. Fierlinger, M.J. Ramsey-Musolf, J.T. Singh, Rev. Mod. Phys. **91,** 015001 (2019)
18. P.A. Butler et al., Nat. Comm. **10**, 2473 (2019)
19. N. Warr, et al., Eur.Phys. J. A **49**, 40 (2013)
20. A. Ostrowski et al., Nucl. Instrum. Methods A **480**, 448–455 (2002)
21. S. Singh, A. Jain, J.K. Tuli, Nucl. Data Sheets **112**, 2851–2886 (2011)
22. I. Wiedenhöver et al., Phys. Rev. Lett. **83**, 2143-6 (1999)
23. J.F.C. Cocks et al., Nucl. Phys. A **645**, 61–91 (1999).
24. W. Nazarewiczet al., Nucl. Phys. A **429,** 269-295 (1984)
25. L.M. Robledo, P.A. Butler, Phys. Rev. C **88**, 051302 (R) (2013)
26. S.E. Agbemava, A.V. Afanasjev, P. Ring, Phys. Rev. C **93**, 044304 (2016)
27. P. Möller, et al, Atom. Data Nucl. Data Tables **94,** 758-780 (2008)
28. L.M. Robledo, G.F. Bertsch, Phys. Rev. C **86**, 054306 (2012)
29. W.J. Vermeer et al, Phys. Rev. C **42**, R1183-1186 (1990)
30. J. Wiśniewski, et al., Phys. Rev. C **96**, 064301 (2017)
31. N. Auerbach, et al., Phys. Rev. C **74**, 025501 (2006)
32. V. Zelevinsky, A. Volya, N. Auerbach, Phys. Rev. C **78**, 014310 (2008)
33. M. Zielinska et al., Eur. Phys. J. A. **52**, 99 (2016)

Three-body description of ^{12}C: From the hyperspherical formulation to the algebraic cluster model and its application to α + ^{12}C inelastic scattering

J. Casal[1,*], *L.* Fortunato[1], *A.* Vitturi[1], and *E. G.* Lanza[2]

[1]Dipartimento di Fisica e Astronomia "G. Galilei" and INFN Sezione di Padova, I-35131 Padova, Italy.
[2]INFN Sezione di Catania, I-95123, Catania, Italy

Abstract. Form factors for $\alpha+{}^{12}$C inelastic scattering are obtained within two theoretical ($\alpha+\alpha+\alpha$) approaches: The hyperspherical framework for three identical bosons, and the algebraic cluster model assuming the D_{3h} symmetry of an equilateral triangle subject to rotations and vibrations. Results show a good agreement, with form factors involving the Hoyle state having a slightly larger extension within the hyperspherical approach. Coupled-channel calculations using these form factors are ongoing.

1 Introduction

The low-lying structure of ^{12}C is still one of the most fascinating open problems in nuclear physics. Alpha-clusterization and the nature of the so-called Hoyle state, which plays a crucial role in nucleosynthesis, have attracted special interest. Microscopic theories (e.g., [1]) support the existence of three-alpha cluster configurations for the ^{12}C nucleus, a fact which justifies the use of cluster models (e.g., [2]) and algebraic methods (e.g., [3]). These approaches, although simpler, are particularly suitable for the description of reaction observables. Experimentally, different probes have been extensively used to access the properties of its ground and excited states. We discuss here the case of α + ^{12}C inelastic scattering studied within two theoretical three-body approaches: The hyperspherical formalism, and the algebraic cluster model. Our goal is to compare form factors for inelastic scattering in these two approaches and set the basis for full coupled-channel calculations.

2 Three-body calculations

As in Refs. [2, 4], the $\alpha + \alpha + \alpha$ problem can be solved within the hyperspherical formalism [5], which has been successfully applied to describe the properties of Borromean nuclei such as the two-neutron halo systems ^{6}He or ^{11}Li, or the weakly bound stable nucleus ^{9}Be [6]. Within this framework, the states of the three-body system can be written as

$$\Psi^{j\mu}(\rho,\Omega) = \rho^{-5/2}\sum_{\beta}\chi_{\beta}^{j}(\rho)\Upsilon_{\beta}^{j\mu}(\Omega), \quad (1)$$

where ρ is the hyper-radius, Ω contains all the angular dependence, functions $\Upsilon_{\beta}^{j\mu}$ are the so-called hyperspherical harmonics, and χ_{β}^{j} are the hyperradial functions to be determined. Here, β is the set of quantum numbers defining each channel. In the case of three zero-spin particles, $\beta \equiv \{K, l_x, l_y\}$, where l_x and l_y are the orbital angular momenta associated to the usual Jacobi coordinates, so that $\boldsymbol{j} = \boldsymbol{l_x} + \boldsymbol{l_y}$ gives the total angular momentum, and K is the so-called hypermomentum. Note that l_x has to be even for identical bosons. Details can be found in Ref. [2]. The hyperangular functions and $\chi_{\beta}^{j}(\rho)$ can be expanded in a given basis,

$$\chi_{\beta}^{j}(\rho) = \sum_{i} C_{i\beta}^{j} U_{i\beta}(\rho). \quad (2)$$

Index i runs over the number of hyperradial excitations included, and coefficients $C_{i\beta}^{j}$ are obtained upon diagonalization of the three-body Hamiltonian with coupling potentials

$$V_{\beta'\beta}^{j\mu}(\rho) = \left\langle \Upsilon_{\beta}^{j\mu}(\Omega) \middle| \sum_{j>i=1}^{3} V_{ij}(\rho,\alpha) \middle| \Upsilon_{\beta'}^{j\mu}(\Omega) \right\rangle + \delta_{\beta\beta'} V_{3b}^{j}(\rho). \quad (3)$$

This contains all binary interactions, integrated over the angular dependence, and a phenomenological three-body force to account for effects not explicitly included in the binary interactions [7]. In this work, calculations are performed with the Ali-Bodmer α-α nuclear potential [8] together with a hard-sphere Coulomb term. Radial functions are expanded in a THO basis [6], and the three-body force is adjusted to reproduce the energies of the first 2_1^+ bound excited state and the 0_2^+ Hoyle state in ^{12}C, which have a well-developed three-body character [1].

The set of coupled differential equations to describe $\alpha + {}^{12}$C inelastic scattering requires the computation of radial form factors involving the projectile-target interaction. Within the hyperspherical three-body framework, this is a four-body problem involving the α-α potential. Between

*e-mail: casal@pd.infn.it

https://doi.org/10.1051/epjconf/201922301008

states labeled i and j, the form factor is simply

$$F_{ij}(\vec{R}) = \langle \Psi_i | \sum_{q=1}^{3} V_{\alpha\alpha}(|\vec{r}_q - \vec{R}|) | \Psi_j \rangle, \quad (4)$$

where $\vec{R}$ is the projectile-target distance and $\vec{r}_q$ is the position of each α within the target. For this purpose, the potential we use is derived from the double folding of two Gaussian densities, adjusted to reproduce the radius of the α particle, with the M3Y nucleon-nucleon interaction [9]. This way, the computed form factors will be consistent with those described in the next section within the formalism of densities and transition densities.

3 Algebraic cluster model

Assuming the D_{3h} symmetry corresponding to an equilateral triangle, the density for the ground-state band for ^{12}C can be writte as [3]

$$\rho_{\text{g.s.}}(\vec{r}, \{\vec{r}_k\}) = \sum_{k=1}^{3} \rho_\alpha(\vec{r} - \vec{r}_k), \qquad \rho_\alpha(\vec{r}) = \left(\frac{d}{\pi}\right)^{3/2} e^{-dr^2}, \quad (5)$$

with $d = 0.56(2)$ fm^{-2} to reproduce the radius of the α particle and $\{\vec{r}_k\}$ at the vertices of the triangle, i.e., $\vec{r}_1 = (\beta, \pi/2, 0)$, $\vec{r}_2 = (\beta, \pi/2, 2\pi/3)$ and $\vec{r}_3 = (\beta, \pi/2, 4\pi/3)$ in spherical coordinates. The radial parameter $\beta = 1.82$ fm ensures the 0_1^+ ground-state radius and $B(E2)$ value to the first 2_1^+ state are reproduced. Expanded in spherical harmonics, it takes the form

$$\rho_{\text{g.s.}}(\vec{r}) = \sum_{\lambda\mu} \rho_{\text{g.s.}}^{\lambda\mu}(r) Y_{\lambda\mu}(\theta, \varphi), \quad (6)$$

where only the multipoles allowed by the D_{3h} symmetry appear in the sum. The 00 term represents the 0_1^+ ground-state density, while others represent the change in density for transitions to higher lying states of the same band, e.g., 20 is the term associated to the 2_1^+ state.

By considering now symmetric vibrations $\Delta\beta^A$ along the radial direction, we can construct the band associated to the Hoyle state as a breathing mode. The transition densities connecting the ground-state band with this A-type band can be obtained in leading order as

$$\delta\rho_{\text{g.s.}\to A}(\vec{r}) = \chi_1 \frac{d}{d\beta} \rho_{\text{g.s.}}(\vec{r}, \beta), \quad (7)$$

with $\chi_1 \simeq 0.247$ to recover the experimental monopole matrix element $M(E0)$. Again, expanding in multipoles one gets

$$\delta\rho_{\text{g.s.}\to A}(\vec{r}) = \sum_{\lambda\mu} \delta\rho_{\text{g.s.}\to A}^{\lambda\mu}(r) Y_{\lambda\mu}(\theta, \varphi). \quad (8)$$

For details about these densities and transition densities, see Refs. [10, 11]. In this case, form factors for $\alpha +^{12}$C inelastic scattering can be obtained following a double folding procedure with the M3Y interaction introduced above,

$$F_{ij}(\vec{R}) = \int \int \rho_\alpha(\vec{r}_1 - \vec{R}) v_{NN}(|\vec{r}_{12}|) \delta\rho^{i\to j}(\vec{r}_2) d\vec{r}_1 d\vec{r}_2. \quad (9)$$

4 Comparison of form factors

The form factors for the $0_1^+ \to 0_2^+$, $0_1^+ \to 2_1^+$ and $2_1^+ \to 0_2^+$ transitions, computed within the two approaches, are shown for comparison in Fig. 1. A rather good agreement is observed, even though they come from very different theoretical approaches. This may indicate that the two models capture essentially the same geometrical properties of ^{12}C. Form factors connecting the bound states ($0_1^+, 2_1^+$) with the Hoyle state (0_2^+) seem to exhibit a larger extension within the hyperspherical approach. This may have implication for the corresponding cross section. Full coupled-channel calculations involving these form factors, as well as those connecting other low-lying states in ^{12}C, are ongoing and will be presented elsewhere.

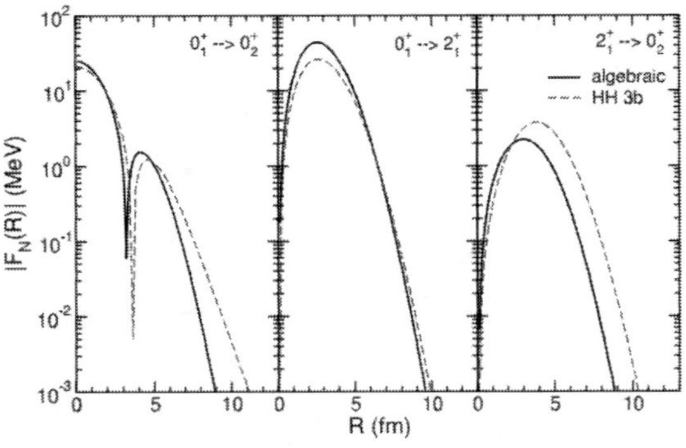

Figure 1. Nuclear form factors for α + ^{12}C inelastic scattering. Comparison between the three-body hyperspherical formalism (dashed) and the algebraic cluster model (solid). See text.

References

[1] M. Chernykh, H. Feldmeier, T. Neff, P. von Neumann-Cosel, A. Richter, Phys. Rev. Lett. **98**, 032501 (2007)
[2] N.B. Nguyen, F.M. Nunes, I.J. Thompson, Phys. Rev. C **87**, 054615 (2013)
[3] V. Della Rocca, R. Bijker, F. Iachello, Nucl. Phys. A **966**, 158 (2017)
[4] S. Ishikawa, Phys. Rev. C **90**, 061604 (2014)
[5] M.V. Zhukov, B.V. Danilin, D.V. Fedorov, J.M. Bang, I.J. Thompson, J.S. Vaagen, Phys. Rep. **231**, 151 (1993)
[6] J. Casal, M. Rodríguez-Gallardo, J.M. Arias, I.J. Thompson, Phys. Rev. C **90**, 044304 (2014)
[7] I.J. Thompson, F.M. Nunes, B.V. Danilin, Comput. Phys. Commun. **161**, 87 (2004)
[8] S. Ali, A.R. Bodmer, Nucl. Phys. **80**, 99 (1966)
[9] G. Bertsch, J. Borysowicz, H. McManus, W.G. Love, Nucl. Phys. A **284**, 399 (1977)
[10] A. Vitturi, J. Casal, L. Fortunato, E.G. Lanza, arXiv:1901.07954 (2019)
[11] A. Vitturi, J. Casal, L. Fortunato, E.G. Lanza, in preparation (2019)

Recent results on heavy-ion induced reactions of interest for neutrinoless double beta decay at INFN-LNS

Manuela Cavallaro[1*], *Luis* Acosta[2], *Clementina* Agodi[1], *Carmen* Altana[1], *Paulina* Amador-Valenzuela[3], *Naftali* Auerbach[4], *Jose* Barea[5], *Jessica I.* Bellone[1], *Roelof* Bijker[6], *Danilo* Bonanno[7], *Thereza* Borello-Lewin[8], *Ismail* Boztosun[9], *Vincenzo* Branchina[1,7], *Sandro* Brasolin[10], *Giuseppe A.* Brischetto[1,11,13], *Oscar* Brunasso[10], *Stefano* Burrello[1], *Salvatore* Calabrese[1,2], *Luciano* Calabretta[1], *Daniela* Calvo[10], *Vittoria* Capirossi[10,12], *Francesco* Cappuzzello[1,11], *Diana* Carbone[1], *Luis E.* Charon Garcia[2], *Efrain R.* Chávez Lomeli[2], *Ruofu* Chen[14], *Irene* Ciraldo[1,11], *Maria* Colonna[1], *Grazia* D'Agostino[1], *Franck* Delaunay[10,12,15], *Nikit* Deshmukh[16], *Haris* Djapo[9], *Gianluigi* De Geronimo[17], *K.* De Los Rìos[2], *Carlo* Ferraresi[18], *Jonas L.* Ferreira[19], *Jacopo* Ferretti[20,21], *Paolo* Finocchiaro[1], *Suna* Firat[9], *Maria* Fisichella[10], *Antonino* Foti[7], *Giuseppe* Gallo[1,11], *Hugo* Garcia-Tecocoatzi[20,22] *Aylin* Hacisalihoglu[23], *Arcadio* Huerta-Hernandez[2], *Jenni* Kotila[24], *Yasemin* Kucuk[9], *Felice* Iazzi[10,12], *Gaetano* Lanzalone[1,25], *José* Antonio Lay[26], *Laura* La Fauci[1,11], *Francesco* La Via[27], *Horst* Lenske[28], *Roberto* Linares[19], *Domenico* Lo Presti[11,7], *Jesus* Lubian[19], *Junbing* Ma[14], *Daniel* Marin-Lámbarri[2], *J.* Mas Ruiz[2], *Nilberto* H. Medina[8], *Djalma* R. Mendes[19], *Paolo* Mereu[10], *Mauricio* Moralles[29], *Lorenzo* Neri[1], *José R.* B. Oliveira[8], *Athena* Pakou[30], *Luciano* Pandola[1], *Horia* Petrascu[31], *Norbert* Pietralla[32], *Federico* Pinna[10,12], *Santo* Reito[7], *Philipp* C. Ries[32], *Marcia R.* D. Rodrigues[8], *Antonio D.* Russo[1], *Giuseppe* Russo[7,11], *Elena* Santopinto[20], *Roberto B.* B. Santos[33], *Leonardo* Serbina[31], *Onoufrios* Sgouros[1], *Marcilei A.* G. da Silveira[33], *Selçuk* O. Solakci[9], *George* Souliotis[34], *Vasileios* Soukeras[1], *Alessandro* Spatafora[1,11], *Domenico* Torresi[1], *Salvatore* Tudisco[1], *Hesiquio* Vargas Hernandez[2], *Ruslan I. M.* Vsevolodovna[20,22], *Jan Song* Wang[14], *Volker* Werner[32], *Yan Yun* Yang[14], *Aydin* Yildirin[9], *Vinicius A.* B. Zagatto[8]

for the NUMEN Collaboration

[1]Istituto Nazionale di Fisica Nucleare, Laboratori Nazionali del Sud, Catania, Italy
[2]Instituto de Fısica, Universidad Nacional Autonoma de Mexico - Mexico City, Mexico
[3]Instituto Nacional de Investigaciones Nucleares - Ocoyoacac, Mexico
[4]School of Physics and Astronomy, Tel Aviv University - Tel Aviv, Israel
[5]Universidad de Concepcion - Concepcion, Chile
[6]Instituto de Ciencias Nucleares, Universidad Nacional Autonoma de Mexico - Mexico City, Mexico
[7]Istituto Nazionale di Fisica Nucleare, Sezione di Catania, Italy
[8]Instituto de Fisica, Universidade de Sao Paulo - Sao Paulo, Brazil
[9]Department of Physics, Akdeniz University - Antalya, Turkey
[10]Istituto Nazionale di Fisica Nucleare, Sezione di Torino, Italy
[11]Dipartimento di Fisica e Astronomia "Ettore Majorana", Università di Catania, Italy
[12]DISAT, Politecnico di Torino, Italy
[13]Centro Siciliano di Fisica Nucleare e Struttura della materia, Italy
[14]Institute of Modern Physics, Chinese Academy of Sciences - Lanzhou, China
[15]LPC Caen, Normandie Universite, ENSICAEN, UNICAEN, CNRS/IN2P3 Caen, France
[16]Nuclear Physics Division, Saha Institute of Nuclear Physics - Kolkata, India
[17]Stony Brook University - Stony Brook, NY, USA
[18]DIMEAS, Politecnico di Torino - Torino, Italy
[19]Instituto de Fisica, Universidade Federal Fluminense - Niteroi, Brazil
[20]Istituto Nazionale di Fisica Nucleare, Sezione di Genova - Genova, Italy
[21]Department of Physics, Yale University - New Haven, CT, USA
[22]Dipartimento di Fisica, Università di Genova - Genova, Italy
[23]Institute of Natural Sciences, Karadeniz Teknik University - Trabzon, Turkey
[24]University of Jyvaskyla - Jyvaskyla, Finland
[25]Università degli Studi di Enna "Kore" - Enna, Italy
[26]Departamento de FAMN, Universidad de Sevilla - Sevilla, Spain
[27]CNR-IMM, Sezione di Catania - Catania, Italy
[28]Department of Physics, University of Giessen - Giessen, Germany
[29]Instituto de Pesquisas Energeticas e Nucleares IPEN/CNEN - Sao Paulo, Brazil
[30]Department of Physics and HINP, University of Ioannina - Ioannina, Greece
[31]IFIN-HH - Magurele, Romania
[32]Institut fur Kernphysik, Technische Universitat Darmstadt - Darmstadt, Germany
[33]Centro Universitario FEI - Sao Bernardo do Campo, Brazil
[34]Department of Chemistry, National and Kapodistrian University of Athens, Athens, Greece

[*] Corresponding author: manuela.cavallaro@lns.infn.it

Abstract. The possibility to use a special class of heavy-ion induced direct reactions, such as double charge exchange reactions, is discussed in view of their application to extract information that may be helpful to determinate the nuclear matrix elements entering in the expression of neutrinoless double beta decay half-life. The methodology of the experimental campaign presently running at INFN - Laboratori Nazionali del Sud is reported and the experimental challenges characterizing such activity are described.

1 Introduction

Charge exchange reactions are versatile tools for both reaction and nuclear structure studies. An especially appealing aspect is the possibility to use them to investigate the spin-isospin-dependent response of nuclei. These studies have a genuine nuclear physics interest, since they reflects a key aspect of nuclear dynamics. However, beyond that, there is an even broader impact of such investigations because of their relevance for nuclear beta decay, thus connecting strong and weak interactions. In this context, the NUMEN and NURE projects at LNS Catania [1-3] aim to promote heavy-ion induced single and double charge exchange reactions in a systematic approach, with the intention to extract information on nuclear matrix elements for single and double charge exchange processes and to identify the possible connections with double beta decay.

In the present paper, the strategy and methodologies of the above-mentioned programmes, the challenges of the experimental activities together with the status of the already performed experimental runs are described and discussed.

2 The motivation

The neutrino is a key particle for astrophysics, nuclear physics, particle physics and cosmology. It is an elementary particle that has only a weak charge, no electric and no colour charge. Thus, neutrino interactions with nuclei are extremely weak and experimental studies related to neutrinos are difficult.

Neutrinos have been extensively studied experimentally and theoretically in the recent 4–5 decades, but some of their fundamental properties are still not well understood and several basic questions about neutrinos remain yet unsolved. Among them, the nature of the neutrino, i.e. whether it is a Majorana particle (neutrino = antineutrino) or a Dirac particle (neutrino ≠ antineutrino) and the absolute mass scale. These questions can be studied well by investigating neutrino-related weak processes in nuclei, such as neutrinoless double beta decays ($0\nu\beta\beta$). If detected, $0\nu\beta\beta$ would provide evidence for the Majorana nature of the neutrino and the non-zero mass [4-9].

The $0\nu\beta\beta$ Nuclear Matrix Elements (NMEs) are very sensitive to nuclear physics involved in $0\nu\beta\beta$ probing the nuclear wave functions of the initial and final states, the spin–isospin correlations, nuclear structure models, nuclear medium effects, the renormalization (quenching) of the effective weak coupling in nuclei, and so on. However, there are no experimental methods to directly measure $0\nu\beta\beta$ NMEs and accurate theoretical calculations of them are hard, leading to discrepancy factors larger than two, as presently reported in literature. Thus experimental inputs relevant to the $0\nu\beta\beta$ responses are useful to help evaluate the $0\nu\beta\beta$ NMEs and to constrain the calculations.

In this context, an experimental campaign [1] has started at the INFN-Laboratori Nazionali del Sud in Catania using the MAGNEX large acceptance magnetic spectrometer [10]. It focuses on the study of heavy-ion induced Double Charge Exchange (DCE) reactions and competing reaction channels involving the nuclei of interest for $0\nu\beta\beta$ decay.

The possible connections of DCE measured cross sections with double beta decay NMEs and consequently with the still unsolved neutrino puzzles are among the goals of the present studies.

3 Heavy ion induced double charge exchange reactions

Important experimental challenges must be addressed to measure heavy ion induced DCE reactions.

Such challenges are related to the request to detect heavy ions with good isotopic separation and energy resolution in a wide angular range, including zero-degree, in order to distinguish transitions to individual states and explore a wide momentum transfer range.

In addition, the rather tiny DCE cross sections (down to few nb) demand for a challengingly high experimental sensitivity, which strongly depend on the rejection capability against unwanted events. These latter may be generated by competing reaction processes, which are rather likely to occur in view of the small DCE cross sections. For this reason high resolution particle identification is a necessary prerequisite for the experiment. In addition, spurious data associated to a wrong determination of detection parameters could casually be misinterpreted as good DCE data. Such background sources can be subtle and require stringent conditions for the experimental set-up together with a very careful characterization of the detection response in order to get rid of and gain the necessary sensitivity for performing meaningful DCE investigations [11].

These results are pursued thanks to the use of the MAGNEX spectrometer, a tool with high performance and flexibility (see Section 4) [12-15].

The accurate description of reactions involving complex nuclei is a demanding task also for nuclear theory. From the side of reaction theory initial and final state interactions of two many-body systems have to be under control and the transition amplitude must be described properly. Reactions involving a few selected degrees of freedom, keeping the bulk of projectile and target essentially intact are the most appropriate tools for those investigations and direct reaction theory can provide the suitable toolbox.

Nuclear structure theory is asked to provide accurate input for the spectroscopy of both projectile and target, even if the main interest is to explore the spectrum of one of the reaction partners only.

Despite of these caveats coherent ion-ion interactions have been used quite successfully in the past to investigate excitations in the quasi-elastic region. Moreover new theoretical developments for the description of the reaction mechanism involved in heavy ion induced reactions have been achieved under specific conditions and others are presently in progress within the NUMEN project [16-20].

4 The experiments

The experimental challenges mentioned in the previous section are faced at INFN-LNS by using the K800 Superconducting Cyclotron to accelerate beams and the MAGNEX large acceptance magnetic spectrometer for the detection of the ejectiles.

The MAGNEX spectrometer is a large acceptance magnetic device made up of a large aperture vertically focusing quadrupole and a horizontally bending dipole magnet. MAGNEX was designed to investigate processes characterized by very low yields and allows the identification of heavy ions with quite high mass ($\Delta A/A \sim 1/160$), angle ($\Delta\theta \sim 0.2°$) and energy resolutions ($\Delta E/E \sim 1/1000$), within a large solid angle ($\Omega \sim 50$ msr) and momentum range ($-14\% < \Delta p/p < +10\%$). High-resolution measurements for quasi-elastic processes, characterized by differential cross sections falling down to tens of nb/sr, were already performed by this setup [21-26]. A crucial feature is the implementation of a technique of trajectory reconstruction, based on differential algebraic techniques, which allows solving the equation of motion of each detected particle to 10th order [27-29]. This is a unique characteristic of MAGNEX, which guarantees the above-mentioned performances and its relevance in the worldwide scenario of heavy-ion physics.

The experimental activity with accelerated beams proposed and presently in progress consists of two main classes of experiments, corresponding to the exploration of the two directions of isospin lowering $\tau^- \tau^-$ and rising $\tau^+ \tau^+$, characteristic of $\beta^-\beta^-$ and $\beta^+\beta^+$ decays, respectively. In particular, the $\beta^+\beta^+$ direction in the target is investigated using an $^{18}O^{8+}$ beam and measuring the (^{18}O,^{18}Ne) DCE transitions, together with other reaction channels involving same beam and target. Similarly, the $\beta^-\beta^-$ direction is explored via the (^{20}Ne,^{20}O) reaction, using a $^{20}Ne^{10+}$ beam and detecting the reaction products of the DCE channel and of the other open channels characterized by same projectile and target.

Exploratory investigations of the two classes of experiments have been already performed, highlighting the strengths and the limiting aspects of the adopted technique and establishing the best working conditions [30-34].

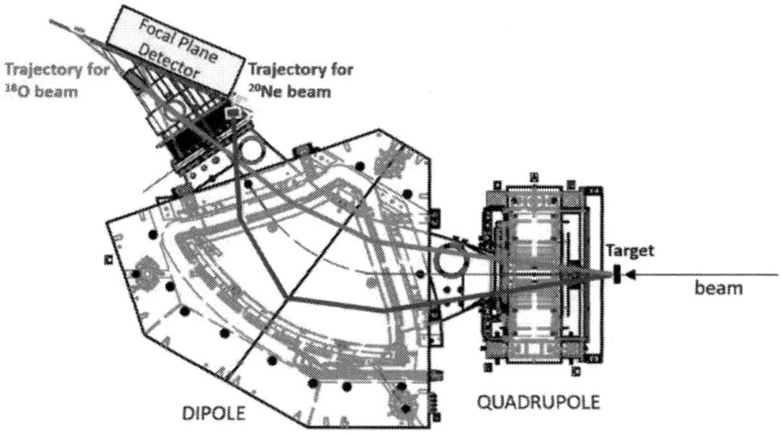

Fig. 1. Drawing of the MAGNEX spectrometer showing two typical trajectories of the ^{18}O and ^{20}Ne beams in the experimental conditions used for the DCE measurements.

4.1 Experiments with ^{18}O beam ($\beta^+\beta^+$ direction)

For the experiments of this class, the reaction channels of our interest are listed below:

- Elastic and inelastic scattering (^{18}O,^{18}O)
- DCE reaction (^{18}O,^{18}Ne)
- Charge-exchange reaction (^{18}O,^{18}F)
- Two-proton pickup reaction (^{18}O,^{20}Ne)
- One-proton pickup reaction (^{18}O,^{19}F)
- Two-neutron stripping reaction (^{18}O,^{16}O)
- One-neutron stripping reaction (^{18}O,^{17}O)

One of the main challenges of such experiments is the measurement at very forward angles, including zero-degree. This is performed by placing the spectrometer with its optical axis at +3° with respect to the beam axis. Thanks to its large angular acceptance, a range $-2° < \theta_{lab} < +9°$ is thus covered. The MAGNEX quadrupole and dipole magnetic fields are set in order that the incident beam, after passing through the magnets, reaches a region besides the FPD. For this class of experiments, in fact, the incident beam ($^{18}O^{8+}$) has higher magnetic rigidity (Bρ) than the ones of the ejectiles of interest (namely ^{18}Ne, ^{18}F, ^{20}Ne, ^{19}F, ^{16}O, ^{17}O with maximum charge state). The beam stops in a specifically designed Faraday Cup, which measures the incident charge in each run, placed in the high-Bρ region besides the FPD.

A drawing of the MAGNEX spectrometer and an example of the typical ^{18}O beam trajectory inside the spectrometer is shown in Fig. 1.

^{116}Sn, ^{76}Se and ^{48}Ti are the targets already explored via (^{18}O,^{18}Ne) reaction at 15 and 22 AMeV in order to study the $^{116}Sn \rightarrow {}^{116}Cd$, $^{76}Se \rightarrow {}^{76}Ge$ and $^{48}Ti \rightarrow {}^{48}Ca$ transitions, respectively, and the competing channels as shown in Fig. 2. The reduction and analysis of the collected data is presently in progress.

4.2 Experiments with ^{20}Ne beam ($\beta^-\beta^-$ direction)

In the class of experiments with $^{20}Ne^{10+}$ beams, the reaction channels we are interested are the following:

- Elastic and inelastic scattering (^{20}Ne,^{20}Ne)
- DCE reaction (^{20}Ne,^{20}O)
- Charge Exchange reaction (^{20}Ne,^{20}F)
- Two-proton stripping reaction (^{20}Ne,^{18}O)
- One-proton stripping reaction (^{20}Ne,^{19}F)
- Two-neutron pickup reaction (^{20}Ne,^{22}Ne)
- One-neutron pickup reaction (^{20}Ne,^{21}Ne)

For these experiments, the incident beam ($^{20}Ne^{10+}$) has a lower magnetic rigidity with respect to the reaction ejectiles of interest. Thus, for a fixed magnetic field setting, the beam will be more bent than the ejectiles of interest. The spectrometer optical axis is typically placed at −3°, thus the covered angular range is $-8° < \theta_{lab} < +3°$. The quadrupole and dipole magnetic fields of MAGNEX are set in order that the $^{20}Ne^{10+}$ beam reaches the low-Bρ region besides the FPD.

A peculiarity of these experiments concerns the treatment of the different charge states of the beam emerging out of the target. The beam components characterized by charge states lower than 10^+, mainly $^{20}Ne^{9+}$ and $^{20}Ne^{8+}$, produced by the interaction of the beam with the electrons of the target material, have a magnetic rigidity which is similar to the ions of interest. Therefore, they enter in the FPD acceptance, causing a limitation in the rate tolerable by the detector.

In order to stop these unwanted ^{20}Ne particles, two aluminium shields are mounted upstream the sensitive region of the focal plane detector.

The shields act on a limited phase space region which stops the 9^+ and 8^+ beams and elastic scattering at very forward angles, but not the other reaction channels generated by these beams. It is known that the charge state distribution of a heavy-ion beam after crossing a material depends on the bombarding energy and on the chemical composition of the target. The targets relevant for NUMEN generate an unwanted charge distribution that can be conveniently changed, minimizing the amount of $^{20}Ne^{9+}$ and $^{20}Ne^{8+}$, by adding an appropriate second foil (post-stripper) downstream of the isotopic target. Recently a specific study of different materials to be used as poststripper has been performed [33].

The systems already experimentally explored using the (^{20}Ne,^{20}O) reaction at 15 AMeV are the ^{116}Cd target (to study the $^{116}Cd \rightarrow {}^{116}Sn$ transition), the ^{130}Te (for the $^{130}Te \rightarrow {}^{130}Xe$) and the ^{76}Ge (for the $^{76}Ge \rightarrow {}^{76}Se$). The data reduction and analysis is in progress.

A scheme of the transitions already studied in the experimental runs is shown in Fig. 2.

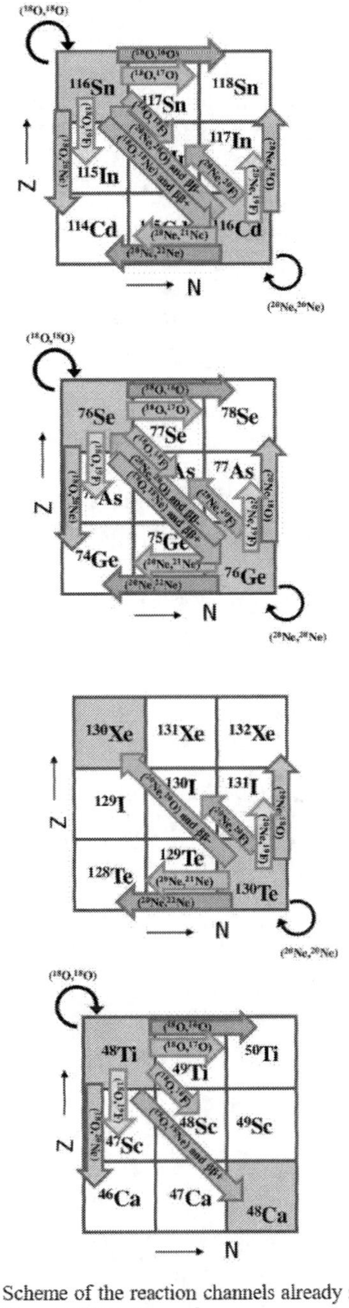

Fig. 2. Scheme of the reaction channels already studied within the NURE and NUMEN experiments.

5 Conclusions

A systematic study of heavy-ion induced double charge exchange reactions and of the other reaction channels characterized by the same projectile and target is in progress at INFN-LNS. The main goal is to investigate the nuclear response to DCE reactions for all the isotopes candidate for $0\nu\beta\beta$ decay to give experimentally driven indication towards the determination of $0\nu\beta\beta$ NMEs.

In the present paper, the focus is on the techniques adopted to set up the MAGNEX spectrometer for the challenging measurements of such suppressed reaction channels. The strategy used to perform the zero-degree measurement, based on an accurate simulation of the ion trajectories along the spectrometer and on the use of two Faraday cups properly designed for each category of experiments, is described.

Acknowledgments

This project has received funding from the European Research Council (ERC) under the European Union's Horizon 2020 research and innovation programme (grant agreement No 714625).

References

1. F. Cappuzzello et al., Eur. Phys. J. A **54**, 72 (2018)
2. M. Cavallaro et al., Proceedings of Science, PoS(BORMIO2017) **015** (2017)
3. F. Cappuzzello et al., J. Phys.: Conf. Ser. **630**, 012018 (2015)
4. H. Ejiri, J. Suhonen, K. Zuber, Physics Reports **797**, 1 (2019)
5. J. Vergados, H. Ejiri, F. Šimkovic, Rep. Progr. Phys. **75**, 106301 (2012)
6. J. Suhonen, O. Civitarese, J. Phys. G: Nucl. Part. Phys. **39**, 124005 (2012)
7. S. Dell'Oro, S. Marcocci, and F. Vissani, Phys. Rev. D **90**, 033005 (2014)
8. J. Engel, J. Menéndez, Rep. Progr. Phys. **60**, 046301 (2017)
9. P. Vogel, Jour. of Phys. G: Nucl. and Part. Phys., **39** 124002 (2012)
10. F. Cappuzzello et al., Eur. Phys. J. A **51**,145 (2015)
11. S. Calabrese et al., in preparation.
12. F. Cappuzzello et al. Eur. Phys. J. A **52**, 167 (2016)
13. C. Boiano et al., IEEE Trans. on Nucl. Science **4723837**, 3563 (2008)
14. M. Cavallaro et al., Nucl. Instr. and Meth. A **700**, 65 (2013)
15. M. Cavallaro et al. Phys. Rev. C **93**, 064323 (2016)
16. H. Lenske et al., Phys. Rev. C **98**, 044620 (2018)
17. E. Santopinto et al., Phys. Rev. C **98** 061601(R) (2018)
18. H. Lenske et al., Prog. Part. Nucl. Phys. **109**, 103716 (2019)
19. J.I. Bellone et al., J. Phys. Conf. Ser. **1056**, 012004 (2018)
20. J.A. Lay et al., J. Phys. Conf. Ser. **1056**, 012029 (2018)
21. F. Cappuzzello et al., Eur. Phys. J. A **52**, 169 (2016)
22. D. Carbone et al., Phys. Rev. C **95**, 034603 (2017)
23. V.A.B. Zagatto et al., Phys. Rev. C **97**, 054608 (2018)
24. M. J. Ermamatov et al., Phys. Rev. C **94**, 024610 (2016)
25. J.R.B. Oliveira et al., Journal of Physics G: Nuclear and Particle Physics **40**, 105101 (2013)
26. F. Cappuzzello et al., Nucl. Instr. and Meth. A **763**, 314 (2014)
27. F. Cappuzzello, et al., Nucl. Instr. and Meth. A **638**, 74 (2011)
28. A. Lazzaro et al., Nucl. Instr. and Meth. A **585**, 136 (2008)
29. A. Lazzaro et al., Nucl. Instr. and Meth. A **602**, 494 (2009)
30. F. Cappuzzello et al., Eur. Phys. J. A **51**, 145 (2015)
31. S. Calabrese et al., Acta Phys. Pol. B **49**, 275 (2018)
32. M. Cavallaro et al., Nucl. Instr. and Meth. B (in press) https://doi.org/10.1016/j.nimb.2019.04.069.
33. M. Cavallaro et al., Results in Phys. **13**, 102191 (2019)
34. A. Spatafora et al., Phys. Rev. C **100**, 034620 (2019)

Comparative study of four reactions at onset of pre-equilibrium emission

Magda Cicerchia[1,2,*], *Fabiana* Gramegna[2], *Daniela* Fabris[3], *Tommaso* Marchi[2], *Marco* Cinausero[2], *Giorgia* Mantovani[1,2], *Meltem* Degerlier[4], *Sandro* Barlini[5], *Maurizio* Bini[5], *Mauro* Bruno[6], *Antonio* Caciolli[1,3], *Alberto* Camaiani[5], Giovanni Casini[5], *Natalia* Cieplicka-Oryńczak[7], *Gianmaria* Collazuol[1,3], *Michela* D'Agostino[6], *Daniele* Dell'Aquila[8], *Bogdan* Fornal[7], *Catalin* Frosin[5], *Nicla* Gelli[5], *Diego* Gruyer[9], *Vladimir* Kravchuk[10], *Silvia* Leoni[11], *Ivano* Lombardo[12], *Daniele* Mengoni[1,3], *Luca* Morelli[6], *Alessandro*Olmi[5], *Pietro* Ottanelli[5], *Gabriele* Pasquali[5], *Silvia* Piantelli[5], *Giacomo* Poggi[5], *Simone* Valdré[5]

[1]Dipartimento di Fisica e Astronomia dell'Università di Padova, Padova, Italy.
[2]INFN Laboratori Nazionali di Legnaro, Legnaro (PD), Italy.
[3]INFN Sezione di Padova, Padova, Italy.
[4]Science and Art Faculty, Physics Department, Nevsehir Haci Bektas Veli Univ., Nevsehir, Turkey.
[5]INFN, Florence Department and University, Physics and Astronomy Department - Florence, Italy.
[6]INFN, Bologna Department and University, Physics and Astronomy Department - Bologna, Italy.
[7]Institute of Nuclear Physics, Polish Academy of Sciences Krakow, Poland.
[8]MSU Department of Physics and Astonomy - Michigan USA.
[9]Grand Accélérateur National d'Ions Lourds, 14076 Caen, France.
[10]National Research Center "Kurchatov Institute" Moscow, Russia.
[11]INFN, Milan Department and University, Physics and Astronomy Department - Milan, Italy.
[12]INFN, Catania Department - Catania, Italy.

Abstract. The study of the emitted particles, comparing pre-equilibrium and thermal components, is a useful tool to examine the nuclear structure of emitters. Possible clustering effects, which may change the expected decay chain probability, could be highlighted on the competition between different reaction mechanisms. The NUCL-EX collaboration (INFN, Italy) has carried out an extensive research campaign on pre-equilibrium emission of light charged particles from hot nuclei. In this framework, the reactions $^{16}O+^{30}Si$, $^{18}O+^{28}Si$, $^{19}F+^{27}Al$ at 7 AMeV and $^{16}O+^{30}Si$ at 8 AMeV have been carried out using the GARFIELD+RCo array at Legnaro National Laboratories. Some anomalies in the α-particle emission channels have been evidenced in the measurement reported above, showing in an exclusive way the observed effects related to the entrance channels. The experimental results are compared to model prediction, for which the same filtering and complete event selection have been applied.

1 Introduction

In the past decades, improvements in the knowledge of the incomplete fusion reactions have been achieved [1–3]; nevertheless, a complete understanding of reaction mechanisms associated with the emission of the particles prior the thermalization, like break-up and pre-equilibrium emissions, is still missing. Frequently, the term *incomplete fusion* has been employed to indicate that, somehow, particles are lost from the projectile and/or target before complete thermalization, which occurs up to the fusion of the remnants.
At bombarding energy above 10 AMeV, even-though usually a complete thermalization occurs, the pre-equilibrium particle emission becomes an increasingly important process as a function of the bombarding energy; such particles are forward focused and emitted in the very early stages of the collision before the attainment of full statistical equilibrium of the compound system [4,5]. Even in the energy region 5-10 AMeV, fast emission processes have been observed when the structure of the projectile plays an important role in the interaction with the target. This projectile break-up mechanism [6–9], as well as the pre-equilibrium, influences the following formation and decay of the hot source.

[*] Corresponding author: cicerchia@lnl.infn.it

 https://doi.org/10.1051/epjconf/201922301010

Since several years, the NUCL-EX collaboration (INFN, Italy) has carried out an extensive research campaign on pre-equilibrium emission of light charged particles from hot nuclei [10–14].

2 The Experiment

In this framework, four reactions $^{16}O+^{30}Si$ at 7 and 8 AMeV, $^{18}O+^{28}Si$ at 7 AMeV and $^{19}F+^{27}Al$ at 7 AMeV, forming the same compound nucleus ($^{46}Ti^*$) in case of complete fusion, were investigated [15,16]. The experiment was carried out at Legnaro National Laboratories (INFN-LNL, Italy) using the GARFIELD+RCo 4π array for charged particles, fully equipped with digital electronics [17]. For sake of comparison, the beam velocity was kept constant (7 AMeV) for the three reactions, since the abundance of pre-equilibrium particles is demonstrated to be dependent on it [18]: in such a way, the non–equilibrium processes are expected to be almost the same. The reaction $^{16}O+^{30}Si$ has been also measured at a beam energy of 8 AMeV to populate the $^{46}Ti^*$ at the same excitation energy of the $^{18}O+^{28}Si$ at 7 AMeV to obtain a similar statistical component. The main characteristics of studied reactions, in case of complete fusion, are reported in Table I.

Table 1. Summary of the main characteristics of the studied reactions.

Entrance Channel	Mass asymmetry	E_{lab} (AMeV)	CN	E^*_{CN} (MeV)
$^{16}O+^{30}Si$	0.30	7	^{46}Ti	88.0
$^{16}O+^{30}Si$	0.30	8	^{46}Ti	98.4
$^{18}O+^{28}Si$	0.22	7	^{46}Ti	98.5
$^{19}F+^{27}Al$	0.17	7	^{46}Ti	103.5

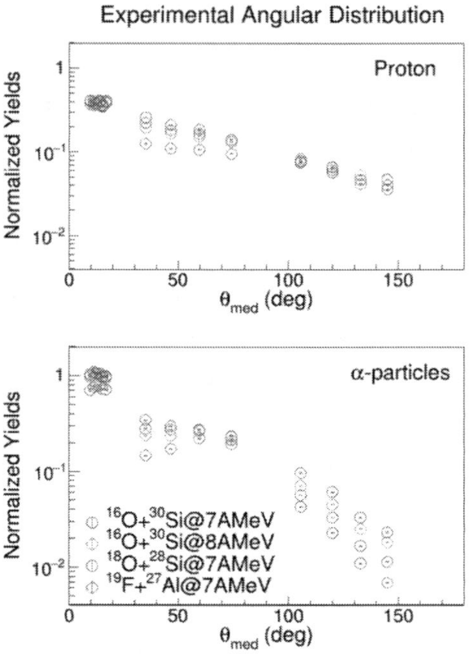

Fig. 1. Experimental angular distribution of protons (upper panel) and α-particles (lower panel) for the four studied reactions. The distributions are normalized to the number of complete events. The error bars are inside the experimental points.

Similarly, the experimental angular distribution depends on v_{cm} (Fig. 1), even-though a strong over-production of LCP (more important in the case of the α-particles) appears at very forward angles (8.8°-17.4°).

3 The Data Analysis

The complete analysis has been performed on an event-by-event basis; a detail description of this analysis is given in Ref. [15]. In the present work, we focus the attention on the complete events ($Z_{tot,detect} = Z_{projectile} + Z_{Target}$). Furthermore, a stricter constrain has been imposed on data: we select the emission of light charged particles (LCP) in coincidence with one and only one fragment with $Z_{frag}>5$ (ER); such events correspond to almost central collisions.

When we compare the experimental global observables (e.g. charge distribution and multiplicities of the emitted light charged particles) of the 4 reactions, a clear dependence of the yields on the excitation energy of the compound nucleus (E_{CN}) and on the center of mass velocity (v_{cm}) is observed.

3.1. The Statistical Code

In order to have a theoretical feedback, the experimental data have been compared with simulations performed with the statistical code GEMINI^{++} [19], which describes the decay of the excited compound nucleus. The simulated events were filtered through a software replica of the experimental setup and, then, selected in the same way of the experimental events [15].

In Fig. 2, the simulated angular distributions of protons and α-particles are shown. The experimental trend and dependence on v_{cm} are reproduced in the angular region from 29.5° to 150.5°, while at very forward angles (8.8°-17.4°) significant deviations from the experimental

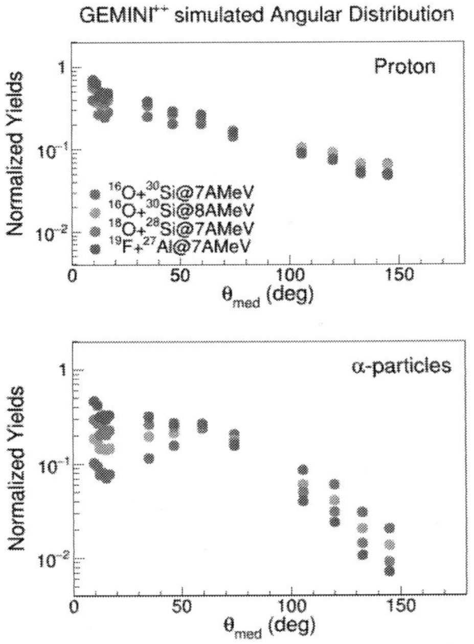

Fig. 2. GEMINI++ simulated angular distribution of protons (upper panel) and α-particles (lower panel) for the four studied reactions. The distributions are normalized to the number of complete events.

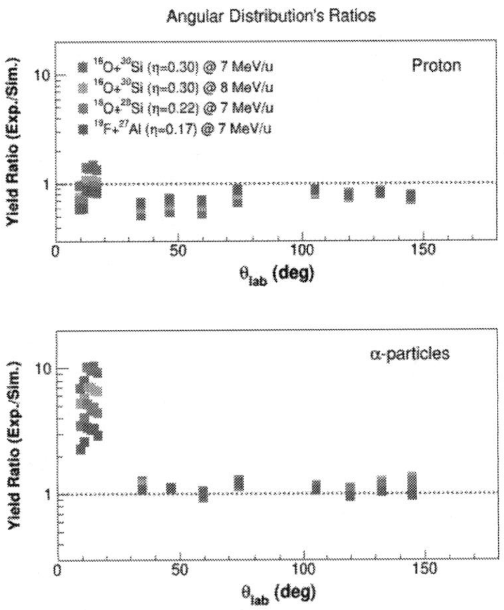

Fig. 3. Ratio of experimental and simulated proton (upper panel) and α-particles (lower panel) yields as a function of the detection angle of the particles for the four studied reactions.

angular distribution are observed. Quantitatively, the divergences between the simulated and experimental angular distributions are shown in Fig. 3, where the ratios between the experimental and simulated α-particles yields are shown as a function of the detection angle of the particles. As it can be observed, in the angular region from 29.5° to 150.5°, the experimental yields of α-particles are compatible with a statistical emission from the compound nucleus.

Otherwise, the observed overproduction of experimental α-particles is present at very forward angles (8.8°-17.4°). According to the literature [18], such over-production of forward focused α-particles should be related to fast emissions from non-equilibrium processes, characterizing the early stage of the reactions; it depends on both the entrance channel mass asymmetry (η) and the beam velocity (v_{beam}). However, in our case we observe some peculiar behaviour: the α-energy spectra are reproduced in shape and the forward missing α-yields are distributed over all the possible energies [15].

Moreover, at variance with what expected, when we compare the results of the two reactions with the same η (same entrance channel: $^{16}O+^{30}Si$), we observe a larger ratio (Exp./GEMINI++) of forward emitted α-particles yields (Fig. 2) in the case of the reaction at the lower v_{beam} (7 AMeV). When comparing the three reactions with the same v_{beam} (7 AMeV), an increase of the ratio at forward angles is seen as the η increases. Despite the small difference in η, this effect seems to be larger than expected, suggesting that the internal structure of the interacting nuclei may also play an important role. For the studied systems, the major part of the forward peaked α-particles is correlated to the exclusive channel with larger Z of residues ; in particular, in the Ca-residue exit channels, a strong inversion of population of 1α + xn and 2p + xn channels is observed with respect to GEMINI++.

Conclusion

We analysed complete events of four reactions having different entrance channels (η) and/or different beam velocity and then different v_{cm}. The observed differences among the four reactions can be ascribed to either entrance channels or structure properties of the reacting partners. Strong dissimilarities between experimental data and statistical model simulation are highlighted especially related to cluster-emission probability. In particular, enhanced pure α-emission has been observed, which, in some cases, become the dominant emission channel at variance with statistical model predictions.

References

1. K. A. Griffioen et al., Phys. Rev. C **37** 2502 (1988).
2. J. Gomez del Campo et al., Phys. Rev. C **60** 021601 (1999); **53** 222 (1996).
3. M. Blann, Phys. Rev. C **31** 1245 (1985).
4. S. J. Luke, Phys. Rev. C **48** 857 (1993).
5. J. Cabrera, Phys. Rev. C **68** 034613-1 (2003).
6. X. Campi et al., Phys. Lett. B **142** 8 (1984).
7. J. Pouliot et al., Phys. Lett. B **299** 210 (1993).
8. D. Shapira et al., Phys. Rev. C **55** 2448 (1997).
9. W. D. M. Rae et al., Phys. Rev. C **30** 158 (1984).
10. T. Marchi et al. - *Nuclear Particle Correlations And Cluster Physics* – Ch. 20 507 (2017);
11. L. Morelli et al., Journ. Of Phys. G **41** 075107 (2014); 075108 (2014).
12. D. Fabris et al., PoS X LASNPA **061** (2013).
13. V.L. Kravchuk, et al. EPJ WoCs **2** 10006 (2010).
14. O. V. Fotina et al., Int. Journ. Mod. Phys. E **19** 1134 (2010).
15. M. Cicerchia, *PhD thesis* (2018), University of Padova (Italy).
16. M. Cicerchia, IL NUOVO CIMENTO **41 C** (2018) 98.
17. M. Bruno et al., EPJ A **49** 128 (2013).
18. P. E. Hodgson, Phys. Rep. **374** 1 (2003).
19. R. J. Charity, Phys. Rev. C **82** 014610 (2010); D. Mancusi et al., Phys. Rev. C **82** 044610 (2010).

New investigations on the $^{32}S(^{3}He,d)^{33}Cl$ reaction at 9.6 MeV bombarding energy

Marco Cinausero[1*], *Daniele* Dell'Aquila[2], *Ivano* Lombardo[2], *Mariano* Vigilante[3,4], *Sandro* Barlini[5,6], *Riccardo* Bolzonella[7,1], *Mauro* Bruno[8,9], *Antonio* Buccola[5,6], *Sara M.* Carturan[7,1], *Giovanni* Casini[6], *Magda* Cicerchia[7,1], *Maurizio* D'Andrea[2], *Meltem* Degerlier[10], *Daniela* Fabris[11], *Leandro* Gasques[12], *Fabiana* Gramegna[1], *Alinka* Lepine-Szily[12], *Giorgia* Mantovani[7,1], *Tommaso* Marchi[1], *Silvia* Piantelli[6], *Valentino* Rigato[1], *Luca* Scomparin[7,1] and *Simone* Valdré[6]

[1] INFN - Laboratori Nazionali di Legnaro, Legnaro, Italy
[2] INFN - Sezione di Catania, Catania, Italy
[3] Università degli Studi di Napoli "Federico II", Napoli, Italy
[4] INFN - Sezione di Napoli, Napoli, Italy
[5] Università degli Studi di Firenze, Firenze, Italy
[6] INFN - Sezione di Firenze, Firenze, Italy
[7] Università degli Studi di Padova, Padova, Italy
[8] Università degli Studi di Bologna, Bologna, Italy
[9] INFN - Sezione di Bologna, Bologna, Italy
[10] Science and Art Faculty, Physics Department, Nevsehir Haci Bektas Veli Univ., Nevsehir, Turkey
[11] INFN - Sezione di Padova, Padova, Italy
[12] Instituto de Fisica da Universidade de São Paulo, 05508-090 São Paulo, Brasil

Abstract. The $^{32}S(^{3}He,d)^{33}Cl$ one-proton transfer reaction is a powerful tool to investigate the spectroscopy of low-lying states in the proton-rich ^{33}Cl nucleus. However, the extraction of firm differential cross-section data at various angles to benchmark and constrain theoretical models is made challenging by the presence of competitive reactions on target contaminants. In this paper we report on a recent measurement using a new generation hodoscope of silicon detectors, capable to detect and identify emitted deuterons down to energies of the order of 2 MeV. The high angular segmentation of our hodoscope combined with a suitable target to control possible contaminants, allowed to unambiguously disentangle the contribution of various states in ^{33}Cl, in particular the 2.352 MeV state lying just few tens of keV above the proton separation energy.

1 Introduction

The properties of low-lying states in the proton rich ^{33}Cl nucleus can be effectively probed by means of the $^{32}S(^{3}He,d)^{33}Cl$ one-proton transfer reaction [1,2]. In particular, the spin and parity (J^{π}) and the spectroscopic factors (C^2S_p) can be determined by the measure of the differential cross-section angular distribution of each individual excited state populated in the residual nucleus and comparing the experimental results with the predictions of theoretical models [1,2]. However, while theoretical calculations based on the shell-model approach have reached a very refined level in the description of (s-d)-shell nuclei [3,4], on an experimental point of view the situation is still unclear. For several low-lying states of ^{33}Cl, in fact, contrasting estimates of the C^2S_p factors are reported in the literature [5] and there is no consensus even in the J^{π} assignement. Indeed, the estimate of the C^2S_p is also relevant in nuclear astrophysics, to detemine the (p,γ) reaction rate in the stars [6]. In this context, we have performed a new study of the $^{32}S(^{3}He,d)^{33}Cl$ reaction at 9.6 MeV incident energy by using a new generation hodoscope of silicon detector, OSCAR [7]. In addition, special care has been taken in the target production to reduce possible contaminants.

2 Experimental details

The experiment has been performed at the CN Van de Graaf accelerator of the Laboratori Nazionali di Legnaro (LNL). A $^{3}He^{++}$ beam was delivered at an energy of 9.6 MeV on a 53 μg/cm^2 $^{nat}Zn^{32}S$ target, 99% enriched in ^{32}S, on a thin (15 μg/cm^2) carbon substrate. The target was manufactured and fully characterized (EDS, XRD and RBS analysis) at LNL, and installed in the center of a circular scattering chamber. A rotating plate hosted two silicon telescopes and an OSCAR telescope [7], placed on the opposite side with respect to the beam direction. Both detectors have two detection stages to enable the particle identification via the ΔE-E technique. The OSCAR telescope was composed by a thin (20 μm) 16-strips Single-Sided Silicon Detector (SSSSD) used as the first detection stage, and 16 individual silicon PIN diode pads (300 μm thick) used as the second detection stage. Each vertical strip of the first detection stage is 3.125 mm wide, while the second stage silicon pads have an active area of 1 cm^2. To determine the angular

* Corresponding author: cinausero@lnl.infn.it

position of each detector with respect to the beam line we performed a dedicated measurement using an optical alignment telescope. Each of the measured positions is determined with an accuracy better than 0.2°. In the OSCAR hodoscope, the correlations between the first and second detection stages can be used to constrain with high precision the impact point on the surface of the detector and therefore the angle of the emitted particles. We define 64 individual pseudo-telescopes, each identified by the correspondence of a particular strip of the SSSSD with a pad in the second detection stage. This method allows to have 64 detection telescopes using only 32 individual electronics channels.

3 Status of the analysis

To reconstruct particles in the OSCAR hodoscope we correlate strips in the first detection stage with pads in the second detection stage. Energy calibration of first and second stages was performed by using the ^{3}He elastic scattering on ^{197}Au at 13 different incident energies. Effects related to electronics were taken into account determining correction factors for each pseudo-telescope and each individual run. This procedure allows to maximize the statistics on the recorded spectra by merging data from different runs without affecting the energy resolution. Average energy resolution of 60 keV [FWHM] is achieved. The particle identification is obtained by means of the ΔE-E technique for each pseudo-telescope. The whole angular range spanned by the detector is divided into 32 different bins by grouping pseudo-telescopes that minimize the θ overlap. This allows to maximize the statistics preserving a good angular resolution. Figure 1 shows the deuteron kinetic energy as a function of the detection angle.

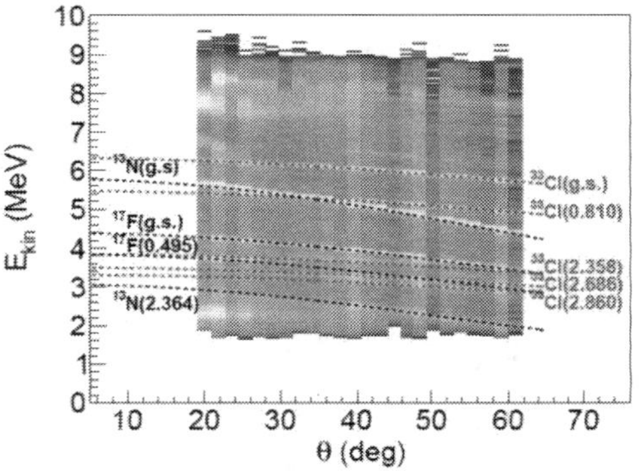

Fig. 1. Experimental deuteron kinetic energy as a function of the detection angle. Red dashed lines show the calculated kinematics lines for the populated ^{33}Cl states. Black lines show the kinematics lines for reactions on target contaminants.

Several lines related to one-proton tranfer in different target nuclei populating different states in the residues are clearly visible in the Figure. Red dashed lines are the result of a kinematical calculation for the ^{32}S(^{3}He,d)^{33}Cl reaction under investigations. Black dashed lines are competing (^{3}He,d) reactions occurring on ^{12}C and ^{16}O contaminants present in the target backing. In this way, the energy peaks related to the states of interest can be clearly identified along the whole spanned angular range. For each angular region, we performed a multi-gaussian fit of the detected peaks in the deuteron E_{kin} spectrum. A typical example of such fits is shown in Figure 2.

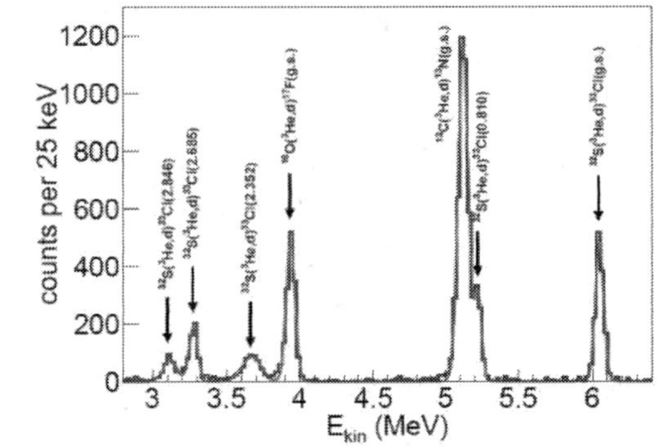

Fig. 2. Experimental deuteron spectrum at θ = 40°. In red is shown the multi-gaussian fit performed. Peaks are identified by labels.

The area under each peak, determined by the fitting procedure, gives the yield of the population of the corresponding state. The absolute cross-section of each ^{33}Cl state can be deduced from these yields by internal normalization to the ^{3}He + ^{nat}Zn Rutherford scattering.

4 Conclusions and perspectives

The ^{32}S(^{3}He,d)^{33}Cl one-proton transfer reaction at 9.6 MeV bombarding energy is investigated by using the new generation hodoscope OSCAR [7]. The good energy and angular resolution achieved by our device allows to disentangle the contribution of reactions of interest from contaminant background. The obtained yields will be used to extract the angular distribution of the differential cross section the identified low-lying states of ^{33}Cl, which will help to solve the experimental ambiguities still present in its spectroscopy.

References

1. G. Inglima *et al.*, Nuovo Cimento A **26**, 211 (1975)
2. R.L. Kozub and D.H. Yongblood, Phys. Rev. C **5**, 413 (1972)
3. B. Wildenthal *et al.*, Phys. Rev. C **4**, 1708 (1971)
4. B. Castel *et al.*, Nucl. Phys. A **162**, 273 (1971)
5. J. Chen and B. Sing, Nucl. Data Sheets **112**, 1393 (2011)
6. C. Iliadis *et al.*, Astr. J. Suppl. Ser. **134,** 151 (2001)
7. D. Dell'Aquila *et al.*, Nucl. Inst. Meth. A **877**, 227 (2018)

EPJ Web of Conferences **223**, 01012 (2019)
NSD2019
https://doi.org/10.1051/epjconf/201922301012

Study of the neutron-rich region in the vicinity of ^{208}Pb via multinucleon transfer reactions

Petra Čolović[1,]* *Andrés* Illana[2], *Suzana* Szilner[1], *Jose-Javier* Valiente-Dobón[2], *Lorenzo* Corradi[2], *Tea* Mijatović[1], *Giovanna* Benzoni[3], *María José* García Borge[4], *Alberto* Boso[5], *Simone* Ceruti[3], *James* Cubiss[6], *Giacomo* de Angelis[2], *Enrico* Fioretto[2], *Franco* Galtarossa[2], *Liam Paul* Gaffney[7], *Maria de La Luz* Jurado-Gomez[8], *Thorsten* Kröll[9], *Tommaso* Marchi[2], *Roberto* Menegazzo[4], *Daniele* Mengoni[4], *Daniel* R. Napoli[2], *Zsolt* Podolyak[10], *Giovanni* Pollarolo[11], *Francesco* Recchia[4] and *Dmitry* Testov[4]

[1]Ruđer Bošković Institute, Zagreb, Croatia
[2]Istituto Nazionale di Fisica Nucleare (INFN), Laboratori Nazionali di Legnaro, Legnaro, Italy
[3]Dipartimento di Fisica, Università di Milano and INFN, Milano, Italy.
[4]Instituto de Estructura de la Materia CSIC, Madrid, Spain
[5]Dipartimento di Fisica e Astronomia, Università di Padova, and INFN, Padova, Italy
[6]Department of Physics, University of York, York, United Kingdom
[7]ISOLDE, CERN, Geneva, Switzerland
[8]Instituto de Física Corpuscular CSIC, Valencia, Spain
[9]Institut für Kernphysik, Technische Universität Darmstadt, Darmstadt, Germany
[10]University of Surrey, Guildford, United Kingdom
[11]Dipartimento di Fisica, Università di Torino and INFN, Torino, Italy

Abstract. The multinucleon transfer reaction mechanism was employed to populate isotopes around the doubly-magic ^{208}Pb nucleus. We used an unstable ^{94}Rb beam on ^{208}Pb targets of different thickness. Transfer channels were studied via the fragment-γ and γ-γ coincidences, by using MINIBALL γ spectrometer coupled to a particle detector. Gamma transitions associated to the different Pb isotopes, populated by the neutron transfers, are discussed in terms of excitation energy and spin. Fragment angular distributions were extracted, and compared with the reaction models.

1 Towards the neutron-rich heavy nuclei

The study of heavy neutron-rich nuclei in the vicinity of the N = 126 shell closure is nowadays in the focus of nuclear physics. The properties of these nuclei are crucial to understand the synthesis of heavy elements. At the same time, this region of the nuclear chart is difficult to access experimentally.

One of the promising tools to reach this heavy neutron-rich region of the nuclear chart is the transfer of many nucleons at energies close to the Coulomb barrier. In fact, in the last decade, the multinucleon transfer (MNT) reactions [1, 2] were extensively used to populate moderately neutron-rich nuclei. Combining the stable (medium-heavy) projectile onto a (heavy) target results in nucleon transfers, with the cross section of each transfer channel governed by optimum Q-value considerations and nuclear form factors. Recently, the total cross sections of different transfer channels were compared for several systems having the same ^{208}Pb target and with projectiles going from neutron-poor to neutron-rich nuclei, i.e., ^{40}Ca, ^{58}Ni, and ^{40}Ar [3]. The neutron-rich (stable) ^{40}Ar beam allowed to get access to the channels involving proton pick-up, leading to the neutron-rich heavy partner. These data were compared with the GRAZING code [4–6]. GRAZING calculates the evolution of the reaction by taking into account, besides the relative motion variables, the intrinsic degrees of freedom of projectile and target, the surface modes and the one-nucleon transfer channels. The relative motion of the system is calculated in a nuclear plus Coulomb field.

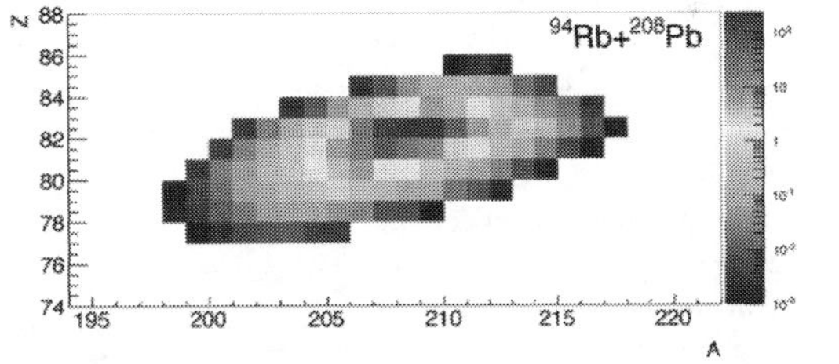

Figure 1. The ^{94}Rb+^{208}Pb reaction cross section distribution [mb] for the heavy partner calculated by the GRAZING code at $E_{\mathrm{lab}}(^{94}\mathrm{Rb})$ = 583 MeV. The calculated cross section includes the effect of neutron evaporation.

*e-mail: petra.colovic@irb.hr

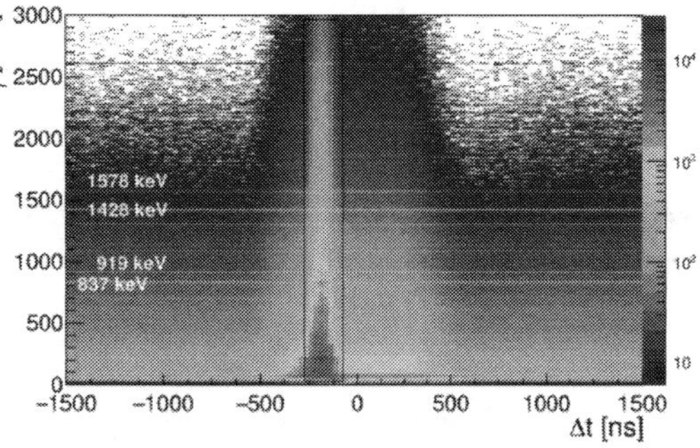

Figure 2. Gamma energy (E_γ) vs. the time difference (Δt) between the particle and γ. The 13 mg/cm^2 ^{208}Pb target was used. The prompt particle-γ coincidences are contained inside the gate of the Δt = 200 ns width. Left from the prompt peak is the isomeric part of the spectrum (delayed region). The strongest background lines are labeled.

The exchange of many nucleons proceeds via a multi-step mechanism of single nucleons.

The same model predicts that by using the (very) neutron-rich unstable beams, the proton pick-up and neutron stripping channels become the dominant ones [7]. Other reactions models, envisaging different mechanisms for the production of heavy neutron-rich nuclei [8, 9], reach the similar conclusions. It was also emphasized that secondary processes, like for example neutron evaporation, may significantly influence the final yield distributions. How to successfully transfer neutrons to the heavy target, while minimizing the influence of the secondary processes, is still an open question (see for example Ref. [10, 11] and references therein).

In order to explore the favorable experimental conditions for the population of the heavy neutron-rich nuclei by the MNT, we performed the measurement of a ^{94}Rb+^{208}Pb reaction at energy close to the Coulomb barrier. This relatively intense neutron-rich beam in the mass region A$\sim$95 has been used, in order to achieve a transfer yield sufficient for the detection, while taking into account also the predicted cross section for the population of the south-west region of the nuclear chart around ^{208}Pb. The prediction of the cross section distribution by the GRAZING code is plotted in Fig. 1. As expected, the largest calculated cross sections are in the neutron transfer channels, being the cross section of the ^{209}Pb larger than the ^{207}Pb one. In this work we will present details of the experimental method and some selected results, concentrating on the neutron transfer channels.

2 The ^{94}Rb+^{208}Pb experiment

The ^{94}Rb radioactive ion beam at E_{lab} = 6.2 MeV·A was delivered by the HIE-ISOLDE facility in CERN onto ^{208}Pb targets of different thickness. Experimental set-up consisted of the MINIBALL γ-array coupled to the Compact-Disk (CD) particle detector. MINIBALL consists of the 24 high-purity germanium crystals, individually encapsulated in triple-cluster cryostats which surround the spherical target chamber in compact geometry [12]. The absolute efficiency of 8% (at 1.3 MeV) is obtained, with the position sensitivity being preserved via six-fold electrical segmentation of the each crystal electrode. Four Double-sided Silicon Strip Detectors (DSSSD) were arranged in the CD configuration, positioned 21 mm downstream of the target with respect to the incoming beam. Front side of each DSSSD is made of 16 annular strips spanning a 24° - 63° angular range.

The identification of reaction products was obtained via their associated γ rays. Fragment-γ coincidences are illustrated in Fig. 2, where the energy, E_γ, is plotted as a function of the time difference between the detected particle and a γ-ray. The same figure also shows how the prompt coincidences (within Δt = 200 ns) can be separated from the the isomeric contribution (on the left side of the prompt region). The spectrum at the right side of the prompt peak illustrates well the complex background. The strongest γ rays, coming from the ^{94}Rb beta-decay chain, were attributed to the ^{94}Sr (837 and 1578 keV lines), ^{94}Y (1428 keV line) and ^{94}Zr (919 keV line). The ^{94}Sr was present also as the beam impurity (10% of the total flux). The background was successfully reduced by subtraction of the random particle-γ coincidences from the prompt and delayed events.

The ^{208}Pb targets of two different thickness were employed, 1 mg/cm^2 and 13 mg/cm^2. We used the thin, 1 mg/cm^2 target, to extract quantitative information relevant for the reaction mechanism, like the angular distributions of different channels. Gamma rays were emitted in flight, thus the Doppler corrections had to be applied for the both Rb-like and Pb-like fragments. The selection of the fragment in the CD detector improved Doppler correction, and the resolution of 1.2% at 1.5 MeV was obtained. These Doppler corrected spectra, for the Rb-like and Pb-like fragments, are plotted in Fig. 3, for a selected energy range. An additional gate was applied to select a prompt coincidence with the Rb-like fragments, and it enabled to obtain a cleaner spectra. In the case of the ^{208}Pb 13 mg/cm^2 thick target, where the Pb-like fragments were stopped, we studied the structure of the strongly populated states via γ-γ coincidences.

3 Results and discussion

The matrix of the fragment energy plotted as a function of the scattering angle, detected with the CD detector, is shown in the Fig. 4. A clear separation of the Rb-like and Pb-like fragments is visible. From this $E-\theta$ matrix we extracted the angular distributions for the total quasi-elastic scattering of the detected Rb-like and Pb-like fragments, and combined them in a single distribution. We remind that we do not have charge and mass identification in the CD detector, thus the experimental distribution, especially around the grazing angle, is a sum of the elastic, inelastic and transfer differential cross sections. Such distribution is a pure Rutherford scattering at the most forward measured angles, which was used to obtain a normalization factor.

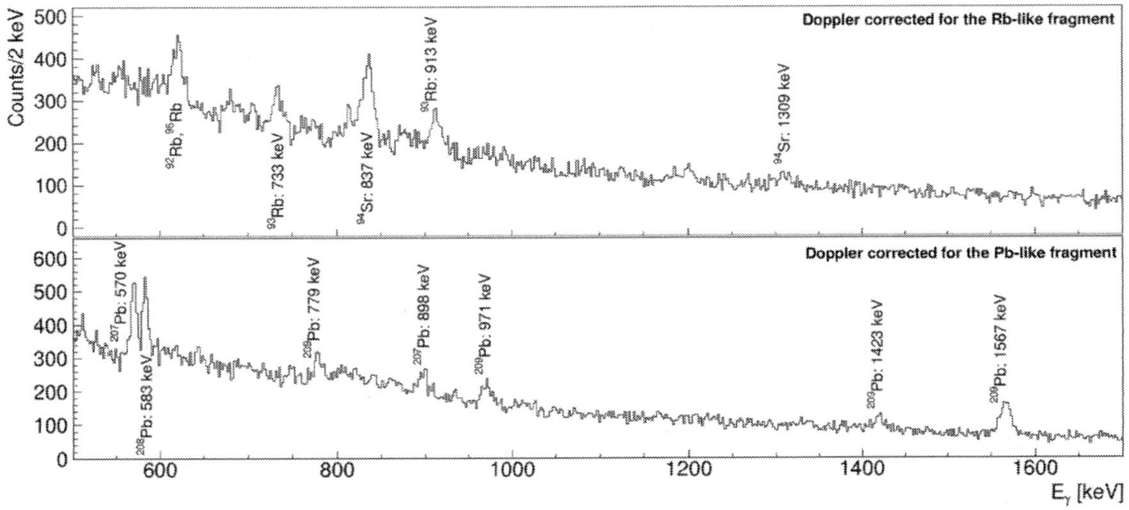

Figure 3. Gamma-spectra Doppler corrected for the Rb-like ($\beta \sim 10\%$), and Pb-like fragments ($\beta \sim 3\%$), is shown on the top, and bottom, respectively. The strongest γ-transitions within a selected energy range are labeled. For the Pb isotopes they correspond to the following transitions: ^{207}Pb, E_γ = 570 keV ($5/2^- \to 1/2^-$) and 898 keV ($3/2^- \to 1/2^-$), ^{208}Pb, E_γ = 583 keV ($5^- \to 3^-$), ^{209}Pb, E_γ = 779 keV ($11/2^+ \to 9/2^+$), E_γ = 971 keV ($3/2^+ \to 5/2^+$), E_γ = 1423 keV ($15/2^- \to 9/2^+$), E_γ = 1567 keV ($5/2^+ \to 9/2^+$).

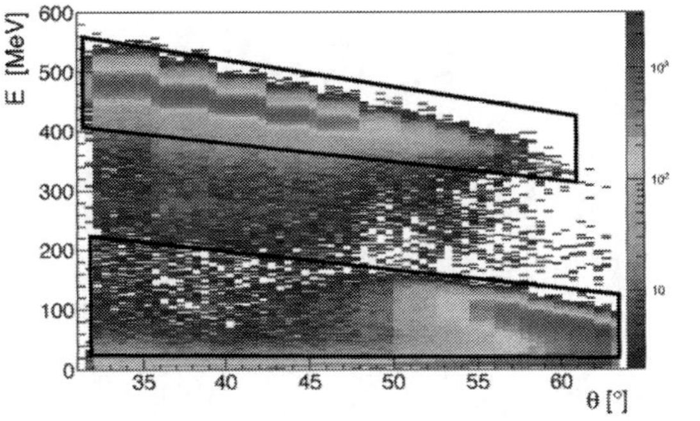

Figure 4. Two-dimensional matrix of angle (θ) vs. the energy (E) measured with the CD detector for the ^{94}Rb+^{208}Pb reaction. The Rb-like (top) and Pb-like (bottom) reaction products are labeled.

The results have been compared with the GRAZING calculations for the entrance channel mass partition. It was found that the measured and calculated differential cross sections agree well at the forward angle and at the quarter point.

In a similar way, by using the fragments in prompt coincidence with γ rays, we constructed angular distributions of strongly excited states, in particular of the 3^- state at 2614.5 keV in ^{208}Pb. The comparison of the obtained distribution with the reaction models, for example, distorted-wave Born approximation calculation, will provide (besides the normalization factor) the interaction potential and will better define the degrees of freedom which are important in the reaction.

In the reaction models the excitation and transfer processes are mediated by the well known single-particle form factors for the fermion degrees of freedom and by the collective form factors, for the vibrational modes. The same degrees of freedom should play a significant role in the distribution of the transfer strength over different states. Preliminary identification of ^{207,208,209}Pb transitions indicates the selective population of excited states (see Fig. 3). In ^{208}Pb the strongest observed transition is the $3^- \to 0^+$, E_γ =2614.5 keV. In one-neutron transfer channels, ^{207}Pb and ^{209}Pb, we observed a significant population of the states of single-particle character. Besides the single-particle character, we also observed the population of the states with the main configuration which can be described as a coupling of the valence neutron (hole) to the ^{208}Pb octupole. The same phenomena, the strong population of the stretched configuration of particle-phonon states, was previously observed in different measured systems [13–16] by using MNT reactions. In neutron transfer channels we measured excitation energies up to ~4 MeV and up to relatively high-spin (i.e. up to ~$12\hbar$ in ^{209}Pb). In the ^{210}Pb, a decay chain from 10^+ to 0^+ was observed, by using the 13 mg/cm^2 thick target and by combining the prompt and delayed transitions in γ-γ coincident analysis. This allowed to measure the decay of the 6^+ and 8^+ isomeric states as well.

Recently, the level scheme of the ^{210}Pb nuclei, populated in the ^{208}Pb+^{208}Pb transfer and deep-inelastic reactions, was established on the basis of prompt and delayed γ-γ coincidences [17], adding previously unobserved states. The neutron rich Pb isotopes were populated by using the fragmentation of the ^{238}U beam [18]. In this measurement, the neutron-rich lead isotopes up to ^{216}Pb were studied by the isomeric-decay γ spectroscopy.

The obtained results were compared with the state-of-the-art shell-model calculations and discussed in terms of the effective three-body interactions and two-body transition operators. These two recent works demonstrate the complementarity of the reaction mechanisms.

4 Summary

We measured the ^{94}Rb+^{208}Pb transfer reaction, by using the neutron-rich unstable beam of the ISOLDE facility. The high-resolution MINIBALL spectrometer, coupled to a position sensitive silicon CD detector, allowed the identification of reaction products via their associated γ rays. The multinucleon transfer reactions were proposed as an efficient reaction mechanism for the production of the moderately neutron-rich nuclei in the vicinity of ^{208}Pb. As we observed a larger cross section in ^{209}Pb than in ^{207}Pb, these preliminary findings indicate that the used mechanism is a very promising tool for studies of the hard-to-reach neutron-rich isotopes around the ^{208}Pb. The more detailed study of the distribution of the transfer strength over different states will shed a light on important degrees of freedom which influence the evolution of reaction.

Acknowledgments

The authors are grateful to the INFN - LNL target laboratory for the excellent targets. This work was partly supported by the HORIZON2020 ENSAR2 Grant Agreement nb. 654002. This work has been supported in part by the Croatian Science Foundation under project no. 7194 and in part under project no.IP-2018-01-1257.

References

[1] L. Corradi, G. Pollarolo, and S. Szilner, J. of Phys. G **36**, 113101 (2009)
[2] R. Broda, J. of Phys. G **32**, R151 (2006)
[3] T. Mijatović et al, Phys. Rev. C **94**, 064616 (2016)
[4] A. Winther, Nucl. Phys. A **572**, 191 (1994)
[5] A. Winther, Nucl. Phys. A **594**, 203 (1995)
[6] Program GRAZING http:/www.to.infn.it/nanni/grazing
[7] C. H. Dasso, G. Pollarolo, and A. Winther, Phys. Rev. Lett. **73**, 1907 (1994)
[8] V. I. Zagrebaev and W. Greiner. Phys, Rev. C **83**, 044618 (2011)
[9] L. Zhu et al., Phys.Lett. B **767**, 437 (2017)
[10] F. Galtarossa et al, Phys. Rev. C **97**, 054606 (2018)
[11] Y. X. Watanabe, Phys. Rev. Lett **115**, 172503 (2015)
[12] N. Warr et al., Eur. Phys. J. A **49**, 40 (2013)
[13] S. Lunardi et al., Phys. Rev. C **76**, 034303 (2007)
[14] S. Szilner et al., Phys. Rev. C **84**, 014325 (2011)
[15] D. Montanari et al., Phys.Lett. B **697**, 288 (2011)
[16] P. Čolović et al., Eur. Phys. J. A **53**, 8 (2017)
[17] R. Broda et al, Phys. Rev. C **98**, 024324 (2018)
[18] A. Gottardo et al., Phys. Rev. Lett. **109**, 162502 (2012)

EPJ Web of Conferences **223**, 01013 (2019) https://doi.org/10.1051/epjconf/201922301013
NSD2019

Study of sub-barrier fusion of ^{36}S+^{50}Ti,^{51}V systems

Giulia Colucci[1,*], *Giovanna* Montagnoli[1,**], *Alberto M.* Stefanini[2,***], *Kouichi* Hagino[3,4,****], *Antonio* Caciolli[1], *Petra* Čolović[5], *Lorenzo* Corradi[2], *Enrico* Fioretto[2], *Franco* Galtarossa[2], *Alain* Goasduff[1], *Jerzy* Grebosz[6], *Marco* Mazzocco[1], *Daniele* Montanari[7], *Concetta* Parascandolo[8], *Fernando* Scarlassara[1], *Marco* Siciliano[9], *Emanuele* Strano[1], *Suzana* Szilner[5] and *Nikola* Vukman[5]

[1]Dipartimento di Fisica e Astronomia, Università di Padova, and INFN, Sez. di Padova, I-35131 Padova, Italy
[2]INFN, Laboratori Nazionali di Legnaro, I-35020 Legnaro (Padova), Italy
[3]Department of Physics, Tohoku University, Sendai 980-8578, Japan
[4]Research Center for Electron Photon Science, Tohoku University, 1-2-1 Mikamine, Sendai 982-0826, Japan
[5]Ruđer Bošković Institute, HR-10002 Zagreb, Croatia
[6]Institute of Nuclear Physics, Polish Academy of Sciences, PL 31-342 Cracow, Poland
[7]IPHC, CNRS-IN2P3, Université de Strasbourg, F-67037 Strasbourg Cedex 2, France
[8]Dipartimento di Fisica, Univ. di Napoli, and INFN, Sez. di Napoli, I-80126 Napoli, Italy
[9]Centre CEA de Saclay, 91190 Gif-sur-Yvette, France

Abstract. A detailed comparative study of the sub-barrier fusion of the two near-by systems ^{36}S+^{50}Ti,^{51}V was performed at the National Laboratories of Legnaro (INFN). Aim of the experiment was the investigation of possible effects of the non-zero spin of the ground state of the ^{51}V nucleus on the sub-barrier excitation function, and in particular on the shape of the barrier distribution. The results show that the two measured excitation functions are very similar down to the level of 20 - 30 μb. The same is observed for the two barrier distributions. Coupled-channels calculations have been performed and are in good agreement with the experimental data. This result indicates that the low-lying levels in ^{51}V can be interpreted in the weak-coupling scheme, that is, $^{51}V(I) = {}^{50}Ti(2^+) \otimes p(1f_{7/2})$.

1 Introduction

Most of the existing near- and sub-barrier fusion studies have concerned systems where both projectile and target are even-even nuclei. This is mainly due to the simpler theoretical treatment when both nuclei have 0^+ ground states. However, interesting effects are expected when odd spins are involved [1]. The ion-ion potential and consequently the height of the Coulomb barrier would be different for each magnetic substate. Thus, every m-substate has to be treated separately, and the coupled-channels (CC) equations solved accordingly. This may produce different fusion cross sections near the barrier and the shape of the barrier distribution should keep memory of the various barriers associated to the m-substates.

In this framework, at the National Laboratory of Legnaro (LNL) a detailed measurement of the fusion excitation functions for the two systems ^{36}S+^{50}Ti,^{51}V has been performed, where no previous data were available. The nucleus ^{50}Ti is spherical and rather stiff because of its closed neutron shell. On the other hand, the ^{51}V nucleus has a large non-zero spin ($7/2^-$) in its ground state and it is also essentially spherical because of its very small quadrupole moment [2].

The aim of the measurement was to identify differences in the fusion excitation function of the two cases, that may possibly be attributed to the non-zero spin (7/2) of the ^{51}V ground state. A different ion-ion potential and consequently a different barrier, is expected for each magnetic substate. Since the nuclei can be treated as spherical, possible effects of the finite spin of the ground state are isolated, without the onset of deformation. This allows to directly compare the two cases before performing detailed CC calculations. By comparing the two systems, we investigated if the shape of the barrier distribution keeps a trace of those different barriers.

2 Experimental procedure

The beam of ^{36}S was provided by the XTU Tandem accelerator of LNL in the energy range of 73 - 100 MeV and with an average intensitiy of 10 pnA. The targets consisted of 50 μg/cm^2 of ^{51}V and $^{50}TiO_2$ (90.3% enriched).

Fusion cross sections have been determined by direct detection of the fusion evaporation residues (ER) at small angles by separating out the beam and beam-like particles using the electrostatic beam deflector [3]. The ER were identified downstream of the deflector by a double Time-of-Flight (ToF) - ΔE - Energy telescope composed of two

*e-mail: giulia.colucci@pd.infn.it
**e-mail: montagnoli@pd.infn.it
***e-mail: Alberto.Stefanini@lnl.infn.it
****e-mail: hagino@nucl.phys.tohoku.ac.jp

 https://doi.org/10.1051/epjconf/201922301013

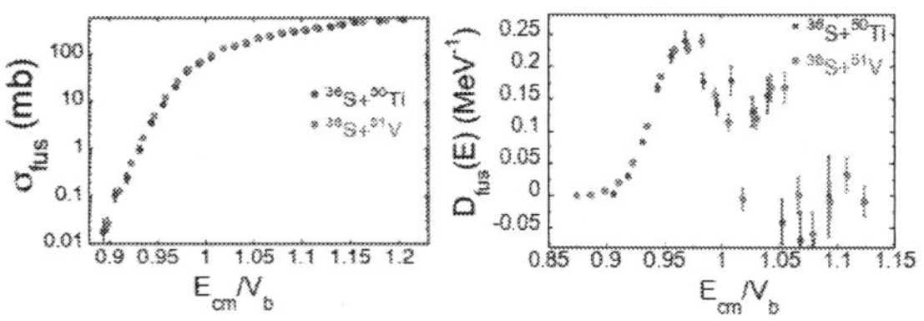

Figure 1. Comparison of the excitation functions (on the left) and barrier distributions (on the right) for the two fusion reactions. The energy scale is normalized to the heights of the two Coulomb barriers.

micro-channel plate time detectors (MCP) followed by a fast ionization chamber (Fast IC) [4] and by a silicon detector placed in the same gas (CH_4) volume of the Fast IC. Four silicon detectors were placed symmetrically around the beam direction at the same scattering angle to monitor the beam and to normalize the fusion yields to the Rutherford scattering cross section. Two ER angular distributions were measured at the energies of 80 and 90 MeV in the range from -6° to +9°.

3 Results

3.1 Excitation function and barrier distributions

The cross sections was measured down to 20 and 30 μb for ^{36}S+^{50}Ti and ^{36}S+^{51}V, respectively. The excitation functions of the two systems are compared in Fig. 1 (left panel), where the error bars are statistical uncertainties, that is, 1 - 2% at high energies and 20 - 30% at sub-barrier energies. The comparison showed a very similar behaviour of the two systems. Thus, in order to put in evidence possible small differences, a comparison of the barrier distributions was performed.

The barrier distributions were obtained using the three-point difference formula [5]. The energy intervals were of $\sim$ 1.5 MeV in order to highlight the structures at energies above and below the barrier. The two barrier distributions are compared in Fig. 1 (right panel). Also in this case, the two shapes are extremely similar. Since no differences are observed between the two systems by the only comparison of the experimental data, a theoretical interpretation is necessary. In this perspective, a coupled-channels analysis was performed.

3.2 Coupled-Channels calculations

The coupled-channels calculations were performed by means of the CCFULL code [6]. The CC calculations for ^{36}S+^{50}Ti included the one-phonon excitation of both the lowest quadrupole vibrational state 2^+ at 1.554 MeV of ^{50}Ti and the first 2^+ state at 3.29 MeV of the ^{36}S. The case of ^{36}S+^{51}V was more tricky. The four magnetic substates m = 1/2, 3/2, 5/2 and 7/2 of the $7/2^-$ ground state of ^{51}V produce different Coulomb barriers that have to be treated individually in the calculations. A modified version of CCFULL was therefore used in order to include the 2^+ excitation in ^{36}S as well as the couplings to the $5/2^-$, $3/2^-$, $11/2^-$, $9/2^-$, and $3/2^-$ states in ^{51}V (for more details see [7]). Despite the slight inconsistency with the structure observed at energies above the main peak of both systems, as shown in Fig. 2 (bottom panels), the CC calculations reproduce very well the two similar excitation functions (Fig. 2 upper panels). This can be interpretated under the weak-coupling approximation, where the low-energy levels of ^{51}V result from the scheme $^{51}V(I) = {}^{50}Ti(2^+) \otimes p(1f_{7/2})$ [8]. The relatively stiff ^{50}Ti (close to the double magic ^{48}Ca) is not significantly influenced by the additional proton to form ^{51}V.

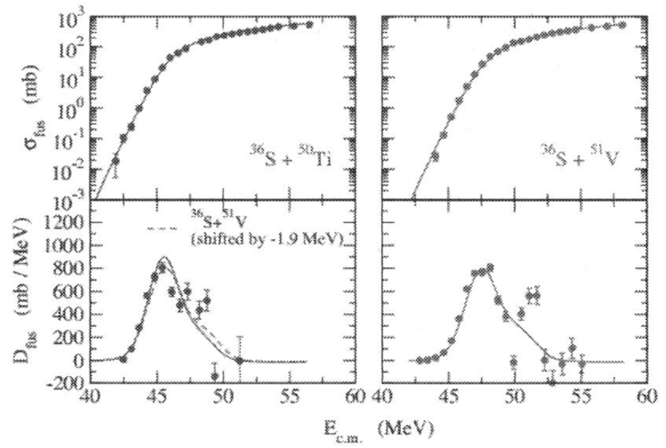

Figure 2. Excitation functions (top) and fusion barrier distributions (bottom) of ^{36}S+^{50}Ti (on the left) and ^{36}S+^{51}V (on the right), compared with the CC calculations described in the text.

4 Summary

The fusion cross sections of the two systems ^{36}S+^{50}Ti,^{51}V show a very similar behaviour down to 20 - 30 μb. A CC analysis was performed in order to highlight differences between the two systems attributable to the non-zero spin ground state of ^{51}V. The CC analysis included the low energy excitations of the ^{36}S and ^{50}Ti,^{51}V nuclei and the results are in very good agreement with the experimental data of both systems. This may be explained in the weak-coupling scheme, where the extra proton in the $f_{7/2}$ shell of the ^{51}V does not significantly affect the sub-barrier fusion.

References

[1] H. Esbensen, Phys. Rev. C **81**, 034606 (2010)
[2] P. Unkel, P. Buch, J. Dembczynski, W. Ertmer and U. Johann, Phys. D **11**, 259 (1989)
[3] G. Montagnoli et al., Phys. Rev. C **97**, 024610 (2018)
[4] G. Colucci et al, Acta Phys. B **50**, 573 (2019)
[5] M. Dasgupta, D. J. Hinde, N. Rowley and A. M. Stefanini, Annu. Rev. Nucl. Part. Sci. **48**, 401 (1998)
[6] K. Hagino, N. Rowley and A.T. Kruppa, Comput. Phys. Commun. **123**, 143 (1999)
[7] G. Colucci et al., Eur. Phys. J. A **55**, 111 (2019)
[8] F. Ballester, E. Casal, J.B.A. England and J.B.A. England, Nucl. Phys. A **513**, 61 (1990)

Systematic Search For Evidence of Tetrahedral and Octahedral Symmetries in Subatomic Physics: Follow-up of the First Identification Case in ^{152}Sm

Jerzy Dudek[1,2,*], *Irene* Dedes[2], *Abdelghafar* Gaamouci[3], *Jie* Yang[2], *Andrzej* Baran[2], *Dominique* Curien[1], *Timo* Dickel[4,5], and *Hua-Lei* Wang[6]

[1]Université de Strasbourg and IPHC, 23, rue du Loess, F-67037 Strasbourg, France, CNRS, UMR7178, F-67037 Strasbourg, France
[2]Instytut Fizyki - Uniwersytet Marii Curie-Skłodowskiej, pl. Marii Curie-Skłodowskiej 1, PL-20031 Lublin, Poland
[3]Laboratory of Theoretical Physics, Faculty of Physics, USTHB, BP 32, El Alia, 16111 Bab Ezzouar, Algiers, Algeria
[4]II. Physikalisches Institut, Justus-Liebig-Universität Gießen, 35392 Gießen, Germany
[5]GSI Helmholtzzentrum für Schwerionenforschung GmbH, 64291 Darmstadt, Germany
[6]School of Physics and Engineering, Zhengzhou University, Zhengzhou 450001, China

Abstract. In a recent article [1] group-theory representation-methods have been combined with the realistic mean-field calculation results to elaborate new, specifically designed methods of experimental identification of the tetrahedral/octahedral symmetries in atomic nuclei. The authors demonstrated that experimental data on ^{152}Sm existing in the literature are fully compatible with the extremely restrictive group-theory criteria of simultaneous presence of tetrahedral and octahedral symmetries, thus identifying these symmetries in subatomic physics for the first time. We discuss theory predictions related to the systematic presence of these symmetries as well as their manifestations throughout the Periodic Table in the form of islands centred around the doubly-magic tetrahedral-symmetry nuclei.

The corresponding theory predictions are discussed in the context of the planned new experiments, which would employ the advanced mass-spectrometry methods [2], in view of the new experimental search criteria [1]. The addressed field of symmetry-research presents particularly promising potentialities in the domain of exotic nuclei studies. Indeed, as it can be demonstrated, in the exact tetrahedral and/or octahedral symmetry limits the corresponding nuclei emit neither E2 nor E1 radiation generating isomeric states with lifetimes which can become much longer than those of the related ground states. This is expected to open the new research strategies for the whole domain of the exotic nuclei studies throughout the Periodic Table.

Keywords: Nuclear structure, Nuclear Platonic symmetries, Exotic nuclei, Shape isomers

1 Introduction: Mechanisms Associated with High-Rank Symmetries

The issue of identifying exotic tetrahedral and octahedral symmetries among possible symmetries of nuclear shapes received increasing attention in recent years, cf. e.g. [3–9] and references therein. The corresponding groups of symmetry are usually denoted in the literature T_d and O_h and sometimes referred to as Platonic or in the nuclear physics context as high-rank. The origins of this interest are numerous and involve fascinating, often unprecedented exotic quantum mechanisms. Due to size limitations, this article presents only a brief description of mechanisms and phenomena whose observation will facilitate finding the evidence of high-rank symmetries in subatomic physics.

1.1 Exotic Degeneracies of Energy Levels

Among the most exotic unprecedented quantum features related to high-rank symmetries in nuclei let us mention the presence of characteristic four-fold degeneracies of the single nucleonic energy-levels. They are predicted to occur in energy spectra generated by the mean-field Hamiltonians invariant under the high-rank symmetry groups [10].

The four-fold degeneracies originate from the characteristic structures of irreducible representations of the two groups. More precisely, the groups T_d and O_h just mentioned describing the forms of geometrical shapes, e.g. molecular ones in a 3D space and/or the tetrahedral and octahedral symmetry properties of the Hamiltonians of systems composed of bosons, have the well known partnership relations with the so-called double point-groups denoted T_d^D and O_h^D, the latter describing the symmetries of the Hamiltonians of the systems composed of Fermions. Whereas T_d^D has two 2-dimensional and one 4-dimensional irreducible representations, the O_h^D has four 2-dimensional and two 4-dimensional irreducible representations. The wave-functions of the states transforming according to the 4-dimensional irreducible representations form quadruplets resulting in the 4-fold degenerate nucleonic energy levels.

It follows that in the exact symmetry limits those levels may lead to 16-fold and 32-fold degeneracies of the

*e-mail: Jerzy.Dudek@iphc.cnrs.fr

particle-hole excited states. Identification of these states in nature together with the identification of the underlying 4-fold degeneracy nucleonic orbitals remains today one of the challenges in the discussed domain of research.

1.2 Unprecedented Structures of Rotational Bands

Another exotic and at the same time fascinating issue related to the structure of tetrahedral-symmetry nuclei are the rotational bands with the totally new, unprecedented properties. Indeed, in contrast to the rotational bands studied traditionally, the latter composed exclusively of the $\Delta I = 2\hbar$ transitions connecting states with either positive or negative parities and composed of either even or odd spin states, the new rotational structures (let us focus on even-even nuclei to start with) are composed of certain even and certain odd spin states, the sequences including both parities and in addition containing degenerate multiplets. For instance, after [11, 12] and references therein, a rotational band built on the tetrahedral symmetry $I^\pi = 0^+$ ground-state forms the spin-parity sequence[1]:

$$\underbrace{I^\pi = 0^+,\, 3^-,\, 4^+,\, \underbrace{(6^+,6^-)}_{\text{doublet}},\, 7^-,\, 8^+,\, \underbrace{(9^+,9^-)}_{\text{doublet}},\, \underbrace{(10^+,10^-)}_{\text{doublet}},\, 11^-,\, \cdots}_{\textbf{Forming a common parabola: Irrep. } \mathbf{A_1}} \tag{1}$$

The above structure in the literature referred to as A_1, can be seen as one example among in total 5 irreducible representations of the T_d-group describing tetrahedral symmetry properties of integer-spin rotational-states of even-even nuclei. The remaining characteristic rotor structures associated with the remaining irreducible representations A_2, E, F_1 and F_2 of the T_d-group have equally exotic structural features, cf. [10, 11, 13] and references therein. For odd-even nuclei the characteristic spin-parity sequences are composed of half-integer spins. Identifying experimentally such very unique structures predicted by group theory can be seen as the strongest argument in favour of the final identification of the symmetries themselves. The first case of such an identification in ^{152}Sm can be found in [1].

1.3 Abundance of Shape Isomers: Isomeric Bands

Another highly exotic feature in the context of the collective rotation of tetrahedral/octahedral-symmetric nuclei is related to the fact that at the exact symmetry limit the electric quadrupole and dipole transition probabilities vanish, $B(E2) = 0$ and $B(E1) = 0$, [4]. It follows that the lowest multipolarity of non-vanishing electric transitions is $\lambda = 3$ (i.e. the octupole transitions) which are orders of magnitude slower as compared to the $E2$ (and/or $E1$) transitions. Recall that the quadrupole transitions dominate in the decay schemes of the "traditional" quadrupole-deformed collectively rotating nuclei and it follows that the states listed in Eq. (1) or analogous ones generated by the irreducible representations (irreps.) A_2, E, F_1 and F_2, are expected to become in fact long-lived isomers whose ensembles placed on the ($E - vs. - I$)-plane form parabolas, [13] and references therein. The condition that isomers form a specific smooth-looking relations of the type $E_{\text{isomer}}(I) \propto I^2$ is unprecedented in the nuclear structure physics and represents a possible focus for the newly established collaborations, cf. discussion accompanying Fig. 1 below.

[1] Spin-parity projected Hartree-Fock calculations demonstrate that the increasing size of the corresponding deformations increases the precision with which the discussed sequences form parabolic E-vs.-I relations.

1.4 Symmetry-Breaking By the Zero-Point Motion

Even though vanishing of the $B(E2)$ and/or $B(E1)$ reduced transition probabilities is a consequence of the exact tetrahedral/octahedral symmetry limits, such conditions are met strictly speaking only at the minimum-points on the potential energy surfaces corresponding to vanishing quadrupole (and other than tetrahedral/octahedral) deformations. However, according to the elementary properties resulting from the collective model of Bohr, deformations are not static and their probability distribution are described by the corresponding collective wave functions.

In particular for the lowest-energy single-minimum states these functions resemble Gaussians and the associated *most probable* (in particular quadrupole) deformations are not vanishing. This implies that there exist non-vanishing, even if small, associated "residual" reduced $B(E2)$ or $B(E1)$ electromagnetic transition probabilities connecting some tetrahedral symmetry states among themselves or with some other states within the spectrum. These probabilities can be estimated theoretically and can be used to identify traces of their presence with the help of the big multi-detector systems such as *AGATA* in Europe or *GRETA* in the USA. The corresponding experimental studies represent a powerful technique of investigations in line with the first tetrahedral/octahedral-symmetry identification-article addressing ^{152}Sm, c.f. [1].

1.5 Symmetry Breaking By Polarisation Effects

Predictions of the high-rank symmetric nuclear configurations are based on the microscopic total energy calculations which suggest a presence of the well pronounced total energy minima corresponding to those exotic shapes which are surrounded by significant potential barriers. In the mean-field calculations such minima originate from the strong single-nucleon energy gaps giving rise to 'tetrahedral magic numbers' – analogues of the well known spherical magic gaps and spherical magic numbers.

However, when adding some nucleons on top of the magic-gap configurations (alternatively subtracting some) we perturb the symmetry similarly to the polarisation of the spherical configurations by extra particle or extra hole configurations. In other words the presence of the extra particles or holes polarises high-rank symmetric doubly-magic cores leading to a partial symmetry breaking.

Similar partial symmetry breaking is expected as the result of the Coriolis polarisation effects induced by an increasing collective rotation. Indeed, the presence of collective angular momentum distinguishes an extra direction is space, the mechanism which breaks the original symmetry induced by the T_d and / or O_h symmetries and increases with increasing spin.

Perhaps paradoxically, both these *symmetry breaking* mechanisms can be helpful in *identifying experimentally* the discussed symmetries – under the condition that they are strong enough to induce small (residual) $B(E2)$ [or $B(E1)$] collective reduced transition probabilities but at the same time sufficiently weak so that they do not destroy the symmetries in question entirely.

1.6 Symmetry Driving Single-Nucleon Orbitals

It turns out that the presence of strong gaps in the single nucleon spectra at high-rank symmetry shapes induces the presence of specific orbitals with particularly strongly down-sloping (usually above the gap) or up-sloping (usually below the gap) orbitals, cf. [14] for certain details. By an appropriate construction of the particle-hole excited states with the particle orbitals selected as down-sloping and hole states as up-sloping one may obtain configurations whose energies decrease by several (up to 10) MeV or so, leading to the final minima at the very significant high-rank symmetry deformations.

This mechanism allows for constructing the particle-hole excited states with strongly increased tetrahedral / octahedral deformations. This may be particularly useful when studying the rotational properties of nuclei in the presence of these symmetries since the structures like the one in Eq. (1) are getting better and better pronounced (e.g. the doublet energies more and more precisely respected) the larger the symmetry-inducing deformation.

1.7 Odd-A Nuclei as Tetrahedral Quantum Rotors

Following semi-analytical results of [15], Coriolis coupling of an odd nucleon to the collective tetrahedral rotor takes a simple form involving mathematical expression with numerical constants referred to as generalised decoupling parameters. These expressions are more and more precisely respected by the full numerical result the stronger the symmetry-preserving deformation. They can be used to facilitate identifying the symmetry-induced mechanisms via comparison with experiment. It follows that employing the generalised decoupling parameters formulae provides a powerful symmetry identification method which is easy in applications.

1.8 Symmetry-Oriented Research of Exotic Nuclei

The states which in contrast to the "traditional" spectroscopic behaviour manifest neither $E2$ nor $E1$ collective transitions, the latter usually dominating the nuclear population and/or decay schemes are expected to form isomers thanks to the tetrahedral/octahedral-symmetry hindrance properties [13]. Their lifetimes, as compared to the life-times of many other states and in particular to the life-times of the ground-states, are expected to be significantly longer. Specifically in very exotic nuclei whose ground-state lifetimes are known to be extremely short their identification is at the limits of the instrumental capacities at present. Under these circumstances a possible existence of series of isomeric bands at low excitation energies opens potentially invaluable possibilities of populating states with sufficiently long lifetimes, possibly longer or much longer than the lifetimes of the ground-states. This would facilitate identifying the exotic nuclear species via long-living excited rather than short living ground-states. The use of the realistic theory-predictions will help optimising the experimental conditions in the specifically designed experiments to come, focussing on exotic and very exotic nuclei.

1.9 New Search Strategies: Combining the Mass Spectrometry and γ-Ray Spectroscopy

In the past, attempts have been undertaken to obtain an experimental evidence of tetrahedral/octahedral symmetries in nuclei by employing mainly the γ-ray spectroscopic methods. The contemporary Ge multi-detector arrays are becoming more and more powerful allowing to eliminate random coincidences and identify weaker and weaker transitions. However, the fact that nothing is known at present about the build-up of the symmetry breaking e.g. via zero-point motion, the extra particle-, or hole-states on top of the magic tetrahedral/octahedral configurations and/or Coriolis coupling in case of collective rotation increases even more the already significant difficulties in approaching the very weak feeding or decay out transitions engaging the high-rank symmetry states.

Under these conditions alternative approaches possibly leading directly to the information about the isomers without directly involving γ-spectroscopic information are extremely welcome. This type of information will be obtained by applying multiple-reflection time-of-flight mass spectrometry (MR-TOF-MS) with highest resolving power as developed at the Justus-Liebig-University, Gießen [2]. This approach will allow to measure isomeric states down to about 200 keV of excitation energies [16]. Moreover, it will offer a possibility of measuring in a broadband mode so that many neighbouring nuclei which are predicted by theory to manifest high-rank symmetry configurations will be measured at the same time.

The nuclei of interest will be produced employing spontaneous fission process. We plan using ^{252}Cf spontaneous fission source installed in a gas-filled stoping cell. The MR-TOF-MS and the gas-cell are parts of the FRS Ion Catcher experiment at GSI, Darmstadt [17]. After thermalisation in the gas, the fission fragments will be extracted with the help of electric field and transported to the MR-TOF-MS. An illustration of the possible zones of research with this particular device compared with the distribution of the doubly-magic tetrahedral nuclei over the Periodic Table are illustrated in Figure 1. The combination of accelerator-independent, broad and stable production together with a very sensitive, universal and broadband detection methods allows long term studies of long time spans when measuring even the very weakly populated states.

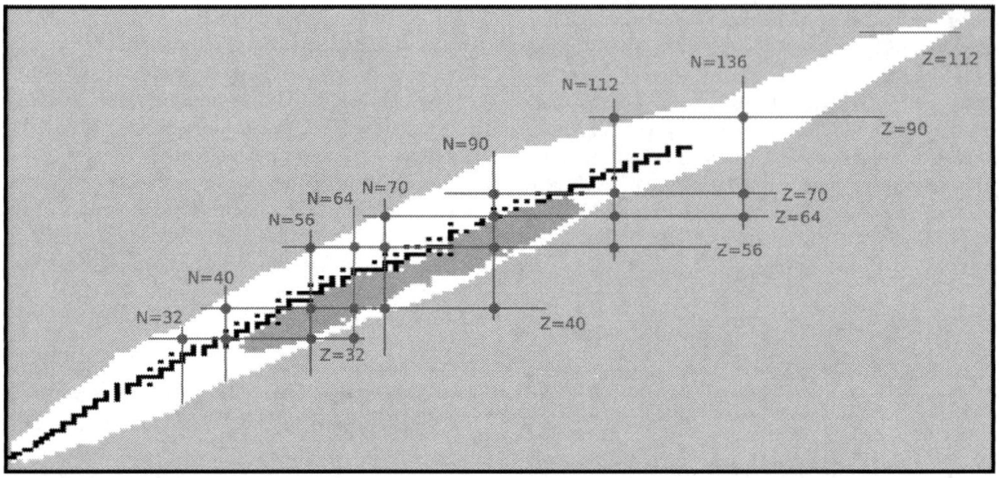

Figure 1. Positions – marked with the full circles – of the doubly-magic tetrahedral nuclei as predicted by theory. There is a relatively broad range of nuclei susceptible to produce the tetrahedral-symmetry configurations *around each central doubly-magic tetrahedral nucleus* as illustrated in figures in Section 2. The range of nuclei produced using ^{256}Cf source in the GSI set-up is marked explicitly.

2 Viewing High-Rank Symmetries as Crypto-Symmetries: Illustrations

At an early stage of theoretical investigations of the tetrahedral and octahedral symmetries in nuclei it became clear that the realistic nuclear mean-field calculations, both the phenomenological ones – based on the Woods-Saxon universal Hamiltonian [18], and the microscopic ones based on the Skyrme Hartree-Fock Bogolyubov Hamiltonians – predict systematic occurrence of the high-rank symmetry configurations in numerous ranges of the Periodic Table. In other words, and in full accordance with the scheme presented in Figure 1, the nuclei for which the high-rank symmetry configuration are predicted are much more numerous then the nuclei with the well known spherical symmetry. And yet, until the very recent announcement [19] of the identification of the high-rank symmetries in a single nucleus ^{152}Sm there was no other case signalled. This situation inspired Dudek and coworkers [19] to introduce the nickname *κρυπτο-συμμετριες* (crypto-symmetries or hidden symmetries) which alludes to the fact that the high-rank symmetries, even though expected to be present allover, have never been seen experimentally – as we believe – mainly because so far nobody was looking for them with the designed methods, and thus the symmetries remain "hidden" (interested reader might compare with the techniques designed in [1]).

In the following we will present briefly certain predictions related to the scheme in Figure 1, in the context of the new approach proposed in Section 1.9. Let us begin by noticing that among 23 doubly-magic tetrahedral nuclei marked in the figure there are only five of them, *viz.* ^{88}Ge, ^{96}Zr, ^{104}Zr, ^{146}Ba and, possibly, ^{154}Gd, which are close to the zone of nuclei populated as fission fragments of ^{252}Cf. All the five can be considered extremely exotic and therefore populated under very unfavourable conditions of statistics.

Consequently it will be of fundamental interest to learn what are the other, i.e. not necessarily *doubly magic* tetrahedral-symmetry nuclei predicted by theory which possibly present strong signatures of the tetrahedral or the octahedral symmetry configurations?

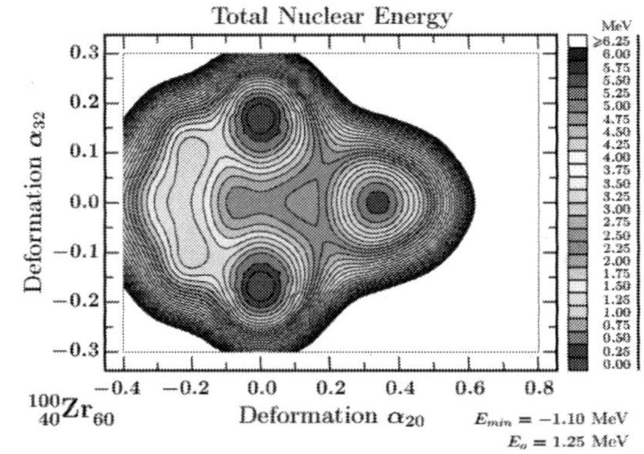

Figure 2. An example of the potential energy surface, here for ^{100}Zr nucleus on the plane spanned by the quadrupole (α_{20}) vs. tetrahedral (α_{32}) deformations using the phenomenological Woods-Saxon realisation of the mean-field approach with the macroscopic-microscopic method of Strutinsky. The symmetric tetrahedral minima at $\alpha_{32} \approx \pm 0.18$ are in direct competition with the strongly deformed axial-symmetry minimum at $\alpha_{20} \approx 0.37$. Similar illustrations apply for numerous nuclei within the fission-fragment population zone marked in orange colour in Figure 1.

To answer this type of questions we have performed systematic calculations using the realistic phenomenological mean-field approach with the deformed potential of the Woods-Saxon type using the so-called "universal" parametrisation of [18]. The Hamiltonian depends on 12 adjustable parameters whose values have been fixed once for all nuclei in the Periodic Table (wherefrom the name "universal" when referring to this parametrisation) using the experimental information about the single-nucleon levels in deformed nuclei. This parameterisation has been shown to describe numerous nuclear structure properties in a very satisfactory fashion, at least on average. Within our analysis, the total energy surfaces have been calculated in the space of deformations including α_{20} and α_{22} (quadrupole deformations), α_{32} (tetrahedral deformation) and α_{40} (hexadecapole deformation) using macroscopic-microscopic method of Strutinsky with the macroscopic energy defined after the so-called Yukawa-folded approach.

Figure 3 shows a selection of the total energy results for Zirconium nuclei with $N \in [56, 76]$ in the form of the one-dimensional cross-sections originating from the results like the ones in Figure 2. Let us notice that actually for all the nuclei displayed the total nuclear energies present tetrahedral-symmetry minima. Given the relatively high potential barriers separating the twin tetrahedral minima like the ones in Figure 2 from one another and comparably important barriers separating the two from the third ones, the large quadrupole-deformation minima, we conclude that according to the calculations there will be numerous nuclear states in the discussed mass range with the tetrahedral symmetry local minima expected to lead to longlived tetrahedral symmetry isomers.

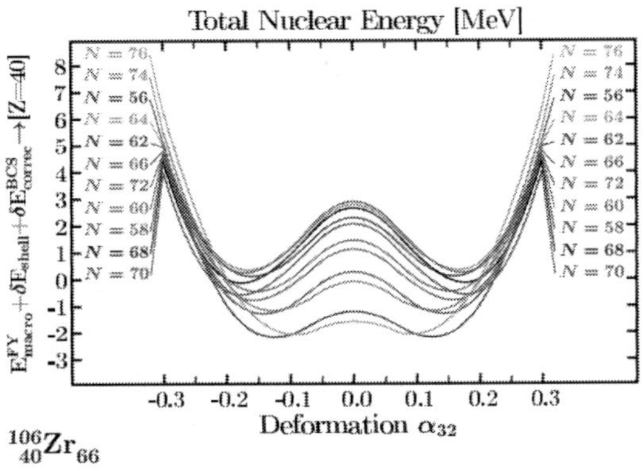

Figure 3. Cross-sections of the total energy surfaces like the one in Figure 2 but for a number of Zirconium isotopes showing the tetrahedral-symmetry minima not only for the magic numbers $N_{T_d} = 56, 64$ and 70 but also for numerous other isotopes. These theory predictions encourage experimental programme briefly announced in Section 1.9.

Analogous results are illustrated in Figure 4 for a series of Gadolinium nuclei. From the results presented for Zirconium and Gadolinium isotopes shown but also from the similar comparisons for other elements we may conclude that the tetrahedral shell effects are very specific in the sense that they do not privilege strongly the tetrahedral doubly magic-number nuclei as compared to their neighbours. In reality, there are several dozens of various isotopes and isotones which, according to the theory predictions should manifest tetrahedral symmetry configurations very likely leading to the isomers. Moreover, let us remind the reader that according to the arguments given in Sections 1.2 and 1.3 each tetrahedral symmetry minimum leads to numerous rotational states expected to lead to the isomers and/or bands composed of isomers.

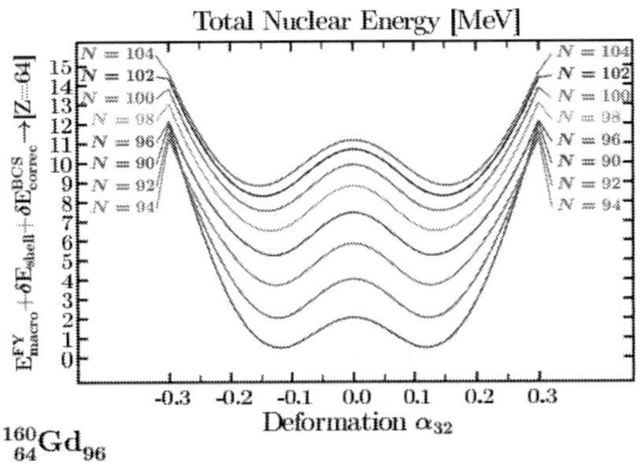

Figure 4. Cross-sections of the total energy surfaces like the ones in Figure 3 but for a number of Gadolinium isotopes showing the tetrahedral-symmetry minima not only for the magic numbers $N_{T_d} = 90$ but also for numerous other isotopes. These theory predictions strongly encourage experimental programme briefly announced in Section 1.9.

3 Summary and Conclusions

In the first part of this article we recalled briefly the numerous exotic and most of the time unprecedented quantum effects expected to accompany the atomic nuclei in their tetrahedral and/or octahedral symmetry configurations. All these effects attracted certain attention in the literature in the past contributing to the growing interest in the exotic symmetries especially in the exotic nuclei. Whereas the quantum effects in question provide strong motivation for this kind of research *per se*, the existence of the hindrance factors which cause vanishing of the usually strong quadrupole and/or dipole electric transitions leads very likely to the novel class of shape isomers opening the new avenues for the studies of exotic nuclei for which ground-state life times are too short for the present day instrumentation.

Group theory considerations recalled briefly in this article lead to the new classes of rotational bands whose members can carry both parities and/or may form doublets, triplets, quadruplets, etc. All these features can be seen as unique experimental identification criteria of the states and phenomena involving these exotic symmetries. The expected existence of many isomeric states with tetrahedral/octahedral symmetries makes it natural to propose the mass spectrometry as one of the most efficient criteria of identification of such nuclear states. In this context we propose applying multiple-reflection time-of-flight mass spectrometry (MR-TOF-MS) in collaboration with the teams from the University of Gießen and from the GSI, Darmstadt in addressing the issue of identifying the presence of the isomers. Let us emphasise that theory predicts the presence of many high-rank symmetry (thus isomeric) states in the area of the Periodic Table covered by the ^{252}CF fission-fragment production zone in the GSI experiment foreseen soon.

Acknowledgements. This work was partially supported by the Polish National Science Centre under Contract No. 2016/21/B/ST2/01227 and Polish-French COPIN-IN2P3 collaboration agreement under project numbers 04-113 and 05-119.

References

[1] J. Dudek, D. Curien, I. Dedes, K. Mazurek, S. Tagami, Y. R. Shimizu, and T. Bhattacharjee; Phys. Rev. **C 97**, 021302(R) (2018)

[2] T. Dickel et al., Nucl. Instrum. Meth. **A777** (2015) 172-188

[3] J. Dudek, A. Góźdź, N. Schunck and M. Miśkiewicz, Phys. Rev. Lett. **88**, 252502 (2002)

[4] J. Dudek, D. Curien, N. Dubray, J. Dobaczewski, V. Pangon, P. Olbratowski, and N. Schunck, Phys. Rev. Lett. **97**, 072501 (2006)

[5] J. Dudek, D. Curien, A. Góźdź, Y. R. Shimizu and S. Tagami, Acta Phys. Polonica **B 44**, 305 (2013)

[6] S. Takami, K. Yabana, and M. Matsuo, Phys. Lett. **B 431** 242 (1998)

[7] M. Yamagami, K. Matsuyanagi, and M. Matsuo, Nucl. Phys. **A 693** 579 (2001)

[8] K. Zberecki, P.-H. Heenen, and P. Magierski, Phys. Rev. **C 79**, 014319 (2009)

[9] J. Zhao, B.-N. Lu, E.-G. Zhao, and S.-G. Zhou, Phys. Rev. **C 95**, 014320 (2017)

[10] M. Hamermesh, *Group Theory and Its Application to Physical Problems*, Dover Publications, Inc., New York, 1962.

[11] S. Tagami, Y. R. Shimizu, and J. Dudek, Phys. Rev. **C 87**, 054306 (2013).

[12] S. Tagami, Y. R. Shimizu, and J. Dudek, J. Phys. G: Nucl. Part. Phys. **42**, 015106 (2015).

[13] J. Dudek, I. Dedes, J. Yang, A. Baran, D. Curien, T. Dickel, A. Góźdź, D. Rouvel and H. L. Wang; Acta Phys. Polonica **B 50**, 685 (2019)

[14] A. Baran nad J. Dudek, Acta Phys. Polonica B Proceedings Supplement, 2019 in print

[15] S. Tagami, Y. R. Shimizu and J. Dudek, Phys. Rev. **C 98**, 024304 (2018)

[16] S. A. S. Andrés et al., Phys. Rev. **C 99** 064313 (2019)

[17] W. R. Plaß et al., Nucl. Instrum. Meth. **B 317** 457 (2013)

[18] The *Universal Woods-Saxon Hamiltonian* and associated, so-called 'universal parametrisation' has been developed in a series of articles:
J. Dudek and T. Werner, J. Phys. G: Nucl. Phys. **4**, 1543 (1978),
J. Dudek, A. Majhofer, J. Skalski, T. Werner, S. Cwiok, and W. Nazarewicz, J. Phys. G: Nucl. Phys. **5**, 1359 (1979),
J. Dudek, W. Nazarewicz, and T. Werner, Nucl. Phys. A **341** 253, (1980),
J. Dudek, Z. Szymański, and T. Werner, Phys. Rev. C **23**, 920 (1981), and has been summarised in
S. Ćwiok, J. Dudek, W. Nazarewicz, J. Skalski, and T. Werner, Comp. Phys. Comm. **46**, 379 (1987).
This parametrisation is being used without modifications by many authors also today.

[19] J. Dudek, N. Schunck, N. Dubray and A. Góźdź, Int. J. Mod. Phys. **E 14** 389 (2005)

Optical Potentials: Microscopic vs. Phenomenological Approaches

Paolo Finelli[1], *Matteo* Vorabbi[2], and *Carlotta* Giusti[3]

[1]Astronomy and Physics Department, University of Bologna (Italy) and INFN section of Bologna (Italy)
[2]TRIUMF, Vancouver (Canada)
[3]Physics Department, University of Pavia (Italy) and INFN section of Pavia (Italy)

Abstract. In this work we study the performances of our microscopic optical potential [1, 2], derived from nucleon-nucleon chiral potentials at fifth order (N^4LO), in comparison with those of a successful non-relativistic phenomenological optical potential in the description of elastic proton scattering data on tin and lead isotopes at energies around and above 200 MeV. Our results indicate that microscopic optical potentials derived from nucleon-nucleon chiral potentials at N^4LO can provide reliable predictions for observables of stable and exotic nuclei, even at energies where the robustness of the chiral expansion starts to be questionable.

1 Introduction

In Ref. [1] we constructed a microscopic optical potential (OP) for elastic proton-nucleus (pA) scattering starting from nucleon-nucleon (NN) chiral potentials derived up to N^3LO in the chiral expansion and we studied the chiral convergence of the NN potential in reproducing the pA scattering observables. The OP was obtained at the first-order term within the spectator expansion of the non-relativistic multiple scattering theory and adopting the impulse approximation and the optimum factorization approximation.

In a subsequent work [2] we adopted the same model to obtain the OP and we studied the chiral convergence of a new generation of NN chiral interactions derived up to N^4LO.

In this contribution we show a sample of our latest work [3] about the predictive power of our microscopic OP derived in Ref. [2] from different chiral potentials at N^4LO and of the successful phenomenological OP (KD) of Ref. [12] in comparison with available data for the observables of elastic proton scattering on different isotopic chains. Results are presented for several proton energies around and above 200 MeV, with the aim to test the upper energy limit of applicability of our OP before the chiral expansion scheme breaks down.

2 Theory

The theoretical justification for the description of the pA optical potential in terms of the microscopical NN interaction has been addressed for the first time by Watson *et al.* [4] and then formalized by Kerman *et al.* (KMT) [5], where the so-called multiple scattering approach to the pA optical potential is expressed by a series expansion of the free NN scattering amplitudes.

In Refs. [1, 2] a microscopic OP V_{opt} was obtained at the first-order term within the spectator expansion of the nonrelativistic multiple scattering theory, corresponding to the single-scattering approximation. The impulse approximation was adopted, where nuclear binding forces on the interacting target nucleon are neglected, as well as the optimum factorization approximation, where the two basic ingredients of the calculations, *i.e.* the nuclear density and the NN t matrix, are factorized. We refer the reader to Refs. [1, 2] for all relevant details and an exhaustive bibliography. In the momentum space, the factorized V_{opt} is obtained as

$$V_{opt}(\mathbf{q}, \mathbf{K}; \omega) \sim \sum_{N=n,p} t_{pN}\left[\mathbf{q}, \frac{A+1}{A}\mathbf{K}; \omega\right]\rho_N(q)\,, \quad (1)$$

where t_{pN} represents the proton-proton (pp) and proton-neutron (pn) free t matrix evaluated at a fixed energy ω, ρ_N the neutron and proton profile density, and the incoming and outgoing projectile momenta $\mathbf{k}$ and $\mathbf{k}'$ are conveniently expressed by the variables $\mathbf{q} \equiv \mathbf{k}' - \mathbf{k}$ and $\mathbf{K} \equiv \frac{1}{2}(\mathbf{k}' + \mathbf{k})$ (see Sect. II of Ref. [1]).

For the neutron and proton densities of the target nucleus we use as in Refs. [1, 2] a Relativistic Mean-Field (RMF) description [6], which has been quite successful in the description of ground state and excited state properties of finite nuclei, in particular in a Density Dependent Meson Exchange (DDME) version, where the couplings between mesonic and baryonic fields are assumed as functions of the density itself [7].

For the NN interaction we use here two different versions of the chiral potentials at fifth order (N^4LO) recently derived by Epelbaum, Krebs, and Meißner (EKM) [8, 9] and Entem, Machleidt, and Nosyk (EMN) [10, 11]. As explained in Ref. [2], the two versions of the chiral N^4LO potentials have significant differences concerning the renormalization procedures and we follow the same

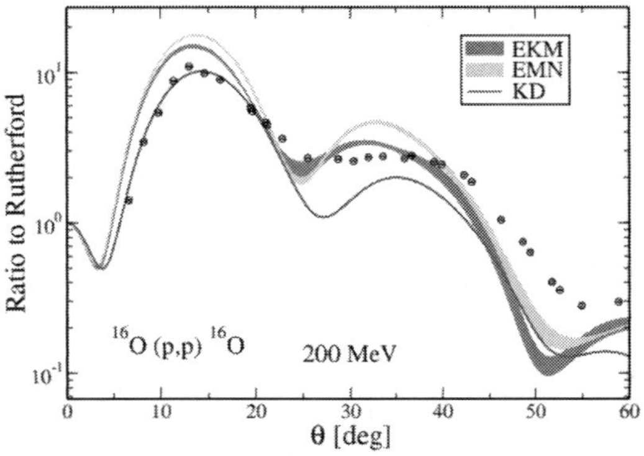

Figure (1) Ratio of the differential cross section to the Rutherford cross section as a function of the center-of-mass scattering angle θ for elastic proton scattering off ^{16}O. Calculations are performed at E = 200 MeV (laboratory energy) with the microscopic OPs derived from the EKM [8, 9](EKM, red band) and EMN [10, 11] (EMN, green band) NN chiral potentials at N^{4}LO and with the phenomenological global OP of Ref. [12] (KD, violet line). The interpretation of the bands is explained in the text. Experimental data from Refs. [13, 14].

prescriptions adopted there. The strategy followed for the EKM potentials [8, 9] consists in a coordinate space regularization for the long-range contributions $V_{\rm long}(\boldsymbol{r})$, by the introduction of $f\left(\frac{r}{R}\right) = \left(1 - \exp\left(-\frac{r^2}{R^2}\right)\right)^n$, and a conventional momentum space regularization for the contact (short-range) terms, with a cutoff $\Lambda = 2R^{-1}$. Five choices of R are available: $0.8, 0.9, 1.0, 1.1$, and 1.2 fm, leading to five different potentials.

On the other hand, for the EMN potentials, a slightly more conventional approach was pursued [10, 11]. A spectral function regularization, with a cutoff $\tilde{\Lambda} \simeq 700$ MeV, was employed to regularize the loop contributions and a conventional regulator function, with Λ = 450, 500, and 550 MeV, to deal with divergences in the Lippmann-Schwinger equation. For all details we refer the reader to Ref. [2].

Since the goal of the present work is to test the predictive power of our microscopic OP in comparison with available experimental data, it is definitely interesting to show the uncertainties on the predictions produced by NN chiral potentials obtained with different values of the regularization parameters. For this purpose, all calculations have been performed with three of the EKM [8, 9] potentials, corresponding to $R = 0.8, 0.9$, and 1.0 fm, and with two of the EMN [10, 11] potentials, corresponding to Λ = 500 and 550 MeV.

In all the next figures the bands give the differences produced by changing R for EKM (red bands) and Λ for EMN (green bands). Thus the bands have here a different meaning than in Ref. [2], where the EKM and EMN NN chiral potentials at N^{4}LO were also used. The aim of Ref. [2] was to investigate the convergence and to assess the theoretical errors associated with the truncation of the chiral expansion and the bands were given to investigate these issues.

We also showed in Ref. [2] that EKM calculations based on different values of R are quite close and consistent with each other (although, as remarked in Ref. [8], larger values of R are probably less accurate due to a larger influence of cutoff artifacts). The same assumption can be made about the EMN potentials: changing the cutoffs does not lead to sizeable differences in the χ^2/datum (see Tab.VIII in Ref. [11]) and it is safe to perform calculations with only two potentials. Because we want to explore elastic scattering at energies around and above 200 MeV, we exclude the EKM potentials with R = 1.1 and 1.2 fm and the EMN potential with Λ = 450 MeV.

In Ref. [1] we compared the results obtained with different versions of EKM and EMN chiral potentials at N^{4}LO for the pp and pn Wolfenstein amplitudes and for the scattering observables of elastic proton scattering off ^{12}C, ^{16}O, and ^{40}Ca nuclei at an incident proton energy of E = 200 MeV.

We show in Fig. 1 the ratio of the differential cross section to the Rutherford cross section for elastic proton scattering off ^{16}O at E = 200 MeV. The results obtained with the EKM and EMN potentials and with the KD optical potential are compared with the experimental data taken from Refs. [13, 14]. The EKM and EMN results for the differential cross section give a reasonable, although not perfect, agreement with data. The experimental ratio is slightly overestimated at lower angles and somewhat underestimated for $\theta \geq 50°$. The differences between the EKM and EMN results are small and not crucial, EKM gives a smaller cross section around the maxima and therefore a somewhat better agreement with the data in this region. The bands, representing the uncertainties on the regularization of the NN chiral potentials, are generally small and not influential for the comparison with data. The KD result gives a good description of the experimental cross section for $\theta \leq 20°$ and underpredicts the data for larger angles.

The results for Sn isotopes at 295 MeV and for ^{120}Sn at 200 MeV are displayed in the upper panel of Fig. 2. In this case all the OPs give qualitatively similar results and a reasonable agreement with data, in particular, for $\theta \leq 20°$. The agreement generally declines for larger angles. KD gives a better description of ^{120}Sn data at 200 MeV, where the EKM and EMN results are a bit larger than the data at the maxima and a bit lower at the minima. We note that ^{120}Sn is included in the experimental database for the KD potential for proton energies up to 160 MeV. At 295 MeV, the microscopic OP gives, in general, a slightly better agreement with the data than KD for all the tin isotopes shown in the figure.

The results for Pb isotopes at 295 MeV and for ^{208}Pb data at 200 MeV are displayed in lower panel of Fig. 2. Also in this case the experimental cross section at 200 MeV is well described by KD, the agreement is better than with the microscopic OP. The experimental database for KD includes ^{208}Pb up to 200 MeV. At 295 MeV a better agreement with data is generally given by the EKM and EMN results, in particular by EMN: for all the three iso-

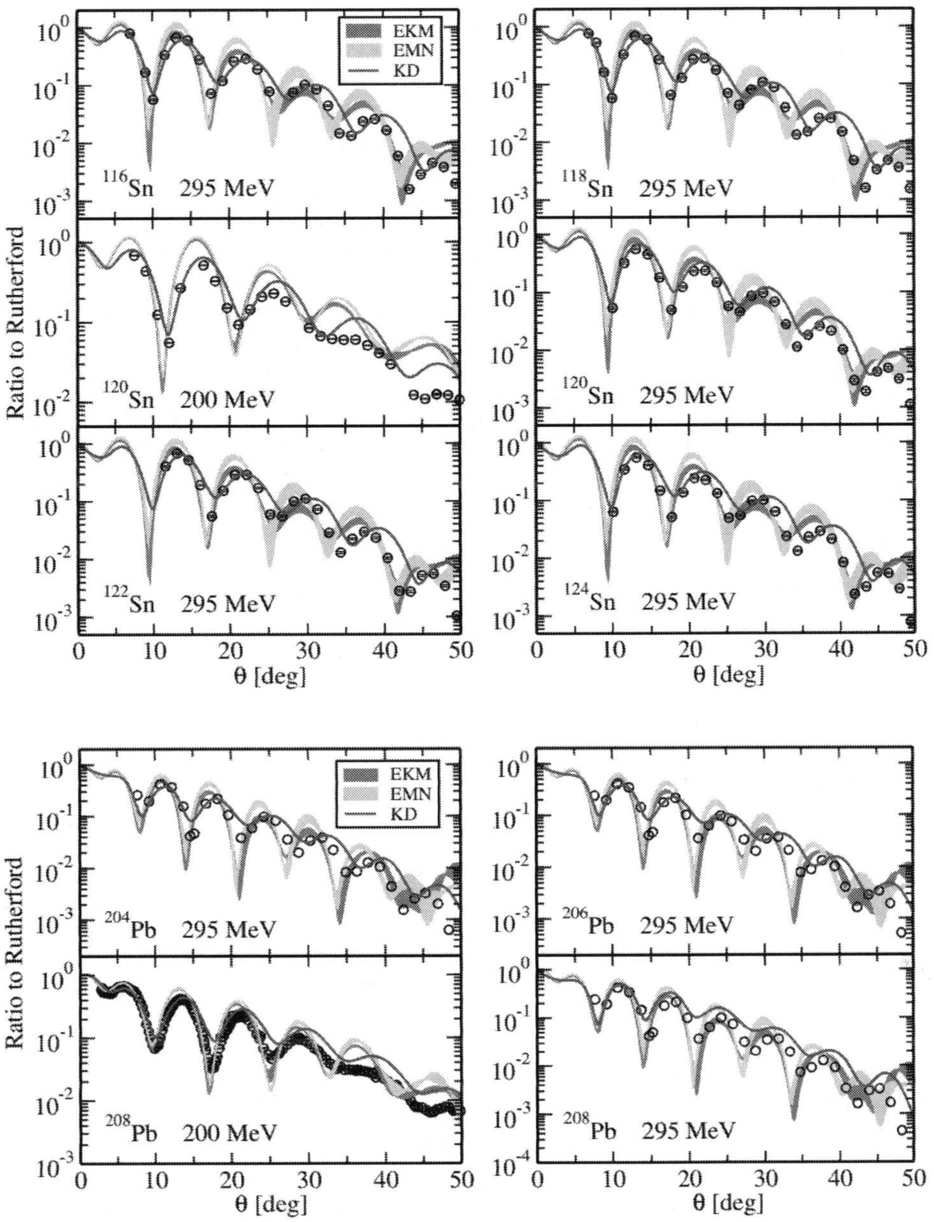

Figure (2) **(Upper panel)** The same as in Fig. 1 for Sn isotopes: ^{120}Sn at E = 200 MeV and ^{116,118,120,122,124}Sn at E = 295 MeV. Experimental data from Refs. [13, 14]. **(Lower panel)** The same as in Fig. 1 for Pb isotopes: ^{208}Pb at E = 200 MeV and ^{204,206,208}Pb at E = 295 MeV. Experimental data from Refs. [13, 14].

topes considered, the two results practically overlap for $\theta \leq 20°$, where they are also very close to the KD result, then they start to separate and the EMN result is a bit larger than the EKM one and in better agreement with data. We point out that the uncertainty bands, that are generally narrow, in this case become larger increasing the scattering angle, when also the agreement with data declines.

The results that we have shown till now indicate that, in comparison with the phenomenological KD potential, our microscopic OP, in spite of the approximations made to derived it, has a comparable and in some cases even better predictive power in the description of the cross sections on the isotopic chains and energy range here considered. KD is able to give a better and excellent description of data in specific situations, in particular, for data included in the experimental database used to obtain the parameters of the phenomenological KD potential and at the lower energies considered. Our microscopic OP is able to give a similar and more homogeneous description of data for all the nuclei of an isotopic chain and, for energies above 200 MeV it gives, in general, a better agreement with data than the phenomenological KD potential.

References

[1] M. Vorabbi, P. Finelli, and C. Giusti, Phys. Rev. C **93**, 034619 (2016)

[2] M. Vorabbi, P. Finelli, and C. Giusti, Phys. Rev. C **96**, 044001 (2017)

[3] M. Vorabbi, P. Finelli, and C. Giusti, Phys. Rev. C **98**, 064602 (2018)

[4] W. B. Riesenfeld and K. M. Watson, Phys. Rev. **102**, 1157 (1956)

[5] A. K. Kerman, H. McManus, and R. M. Thaler, Annals Phys. **8**, 551 (1959)

[6] T. Niksic, N. Paar, D. Vretenar, and P. Ring, Computer Physics Communications **185**, 1808 (2014)

[7] T. Niksic, D. Vretenar, P. Finelli, and P. Ring, Phys. Rev. C **66**, 024306 (2002)

[8] 'E. Epelbaum, H. Krebs, and U. G. Meissner, Phys. Rev. Lett. **115** 122301 (2015)

[9] E. Epelbaum, H. Krebs, and U. G. Meissner, Eur. Phys. J. A **51** 53 (2015)

[10] D. R. Entem, N. Kaiser, R. Machleidt, and Y. Nosyk, Phys. Rev. C **91** 014002 (2015)

[11] D. R. Entem, R. Machleidt, Y. Nosyk, Phys. Rev. **C96** 024004 (2017)

[12] A. J. Koning and J. P. Delaroche, Nucl. Phys. A **713**, 231 (2003)

[13] `http://www.nndc.bnl.gov/exfor/exfor.htm`

[14] `http://www.physics.umd.edu/enp/jjkelly/`

Structural investigation of neutron-deficient Pt isotopes: the case of ^{178}Pt

Christoph Fransen[1,*], *Franziska* Mammes[1], *Robert* Bark[2], *Thomas* Braunroth[1], *Zinhle* Buthelezi[2], *Alfred* Dewald[1], *Tshepo* Dinoko[2], *Siegie* Förtsch[2], *Matthias* Hackstein[1], *Jan* Jolie[1], *Pete* Jones[2], *Elena* Lawrie[2], *Julia* Litzinger[1], *Claus* Müller-Gatermann[1], *Richard* Newman[3], *Nima* Saed-Samii[1], *John F.* Sharpey-Schafer[2,4], *Obed* Shirinda[2], *Ricky* Smit[2], *Nigel* Warr[1], *Mathis* Wiedeking[2], and *Karl-Oskar* Zell[1]

[1]Institut für Kernphysik, Universität zu Köln, Zülpicher Strasse 77, 50937 Cologne, Germany
[2]iThemba LABS, National Research Foundation, P.O.Box 722, Somerset West, 7129, South Africa
[3]Department of Physics, Stellenbosch University, Matieland, Stellenbosch 7602, South Africa
[4]Department of Physics, University of the Western Cape, Robert Sobukwe Road, Bellville 7535, South Africa

Abstract. Lifetime measurements with the recoil distance Doppler-shift technique have been performed to determine yrast $E2$ transition strengths in ^{178}Pt. The experimental data are related to those on neighboring Pt isotopes, especially recent data on ^{180}Pt, and compared to calculations within the interacting boson model and a Hartree-Fock Bogoliubov approach. These models predict prolate deformed ground states in Pt isotopes close to neutron midshell consistent with the experimental findings. Further, evidence was found that the prolate intruder structure observed in neutron-deficient Hg isotopes that is minimum in energy in ^{182}Hg becomes the ground state configuration in ^{178}Pt and neighboring ^{180}Pt with nearly identical transition quadrupole moments. The new data on ^{178}Pt are further discussed in the context of the systematics along the Pt isotopic chain with respect to a possible sharp shape transition towards a weakly deformed or a quasi-vibrational ground state whereas the prolate structure increases in energy in ^{174,176}Pt and becomes an intruder configuration similar to, for example, ^{180,182}Hg.

1 Introduction

Neutron deficient nuclei in the $A = 180$ mass region close to the $Z = 82$ main shell closure are the subject of numerous experimental and theoretical studies. Here, challenges are the understanding of both the shape evolution for isotopic chains where the proton number is not far from magicity and the phenomenon of shape coexistence that is very predominant in Pb and Hg isotopes close to neutron midhell (see, e.g., [1] for a general overview and [2, 3] for experimental data on Pb and Hg isotopes, respectively). In Hg isotopes close to neutron midshell, i.e., around $N = 104$, clear evidence was found for a weakly oblate ground state configuration with quite constant energy systematics and a prolate intruder band with a parabolic trend of the level energies with respect to the neutron number where minimum energies of the latter are reached for ^{182}Hg [4]. In a very recent work on ^{178}Hg [5] the expected shift of the prolate deformed structure toward higher excitation energies for decreasing neutron number was proven from $E2$ transition strengths. In addition, it turned out that an intermediate deformation between normal and superdeformation has been attributed to the prolate intruders [5]. The latter is a new phenomenon in the chain of the Hg isotopes.

Due to these experimental findings that are also supported by different theoretical approaches, an investigation of the neighboring neutron deficient Pt isotopic chain is of enhanced interest. A detailed analysis of ^{180}Pt including the measurement of $E2$ transition strengths between the lowest yrast levels indicates an axially deformed shape with slight γ softness [6]. It is important to stress that the deformation found for the ground state configuration in ^{180}Pt is comparable to the deformation of the intruder states in ^{182}Hg as discussed in [6]. In that framework it should be noted that neighboring ^{176,178}Os exhibit signatures for nuclei at the critical point of the transition from spherical to axialsymmetric rotor nuclei (X(5)) [7]. Nevertheless, ^{180}Pt does not follow the expected trend for transition quadrupole moments in the close vicinity of X(5) nuclei and approaching the $N = 82$ shell closure. In fact, the rather large quadrupole deformations of both ^{180}Pt [6] and ^{182}Pt [8] indicate that this deviation from the expected trend is a general effect, which might be correlated to a weakening of the $Z = 82$ shell closure.

For a decreasing neutron number in the lighter Pt isotopes, an evolution from a rotor-like ground state configuration towards a shape coexistence of a quasi-vibrational structure and a well-deformed rotor intruder configuration at very low spins is expected. This was already discussed by Dracoulis and coworkers in an older publication [9] and is partly supported by the level schemes. The authors of [9] point out that the yrast bands of ^{176,178}Pt show a coexistence of a well deformed 'intruder' configuration in ^{176}Pt, that is falling in energy to be the main component

*e-mail: fransen@ikp.uni-koeln.de

 https://doi.org/10.1051/epjconf/201922301016

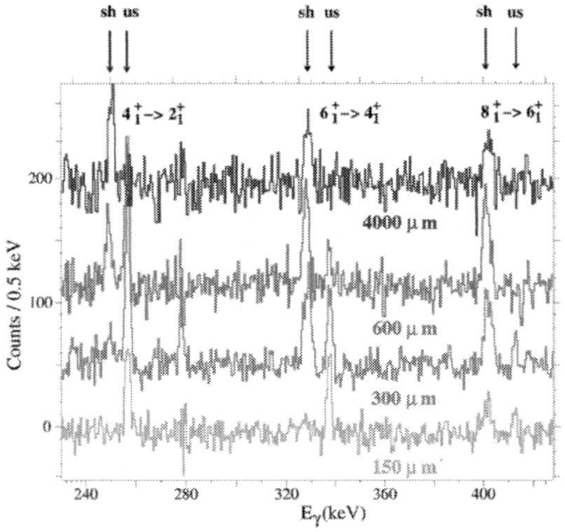

Figure 1. γ-ray spectra of ^{178}Pt for four different target–degrader distances energy-gated on the $2_1^+ \to 0_1^+$ transition. The labels "us" and "sh" denote the unshifted and shifted components of the respective γ-ray lines.

in the ground state of the heavier Pt isotopes as is the case for ^{180}Pt and as was observed by [6]. In addition, the very similar and nearly constant transition quadrupole moments within the yrast band of ^{180}Pt and between the higher yrast states in ^{180,182}Hg that belong to the prolate intruder configuration of $Q_t \approx 8eb$ further support this picture. The prolate ground state configuration of ^{180}Pt is also evident from theoretical approaches, e.g., with the Interacting Boson Model (IBM) Hamiltonian derived from the Gogny energy density functional [10] and with the IBM with configuration mixing (IBM-CM) and a self-consistent mean-field Hartree-Fock-Bogoliubov (HFB) approach, starting from the Gogny-D1S interaction [11]. Even though these calculations explain the experimental observations very well, there are deviating results for the lighter even-even Pt isotopes from ^{178}Pt down to ^{172}Pt as pointed out in [11]: The IBM-CM calculation predicts a smooth transition to a weakly prolate deformation for decreasing neutron number towards ^{172}Pt. In contrast, the self-consistent mean-field HFB approach hints at a transition from a prolate structure in ^{178}Pt via a coexistence of prolate deformation and a γ-soft structure towards spherical shape in ^{172,174}Pt. Before this work experimental data were too sparse to clarify this situation.

As ^{178}Pt is expected to be just at the edge of the aforementioned transition towards a suggested shape coexistence an investigation of this nucleus with respect to the precise determination of yrast $E2$ transition strengths is of high interest and was done in this work. So far, for ^{178}Pt level lifetimes and thus $E2$ transition strengths were only known for the 4_1^+ and 6_1^+ states from an experiment with a single Compton-suppressed Ge detector where six NaI(Tl) detectors were used as a multiplicity trigger [9]. The authors of this publication also give a (tentative) lifetime value for the 8_1^+ state but state that the respective γ-ray line is contaminated. Due to the fact that no gating on populating transitions of the levels of interest in ^{178}Pt was possible the results from [9] might be influenced by unobserved delayed sidefeeding. The lifetime of the 2_1^+ state is well known from a more recent fast-timing experiment [12].

2 Experimental details

To populate excited states in ^{178}Pt, the ^{98}Mo(^{84}Kr,4n)^{178}Pt reaction was used. The ^{84}Kr beam was provided by the K=200 separated sector cyclotron (SSC) at iThemba LABS, South Africa. The target consisted of an enriched, self-supporting 0.9 mg/cm^2 ^{98}Mo foil. After the fusion-evaporation reaction, the reaction products were stopped in a 15 mg/cm^2 gold foil. During the flight in vacuum the ^{178}Pt nuclei had a mean velocity of about 4.2% of the speed of light. Target and stopper foils were mounted in the Cologne coincidence plunger device [13] to employ the Recoil Distance Doppler-shift (RDDS) method for the determination of level lifetimes. This plunger device was used for the first time at iThemba LABS. The target-to-stopper distance was monitored by measuring the capacitance between the foils. Beam-induced drifts were compensated with a feedback system using a piezoelectric device. The plunger device was coupled to the AFRODITE Ge-detector array [14]. In this experiment, it comprised of 8 escape suppressed high-purity germanium (HPGe) Clover detectors mounted in two rings with average angles of 45 and 135 degrees with respect to the beam axis. Furthermore, a low energy photon spectrometer (LEPS) detector was mounted at an angle of 90 degrees with repect to the beam axis.

Data were taken at 14 target-to-stopper distances ranging from nearly electrical contact of the foils up to 7.2 mm. Each distance was measured for about 10 hours with an average ^{84}Kr beam current of 2.0–2.5 pnA. The reaction cross section for the 4n channel to ^{178}Pt was only of the order of 1 mb. Nevertheless, it was possible to analyze the data by gating on the Doppler-shifted component of each direct feeding transition of the respective level of

Table 1. Results of the RDDS experiment on the lowest yrast states in ^{178}Pt giving the level energies E_{lev} (from [16]), the level lifetimes and the $E2$ decay transition strengths $B(E2; I_1^+ \to J_1^+)$ ($J = I - 2$) and transition quadrupole moments derived from these observables. The more precise lifetime value of the 2_1^+ state labeled with "*" was taken from [12] and used for the determination of the respective $E2$ transition strength.

I_1^+	E_{lev} (keV)	$\tau\,(I_1^+)$ (ps)	$B(E2; I_1^+ \to J_1^+)$ (W.u.)	Q_t (eb)
2_1^+	170.3	445 (100)		
		412 (30)*	143^{+11}_{-10}	$6.53^{+0.26}_{-0.22}$
4_1^+	427.4	40.8 (24)	259^{+16}_{-14}	$7.36^{+0.23}_{-0.21}$
6_1^+	765.2	11.9 (11)	245^{+25}_{-21}	$6.81^{+0.35}_{-0.29}$
8_1^+	1178.3	3.79 (50)	289^{+44}_{-34}	$7.24^{+0.55}_{-0.42}$
10_1^+	1661.3	1.84 (82)	276^{+222}_{-85}	$7.0^{+2.8}_{-1.1}$

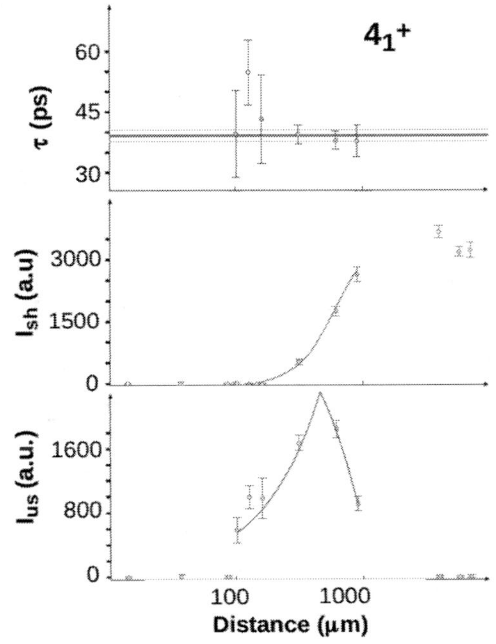

Figure 2. τ-curve of the 4_1^+ state in ^{178}Pt determined from a gate on the flight component of the feeding $6_1^+ \rightarrow 4_1^+$ transition (upper panel). The lower two panels show the intensities of the shifted (I_{sh}, middle panel) and unshifted (I_{us}, lower panel) components.

interest to exclude contributions from unoberved delayed sidefeeding.

The listmode data with $\gamma\gamma$ coincident events between all Clover HPGe detectors data were sorted into four $\gamma\gamma$ matrices for all detector-angle combinations for each distance. Each Clover was used as a single HPGe detector, i.e., the spectra of the respective four HPGe segments were summed up including an addback of Compton scattered γ-rays between the segments of the respective Clover. An example for the γ-ray spectra of this measurement showing the unshifted and Doppler-shifted components of the lowest yrast transitions in coincidence to the $2_1^+ \rightarrow 0_1^+$ γ-ray transition for four target to stopper distances is shown in Fig. 1.

The lifetimes were determined with the well established Differential Decay Curve Method (DDCM) [13, 15]. For $\gamma\gamma$-coincidence data the relation between the lifetime $\tau_{(D)}$ for a distance D between target and stopper is given by

$$\tau_{(D)} = \frac{\{Y_{sh}, X_{us}\}_{(D)}}{\frac{d}{dt}\{Y_{sh}, X_{sh}\}_{(D)}}, \qquad (1)$$

where $\{Y_{sh}, X_{us}\}$ and $\{Y_{sh}, X_{sh}\}$ are the intensities of the Doppler shifted (sh) and unshifted (us) components of a decay transition X, determined in a direct gate on the Doppler shifted component of a direct feeding transition Y. The decay curve $\{Y_{sh}, X_{sh}\}_{(D)}$ was obtained by fitting it with piecewise continuously differentiable second order polynomials. As an example, Fig. 2 depicts the intensities of the shifted (I_{sh}, middle panel) and unshifted (I_{us}, lower panel) components of the $4_1^+ \rightarrow 2_1^+$ decay transition in the direct gate on the shifted component of the feeding $6_1^+ \rightarrow 4_1^+$ transition including polynomial fits for all relevant distances, i.e., in the sensitive range for the 4_1^+ lifetime. The upper panel of Fig. 2 shows the lifetime values deduced via equation (1).

From the data of this experiment we were able to determine lifetimes of the $2_1^+, 4_1^+, 6_1^+, 8_1^+$, and 10_1^+ yrast states in ^{178}Pt which are listed in Tab. 1 together with the yrast $E2$ transition strengths and transition quadrupole moments Q_t deduced from the level lifetimes. The lifetime of the 2_1^+ state was only determined in this work with a large uncertainty due to the low intensity of the $2_1^+ \rightarrow 0_1^+$ γ-ray transition because of its rather large conversion coefficient and a contamination of the Doppler-shifted γ-ray line observed under backward angle with a γ-ray line from ^{181}Ta. Therefore, the more precise value for the 2_1^+ lifetime from a fast-timing experiment of C.B. Li and coworkers [12] is given, too, and used for the determination of the $E2$ transition strength and Q_t value. The fact that in the work of Dracoulis and coworkers [9] significantly longer lifetimes are given for the 4_1^+ and 6_1^+ states of $\tau(4_1^+) = 54.1(46)$ ps and $\tau(6_1^+) = 15.7(12)$ ps supports that these results are considerably influenced by unobserved delayed sidefeeding.

3 Discussion

Fig. 3 displays the kinetic moments of inertia for the lower yrast states up to the 10_1^+ state in ^{174,176,178,180}Pt and ^{180}Hg as a function of the frequency $\omega(I_1^+ \rightarrow (I-2)_1^+)$. A comparison of these data gives hints at a very similar structure of the prolate intruder configuration in ^{180}Hg and the ground state configuration of ^{178,180}Pt: the high absolute values of the latter and only a slight increase within the yrast band can be attributed to a quite stable deformation. The small but still recognizable decrease for the 2_1^+ state in ^{178}Pt might hint at the fact that this nucleus is right at the edge of a change towards a different structure of the lowest states for lighter Pt isotopes. In ^{176}Pt a clear decrease of

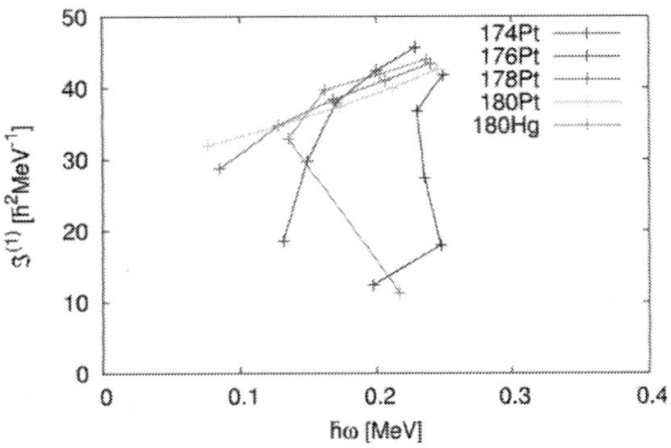

Figure 3. Kinetic moments of inertia of yrast levels of $^{174-180}$Pt and ^{180}Hg up to the 10_1^+ states as a function of the frequency $\omega(I_1^+ \rightarrow (I-2)_1^+)$.

the kinetic moments of inertia is observed for both the 2_1^+ and 4_1^+ where for ^{174}Pt this happens already for yrast states below the 10_1^+. This is consistent with a different shape of the lower yrast states as compared to the rotational picture for ^{178,180}Pt and the higher yrast states in ^{174,176}Pt, ^{180}Hg and neighboring Hg isotopes as discussed in detail in [5]. This is also evidenced by a comparison of the transition quadrupole moments for ^{178,180}Pt and ^{180,182}Hg shown in Fig. 4. For ^{174}Pt no such data are available to date and for ^{176}Pt the older data from [9] could be questionable due to possible unobserved delayed sidefeeding similar to the data for ^{178}Pt from the same work.

From mixing calculations for Hg isotopes with $A \geq 180$ it turned out that the wave function of the 2_1^+ state contains a significant part of the prolate intruder configuration while the ground state is weakly oblate deformed. This reduces the transition probability of the $2_1^+ \rightarrow 0_1^+$ transition as pointed out in [5] (see also [3, 4, 18]).

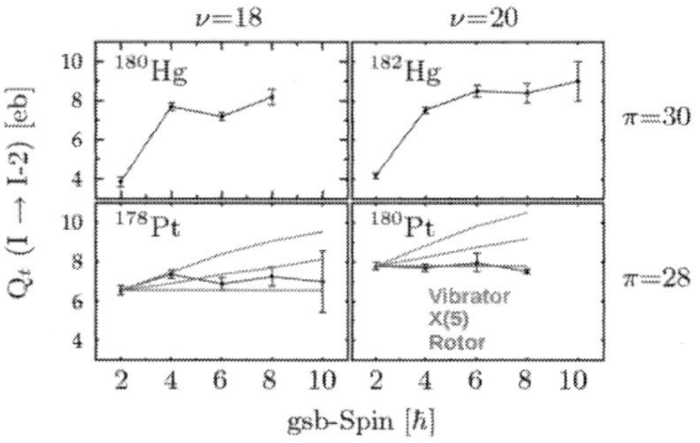

Figure 4. Transition quadrupole moments between the lowest yrast states in ^{180}Hg [4], ^{182}Hg [17], ^{178}Pt (this work and [12]) and ^{180}Pt [6, 19].

For states higher than the 4_1^+ state the constant transition quadrupole moments between the corresponding yrast states of $Q_t \approx 8eb$ give evidence for a rotational structure. The very similar Q_t values within the yrast bands of both ^{178,180}Pt down to the ground state and the agreement with the expectation values for a rotor nucleus from a geometrical model prove that the prolate structure observed as an intruder structure in neutron deficient Hg isotopes represents the ground state configuration in ^{178}Pt in agreement with the data on the kinetic moments of inertia. However, it should be noted that the systematically slightly smaller yrast Q_t values in ^{178}Pt hint for a slight decrease of the collectivity of the yrast states as compared to ^{180}Pt.

Nevertheless, the approaches made for ^{180}Pt in [6] within the IBM with extended Q formalism (see [20]) and the general collective model (GCM) [21] allowing to interprete this nucleus as a rotor nucleus with slight γ softness apply for ^{178}Pt, too, due to its very similar characteristics. This is also in agreement with calculations with, e.g., the IBM-CM and the self-consistent mean field HFB approach [11] that were alluded in section 1. These predict a rotor like structure of ^{178}Pt and a rapid structural change is expected for lighter Pt isotopes where contradicting theoretical approaches exist to date that cannot be tested in detail due to sparse experimental data. This highly motivates a measurement of yrast $E2$ transition strengths in the lighter Pt isotopes, especially in ^{174,176}Pt which are expected to be just at the edge of this change towards a potential shape coexistence in ^{176}Pt as is supported by the kinetic moments of inertia of the lowest yrast states (see Fig. 3).

This work was supported by the Deutsche Forschungsgemeinschaft (DFG) under Contracts No. FR 3276/1-1 and DE 1516/3-1 and in part by the National Research Foundation of South Africa.

References

[1] K. Heyde, J.L. Wood, Rev. Mod. Phys. **83**, 1467 (2011)
[2] A.N. Andreyev et al., Nature **405**, 430 (2000)
[3] L.P. Gaffney et al., Phys. Rev. C **89**, 024307 (2014); Erratum Phys. Rev. C **89**, 059905 (2014)
[4] T. Grahn et al., Phys. Rev. C **80**, 014324 (2009)
[5] C. Müller-Gatermann et al., Phys. Rev. C **99**, 054325 (2019)
[6] C. Müller-Gatermann et al., Phys. Rev. C **97**, 024336 (2018)
[7] A. Dewald et al., J. Phys. (London) G **31**, S1427 (2005)
[8] K.A. Gladnishki et al., Nucl. Phys. A **877**, 19 (2012)
[9] G.D. Dracoulis et al., J. Phys. G: Nucl. Phys. **12**, L97 (1986)
[10] K. Nomura et al., Phys. Rev. C **83**, 014309 (2011)
[11] J.E. Garcia-Ramos, K. Heyde, L.M. Robledo, R. Rodriguez-Guzman, Phys. Rev. C **89**, 034313 (2014)
[12] C.B. Li et al., Phys. Rev. C **90**, 047302 (2014)
[13] A. Dewald, O. Möller, P. Petkov, Prog. Part. Nucl. Phys. **67**, 786 (2012)
[14] R.A. Bark et al., J. Phys. G **31**, S1747 (2005)
[15] A. Dewald, S. Harissopulos, P. von Brentano, Z. Phys. A Atomic Nuclei **334**, 163 (1989)
[16] E. Achterberg, O. A. Capurro, G. V. Marti, Nucl. Data Sheets **110**, 1473 (2009)
[17] M. Scheck et al., Phys. Rev C **81**, 014310 (2010)
[18] N. Bree et al., Phys. Rev. Lett. **112**, 162701 (2014)
[19] C. Müller-Gatermann et al., Nucl. Inst. Meth. Phys. Res. A **920**, 95 (2019)
[20] D.D. Warner, R.F. Casten, Phys. Rev. C **28**, 1798 (1983)
[21] G. Gneuss, W. Greiner, Nucl. Phys. A **171**, 449 (1971)

EPJ Web of Conferences **223**, 01017 (2019) https://doi.org/10.1051/epjconf/201922301017
NSD2019

Decoherence of collective motion in warm nuclei

S. Frauendorf[1,*], *C. M.* Petrache[2], *R.* Schwengner[3], and *K.* Wimmer[4]

[1]University Notre Dame, USA
[2]University Paris Sud & CSNSM-CNR, France
[3]Helmholtz Center Dresden Rossendorf, Germany
[4]The University of Tokyo, Japan

Abstract. Collective states in cold nuclei are represented by a wave function that assigns coherent phases to the participating nucleons. The degree of coherence decreases with excitation energy above the yrast line because of coupling to the increasingly dense background of quasiparticle excitations. The consequences of decoherence are discussed, starting with the well studied case of rotational damping. In addition to superdeformed bands, a highly excited oblate band is presented as a new example of screening from rotational damping. Suppression of pair correlation leads to incoherent thermal M1 radiation, which appears as an exponential spike (LEMAR) at zero energy in the γ strength function of spherical nuclei. In deformed nuclei a Scissors Resonance appears and LEMAR changes to damped magnetic rotation, which is interpreted as partial restoration of coherence.

1 Introduction

The study of various quantal phenomena in cold nuclei, i. e. near the yrast line, is the major thrust of the nuclear structure community. The study of hot nuclei by means of heavy ion reactions is another active field. In this talk I address the properties of "warm nuclei" which are excited by several MeV above the yrast line (see Fig. 1). My focus is the disappearance of quantal phenomena characterizing cold nuclei, which is also referred to as the transition from order to chaos. I call this "decoherence of collective motion" alluding to condensed matter physics where the analog loss of coherence is observed when the system's temperature increases (for example the disappearance of the pair condensate in superconductors/superfluids). Compared to macroscopic systems, nuclei are small enough for finding "exact" solutions by numerical diagonalization of a model Hamiltonian in a finite basis. Such simulations of the nuclear dynamics can be considered as the theoretician's experiment for warm nuclei. The challenge is to distill simple features from the huge set of data generated by the computer.

The properties of warm nuclei are an important input into network calculations for the stellar processes as well as for applications to nuclear engineering. One example are the strength functions for γ emission below the neutron emission threshold.

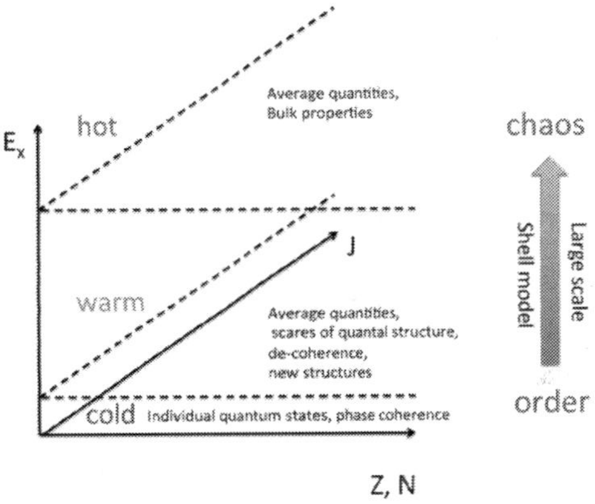

Figure 1. The region of warm nuclei.

2 Collectivity vs. Coherence

Discussing cold nuclei, the termini "collective" or "collectivity" indicate the presence of a "collective wave function" that is carried by a large number a nucleons. In the context of warm nuclei one needs sharpen terminology. A collective wave function implies the existence of definite phase relations between the constituents, which is called coherence in condensed matter physics. "Collective" implies that many constituents move in an organized way. As illustrated in Fig. 2, collectivity does not necessarily imply coherence and vice versa. Heavy ion fusion is a collective process, but not coherent. Magnetic rotation is carried by few nucleons, not very collective but coherent. The fixed phase relations between the nucleons lead to the emergence of the same $I(I+1)$ energy spectrum that is characteristic for the highly collective rotation of superdeformed nuclei.

The theory of superconductivity introduces the "coherence length", which quantifies the degree of coherence.

*e-mail: sfrauend@nd.edu

https://doi.org/10.1051/epjconf/201922301017

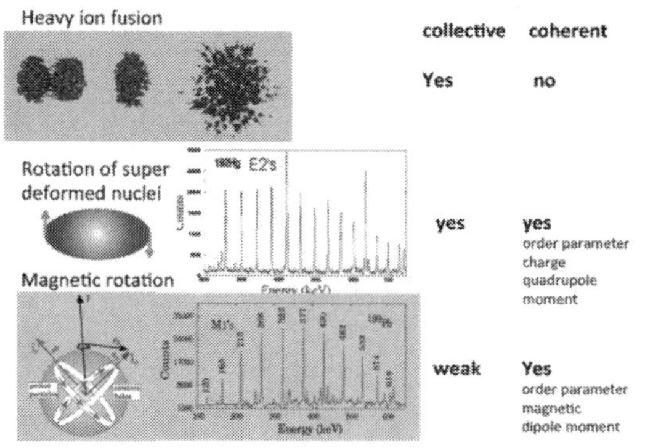

Figure 2. Relation between collectivity and coherence.

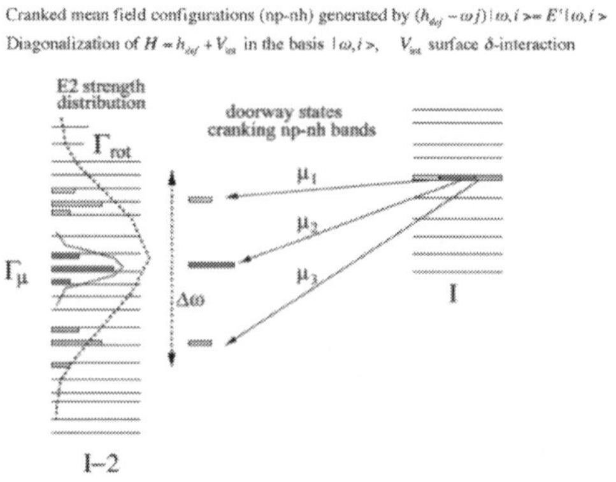

Figure 3. Fragmentation of the E2-transition strength by band mixing. From Ref. [4].

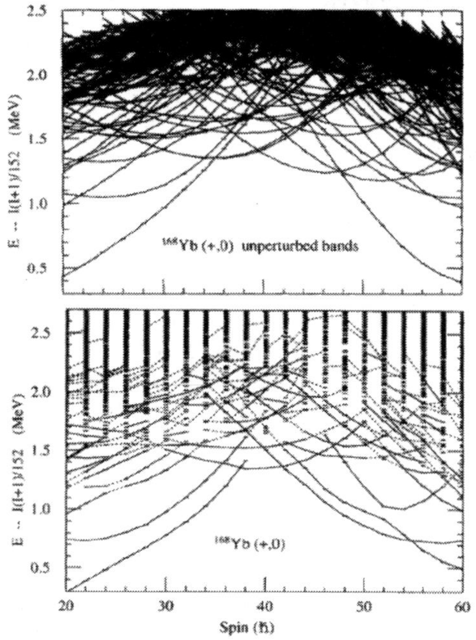

Figure 4. Reduction of the E2 transition strength by rotation damping. The short horizontal lines show the energies of the states. Upper panel (before mixing): the full lines trace individual ph-configurations which represent discrete rotational bands. They are connected by transitions with large $B(E2, I \rightarrow I-2)$ values (assumed to be 1). Lower panel (after mixing): the full lines connect mixed states for which $B(E2, I \rightarrow I-2) > 0.7$ and dotted lines for $0.7 > B(E2, I \rightarrow I-2) > 0.5$. From Ref. [6].

Loosely speaking, it is the size of a Cooper pair. The quantal phenomena, as the Meissner effect, Josephson effect or flux quantization, emerge as consequences of the existence of a wavefunction of the pair condensate. The coherence length is the minimal length on which the wave function lives. Its phase cannot increase more than 2π within this interval. By the Uncertainty Principle, this sets the upper limit of the momentum of the wave function and, as a consequence, it limits the super current density.

In analogy, Frauendorf [1] introduced the coherence angle $\Delta\psi$ for nuclear rotation as the width of the overlap between two mean field states of different orientation. The complementary coherence length ΔI counts how many times $\Delta\psi$ fits into the angle range of 2π (π in case of $\mathcal{R}(\pi) = 1$ symmetry), which limits the angular momentum range of a rotational band by the Uncertainty Principle. For the superdeformed bands $\Delta\psi = 5°$ and $\Delta I = \pi/\Delta\psi = 36$. For well deformed nuclei $\Delta\psi = 8°$ and $\Delta I = \pi/\Delta\psi = 22$. For the weakly deformed magnetic bands $\Delta\psi = 25°$ and $\Delta I = 2\pi/\Delta\psi = 14$. The angular momentum coherence length ΔI correlates well with the number of observed transitions for the three types of rotational bands (see two examples in Fig. 2). Further details can be found in Refs. [2, 3].

There, the coherence of the axial β vibration is discussed as well. The coherence length $\Delta\beta$ decreases with the number of crossings between the single particle states as functions of the deformation β. For the well deformed nuclei in the center of the rare earth region, $\Delta\beta$ is too large to support the first vibrational excitation with one node. In the transitional region around $N = 90$ it is short enough to support one node. This correlates with the experimental evidence for the respective absence or presence of a collective β vibrational excitation.

In general, coherence becomes weaker with increasing excitation energy, which leads to phase transitions from ordered to more chaotic phases in extended systems. The coherence length increases with warming up nuclei, and the quantal collective modes of cold nuclei dissolve. In the following the decoherence of the quantal rotational mode will be discussed in some detail.

2.1 Rotational Damping

The coherence of rotational motion is manifest by the appearance of rotational bands, which are characterized by a smooth increase of the energy with angular momentum and strong intra-band E2 transitions. The latter are detected by means of $\gamma-\gamma$ coincidence measurements, which are THE TOOL to identify the members of a band. As

 https://doi.org/10.1051/epjconf/201922301017

such they probe the coherence length ΔI in angular momentum.

The Copenhagen/Milano/Kyoto collaboration carefully studied the progressive decoherence of quantal nuclear rotation, which they called "Rotational Damping". Recently Leoni and Lopez-Martens presented the work in a concise review [4]. The theoretical analysis starts from the single particle configurations in a rotating deformed potential described by the Cranked Shell Model. The various configurations, diabatically continued as functions of the rotational frequency, well account for the discrete rotational bands observed in the region of about 1 MeV above the yrast line (e. g. see [5]).

The particles are assumed to interact by a surface δ-interaction, and the resulting Hamiltonian is numerically diagonalized. Because of the exponential decrease of the level spacing with excitation energy, there appears strong band mixing, which is illustrated in Fig. 3. The state I is composed of three configurations (yellow, pink, green). As each of which has a somewhat different moment of inertia a strong E2- transition μ_i arrives at different energy at spin $I-2$. The band mixing causes a fragmentation of each configuration by Γ_μ, which generates a fragmentation of the strong E2 strength over states within Γ_{rot}, called "rotational damping".

The fragmentation dilutes the strong E2-transitions that serve as filter to thread a rotational band, which is illustrated in Fig. 4. The intra band $B(E2, I \rightarrow I-2)$ values of the pure bands in the upper panel are normalized to 1. The $B(E2I \rightarrow, I-2)$ values of the strongest transitions between the mixed states rapidly fall below 0.5. Together with the finite resolution by the detectors this sets a limit to identify discrete rotational bands. Still one finds a correlation between the unresolved E2 transitions in form of enhanced $\gamma-\gamma$ coincidence probability along ridges in the $E_\gamma - E_\gamma$ plane. The number N_{ridge} of resolved ridges can be taken as a measure of the coherence length $\Delta I = N_{ridge} \times 2$ of warm nuclei. The typical number $N_{ridge} = 2, 3$ give $\Delta I = 4, 6$ to be compared with $\Delta I = 22$ in cold rare earth nuclei.

3 Islands of order in the sea of chaos

Not all excited bands are rotationally damped. The upper panels of Fig. 5 illustrate the well studied case of superdeformed bands. They are populated at high spin, where they make up the yrast region. They dispose of their angular momentum by emitting E2 radiation, while becoming excited with respect to the normal deformed states. Around $I = 10$ they are 5-8 MeV above yrast line, embedded in the dense environment of rotationally damped normal deformed states. Their mixing with the background state is suppressed by a potential well in the deformation degrees of freedom. The tiny admixtures do not disturb the regular band pattern and their decay branches cannot compete with the strong intra-band E2 transitions. At the lowest spins the coupling increases and the band de-excites via fragmented path ways.

The lower left panel of Fig. 5 presents ^{137}Nd as a new case of a regular band that extends 5 MeV above the yrast

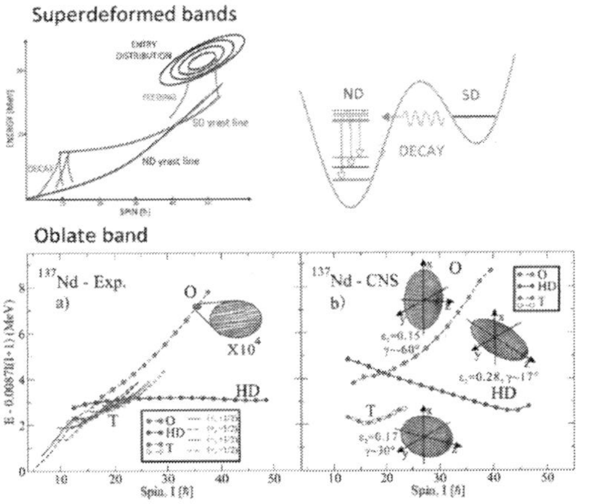

Figure 5. Examples of bands screened from rotational damping. Upper panels: Superdeformed band separated by a potential well from the damped normal deformed states. From Ref. [4]. Lower panels: Oblate band in ^{137}Nd screened from the damped highly deformed states by re-occupation of many orbitals. Left: experimental energies. Right: CNS calculations. From Ref. [7].

line well in to the damping region. The lower right panel shows calculations in the framework of the Cranked Nilsson Strutinsky (CNS) model, which well reproduce the experimental band structure. There are three classes of states. For $I < 20$ the yrast region consist of weekly deformed triaxial (T) bands, for $I > 20$ it consist of highly deformed (HD) , slightly triaxial bands. The third class are oblate (O) configurations to which we assign the highly nonyrast band. In contrast to the superdeformed bands the CNS calculation does not give a potential well between the O and HA configurations. We attribute the suppression of the coupling to the substantial structural difference. Relative to the $Z = 50$, $N = 82$ core the states are built on the lowest configurations
HD: $[\pi(h^4_{11/2}(sdg)^6) \times \nu(h^{-4}_{11/2}h^2_{9/2}i^1_{13/2}(sdg)^{-8})]$
O: $[\pi(h^2_{11/2}(sdg)^8) \times \nu(h^{-1}_{11/2}(sdg)^{-4})]$.
The O band acts like a funnel. Any feeding transition is followed by fast E2 transitions down to the region of triaxial bands, where O dissolves.

4 Low Energy Magnetic Radiation

Using their new analyzing technique of the γ quasi continuum, the Oslo group discovered an enhancement of the γ strength function (GSF) below the transition energy of 2 MeV [8]. The observation came as a surprise, because the dipole excitations from the ground state do not show such an enhancement. It was also met with skepticism, because it was on variance with the generally accepted Brink-Axel hypothesis, which states that GSF depends only on the transition energy E_γ but not on the energy of the initial state. Later experiments demonstrated the systematic appearance of the enhancement (see e. g. Appendix II of Ref. [9]).

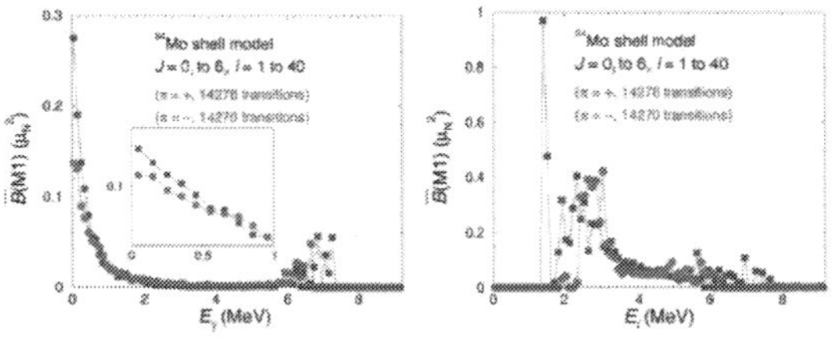

Figure 6. (Color online) Left: Average $\overline{B}(M1)$ values in 100 keV bins of E_γ and averaged over all E_i bins for positive-parity (blue squares) and negative-parity (red circles) states in ^{94}Mo. The inset shows the low-energy part in logarithmic scale. Right: Average $\overline{B}(M1)$ values in 100 keV of E_f averaged over the $E_\gamma <$ 5 *MeV* bins. From [10].

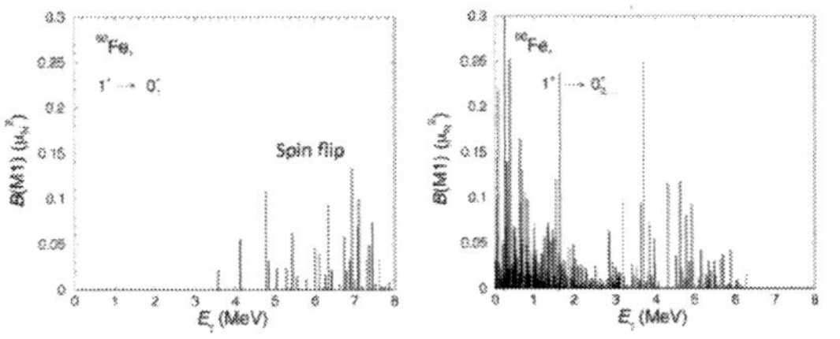

Figure 7. Reduced transition probabilities $B(M1)$ of all transitions ending in the ground state 0_1^+ and the first excited state 0_2^+ of ^{60}Fe. Generated from the Shell Model calculations of Ref. [11].

Shell Model Calculations for ^{94}Mo and neighbors with a $Z = 28$, $N = 40$ core provided an explanation of the phenomenon [10]. The authors calculated the $B(M1)$ for all allowed transitions between the lowest 40 states of given angular momentum $I = 0 - 6$. The large set of data was analyzed as follows (present version, the original is a special case). The values were collected into $[E_\gamma, E_i]$ bins of 100 keV width. Average $\overline{B}(M1)$ values were obtained by dividing the cumulative values by the number of transition allowed for each bin and taking the mean value over a range of the initial energies E_i of the transitions. As seen in Fig. 6, there is a strong enhancement of the average $\overline{B}(M1)$ values at $E_\gamma = 0$ which falls off approximately exponentially with a constant $1/T_B$, where $T_B = 0.49$ MeV. The exponential decrease is retained by GSF, which was calculated by

$$f_{M1}(E_\gamma) = \frac{16\pi}{9(\hbar c)^3}\overline{B}(M1, E_\gamma)\rho(E_i), \quad (1)$$

where the level density $\rho(E_i)$ was obtained from the number of initial states in the E_i bins. The authors of Ref. [10] associated the experimentally observed enhancement of the GSF below 2 MeV with the calculated spike at $E_\gamma = 0$, which they called Low Energy Magnetic Radiation (LEMAR). The presence of exponential LEMAR spike in weakly deformed nuclei was confirmed by subsequent Shell Model studies along the same line for the Fe isotopes [11] and the nuclei in the fp-shell [9, 12, 13] (and earlier work cited).

The low-energy transitions appear only when they end in excited states (see right panel of Fig. 6 and Fig. 7). The analysis of the initial and final states showed that transitions involve a reorientation of high-j orbitals within one and the same partition of protons and neutrons (e. g. in ^{94}Mo: $\pi g_{9/2}^2 \times \nu(d_{5/2}^1 g_{7/2}^1)$, and $\pi(p_{1/2}^{-1} g_{9/2}^3) \times \nu(d_{5/2}^2 g_{7/2}^1 g_{9/2}^{-1})$, and more). The reorientation of a high-j orbital has a large transitional magnetic moment of $\vec{\mu}_{12} = g_j(\vec{j}_2 - \vec{j}_1)$ which is reflected by the $B(M1)$ value. The transitional magnetic moment is particularly large when the reorientation involves protons and neutrons such that they enhance each other (like in the case of magnetic rotation the simultaneous reorientation high-j protons and high-j neutron holes [1–3]). As illustrated in Fig. 8, without residual interaction the states in one partition that differ by the mutual orientation of the occupied j-orbital have all the same energy, and no radiative transitions appear. The mixing of these states by the residual interaction generates a splitting of the complex eigenstates, among which the low-energy radiative transitions appear. A calculation for ^{94}Mo with artificially reduced scale of the residual interaction by a factor of 2 gives $T_B = 0.24$ MeV which is one half of $T_B = 0.49$ MeV found with the full interaction strength. This demonstrates that the width of the LEMAR spike is determined by the mixing.

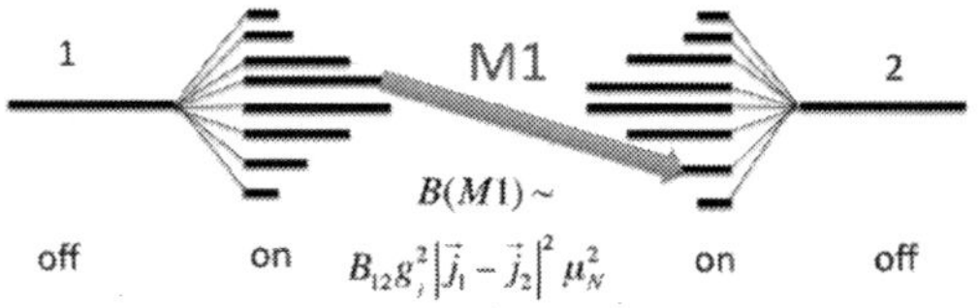

Figure 8. Two states 1, 2 within one and the same partition that are related by mutual re-alignment of the spins of the active high-j orbitals are connected by a large M1 matrix element. Without residual interaction (off) the two states have the same energy → no radiation. The residual interaction mixes the states in a chaotic way (on), which generates energy differences between them and fragments the M1 strength → radiation.

The fragmentation of the two states reduces their contribution to $B(M1) \sim B_{12}\left|g_j(\vec{j}_2 - \vec{j}_1)\right|^2$, where B_{12} is a stochastic factor that measures the probability of the pair [1 2] contributing to the transition between the two mixed states. The stochastic factor can be estimated assuming random mixing. The average probability of one of the basis states to be present in the complex state is $\propto 1/N$ with N being the total number of basis states contributing to the mixed state. In the chaotic regime the residual interaction is much larger than the level distance d. Therefore $N \propto 1/d$ and $B_{12} \propto 1/(d_i d_f) \propto \rho_i \rho_f$. At the energies of interest, the level densities ρ_i and ρ_f are well reproduced by the "constant temperature" expression $\rho \propto \exp[E^*/T_L]$, which gives $B_{12} \propto \rho_i^2 \exp[-E_\gamma/T_L]$. In fact, the Shell Model calculations of Ref. [10, 11] are consistent with an exponential fall-off of the GSF given by $T_B = T_L$, where

T_L is taken from the exponential increase of the level density in the same calculation (see Ref.[14] for details).

Zelevinsky and co-workers [15] discussed the relation of chaotic mixing of the basis states and the statistical concept of an ensemble at a temperature T. From this perspective, one may associate T_B and T_L with the ensemble temperature T, and the appearance of the LEMAR spike may interpreted as follows. In the ground state the nucleons on the high-j orbitals are frozen by the pair correlations which couple them to zero spin pairs. To generate an M1 transition, one has to break such a pair which costs 2Δ of energy. Warming up the nucleus destroys the pair correlations (decoherence). Now the nucleons on the high-j orbitals are free to re-orient. Agitated by the warm environment they re-orient emitting radiation with characteristic probability $\propto \exp[-E_\gamma/T]$ per gamma quant. Thus, the LEMAR spike represents incoherent thermal radiation of an ensemble of uncoupled magnetic dipoles generated by the nucleons on the high-j orbitals.

LEMAR boosts the rates of (n,γ) reactions. We continued our study of nuclides around ^{132}Sn (details see [16]) which are a bottle neck in the r-process of element synthesis in violent stellar events [17]. A LEMAR spike with a GSF comparable to the Mo isotopes was found for all nuclides with at least 3 particles/holes with respect to the $Z = 50$ and $N = 82$ core. Its existence north-east of ^{132}Sn is at variance with the conjecture [10, 16] that it appears only in regions where Magnetic Rotation is predicted [16].

5 Deformation

With N and Z moving away from closed shells deformation sets in, which represents a correlation between nucleons. The impact on the M1 GSF was studied in Ref. [11] for the ^{60,64,68}Fe isotopes by means of Shell Model calculations using $Z = 20$ and $N = 28$ as a core. The calculations account for the increase of quadrupole collectivity in the considered isotopic chain. For ^{60,68}Fe the respective experimental $B(E2, 2_1^+ \to 0_1^+) = 190,\ 450(eb)^2$ and the calculated $B(E2, 2_1^+ \to 0_1^+) = 368,\ 406(eb)^2$. The values are consistent with the transitional character of the ground state bands of all considered isotopes.

Fig. 9 shows the GSF for ^{60,68}Fe. The shape changes from monomodal (only the LEMAR spike) at $N = 34$, four neutrons above the closed shell, to bimodal at $N = 42$, the middle of the fpg shell. We interpret the bump in the range from about 2 to 5 MeV as the Scissors Resonance (SR) built on excited states. The integrated strength

$$\Sigma\overline{B}(M1) = \frac{9(\hbar c)^3}{16\pi}\int_{E_{\gamma 1}}^{E_{\gamma 2}} f1(E_\gamma)dE_\gamma \qquad (2)$$

deduced from the strength functions below 5 MeV varies by only 8% at most from an average of 9.80 μ_N^2. For the near spherical ^{60}Fe it is mainly located in the LEMAR spike. For the deformed ^{68}Fe, more than one half is shifted to the SR.

Taking GSF for ^{64}Fe into consideration, the authors of Ref. [11] concluded that the bimodal structure gradually develops with increasing N by shifting strength from the LEMAR spike to the SR while the sum stays nearly constant. The gradual development of the SR and the shift of the integrated strength when entering a shell has been confirmed experimentally for odd-A Sm isotopes [18]. Systematic calculations using the full pf shell as configuration space for the Ge isotopes [19] and the Ga isotopes [9] show that near the closed shells $N = 28$ and 50 only the LEMAR spike exists. Going into the open shell from below and above the SR develops, first as a shoulder, and in the middle, where the deformation has a maximum, it becomes a distinct resonance.

Fig. 10 illustrates how the development of the bimodal LEMAR-SR structure can be explained by the on-

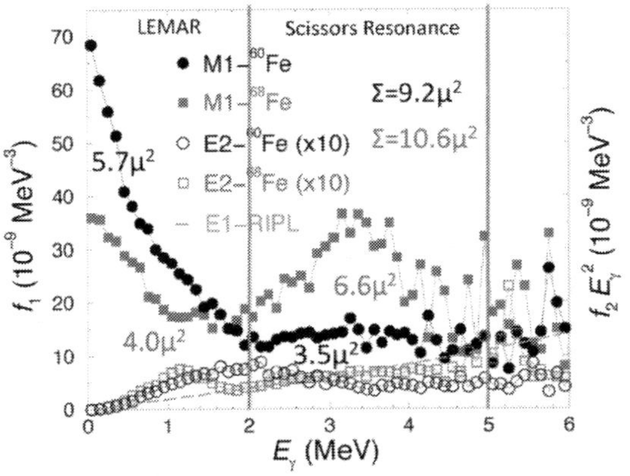

Figure 9. (Color online) Calculated M1 strength functions (f_1) and E2 strength functions $\left(f_2 E_\gamma^2\right)$ in ^{60,68}Fe. Note the factor of 10 for E2. The numbers quote the integrated strength (2) where vertical lines indicate the respective integration regions. For ^{68}Fe the sum of the M1 strength from the ground state to all 1^+ states is $\sum B(M1, 0_1^+ \to 1^+) = 1.7\mu^2$. Adapted from Ref. [11].

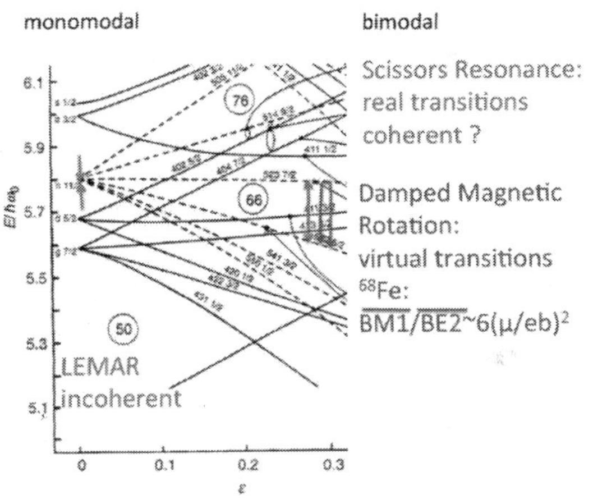

Figure 10. (Color online) Transitions that generate the LEMAR spike and the Scissors Resonance. The energy scale is $\hbar\omega_0$=7.7 MeV for A =153.

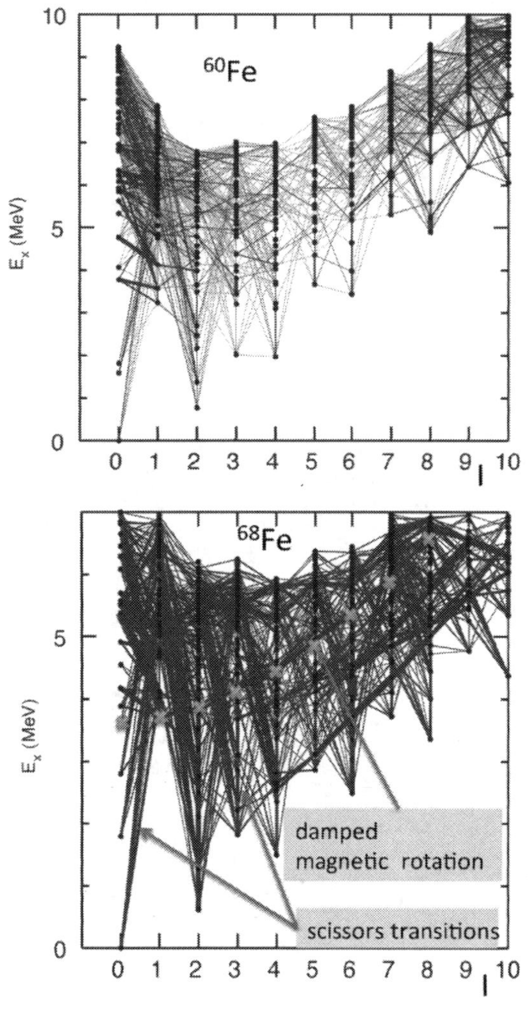

Figure 11. (Color online) Transitions with $B(M1) > 0.1\mu_N^2$ (blue lines) between the calculated positive-parity states (black dots) in ^{60,68}Fe. The widths of the lines are proportional to the $B(M1)$ values.

set of quadrupole deformation. The mechanism causing large $B(M1)$ values is reorientation of the high-j single-particle angular momenta. Without deformation this occurs between the various configurations of the nucleons in an incompletely filled j-shell ($h_{11/2}$ in the figure), which generates the LEMAR spike as discussed above. The magnetic (m) substates of the high-j multiplet split with the onset of deformation. Reorientation occurs in two different ways. First, there are the real transitions between the m-substates. Particle-hole excitations of this type are known to generate the SR [20]. As seen in Fig. 10, the splitting between the m-states corresponds to the position of the SR for $\varepsilon = 0.3$. Second, there are the virtual excitations from the occupied to the empty m-states. These are known generating the rotational motion of deformed nuclei. Along a rotational band the angular momentum increases by a coherent gradual alignment of the spins of the nucleons in the deformed orbitals, which is described by the virtual particle-hole excitations between the m-states.

With increasing deformation, the LEMAR spike changes from incoherent thermal-like radiation to partially coherent rotational radiation. The ratio $B(M1)/B(E2) \sim 6(\mu/eb)^2$ is characteristic for magnetic rotation [1–3]. As shown in Fig. 11, the M1 transitions between the states in ^{60}Fe are chaotic. For ^{68}Fe they acquire a certain preference along the curves with $E_\gamma = I/\mathcal{J}$, which is indicated by sequence of blue crosses in Fig. 11. The moment of inertia $\mathcal{J} \approx 14$ MeV^{-1} is close to the rigid body value $\mathcal{J}_{rig} \approx 17$ MeV^{-1}. The preference is seen as the fuzzy threads, which signal the onset of rotational coherence in form of damped magnetic rotation. The second preference are the strong transitions $I \to I \pm 1$ with an energy $E_\gamma \approx 3MeV$, which generate the SR. It is not clear if their presence indicates vibrational coherence, which the concept of a collective Scissors Mode implies. The may represent incoherent radiation from an ensemble of magnetic dipoles, each of which oscillates independently in the deformed nuclear potential.

S. F. acknowledges support by the DOE Grant DE-FG02-95ER4093.

References

[1] S. Frauendorf, Rev. Mod. Phys. **73** 463 (2001)
[2] S. Frauendorf, AIP Conf. Proc **1753**, 030001 (2016)
[3] S. Frauendorf, Phys. Scr. **93**, 043003 (2018)
[4] S. Leoni, A. Lopez-Martens, Phys. Scr. **91**, 063009 (2016)
[5] S. Frauendorf, Phys. Scr. **93** 043003, (2018)
[6] M. Matsuo *et al.*, Nucl. Phys. A **617**, 1 (1997)
[7] C. M. Petrache *et al.*, Phys. Rev. C **99**, 041301 (2019)
[8] M. Guttormsen *et al.*, Phys. Rev. C **71** 044307 (2005)
[9] J. E. Mitbø *et al.*, Phys. Rev. C **98**, 064321 (2018)
[10] R. Schwengner *et al.*, Phys. Rev. Lett. **111** 232504 (2013)
[11] R. Schwengner *et al.*, Phys. Rev. Lett. **118**, 092502 (2017)
[12] S. Karampagia *et al.* Phys. Rev. C **95**, 024322 (2017)
[13] K. Sieja, Phys. Rev. C **98**, 064312 (2018)
[14] S. Frauendorf *et al. Proc. Int. Conf. on Fission and Properties of Neutron-rich Nuclei, 2016, Sanibel Island, Florida* World Science Pub. Singapore 2017 37-46; arXiv 1907.00398
[15] V. Zelevinsky, Ann. Rev. Nucl. Part. Sci. **46**, 237 (1996); V. Zelevinsky *et al.*, Phys. Rep. **391**, 311 (2004)
[16] S. Frauendorf *et al.*, EPJ Web of Conferences **93** 04002 (2015), arXiv 1907.02641
[17] S. Frauendorf, K. Wimmer, to be published
[18] F. Naqvi *et al.*, Phys. Rev. C **99**, 054331 (2019)
[19] S. Frauendorf, R.Schwengner, to be published
[20] I. Hamamoto and S. Åberg, Phys. Lett. B **145**, 164 (1984)

Chiral three-body force and monopole properties of shell-model Hamiltonian

Tokuro Fukui[1,*], *Luigi* Coraggio[1], *Luca* De Angelis[1], *Angela* Gargano[1], *Nunzio* Itaco[1,2], *Yuanzhuo* Ma[3], and *Furong* Xu[3]

[1]Istituto Nazionale di Fisica Nucleare, Sezione di Napoli, Complesso Universitario di Monte S. Angelo, Via Cintia I-80126 Napoli, Italy
[2]Dipartimento di Matematica e Fisica, Università degli Studi della Campania "Luigi Vanvitelli", viale Abramo Lincoln 5 - I-81100 Caserta, Italy
[3]School of Physics, and State Key Laboratory of Nuclear Physics and Technology, Peking University, Beijing 100871, China

Abstract. So far, the nature of three-nucleon forces (3NFs) derived by the chiral effective field theory has been intensively investigated by various theoretical approaches. In this work, to address the chiral 3NF within the shell-model framework, three-body matrix elements are formulated in terms of the harmonic oscillator basis functions, by adopting the nonlocal regulator. We perform a benchmark test for p-shell nuclei in order to confirm our framework. Then we show that the contribution of the 3NF to the monopole component of the effective shell model Hamiltonian plays an essential role to account for the shell evolution of fp-shell nuclei.

1 Introduction

The realistic shell model (RSM) employing a realistic nuclear force is one of the theoretical tools describing the nuclear structure microscopically and exploring the nature of the nuclear force. As regards obtaining the state-of-the-art nuclear force, the chiral effective field theory [1] has a great advantage that many-body forces appear on an equal footing.

Zuker suggested by a phenomenological way that the RSM Hamiltonian consisting of a realistic two-nucleon force (2NF) needs to be corrected due to the absence of three-nucleon forces (3NFs) [2]. In particular, the correction to its monopole component, which is responsible for the evolution of the spherical mean field, was found to be crucial. Indeed, it was shown that, by the RSM calculations, 3NFs play a significant role to explain the neutron drip line of the oxygen isotope [3]. The effect of the chiral 3NF on the Ca isotopes was also investigated [4], but how its monopole correction is important in this region remains unclear.

We aim to clarify the 3NF contribution to the monopole Hamiltonian for fp-shell nuclei including the Ca isotopes. To this end, we derive new formalism of the three-body matrix element and perform a benchmark test for p-shell nuclei.

2 Theoretical Framework

Within the RSM framework, we separate the Hilbert space into two parts, the valence-nucleon and inert-core parts, and explicitly address the valence nucleons only. First, we need to compute the three-body matrix element with the harmonic oscillator (HO) basis functions as follows:

(1) Transformation from the three-body JT-coupled HO states into the Jacobi-HO states [5], by Talmi transformation [6–11], with the separation of the center-of-mass and relative motions.

(2) Antisymmetrization of the Jacobi-HO states by diagonalizing the three-body antisymmetrizer [12, 13].

(3) Evaluation of the three-body matrix element regularized by the nonlocal regulator with the Jacobi-HO basis at the chiral next-to-next-to-leading order (N^2LO) [14].

The steps (1) and (2) are essentially same as those reported in Ref. [15], whereas the step (3) is newly developed by the present work [16]. Next, the three-body operator appearing in the Hamiltonian is reduced to the one- and two-body operators by the normal-ordering approximation [17]. Then, in order to incorporate the chiral two-body interaction at next-to-next-to-next-to-leading order [14] and the normal-ordered three-body interaction into the RSM, we adopt the many-body perturbation theory [18], and thus obtain the effective Hamiltonian associated with the valence-nucleon space.

The further description of our theoretical framework is reported in Refs. [16, 19], as well as the model spaces and parameters adopted for the calculations.

3 Results

3.1 Benchmark test for p-shell nuclei

Figure 1 shows the energy spectra relative to the ground state (g.s.) for the selected p-shell nuclei, ^{6}Li and ^{10}B. Our results obtained by the RSM are compared with the experimental data (Exp) [20] and those by the *ab initio* no-core shell model (NCSM) [21, 22]. One finds from Figs. 1(a) and (c) that, even with the 2NF only, the RSM gives the

*e-mail: fukui@na.infn.it

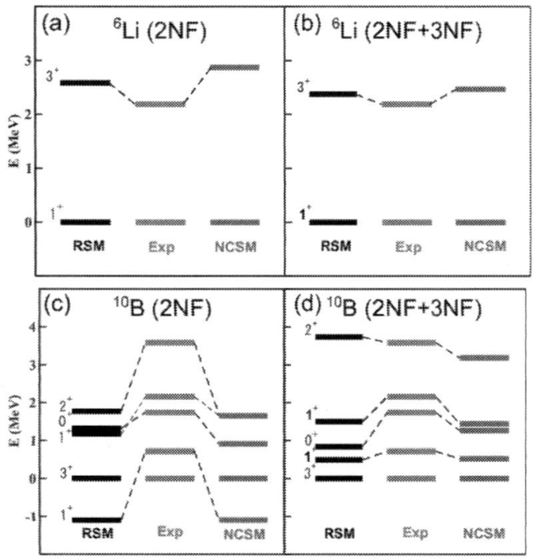

Figure 1. The comparison of the low-lying energy spectra of ^{6}Li and ^{10}B computed by the RSM and NCSM, together with the experimental data. See text for details.

low-lying spectra comparable to those by the NCSM, although they are not consistent with the data. Indeed, both models at the 2NF level cannot provide the correct g.s. for ^{10}B. Then, as shown in Figs. 1(b) and (d), the inclusion of the 3NF plays a significant role on the spectra and improves the levels consistent with the data, with the correct g.s. of ^{10}B. The similar results are obtained also for ^{8}Li, ^{8}Be, ^{8}B, ^{11}B, ^{12}C, and ^{13}C [16].

In Fig. 2(a), the consistency between the RSM and NCSM at the 2NF level is also confirmed from the g.s. energy $E_{\rm g.s.}$ of the p-shell nuclei as a function of the mass number A. We select the systems with the neutron number equal to the proton number. The 2NF only is not enough to gain $E_{\rm g.s.}$ consistent with the experimental data. The NCSM with the 3NF significantly improves the results, which coincide with the data, while the RSM with the 3NF remains overestimate the data, represented in Fig. 2(b). This may suggest that the normal-ordering approximation

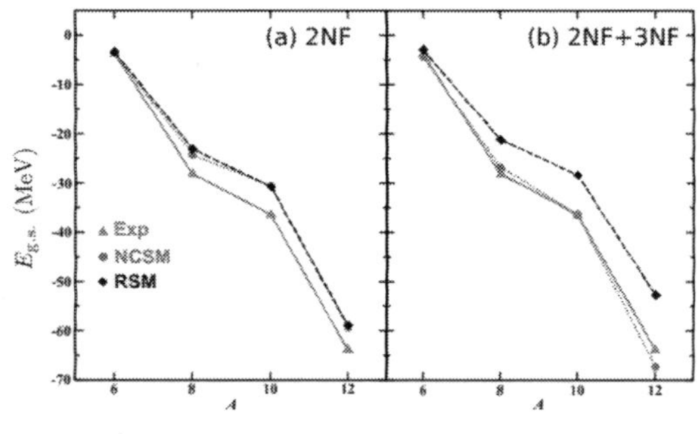

Figure 2. The g.s. energy for the selected p-shell nuclei obtained with (a) the 2NF only and (b) 2NF plus 3NF in both models.

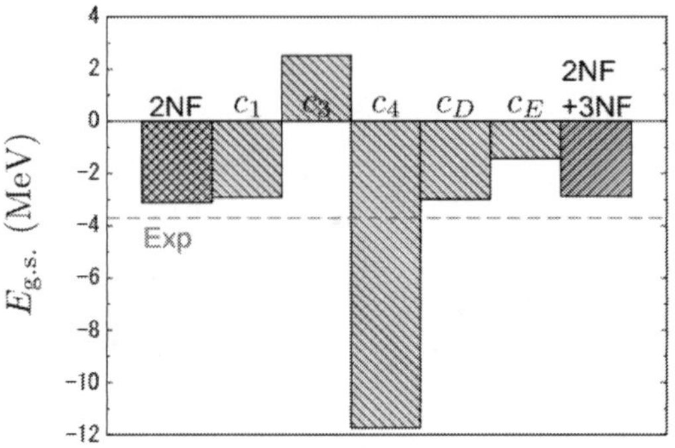

Figure 3. The individual contribution of each term of the 3NF to $E_{\rm g.s.}$ of ^{6}Li. See text for details.

is not sufficient for $E_{\rm g.s.}$, although it works for the relative energy spectra. As mentioned above, we adopt the normal-ordering approximation for the three-body operator in the RSM, whereas the NCSM explicitly addresses it without the approximation. It is required to make the approximation more precise.

The chiral 3NF at N^{2}LO can be decomposed into five terms associated with each low-energy constant (LEC). The two-pion exchange term involves the three LECs, c_1, c_3, and c_4, while the one-pion exchange plus contact term (the contact term) has the LEC c_D (c_E). Figure 3 expresses the individual contribution of each of those terms to $E_{\rm g.s.}$ of ^{6}Li. The experimental value is shown by the dashed line and the calculated result only with the 2NF is given by the cross-hatched bar. We see that the c_4 term (declining-hatched bar) has the largest contribution to $E_{\rm g.s.}$. This is consistent with the fact that the two-pion exchange term is the most dominant component in the chiral 3NF at N^{2}LO (see, for example, Ref. [23]). It is also clearly exhibited that there is a large coherent effect when all the terms exist, since an incoherent sum of each c_i contribution, which is always attractive except for the c_3 one, does not reproduce the full result (soaring-hatched bar).

3.2 Shell evolution on fp-shell nuclei

In the RSM calculations for the fp-shell nuclei, we adopt the single-particle energy corrected by the 3NF [19], and

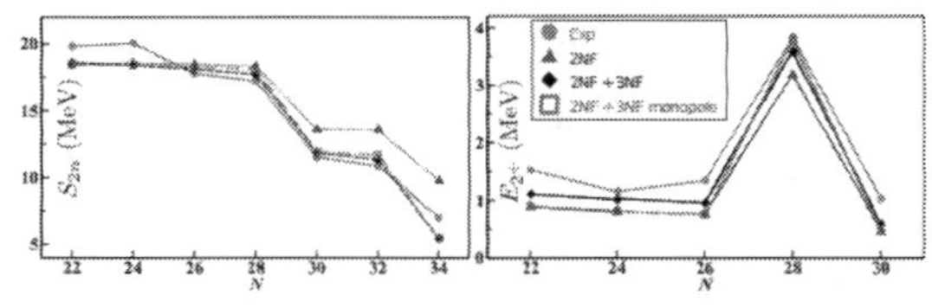

Figure 4. The comparison of the measured data [20] and RSM results on the two-neutron separation energy S_{2n} and the 2$^+$ excitation energy E_{2^+} for the Ca isotopes. Three types of the RSM calculations are performed: the 2NF only (triangle), 2NF plus 3NF (filled square), and 2NF with the monopole correction by the 3NF (open square).

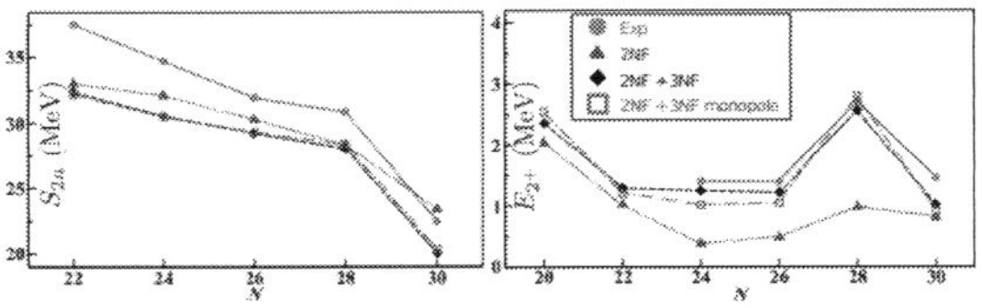

Figure 5. Same as Fig. 4 but for the Ni isotopes.

thus, in the following we discuss the 3NF effect on the two-body term. Figure 4 shows the two-neutron separation energy S_{2n} and the 2^+ excitation energy E_{2^+} as a function of the neutron number N for the Ca isotopes. One sees that experimental S_{2n} (circle) suddenly drops from $N = 28$ to 30 owing to filling the neutron $0f_{7/2}$ orbit. The RSM performed with the 2NF plus 3NF (filled square) satisfactory simulates this behavior, as well as the 2NF calculation with the monopole correction by the 3NF (open square), in contrast to the calculation only with 2NF (triangle) failing to explain this sudden drop. This clearly presents that the 3NF plays an important role to be responsible for the shell evolution on the Ca isotopes, and the 3NF contribution to the monopole Hamiltonian is essential. For E_{2^+}, all the calculations give the similar results, which are slightly underestimated compared to the measured data. However, the experimental E_{2^+} at $N = 28$ coincides well with the RSM results by the 3NF and 2NF with the monopole correction, while the result of the 2NF only remains lower than the data. Thus we find that the 3NF is substantial to depict the closure property.

We show in Fig. 5 the comparison of the RSM results with the measured data for S_{2n} and E_{2^+} of the Ni isotopes. The RSM globally underestimates the measured S_{2n} but the 3NF result well reproduces the drop of S_{2n} at $N = 30$, the result of the 2NF with the monopole correction as well. In contrast, the 2NF calculation cannot describe this feature. The above consequence obtained from S_{2n} is qualitatively consistent with that for the Ca isotopes. For E_{2^+}, however, we obtain results very different from the Ca case. The RSM only with the 2NF gives smaller values of E_{2^+}, specifically at $N = 28$, compared to the other two calculations, which provide E_{2^+} consistent with the measured data. Thus we find the 3NF responsible for the closure property of the Ni isotopes is more prominent than that of the Ca case. This can be interpreted by scrutinizing the effective single-particle energy (ESPE).

The ESPE is the evolved single-particle energy due to the monopole interaction, which is the angular-momentum averaged two-body interaction (see, for instance, Ref. [24]). Using the ESPE, we are able to elucidate how the single-particle structure evolves as the number of neutrons increases.

Figure 6 displays the neutron ESPEs relevant for the Ca isotopes as a function of N. The distance between each ESPE is almost constant as N increases, and they are well separated from each other. In particular, we see the large gap between the $0f_{7/2}$ and $1p_{3/2}$ ESPEs, which accounts for the closure property at $N = 28$. The inclusion of the

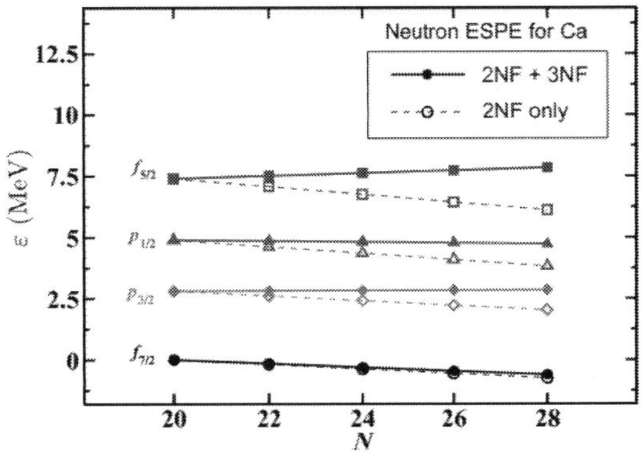

Figure 6. The neutron ESPEs calculated with the 2NF plus 3NF (filled symbols) and the 2NF only (open symbols), for the Ca isotopes as a function of N.

3NF enlarges this gap by about 0.7 MeV at $N = 28$, although each ESPS remains isolated from the others. This induces the more substantial shell closure.

The neutron and proton ESPEs of the Ni isotopes as a function of N are presented in Fig. 7. We find that the neutron $0f_{7/2}$ ESPE is well separated from the other three ESPEs. This makes the shell closure stronger on ^{56}Ni. In contrast, the grouping of the neutron ESPEs of the $1p_{3/2}$, $1p_{1/2}$, and $0f_{5/2}$ orbits enhances the collectivity reducing the closure. The similar feature is observed also in the proton ESPEs. Consequently, due to the interplay of these op-

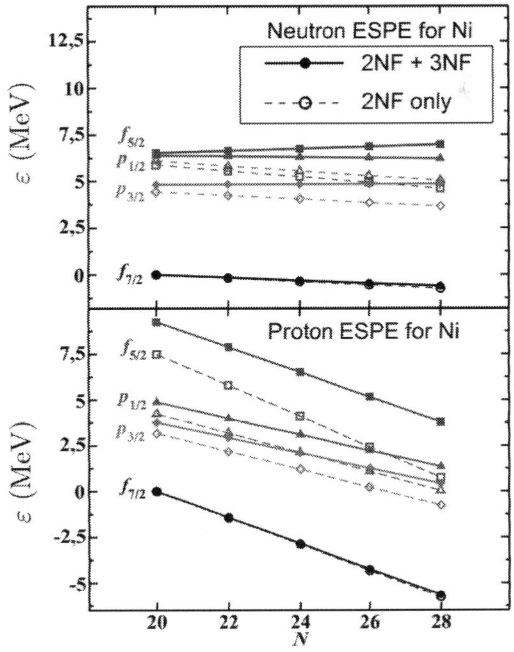

Figure 7. The neutron and proton ESPEs calculated with the 2NF plus 3NF (filled symbols) and the 2NF only (open symbols), for the Ni isotopes as a function of N.

posite effects, the shell closure on ^{56}Ni is slightly smaller than that on ^{48}Ca, evidenced by the value of E_{2^+} of ^{56}Ni that is about 1 MeV smaller than that of ^{48}Ca.

When we include the 3NF, the gap between the $0f_{7/2}$ and $1p_{3/2}$ orbits increases by about 1 MeV in both neutron and proton ESPEs. Moreover, the $0f_{5/2}$ orbit of the proton ESPE is strongly enhanced by the 3NF at $N = 28$. Therefore, the 3NF effect observed on E_{2^+} of the Ni isotopes is more distinguished than that of the Ca case.

This fact is also understood by the property of the two-pion exchange contributions. As presented in Fig. 3 as an example, the 3NF is generally dominated by the c_4 term. However, due to the charge conservation, the c_4 term vanishes when the three particles are identical. Naively, such situation in the Ca isotopes is relatively important than that in the Ni case, and thus we can explain the different behavior of the 3NF on the Ca and Ni isotopes.

4 Perspectives

We require a prescription beyond the normal-ordering approximation, which is equivalent to the first-order term in the perturbative expansion of the interaction. The inclusion of the second-order terms in addition to the normal-ordering approximation would modify the single-particle energies. We expect, therefore, that the second-order contribution meliorates the simulation of $E_{\text{g.s.}}$, as well as S_{2n} and E_{2^+}. In order to systematically investigate the contribution of the 3NF, we plan to apply the present framework to heavier systems, such like candidates of the neutrinoless double-beta decay. The implementation of the chiral 3NF into the Gamow shell model [25–29] is also interesting since both 3NF and continuum effects are expected to be crucial for neutron-rich nuclei near drip lines. Indeed, the oxygen drip line is explained by a repulsive property of the 3NF [3], while the continuum degrees of freedom act as a attractive nature [30]. The interplay of the 3NF and the continuum effects needs to be clarified.

References

[1] S. Weinberg, Phys. A **96**, 327 (1979)
[2] A.P. Zuker, Phys. Rev. Lett. **90**, 042502 (2003)
[3] T. Otsuka, T. Suzuki, J.D. Holt, A. Schwenk, Y. Akaishi, Phys. Rev. Lett. **105**, 032501 (2010)
[4] J.D. Holt, J. Menéndez, J. Simonis, A. Schwenk, Phys. Rev. C **90**, 024312 (2014)
[5] R. Roth, A. Calci, J. Langhammer, S. Binder, Phys. Rev. C **90**, 024325 (2014)
[6] I. Talmi, Helv. Phys. Acta **25**, 185 (1952)
[7] T.A. Brody, M. Moshinsky, *Tables of Transformation Brackets* (Monografias del Instituto de Fisica, Mexico, 1960)
[8] M. Moshinsky, T.A. Brody, Rev. Mex. Fís. **9**, 181 (1960)
[9] L. Trlifaj, Phys. Rev. C **5**, 1534 (1972)
[10] B. Buck, A.C. Merchant, Nucl. Phys. A **600**, 387 (1996)
[11] G.P. Kamuntavičius, R.K. Kalinauskas, B.R. Barrett, S. Mickevičius, D. Germanas, Nucl. Phys. A **695**, 191 (2001)
[12] P. Navrátil, B.R. Barrett, W. Glöckle, Phys. Rev. C **59**, 611 (1999)
[13] P. Navrátil, G.P. Kamuntavičius, B.R. Barrett, Phys. Rev. C **61**, 044001 (2000)
[14] R. Machleidt, D. Entem, Phys. Rep. **503**, 1 (2011)
[15] P. Navrátil, Few-Body Syst. **41**, 117 (2007)
[16] T. Fukui, L. De Angelis, Y.Z. Ma, L. Coraggio, A. Gargano, N. Itaco, F.R. Xu, Phys. Rev. C **98**, 044305 (2018)
[17] R. Roth, S. Binder, K. Vobig, A. Calci, J. Langhammer, P. Navrátil, Phys. Rev. Lett. **109**, 052501 (2012)
[18] L. Coraggio, A. Covello, A. Gargano, N. Itaco, T. Kuo, Ann. Phys. **327**, 2125 (2012)
[19] Y.Z. Ma, L. Coraggio, L. De Angelis, T. Fukui, A. Gargano, N. Itaco, F.R. Xu, Phys. Rev. C **100**, 034324 (2019)
[20] *Data extracted by using the nndc on-line data service from the ensdf database, file revised as of march 15, 2019*, `https://www.nndc.bnl.gov/ensdf`
[21] P. Navrátil, V.G. Gueorguiev, J.P. Vary, W.E. Ormand, A. Nogga, Phys. Rev. Lett. **99**, 042501 (2007)
[22] P. Maris, J.P. Vary, P. Navrátil, Phys. Rev. C **87**, 014327 (2013)
[23] H. Krebs, A.M. Gasparyan, E. Epelbaum, Phys. Rev. C **98**, 014003 (2018)
[24] A. Umeya, K. Muto, Phys. Rev. C **74**, 034330 (2006)
[25] R. Liotta, E. Maglione, N. Sandulescu, T. Vertse, Physics Letters B **367**, 1 (1996)
[26] N. Michel, W. Nazarewicz, M. Płoszajczak, K. Bennaceur, Phys. Rev. Lett. **89**, 042502 (2002)
[27] R. Id Betan, R.J. Liotta, N. Sandulescu, T. Vertse, Phys. Rev. C **67**, 014322 (2003)
[28] N. Michel, W. Nazarewicz, M. Płoszajczak, J. Okołowicz, Phys. Rev. C **67**, 054311 (2003)
[29] R. Id Betan, R.J. Liotta, N. Sandulescu, T. Vertse, Phys. Rev. Lett. **89**, 042501 (2002)
[30] K. Fossez, J. Rotureau, N. Michel, W. Nazarewicz, Phys. Rev. C **96**, 024308 (2017)

EPJ Web of Conferences **223**, 01019 (2019) https://doi.org/10.1051/epjconf/201922301019
NSD2019

Particle-Hole Cluster Shell Model States in ^{8}Be, a Challenge for Ab-Intio Nuclear Theories

Moshe Gai[1,*]

[1]LNS at Avery Point, University of Connecticut, 1084 Shennecossett Rd, Groton, CT 06340, USA

Abstract. We propose the first application of the Cluster Shell Model (CSM) of Della Rocca and Iachello to particle-hole (p-h) states in ^{8}Be. We demonstrate a few essential features of the CSM in ^{8}Be: 1) All predicted p-h states of the CSM, and only the predicted p-h states, are observed near thresholds and up to 19.5 MeV in ^{8}Be. 2) The states are observed in the predicted order, with positive parity states below negative parity states. 3) Some of the p-h states are already known to have the rotational structure predicted for the deformed p-h states. 4) The rotational structures observed at high excitations in the p-h bands in ^{8}Be, resemble the ground state bands of ^{8}Be, ^{9}Be and ^{9}B, with similar moment of inertia. Based on these observations we contemplate new measurements of the spectroscopy of ^{8}Be at energies above 19.5 MeV, where our knowledge of states in ^{8}Be is scarce. We examine the observed B(M1)s and B(E2)s in these nuclei and contemplate a measurement of the B(E2) of the isobaric Analog transition in ^{8}B. We discuss the observed rotational structure in ^{8}Be as a challenge to ab-initio calculations that searched for "emerging rotational structures at high excitations" in beryllium nuclei, and reveal rotational structure at high excitations in ^{10}Be and ^{12}Be but not in ^{8}Be.

1 Introduction

The nuclei ^{12}C and ^{8}Be have currently attracted much attention. On one hand the algebraic cluster model (ACM) of Bijker and Iachello [1], presents some of the most interesting application of geometrical point group symmetries to cluster states, and on the other hand ab-initio theories such as the no-core shell model (NCCI) [2] and the quantum Monte Carlo (GFMC) [3], can be applied to these light nuclei as testing grounds. But, perhaps one of the most intriguing aspect of these new theoretical development is the prediction and observation in ^{12}C of the mixed parity ground state rotational band including the 4^+ and 4^- parity doublet, and the states of $J^\pi = 0^+, 2^+, 3^-, 4^\pm$ and 5^- [4].

1.1 Ab-Initio No Core Shell Model

Ab-initio no-core shell model (NCCI) [2] and the quantum Monte Carlo (GFMC) [3] calculations reached an impressive level of accomplishment where they can be compared to the spectra of light nuclei at very high excitations, as shown in Fig. 1. However, these ab-initio calculations appear at the present time to have difficulties with calculating negative parity states. Of interest for this paper is the study of the "emergence of rotational bands at high excitation energies" in ab-initio no core shell model calculations [5] as shown in Fig. 2. We note that these ab-initio calculations reveal the emergence of positive parity rotational bands at high excitations in ^{10}Be and ^{12}Be, but not in ^{8}Be.

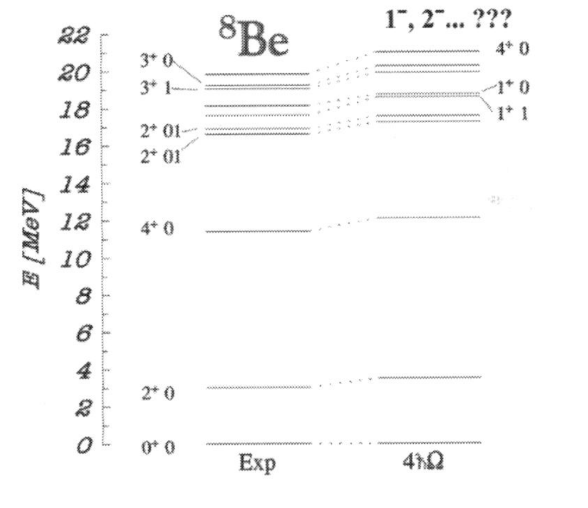

Figure 1. Ab-initio shell model calculations [2] of (only) positive parity states extended to very high excitations in ^{8}Be.

1.2 The Cluster Shell Model

The ACM [1] was recently extended to describe single particle motion in molecular orbits within the cluster shell model (CSM) of Della Rocca and Iachello [6]. In Fig. 3 we show the Nilsson-like (neutron) single particle states in ^{9}Be as a function the two alpha-particles separation distance (β), as predicted by the CSM [6]. However, while the CSM is a Nilsson-like model, it predicts molecular single particle orbits, hence the occurrence of rotational bands

*e-mail: moshe.gai@uconn.edu

https://doi.org/10.1051/epjconf/201922301019

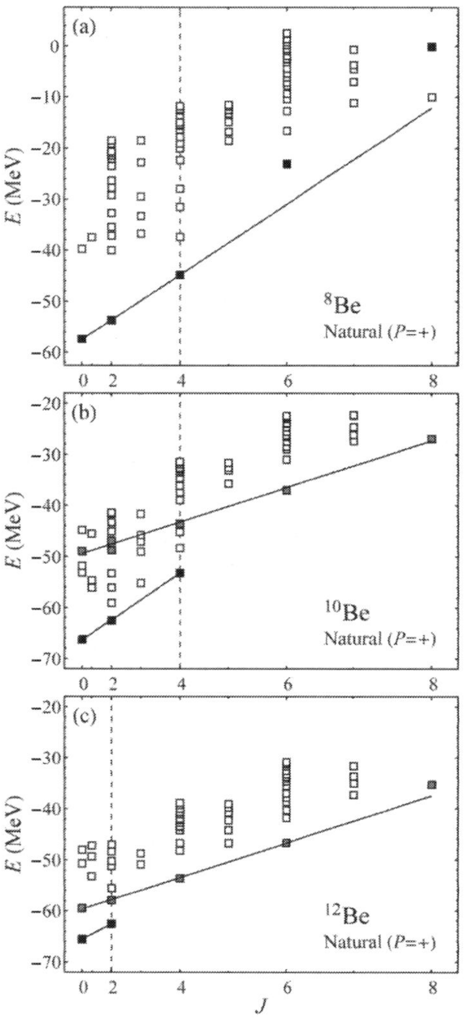

Figure 2. Excited states rotational bands of (only) positive parity states revealed in ab-initio no-core configuration interaction calculations [5] of the beryllium isotopes. Note the emergence of such rotational bands at high excitations in ^{10}Be and ^{12}Be, but not in ^{8}Be.

with parity doublets [6], which are not predicted in the Nilsson Model. In addition the CSM predicts similarities and relationships between rotational bands in ^{9}Be and ^{8}Be [6], which are not evident in the Nilsson model.

2 CSM p-h states in ^{8}Be

The lowest particle-Hole (p-h) states in ^{8}Be arise in the CSM as holes in the K$^{\pi}$= 1/2^{-} and 1/2^{+} proton or neutron molecular orbits, excited to the K$^{\pi}$ = 3/2^{-} orbit. Hence the lowest p-h sates in ^{8}Be are expected to be with K$^{\pi}$ = 2^{+}, 1^{+}, 2^{-} and 1^{-}, with the positive parity states at lower excitations. In addition, both isospin 0 and 1 are predicted, corresponding to neutron (^{9}Be like) and proton (^{9}B like) excitations. In Fig. 4 we show all the known states in ^{8}Be below 19.5 MeV, we also add three selected states above 20 MeV. In the energy range below 19.5 MeV, aside from the ground state $\alpha + \alpha$ band, all p-h states predicted by the CSM are observed at these high excitations, and only the predicted states are observed. The predicted order of the negative parity states above the positive parity is also observed. Hence, we tentatively denote the states by the K$^{\pi}$ classification of the CSM.

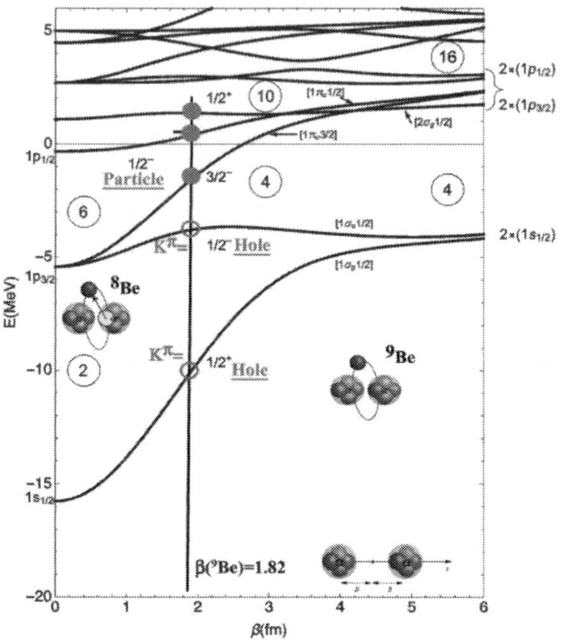

Figure 3. Nilsson like single (neutron) particle states in ^{9}Be predicted by the CSM [6] as a function of the separation distance of the alpha-particles (β). The particle (hole) states in ^{8}Be are indicated by full (open) circles.

The phenomenological agreement with the CSM of the known eigen states of ^{8}Be up to 19.5 MeV, encourage us on one hand to consider measurements of the spectroscopy of ^{8}Be above 20 MeV [7], in order to clarify the spectroscopy of ^{8}Be above 20 MeV [8, 9]. On the other hand, the observed spectrum gives very strong motivation to calculations of p-h states in 8Be, in the frame of the CSM.

Two additional important features of p-h CSM states in ^{8}Be are highlighted: First, since the p-h states are deformed, a rotational band is predicted on top of each p-h state. Second, the rotational band structure is predicted to be very similar to the ground state rotational bands observed in ^{9}Be and ^{9}B. In addition, the ground state rotational bands in ^{9}Be and ^{9}B were already demonstrated [6] to be similar to the K$^{\pi}$ = 0^{+} ground state rotational band of ^{8}Be, also shown in Fig. 4. Hence, the p-h bands in ^{8}Be, the g.s. rotational bands in ^{9}Be and ^{9}B, the g.s. rotational band in ^{8}Be, are all predicted to have very similar properties including similar moment of inertia and related B(E2) and B(M1) values.

https://doi.org/10.1051/epjconf/201922301019

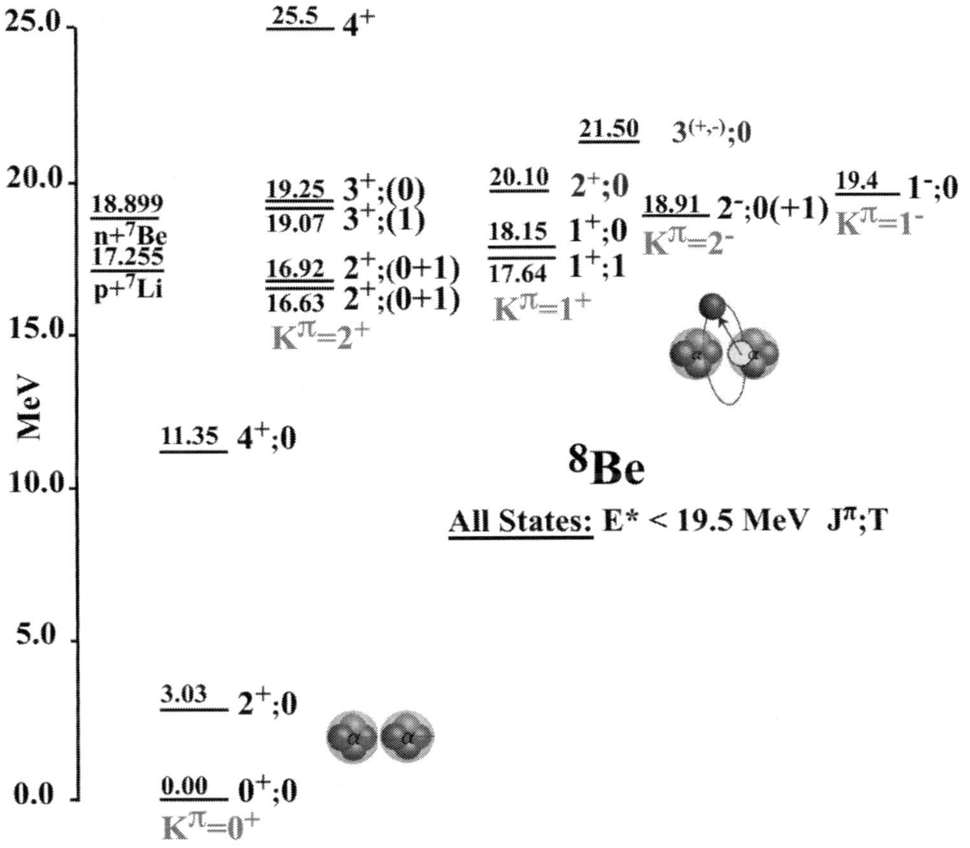

Figure 4. Measured states in ^{8}Be up to 19.5 MeV, with the (tentative) K^π assignments in the CSM indicated in red.

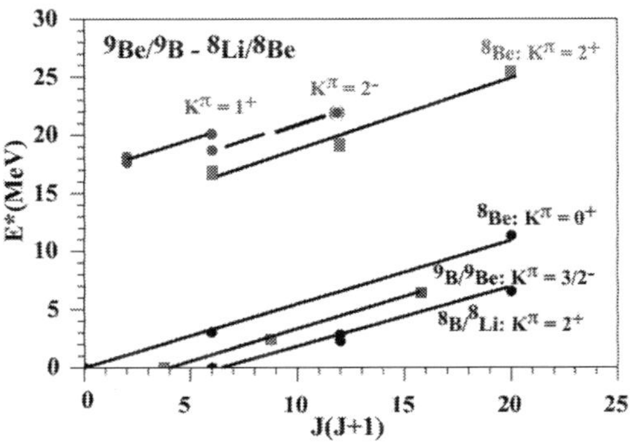

Figure 5. Rotational band structures in ^{8}Be and the isobaric Analog ^{8}B and ^{8}Li, and ^{9}Be and ^{9}B. The tentative 3^- state at 21.5 MeV is from [9]

In Fig. 5 we show the $K^\pi = 0^+$ g.s. rotational band in ^{8}Be, the $K^\pi = 3/2^-$ g.s. rotational bands in ^{9}Be and ^{9}B, the $K^\pi = 2^+$ (and $K^\pi = 1^-$ and 2^-) p-h bands in ^{8}Be, and the $K^\pi = 2^+$ g.s. rotational band in a ^{8}Li (and ^{8}B), the isobaric-analog of the $K^\pi = 2^+$ p-h band in ^{8}Be. The similarity of the observed moments of inertia is striking (with differences smaller or equal to 7%).

In Table I we compare the measured B(M1) in the $K^\pi = 3/2^-$ g.s. rotational bands in ^{9}Be and the $K^\pi = 2^+$ g.s. rotational band in ^{8}Li, the isobaric-analog of the $K^\pi = 2^+$ p-h band in ^{8}Be. The similarity of the B(M1) is indeed striking. We however note that when including the Clebsch-Gordon coefficients, slightly different B(M1)s are predicted due to the different spins involved.

Table 1. B(M1)s in ^{8}Li (isobaric analog of ^{8}Be) and ^{9}Be

$E_i \to E_f$ (MeV)	$J^\pi;T \to J^\pi;T$	B(M1) (W.u.)
2.26→ 0.0	$3^+;1 \to 2^+;1$	0.29±0.12
2.4294→ 0.0	$\frac{5}{2}^-;\frac{1}{2} \to \frac{3}{2}^-;\frac{1}{2}$	0.3±0.03

Table 2. B(E2)s in ^{8}Be and ^{9}Be

$E_i \to E_f$ (MeV)	$J^\pi;T \to J^\pi;T$	B(E2) (W.u.)
11.35→ 3.03	$4^+;0 \to 2^+;0$	25±8.4
2.4294→ 0.0	$\frac{5}{2}^-;\frac{1}{2} \to \frac{3}{2}^-;\frac{1}{2}$	24.4±1.8
6.38→0.0	$\frac{7}{2}^-;\frac{1}{2} \to \frac{3}{2};\frac{1}{2};\frac{1}{2}$	8.5±3.7

The similarity of the isobaric-Analog transition in ^{8}Li and ^{9}Be, shown in Table I, encourage us to consider the B(E2) of these nuclei. A strong B(E2: $3^+ \rightarrow 2^+$) in ^{8}B or ^{8}Li, will further demonstrate the rotational character of the isobaric Analog p-h band in ^{8}Be shown in Fig. 5. In Table II we list measured B(E2) in ^{8}Be and ^{9}Be. Based on the similarity of the rotational structure in the beryllium isotopes we expect the B(E2: $3^+ \rightarrow 2^+$) in ^{8}B or ^{8}Li to be approximately 25 W.u. We are engaged in developing a proposal to measure the Coulomb excitation of the ^{8}B to measure the B(E2: $3^+ \rightarrow 2^+$).

3 Conclusions

We identified a rotational structure in ^{8}Be at very high excitations above 16 MeV. The observed structure is indicative of particle-hole states of the cluster shell model (CSM). They arise from a particle and a hole moving in molecular orbits. The description of molecular states requires many $\hbar\omega$ model space and as such they pose a challenge to ab-intio no core shell model calculations, that searched for the "emergence" (in NCCI calculations) of such rotational structure at high excitations in ^{8}Be.

Acknowledgement

The material presented in this paper is based upon work supported by the U.S. Department of Energy, Office of Science, Office of Nuclear Physics, Award Numbers: DE-FG02-94ER40870.

References

[1] R. Bijker and F. Iachello, Ann. Phys. **298**, 334 (2002).
[2] P. Navratil and B.R. Barrett, Phys. Rev. C **57**, 3119 (1998), Petr Navratil and W. Erich Ormand, Phys. Rev. C **68**, 034305 (2003).
[3] S. Pastore, Steven C. Pieper, R. Schiavilla, and R.B. Wiringa, Phys. Rev. C **87**, 035503 (2013).
[4] D.J. Marin-Lambarri, R. Bijker, M. Freer, M. Gai M, Tz. Kokalova, D.J. Parker, and C. Wheldon, Phys. Rev. Lett. **113**, 012502 (2014).
[5] P. Maris, M.A. Caprio, and J.P. Vary, PRC **91**, 014310 (2015).
[6] V. Della Rocca, and F.Iachello, Nucl. Phys. A **973**, 1 (2018).
[7] Workshop of the ISOLDE Solenoidal Spectrometer (ISS), August 27-28, 2019, Liverpool, UK.
[8] D.R. Tilley, J.H. Kelley, J.L. Godwin, D.J. Millener, J. Purcell, C.G. Sheu and H.R. Weller, Revised Manuscript: August 4, 2008, References closed: March 31, 2004, A=8 compilation: http://tunl.duke.edu.
[9] P. R. Page, Phys. Rev. C **72**, 054312 (2005).

Recent applications of the subtracted SRPA approximation

Danilo Gambacurta[1,*]

[1]Extreme Light Infrastructure - Nuclear Physics (ELI-NP), Horia Hulubei National Institute for Physics and Nuclear Engineering, 30 Reactorului Street, RO-077125 Magurele, Jud. Ilfov, Romania

Abstract. The Second Random Phase Approximation (SRPA) is a natural extension of the Random Phase Approximation obtained by introducing more general excitation operators where two particle-two hole configurations, in addition to the one particle-one hole ones, are considered. Only in the last years, large-scale SRPA calculations, without usually employed approximations have been performed. The SRPA model corrected by a subtraction procedure designed to cure double counting issues and the related instabilities has been recently implemented and applied in the study of different physical cases. We report here on some of the most recent results obtained by using this model. In particular, results on the dipole strength ^{48}Ca and on a systematic study of the isoscalar giant quadrupole resonance in spherical nuclei will be shown and discussed.

1 Introduction

The random–phase–approximation (RPA) model provides a microscopic description of the nuclear collective excitations constructed as superpositions of 1 particle–1 hole (1p1h) configurations. This approach is able to provide the global features of Giant Resonances (GR) such as the centroid energy, the total strength and the corresponding energy-weighted sum rules. However, it is not suited to describe other important properties as for example the GR's fine structure and the spreading width, generated by the coupling between 1p1h configurations with more complex degrees of freedom. Approaches aimed at the description of this coupling are for example the particle-phonon coupling models based on the so-called time-blocking approximation [1, 2] and the relativistic quasiparticle time-blocking approxi- mation [3, 4], the quasiparticle-phonon model [5, 6], particle-phonon (or quasiparticle-phonon) coupling models [7, 8].

The Second RPA (SRPA) model is a natural extension of RPA where a more general description of the nuclear excitations is considered and provides a valuable tool for the prediction of spreading widths and fine structure properties due to the introduction of 2 particle-2 hole (2p2h) configurations [9–12]. The SRPA model has been recently improved by using the so called subtraction procedure [13, 14] designed to handle the problem of the double counting of correlations within energy–density–functional (EDF) based model which go beyond the RPA level. This procedure cures some of the drawbacks and the limitations of the SRPA model formulated in the EDF framework providing thus a robust and stable theoretical tool for a beyond–mean–field description of the excitation spectra of many–body systems. In this work, we briefly report about some recent applications of the SRPA model based on the subtraction procedure that we will refer at as SSRPA.

In Section 2, we analyze the dipole response in ^{48}Ca. In particular we show how the SSRPA provides a satisfactory description of the low–lying dipole spectrum (below 10 MeV) and of the giant dipole resonance (GDR).

In Section 3 we present the systematic trends obtained for the centroids and the widths in the SSRPA model for the isoscalar giant quadrupole resonances for different nuclei, from ^{30}Si to ^{208}Pb. In the same section we also show a detailed comparison of the strength distributions for ^{90}Zr. Comparison with the RPA results and with experimental spectra are also shown. Finally in Section 4 some conclusions are drawn.

2 Dipole response in ^{48}Ca

Details on the SRPA and SSRPA formalism can be found in Refs [10, 11, 13]. In the following analysis, we used a cutoff of 100 MeV in building the 1ph configurations, ensuring a full preservation of the isoscalar and isovector Energy Weighted Sum Rules (EWSRs). Results are illustrated with two Skyrme parametrizations, SGII [15] and SLy4 [16] and more detailes can be found in Ref. [17]. Deviations of less than 1% are found in both cases in RPA. In the SRPA and SSRPA calculations, a cutoff of 60 MeV in the 2p2h configurations is used. The dipole response below the neutron threshold in ^{48}Ca has been measured with the (γ, γ') reaction [18]. The RPA predictions fail in reproducing such a low-lying strength either because the lowest RPA energies are higher than 10 MeV or because the RPA model cannot provide the observed fragmentation. A fragmented SRPA dipole strength below 10 MeV was indeed found [11]. However, the SRPA $B(E1)$ transition probability, integrated up to 10 MeV, was found to be definitely much larger than the experimental value. We

*e-mail: danilo.gambacurta@eli-np.ro

 https://doi.org/10.1051/epjconf/201922301020

show here that problem of overestimation of the transition probability is strongly reduced in SSRPA, providing thus a rather good description of the experimental data. Some states having strong 1ph components, which are too much shifted downwards by the standard SRPA, are pushed to higher energies by the subtraction, reducing in this way the $B(E1)$ value below 10 MeV.

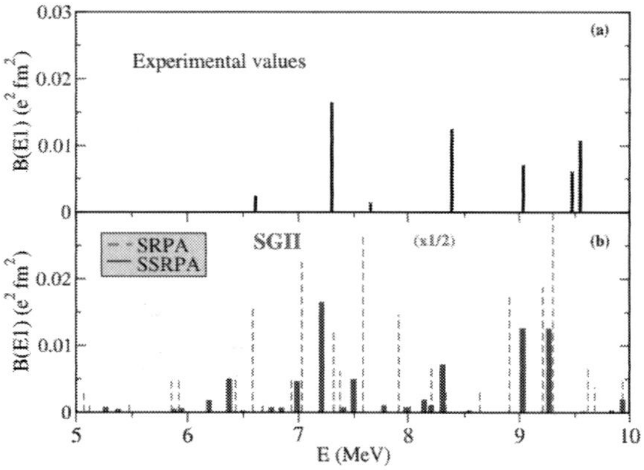

Figure 1. Low-lying dipole strength in ^{48}Ca : (a) Experimental $B(E1)$ values [18]; (b) Theoretical predictions for the transition probabilities $B(E1)$ calculated with the standard SRPA (dashed red bars; the values have been divided by 2) and with the SSRPA (blue thick bars).

In Figure 1(a) we show the experimental transition probabilities $B(E1)$ [18]. The corresponding theoretical results are displayed in Figure 1(b). One observes that the total strength provided by the standard SRPA is much higher than the experimental one as already found and discussed [12]. The theoretical prediction given by the SS-RPA model, on the other side, is in a very satisfactory agreement with respect to the experimental data. The fragmentation of the states follows quite well the experimental distribution in the energy position of the main peaks. The total $BE(1)$ integrated up to 10 MeV is 0.126 (0.078) e^2 fm^2 in the SLy4 (SGII) case to be compared with the experimental value 0.068 e^2 fm^2.

We move now to the energy region between 10 and 25 MeV, where the GDR is located. Figure 2 shows the strength distributions calculated with the parametrization SGII in RPA, SRPA, and SSRPA, compared with the experimental distributions [19]. Results with the SGII interaction are shown but a similar behaviour is found also with SLy4. To make simpler the comparison, a folding with a Lorentzian having a width of 0.25 MeV is performed. This folding produces some artificial spreading in the RPA case. For both SRPA and SSRPA, a more pronounced spreading is found corresponding to a physical width produced by the extremely dense distribution generated by the 2p2h configurations. It can be observed that the width is indeed in a satisfactory agreement with the experimental distribution. Experimental centroid energy E_C and width Γ are equal to 18.9 ± 0.2 and 3.9 ± 0.4 MeV, respectively [19]. We have computed the theoretical centroid energies and widths using the expressions

$$E_C = \frac{m_1}{m_0}, \quad \Gamma_C = \sqrt{m_2/m_0 - (m_1/m_0)^2},$$

where m_k represents the moment of order k integrated in the energy region of interest.

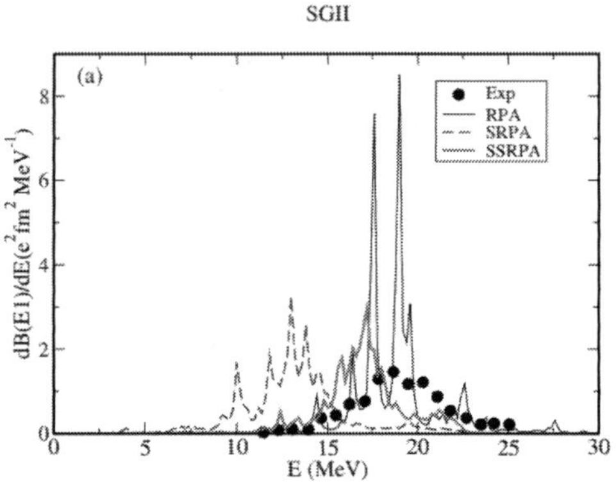

Figure 2. Dipole strength distributions in ^{48}Ca evaluated with RPA (solid black line), SRPA (blue dashed line), and SSRPA (thick red line), compared with the experimental distributions (black circles) from Ref. [19]. The SGII Skyrme interaction is used.

The theoretical centroid energies are 18.0, 13.4, and 16.2 MeV with RPA, SRPA, and SSRPA, respectively. The widths are equal to 2.5 and 2.1 MeV with SRPA and SSRPA, respectively. The parametrization SLy4 provides less satisfactory results than the Skyrme interaction SGII. In particular, the SSRPA centroid energy is lower than the experimental centroid by more than 2.5 MeV. Figure 3 displays in the two panels the SSRPA strength distributions obtained with SGII (a) and SLy4 (b), shifted upwards respectively by 1.5 (a) and 2.7 (b) MeV, together with the RPA strength distributions and the experimental data. This plot clearly illustrates how well the width of the resonance is described and shows the significant improvement with respect to RPA.

3 Isoscalar GQR: a systematic study

In this section we analyze the isoscalar GQR in several closed shell nuclei. More details can be found in Ref. [20]. It is quite clear, that the SRPA approach, especially if performed within the EDF approach, i.e. with the use of effective interactions, is plagued by double–counting and instabilities issues[21, 22]. We underline that these issues affects mainly the centroid of the strength distribution but not its width, so the SRPA and SSRPA widths are indeed very similar. However, as the SRPA centrids are quite far from th experimental ones, in the following we will show only the comparison between the RPA and SSRPA results. The SLy4 interaction is used. The single–particle space

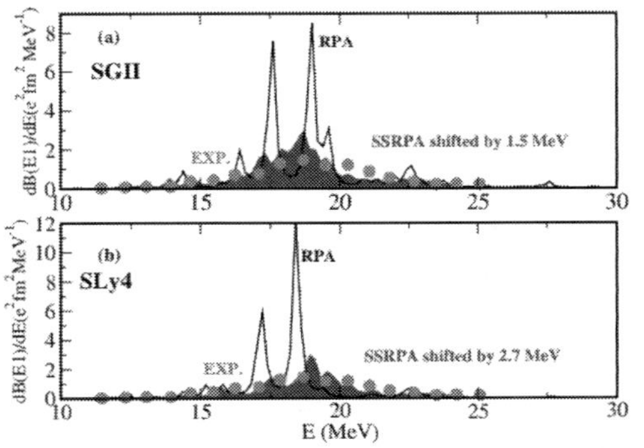

Figure 3. (a) SSRPA results for ^{48}Ca shifted by 1.5 MeV (blue area) compared with the RPA strength (black line) and with the experimental values (red circles), obtained with the parametrization SGII. (b) Same as in (a) but the results are obtained with the parametrization SLy4 and the shift of the SSRPA spectrum is larger (2.7 MeV).

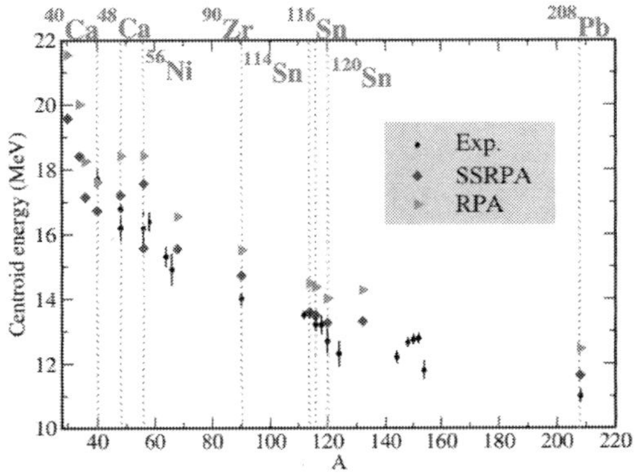

Figure 4. Centroids of the isoscalar GQR: the experimental data in black circles are compared with the SSRPA (RPA) predictions are plotted as blue diamonds (magenta triangles).

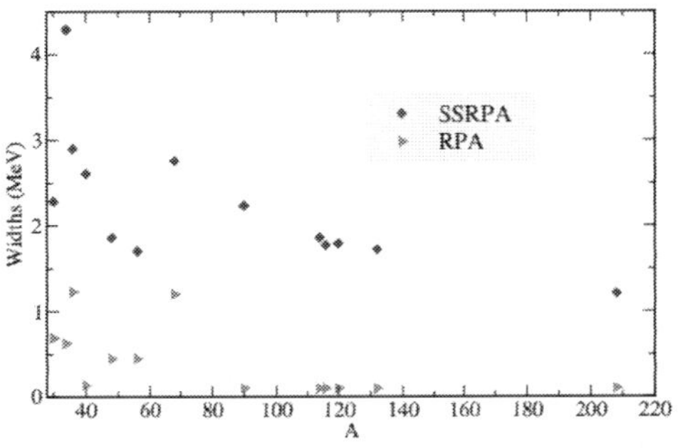

Figure 5. Theoretical widths calculated with the fit of a Lorentzian distribution within RPA and SSRPA models.

is chosen large enough to assure that the EWSR are preserved within 1%. For the 2p2h space in the SSRPA calculations, we use a cutoff of 60 MeV for medium–mass nuclei (up to ^{68}Ni) and of 50 MeV for the heavy ones (up to ^{208}Pb). We checked that these cutoff values provide stable results. Centroids and widths of the IS GQRs were computed by fitting a Lorentzian distribution. This adjustment was done on curves obtained by folding the discrete spectra with narrow Lorentzian distributions. In the SSRPA case, the folding done with very narrow distributions follows fairly well the extremely dense SSRPA spectra and no artificial effects induced by the performed folding are observed when the Lorentzian distribution is fitted. On the other side, in those RPA cases where there is a unique dominant peak, the width extracted in this way turns out to be equal to the width of the folding Lorentzian. We used in these cases very narrow folding Lorentzian (100–keV width) not to induce any artificial spreading effect.

The systematic trend provided by the SSRPA model for the centroids is shown in Fig. 4 where SSRPA results (blue diamonds) are compared with the corresponding RPA results (magenta triangles) and with the experimental data (black circles) represented in the figure with the corresponding error bars. Nuclei for which a comparison between our theoretical predictions and the corresponding experimental data may be done are identified in the figure by vertical dotted red lines. We observe that the SSRPA centroids are systematically located at lower energies than the RPA values. For most of the cases where the experimental centroids are available, we observe that the SSRPA energies are in better agreement with the experimental values than the RPA centroids, which in general overestimate the data, as can be seen in Fig. 4.

We expect that the description of the widths is strongly modified in the SSRPA model, compared to the RPA case, because an additional spreading effect is explicitly taken into account (in addition to the single–particle Landau damping which is already present in RPA) owing to the coupling between 1p1h and 2p2h configurations. Figure 5 displays the SSRPA and RPA widths. We specify once more that our predictions of the total widths contain the Landau–damping effect (RPA and SSRPA) and the spreading contribution (SSRPA). The escape width is missing. Even if the escape width is expected to be less important than the spreading width, it may modify some of our predictions, especially for light nuclei. We observe that the SSRPA widths are, as expected, systematically larger than the RPA ones.

As far as width, fine structure and fragmentation are concerned, we dedicate a more focused discussion to a specific case, namely ^{90}Zr. High–resolution (p, p') spectra are available [23]. Energy resolutions of $\sim$ 40 keV could be achieved and the fine structure of the excitation spectra could be examined. We present in Fig. 6(c) the

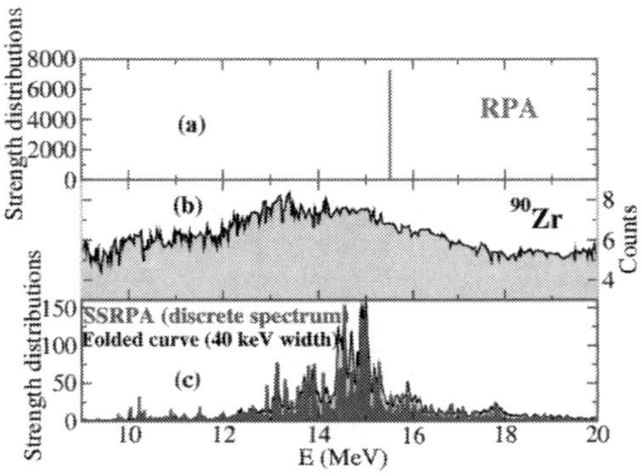

Figure 6. (a) RPA strength distributions, (b) Experimental spectrum [23] and (c) SSRPA strength distributions for the isoscalar GQR for ^{90}Zr. For the cases of the RPA and SSRPA discrete spectra, units are e^2 fm^4. For the SSRPA folded case, units are e^2 fm^4 MeV^{-1}.

SSRPA strength distribution (violet bars) for the nucleus ^{90}Zr. To better compare it with the corresponding experimental spectrum (b), a folded curve is also plotted (black solid line and grey area), obtained by folding the discrete distribution with a Lorentzian of width equal to 40 keV, which corresponds to the experimental energy resolution. The RPA centroid is larger by more than 1 MeV compared to the experimental value (Fig. 4). The SSRPA prediction is located at lower energies, in better agreement with data. We observe that the folded curve follows well the fine structure provided by the discrete spectrum. The RPA strength distribution is also shown in Fig. 6(a). Our RPA prediction displays a unique dominant peak, whereas the SSRPA strength distribution is much more fragmented and extends over a larger energy region where the experimental data are spread. We note however that the experimental strength is distributed in a broader region than the theoretical one. This is probably related to missing effects in our theoretical model, such as the inclusion of higher–order configurations or, to a less extent, to the coupling with the continuum which is neglected here.

4 Conclusions

We have discussed two recent applications of the SSRPA model. First we analyze the dipole response in ^{48}Ca and we show how the SSRPA is able to provide a better description with respect the RPA regarding both the low-lying excited states and the width of the GDR. Afterwards, we have shown a systematic analysis of the isoscalar GQR in several spherical nuclei. Also in this case the SSRPA ia able to provide a better description of the centroid energies and widths with respect the RPA. We also discussed the fine structure properties described by the SSRPA for the specific case of ^{90}Zr. A significant improvement of the description of the spreading width is found in SSRPA compared to RPA (where a single dominant peak is predicted) and fragmented strengths are obtained.

References

[1] S.P. Kamerdzhiev, G.Y. Tertychny, V.I. Tselyaev, Phys. Part. Nucl. 28, 134 (1997).
[2] V.I. Tselyaev, Phys. Rev. C 75, 024306 (2007).
[3] E. Litvinova, P. Ring, and V. Tselyaev, Phys. Rev. C 78, 014312 (2008).
[4] I.A. Egorova and E. Litvinova, Phys. Rev. C 94, 034322 (2016).
[5] V.G. Soloviev, *Theory of Atomic Nuclei: Quasiparticles and Phonons* (Institute of Physics, Bristol and Philadelphia, 1992).
[6] A.P. Severyukhin, S. Aberg, N.N. Arsenyev, and R.G. Nazmitdinov, Phys. Rev. C 95, 061305(R) (2017).
[7] G. Colò, H. Sagawa, N. Van Giai, P.F. Bortignon, and T. Suzuki, Phys. Rev. C 57, 3049 (1998).
[8] Y.F. Niu, G. Colò, E. Vigezzi, C.L. Bai, and H. Sagawa, Phys. Rev. C 94, 064328 (2016).
[9] P. Papakonstantinou and R. Roth, Phys. Rev. C **81**, 024317 (2010)
[10] D. Gambacurta, M. Grasso, and F. Catara, Phys. Rev. C **81**, 054312 (2010)
[11] D. Gambacurta, M. Grasso, and F. Catara, Phys. Rev. C **84**, 034301 (2011)
[12] D. Gambacurta, M. Grasso, V. De Donno, G. Co', and F. Catara, Phys. Rev. C **86**, 021304(R), (2012)
[13] D. Gambacurta, M. Grasso, and J. Engel, Phys. Rev. C **92**, 034303 (2015)
[14] D. Gambacurta and M. Grasso, Eur. Phys. J. A **52**, 198 (2016)
[15] N. V. Giai, H.. Sagawa, Phys. Lett. B 106, 379 (1981); Nucl. Phys. A **371**, 1 (1981)
[16] E. Chabanat, P. Bonche, P. Haensel, J. Meyer, R. Schaeffer, Nucl. Phys. A **627**, 710 (1997); *ibid* A **635**, 231 (1998); *ibid* **643**, 441 (1998)
[17] D. Gambacurta, M. Grasso, and O. Vasseur, Phys. Lett. B **777**, 163 (2018)
[18] T. Hartmann, J. Enders, P. Mohr, K. Vogt, S. Volz, and A. Zilges, Phys. Rev. C **65**, 034301 (2002)
[19] J. Birkhan *et al.*, Phys. Rev. Lett. **118**, 252501 (2017)
[20] O. Vasseur, D. Gambacurta, and M. Grasso,Phys. Rev. C **98**, 044313, (2018)
[21] V. I. Tselyaev, Phys. Rev. C 88, 054301 (2013).
[22] P. Papakonstantinou, Phys. Rev. C 90, 024305 (2014).
[23] A. Shevchenko, et al., Phys. Rev. Lett. **93**, 122501 (2004)

Intertwined quantum phase transitions in the Zr chain

N. Gavrielov[1,*], *A.* Leviatan[1,**], and *F.* Iachello[2,***]

[1]Racah Institute of Physics, The Hebrew University, Jerusalem 91904, Israel
[2]Center for Theoretical Physics, Sloane Physics Laboratory, Yale University, New Haven, Connecticut 06520-8120, USA

Abstract. We introduce the notion of intertwined quantum phase transitions (IQPTs), for which a crossing of two configurations coexists with a pronounced shape-evolution of each configuration. A detailed analysis in the framework of the interacting boson model with configuration mixing, provides evidence for this scenario in the Zr isotopes. The latter exhibit a normal configuration which remains spherical along the chain, but exchanges roles with an intruder configuration, which undergoes first a spherical to prolate-deformed [U(5)→SU(3)] QPT and then a crossover to γ-unstable [SU(3)→SO(6)].

Quantum phase transitions (QPTs) are qualitative changes in the properties of a physical system, induced by a variation of parameters that appear in the quantum Hamiltonian. In nuclear physics, most of the attention in the study of QPTs, has been devoted to shape phase transitions in a single configuration, described by a single Hamiltonian, $\hat{H} = (1-\xi)\hat{H}_1 + \xi\hat{H}_2$, where ξ is the control parameter. As ξ changes from 0 to 1, the eigenvalues of the system change from those of $\hat{H}_1$ to those of $\hat{H}_2$. For sake of clarity, we denote these phase transitions Type I. The latter have been observed in the neutron number 90 region, *e.g.*, for Nd-Sm-Gd isotopes [1].

A different type of phase transitions occurs when two (or more) configurations coexist [2]. In this case, the quantum Hamiltonian has a matrix form with entries: $\hat{H}_A(\xi^A)$, $\hat{H}_B(\xi^B)$, $\hat{W}(\omega)$, where the indices A, B denote the two configurations and $\hat{W}$ denotes their coupling. We call for sake of clarity these phase transitions Type II [3], to distinguish them from those of a single configuration. Type II QPTs have been observed in nuclei near shell closure, *e.g.*, in the light Pb-Hg isotopes [4], albeit with strong mixing between the two configurations. In the present contribution, we explore a situation where in parallel to the crossing, each configuration maintains its purity and its own shape-evolution with nucleon number. We refer to such a scenario as intertwined quantum phase transitions (IQPTs) in the sense that Type I and Type II coexist, and show empirical evidence for it in the Zr chain [5].

The $_{40}$Zr isotopes have been recently the subject of several experimental [6–10] and theoretical investigations, including mean-field based methods [11–13] and the Monte-Carlo shell-model (MCSM) [14]. We adapt here the algebraic approach of the Interacting Boson Model (IBM) [15], with bosons representing valence nucleon pairs counted from the nearest closed shells. This provides a simple tractable framework, where phases of quadrupole shapes: spherical, prolate-deformed and γ-unstable correspond to U(5), SU(3) and SO(6) dynamical symmetries.

To be specific, we use the configuration mixing model (IBM-CM) of [16], and write the Hamiltonian not in matrix form, but rather in the equivalent form

$$\hat{H} = \hat{H}_A^{(N)} + \hat{H}_B^{(N+2)} + \hat{W}^{(N,N+2)}\,, \tag{1}$$

where $\hat{O}^{(N)} = \hat{P}_N^\dagger \hat{O}\hat{P}_N$ and $\hat{O}^{(N,N')} = \hat{P}_N^\dagger \hat{O}\hat{P}_{N'}$, for an operator $\hat{O}$, with $\hat{P}_N$, a projection operator onto the $[N]$ boson space. Here $\hat{H}_A^{(N)}$ represents the so-called normal (N boson space) configuration and $\hat{H}_B^{(N+2)}$ represents the so-called intruder ($N+2$ boson space) configuration, which we have assumed, as in [17] where a similar calculation was done for the $_{42}$Mo isotopes, to be a proton excitation across the subshell closure at proton number 40 (see Fig. 1 of [17]). The explicit form of these Hamiltonians is

$$\hat{H}_A = \epsilon_d^{(A)}\hat{n}_d + \kappa^{(A)}\hat{Q}_\chi\cdot\hat{Q}_\chi\,, \tag{2a}$$
$$\hat{H}_B = \epsilon_d^{(B)}\hat{n}_d + \kappa^{(B)}\hat{Q}_\chi\cdot\hat{Q}_\chi + \kappa'^{(B)}\hat{L}\cdot\hat{L} + \Delta_p\,, \tag{2b}$$
$$\hat{W} = \omega\,[\,(d^\dagger\times d^\dagger)^{(0)} + (s^\dagger)^2\,] + \text{H.c.}\,, \tag{2c}$$

where the quadrupole operator is defined as $\hat{Q}_\chi = d^\dagger s + s^\dagger\tilde{d} + \chi(d^\dagger\times\tilde{d})^{(2)}$ and $\hat{n}_d$ is the d-boson number operator. In Eq. (2b), Δ_p is the off-set between the normal and intruder configurations, where the index p denotes the fact that this is a proton excitation. The resulting eigenstates $|\Psi; L\rangle$ with angular momentum L, are linear combinations of the wave functions, Ψ_A and Ψ_B, in the two spaces $[N]$ and $[N+2]$,

$$|\Psi; L\rangle = a\,|\Psi_A; [N], L\rangle + b\,|\Psi_B; [N+2], L\rangle\,, \tag{3}$$

with $a^2 + b^2 = 1$.

By employing the IBM-CM framework described above, we have calculated the spectra and other observables of the entire chain of Zr isotopes, from neutron number 52 to 70. The values of the Hamiltonian parameters, obtained by a global fit to energy and $E2$ data, are

*e-mail: noam.gavrielov@mail.huji.ac.il
**e-mail: ami@phys.huji.ac.il
***e-mail: francesco.iachello@yale.edu

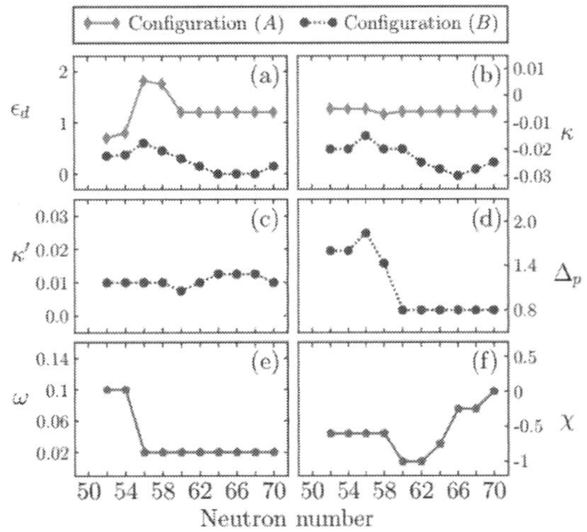

Figure 1. Parameters of the IBM-CM Hamiltonians, Eq. (2) are in MeV and χ is dimensionless. Configurations A and B correspond to boson spaces $[N]$ and $[N+2]$, respectively, with $N=1,2,\ldots,8$ ($N=7,6$), for neutron number 52-66 (68,70).

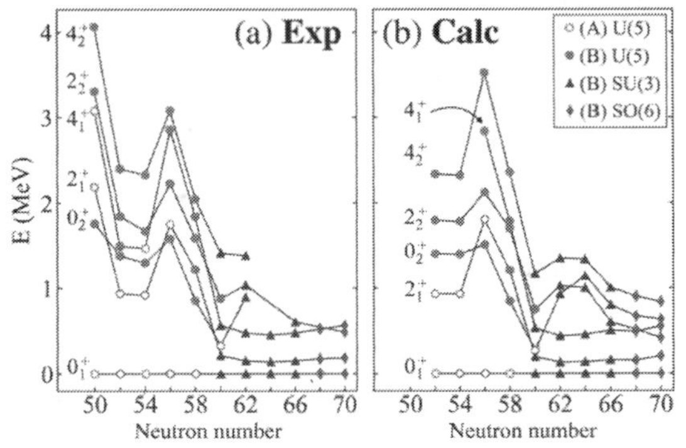

Figure 2. Comparison between (a) experimental [8, 21] and (b) calculated energy levels $0_1^+, 2_1^+, 4_1^+, 0_2^+, 2_2^+, 4_2^+$. Empty (filled) symbols indicate a state dominated by the normal A-configuration (intruder B-configuration), with assignments based on the decomposition of Eq. (3). The shape of the symbol [○, ▽, ◇], indicates the closest dynamical symmetry [U(5), SU(3), SO(6)] relevant to the level considered. Note that the calculated values start at neutron number 52, while the experimental values include the closed shell at 50. Adapted from [5].

shown in Fig. 1. It should be noted that beyond the middle of the shell, at neutron number 66, bosons are replaced by boson holes [15]. Apart from some fluctuations due to the subshell closure at neutron number 56 (the filling of the $2d_{5/2}$ orbital), the values of the parameters are a smooth function of neutron number and, in some cases, a constant. A notable exception is the sharp decrease by 1 MeV of the energy off-set parameter Δ_p beyond neutron number 56. Such a behavior was observed for the Mo and Ge chains [17, 18] and, as noted in [17], it reflects the effects of the isoscalar residual interaction between protons and neutrons occupying the partner orbitals $1g_{9/2}$ and $1g_{7/2}$, which is the established mechanism for descending cross shell-gap excitations and onset of deformation in this region [19, 20]. The $E2$ operator reads $\hat{T}(E2) = e^{(A)}\hat{Q}_\chi^{(N)} + e^{(B)}\hat{Q}_\chi^{(N+2)}$, where $\hat{Q}_\chi^{(N)} = \hat{P}_N^\dagger \hat{Q}_\chi \hat{P}_N$, $\hat{Q}_\chi^{(N+2)} = P_{N+2}^\dagger \hat{Q}_\chi \hat{P}_{N+2}$ and $\hat{Q}_\chi$ is the same operator as in the Hamiltonian (2). Here $e^{(A)}=0.9$ and $e^{(B)}=2.24$ (W.u.)$^{1/2}$ are the boson effective charges.

In Fig. 2 we show a comparison between experimental and calculated levels. One can see here a rather complex structure. In the region between neutron number 50 and 56, there appear to be two configurations, one spherical (seniority-like), (A), and one weakly deformed, (B), as evidenced by the ratio $R_{4/2}$, which is at 52-56, $R_{4/2}^{(A)} \cong 1.6$ and $R_{4/2}^{(B)} \cong 2.3$. From neutron number 58, there is a pronounced drop in energy for the states of configuration B and at 60, the two configurations exchange their role indicating a Type II QPT. At this stage, the intruder configuration (B) appears to be at the critical point of a U(5)-SU(3) Type I QPT, as evidenced by the low value of the excitation energy of the first excited 0^+ state of this configuration (the 0_3^+ state in ^{100}Zr shown in Fig. 4). The same situation is seen in the $_{62}$Sm and $_{64}$Gd isotopes at neutron number 90 [15]. Beyond neutron number 60, the intruder configuration (B) is strongly deformed, as evidenced by the small value of the excitation energy of the state 2_1^+, $E_{2_1^+} = 139.3$ keV and by the ratio $R_{4/2}^{(B)} = 3.24$ in ^{104}Zr. At still larger neutron number 66, the ground state band becomes γ-unstable as evidenced by the close energy of the states 2_2^+ and 4_1^+, $E_{2_2^+} = 607.0$ keV, $E_{4_1^+} = 476.5$ keV, in ^{106}Zr, and especially by the recent results $E_{4_1^+} = 565$ keV and $E_{2_2^+} = 485$ keV in ^{110}Zr [8], a signature of the SO(6) symmetry. In this region, the ground state configuration undergoes a crossover from SU(3) to SO(6).

The above spectral analysis signals the presence of coexisting Type I and Type II QPTs, which is the defining property of IQPTs. In order to understand the nature of these phase transitions, one needs to study the behavior of the order parameters. The latter are given by

$$\frac{\langle \hat{n}_d \rangle_A}{\langle \hat{N} \rangle_A}\ , \quad \frac{\langle \hat{n}_d \rangle_B}{\langle \hat{N} \rangle_B}\ , \quad \frac{\langle \hat{n}_d \rangle_{0_1^+}}{\langle \hat{N} \rangle_{0_1^+}}\ . \tag{4}$$

They involve the expectation value of $\hat{n}_d$ in the ground state wave function, $|\Psi; L=0_1^+\rangle$ and in its Ψ_A and Ψ_B components (3), normalized by the respective boson numbers, $\langle \hat{N} \rangle_A = N$, $\langle \hat{N} \rangle_B = N+2$, $\langle \hat{N} \rangle_{0_1^+} = a^2 N + b^2 (N+2)$. Here $\langle \hat{n}_d \rangle_A$ and $\langle \hat{n}_d \rangle_B$ portray the shape-evolution in configuration (A) and (B), respectively, and $\langle \hat{n}_d \rangle_{0_1^+} = a^2 \langle \hat{n}_d \rangle_A + b^2 \langle \hat{n}_d \rangle_B$ contains information on the normal-intruder mixing. Fig. 3(a) shows the evolution of the order parameters involving $\langle \hat{n}_d \rangle_A$ and $\langle \hat{n}_d \rangle_B$ in dotted lines, and $\langle \hat{n}_d \rangle_{0_1^+}$ in solid line. Configuration (A) is seen to be spherical for all neutron numbers considered. In contrast, configuration (B) is weakly-deformed for neutron number 52-58. One can see here clearly a jump between neutron number 58 and 60 from configuration (A) to configuration (B), indicating a 1st order Type II phase transition [3], a further increase

https://doi.org/10.1051/epjconf/201922301021

at neutron numbers 60-64 indicating a U(5)-SU(3) Type I QPT, and, finally, there is a decrease at neutron number 66, due in part to the crossover from SU(3) to SO(6) and in part to the shift from boson particles to boson holes after the middle of the major shell 50-82. $\langle \hat{n}_d \rangle_{0_1^+}$ is close to $\langle \hat{n}_d \rangle_A$ for neutron number 52-58 and coincides with $\langle \hat{n}_d \rangle_B$ above 60, indicating a high degree of purity and small configuration-mixing, with the exception of a narrow transition region. Indeed, the ground state wave function (3) has $a^2 = 98.2\%$, $b^2 = 87.2\%$ and $b^2 = 99.9\%$ for ^{98}Zr, ^{100}Zr and ^{102}Zr, respectively. These conclusions are stressed by an analysis of other observables, in particular, the $B(E2)$ values. As shown in Fig. 3(b), the calculated $B(E2)$'s agree with the empirical values and follow the same trends as the respective order parameters.

Further evidence can be obtained from an analysis of the isotope shift $\Delta\langle \hat{r}^2 \rangle_{0_1^+} = \langle \hat{r}^2 \rangle_{0_1^+;A+2} - \langle \hat{r}^2 \rangle_{0_1^+;A}$, where $\langle \hat{r}^2 \rangle_{0_1^+}$ is the expectation value of $\hat{r}^2$ in the ground state 0_1^+. In the IBM-CM the latter is given by $\langle \hat{r}^2 \rangle = r_c^2 + \alpha N_v + \eta[\langle \hat{n}_d^{(N)} \rangle + \langle \hat{n}_d^{(N+2)} \rangle]$, where r_c^2 is the square radius of the closed shell, N_v is half the number of valence particles, and η is a coefficient that takes into account the effect of deformation [15, 24]. $\Delta\langle \hat{r}^2 \rangle_{0_1^+}$ depends on two parameters, $\alpha = 0.235$, $\eta = 0.264$ fm^2, whose values are fixed by the procedure of Ref [24]. $\Delta\langle \hat{r}^2 \rangle_{0_1^+}$ should increase at the transition point and decrease and, as seen in Fig. 3(c), it does so, although the error bars are large and no data are available beyond neutron number 60. Similarly, the two-neutron separation energies S_{2n} can be written as [15], $S_{2n} = -\tilde{A} - \tilde{B}N_v \pm S_{2n}^{\text{def}} - \Delta_n$, where S_{2n}^{def} is the contribution of the deformation, obtained by the expectation value of the Hamiltonian in the ground state 0_1^+. The + sign applies to particles and the − sign to holes, and Δ_n takes into account the neutron subshell closure at 56, $\Delta_n = 0$ for 50-56 and $\Delta_n = 2$ MeV for 58-70. The value of Δ_n is taken from Table XII of [25] and $\tilde{A} = -16.5$, $\tilde{B} = 0.758$ MeV are determined by a fit to binding energies of ^{92,94,96}Zr. The calculated S_{2n}, shown in Fig. 3(d), displays a complex behavior. Between neutron number 52 and 56 it is a straight line, as the ground state is spherical (seniority-like) configuration (A). After 56, it first goes down due to the subshell closure at 56, then it flattens as expected from a 1st order Type I QPT (see, for example the same situation in the $_{62}$Sm isotopes [15]). After 62, it goes down again due to the increasing of deformation and finally it flattens as expected from a crossover from SU(3) to SO(6).

We note that our calculations describe the experimental data in the entire range $^{92-110}$Zr very well. Here we show only two examples, ^{100}Zr and ^{110}Zr. ^{100}Zr is near the critical point of both Type I and Type II QPT and yet our description of energy levels and $B(E2)$ values is excellent, Fig. 4(a)-(b). The ground state band, configuration (B), appears to have features of the so-called X(5) symmetry [26], while the spherical configuration (A) has now become the excited band 0_2^+. ^{110}Zr, Fig. 4(c)-(d), appears instead to be an excellent example of SO(6) symmetry [15], although few experimental data are available. In general, the current results resemble those obtained in the MCSM [14], however, there are some noticeable differences. Specifically,

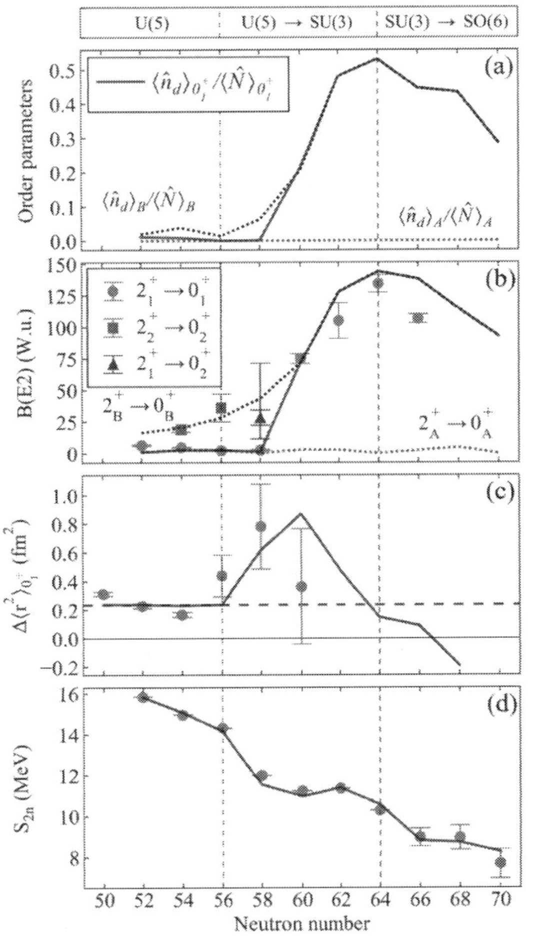

Figure 3. Evolution of order parameters and of observables along the Zr chain. Symbols (solid lines) denote experimental data (calculated results). (a) Order parameters, Eq. (4). Notation of lines is explained in the text. (b) $B(E2)$ values in Weisskopf units (W.u.). Data taken from [6, 7, 9, 10, 21]. Dotted lines denote calculated $E2$ transitions within a configuration. (c) Isotope shift, $\Delta\langle \hat{r}^2 \rangle_{0_1^+}$ in fm^2. Data taken from [22]. The horizontal dashed line at 0.235 fm^2 represents the smooth behavior in $\Delta\langle \hat{r}^2 \rangle_{0_1^+}$ due to the $A^{1/3}$ increase of the nuclear radius. (d) Two-neutron separation energies, S_{2n}, in MeV. Data taken from AME2016 [23]. Adapted from [5].

the replacement γ-unstable → triaxial and the inclusion of more than two configurations in the MCSM. The spherical state in ^{100}Zr is identified in the MCSM as 0_4^+, in contrast to 0_2^+ in the current calculation and the data. Both calculations show a large jump in $B(E2; 2_1^+ \to 0_1^+)$, between ^{98}Zr and ^{100}Zr, typical of a 1st order QPT. This is in contrast with mean-field based calculations [11–13], which due to their character smooth out the phase transitional behavior, and show no such jump at the critical point of the QPT (see Fig. 2 of [10]). The observed peak in $B(E2; 2_1^+ \to 0_1^+)$ for ^{104}Zr, is reproduced here but not by the MCSM.

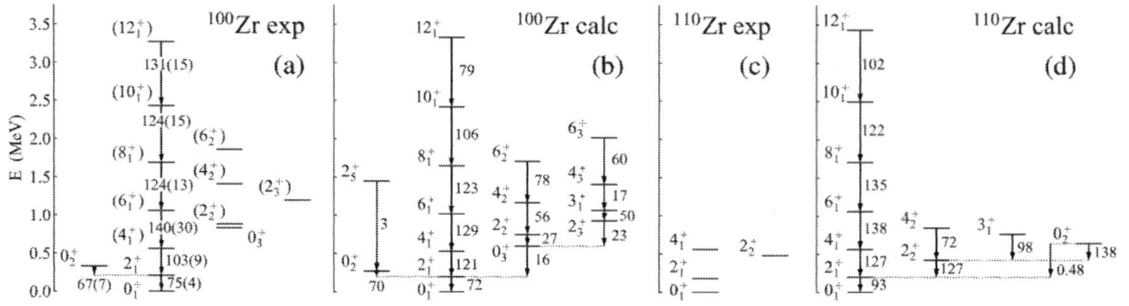

Figure 4. Experimental and calculated energy levels in MeV and $E2$ rates in W.u. for ^{100}Zr and ^{110}Zr. Adapted from [5].

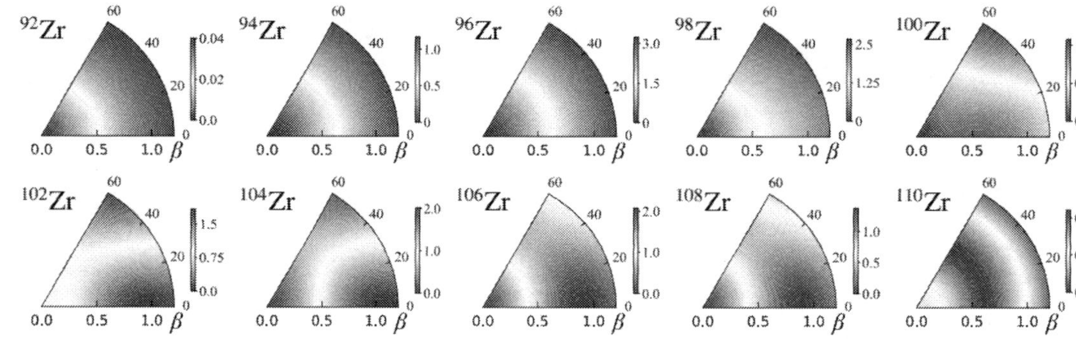

Figure 5. Contour plots in the (β, γ) plane of the classical potential surface for the $^{92-110}$Zr isotopes. Adapted from [5].

The algebraic approach allows both a quantum and a classical analysis of QPTs. Classical potential surfaces are obtained by the method of matrix-coherent-states [3]. As seen in Fig. 5, the calculated surfaces confirm the quantum results, as they show a transition from spherical ($^{92-98}$Zr), to a flat-bottomed potential at ^{100}Zr, to axially deformed ($^{102-104}$Zr), and finally to γ-unstable ($^{106-110}$Zr).

This work was supported in part by U.S. DOE under Grant No. DE-FG02-91ER-40608 and by the US-Israel Binational Science Foundation Grant No. 2016032.

References

[1] P. Cejnar, J. Jolie and R. F. Casten, Rev. Mod. Phys. **82**, 2155 (2010).
[2] K. Heyde and J. L. Wood, Rev. Mod. Phys. **83**, 1467 (2011).
[3] A. Frank, P. Van Isacker and F. Iachello, Phys. Rev. C **73**, 061302(R) (2006).
[4] J. E. García-Ramos and K. Heyde, Phys. Rev. C **89**, 014306 (2014); Phys. Rev. C **92**, 034309 (2015).
[5] N. Gavrielov, A. Leviatan and F. Iachello, Phys. Rev. C **99**, 064324 (2019).
[6] C. Kremer *et al.*, Phys. Rev. Lett. **117**, 172503 (2016).
[7] S. Ansari *et al.*, Phys. Rev. C **96**, 054323 (2017).
[8] N. Paul *et al.*, Phys. Rev. Lett. **118**, 032501 (2017).
[9] W. Witt *et al.*, Phys. Rev. C **98**, 041302(R) (2018).
[10] P. Singh *et al.*, Phys. Rev. Lett. **121**, 192501 (2018).
[11] J.-P. Delaroche *et al.*, Phys. Rev. C **81**, 014303 (2010).
[12] H. Mei *et al.*, Phys. Rev. C **85**, 034321 (2012).
[13] K. Nomura *et al.*, Phys. Rev. C **94**, 044314 (2016).
[14] T. Togashi, Y. Tsunoda, T. Otsuka and N. Shimizu, Phys. Rev. Lett. **117**, 172502 (2016).
[15] F. Iachello and A. Arima, *The Interacting Boson Model*, (Cambridge Univ. Press, Cambridge, 1987).
[16] P. D. Duval and B. R. Barrett, Phys. Lett. B **100**, 223 (1981); Nucl. Phys. A **376**, 213 (1982).
[17] M. Sambataro and G. Molnár, Nucl. Phys. A **376**, 201 (1982).
[18] P. D. Duval *et al.*, Phys. Lett. B **124**, 297 (1983).
[19] P. Federman and S. Pittel, Phys. Rev. C **20**, 820 (1979).
[20] K Heyde *et al.*, Phys. Lett. B **155**, 303 (1985).
[21] Evaluated Nuclear Structure Data File (ENSDF), https://www.nndc.bnl.gov/ensdf/.
[22] I. Angeli and K. P. Marinova, At. Data Nucl. Data Tables **99**, 69 (2013).
[23] M. Wang *et al.*, Chinese Phys. C **41**, 030003 (2017).
[24] S. Zerguine *et al.*, Phys. Rev. C **85**, 034331 (2012).
[25] J. Barea and F. Iachello, Phys. Rev. C **79**, 044301 (2009).
[26] F. Iachello, Phys. Rev. Lett. **87**, 052502 (2001).

Preliminary results of lifetime measurements in neutron-rich ^{53}Ti

Alina Goldkuhle[1]*, *Christoph* Fransen[1], *Alfred* Dewald[1], *Navin* Alahari[2], *Marcel* Beckers[1], *Benedikt* Birkenbach[1], *Andrey* Blazhev[1], *Thomas* Braunroth[1], *Emmanuel* Clément[2], *Gilles* De France[2], *Jérémie* Dudouet[3,4], *Jürgen* Eberth[1], *Herbert* Hess[1], *Bertrand* Jacquot[2], *Yung-Hee* Kim[5], *Antoine* Lemasson[2], *Silvia Monica* Lenzi[6,7], *Hongjie* Li[2], *Joa* Ljungvall[3], *Julia* Litzinger[1], *Caterine* Michelagnoli[2,6,7], *Claus* Müller-Gatermann[1], *Daniel Ricardo* Napoli[8], *Bondili Sreenivasa* Nara Singh[9], *Rosa Maria* Perez-Vidal[10], *Damian* Ralet[3,11,12], *Peter* Reiter[1], *Maurycy* Rejmund[2], *Andreas* Vogt[1], *Nigel* Warr[1], *Karl Oskar* Zell[1], *Magda* Zielińska[13], and the AGATA-Collaboration

[1]Institut für Kernphysik, Universität zu Köln, 50937 Köln, Germany
[2]GANIL, CEA/DRF-CNRS/IN2P3, BP 55027, 14076 Caen Cedex 5, France
[3]CSNSM, Université Paris-Sud, CNRS/IN2P3, Université Paris-Saclay, F-91405 Orsay, France
[4]Université de Lyon, Université de Lyon 1, CNRS/IN2P3, IPN-Lyon, F-69622 Villeurbanne, France
[5]Institut Laue-Langevin, BP 156, 38042 Grenoble Cedex 9, France
[6]INFN Sezione di Padova, I-35141 Padova, Italy
[7]Dipartimento di Fisica e Astronomia dell'Università di Padova, I-35131 Padova, Italy
[8]Laboratori Nazionali di Legnaro, INFN, I-35020 Legnaro, Italy
[9]Nuclear Physics Group, Schuster Laboratory, University of Manchester, Manchester, UK
[10]GSI, Helmholtzzentrum für Schwerionenforschung GmbH, 64291 Darmstadt, Germany
[11]Institut für Kernphysik, Technische Universität Darmstadt, 64289 Darmstadt, Germany
[12]Instituto de Físcia Corpuscular, CSIC-Universidad de Valencia, E-46071 Valencia, Spain
[13]Irfu, CEA, Université Paris-Saclay, F-91191 Gif-sur-Yvette, France

Abstract. To study the nuclear structure of neutron-rich titanium isotopes, a lifetime measurement was performed at the Grand Accélérateur National d'Ions Lourds (GANIL) facility in Caen, France. The nuclei were produced in a multinucleon-transfer reaction by using a 6.76 MeV/u ^{238}U beam. The Advanced Gamma Tracking Array (AGATA) was employed for the γ-ray detection and target-like recoils were identified event-by-event by the large-acceptance variable mode spectrometer (VAMOS++). Preliminary level lifetimes of the $(5/2^-)$ to $13/2^-$ states of the yrast band in the neutron-rich nucleus ^{53}Ti were measured for the first time employing the recoil distance Doppler-shift (RDDS) method and the compact plunger for deep inelastic reactions. The differential decay curve method (DDCM) was used to obtain the lifetimes from the RDDS data.

1 Introduction

The nuclear structure in the region of neutron-rich Ti isotopes is of enhanced interest due to the observed features hinting at subshell closures at $N = 32, 34$ [1]. For example, in neutron-rich $^{50-56}$Ti isotopes the staggering of the excitation energies for the 2_1^+ state points to emergence of subshells. In case of ^{53}Ti, the excited states were reported by prompt γ-ray spectroscopy following β decay [2]. The knowledge was extended by studies performed at the Gammasphere spectrometer [3] to identify the ^{53}Ti yrast cascade for the first time and to determine the excitation energy of states with spins up to $I^\pi = 21/2^-$. In addition, the experimental yrast structure was compared with shell-model calculations where the experimental data are particularly well described by calculations using the GXPF1 and GXPF1A interactions [3]. For the level energies this is directly related to the fact that the $\nu 1f_{5/2}$ and $\nu 1p_{1/2}$ orbitals are well separated from the $\nu 1p_{3/2}$ orbital, a fact that also gives rise to the subshell closure at $N = 32$. In contrast, calculations with the FPD6 Hamiltonian do not predict gaps at $N = 32, 34$. In order to verify the previous findings on shell closures, lifetimes of excited states are of great importance. From these, transition probabilities can be determined which represent important quantities to test the shell structure. In this work preliminary data of lifetimes for the $(5/2^-)$ to $13/2^-$ states of the yrast band of ^{53}Ti are presented for the first time.

*e-mail: agoldkuhle@ikp.uni-koeln.de

2 Experimental Details

Excited states in neutron-rich titanium isotopes were populated as products of a multinucleon-transfer reaction induced by a ^{238}U beam at an energy of $E(^{238}\text{U}) = 1608.9$ MeV (= 6.76 MeV/u) at the Grand Accélérateur National of d'Ions Lourds (GANIL) in Caen, France. The experimental setup consisted of the compact plunger for lifetime measurements of excited states produced in deep inelastic reactions [4] in combination with the Advanced Gamma Tracking Array (AGATA) [5] γ-ray spectrometer consisting of 29 36-fold segmented capsules in 10 cryostats. A 1.5 mg/cm^2 enriched ^{50}Ti plunger target was used, on which a natural copper fronting with a thickness of 0.4 mg/cm^2 for heat conduction was evaporated. A ^{nat}Mg degrader foil with a thickness of 3.2 mg/cm^2 was placed downstream from the target. Both the target and degrader foils were placed at an angle of 45° with respect to the incoming beam which is close to the grazing angle of the multinucleon transfer reaction. Since the expected lifetimes in the neutron-rich titanium isotopes are a few picoseconds, the recoil distance Doppler-shift (RDDS) method [6] was employed. The target-like reaction products were detected and identified in the large-acceptance variable mode spectrometer VAMOS++ [7–9]. These products exit the target foil with an average velocity v_T resulting $\beta_T = v_T/c = 12.7(2)\,\%$. The Mg foil degrades the energy of the ions to 3.68 MeV/u resulting in an average velocity of $\beta_D = 11.7(2)\,\%$ that was measured directly by VAMOS++. β_T was deduced from the experimental Doppler-shift between the two components of the transitions:

$$E_{\mathrm{sh}}(\Delta\beta, \theta) = E_{\mathrm{us}} \cdot \frac{\sqrt{1 - \Delta\beta^2}}{1 - \Delta\beta \cdot \cos(\theta)}, \quad (1)$$

where E_{sh} is the energy of the fast component, i.e. the emission before the degrader, and E_{us} is the intrinsic γ-ray energy of the degraded component, i.e. the emission after the degrader, which depends on the velocity difference $\Delta\beta = \beta_T - \beta_D$ as well as the polar angle θ. During the experiment beam induced structural changes of the target prevented a determination of the distances between target and degrader, which are essential for a lifetime analysis. This problem could be solved by determining effective distances using precisely known lifetimes of ^{46}Ti [10]. Detailed information about the distance determination will be given in Ref. [11]. A total of six target-to-degrader distances, ranging from 130 µm and 1000 µm, were set during the experiment, each with a measurement time of about 24 hours.

3 Data analysis and results

The reaction products were completely identified on an event-by-event basis using VAMOS++. The matrix of the energy loss obtained in the sum of the first three rows of the ionization chamber versus the total energy measured in the sum of all rows was used for the Z identification. The mass A and mass over charge (A/Q) were obtained from the time-of-flight, the total energy and the magnetic rigidity ($B\rho$). Further details will be given in Ref. [11]. After the particle identification Doppler-corrected γ-ray singles spectra could be generated for each nucleus by means of corresponding Z and A gates. The differential decay curve method (DDCM) [12] was employed for the lifetime analysis using the γ-ray singles spectra. Figure 1 shows the γ-ray statistics for ^{53}Ti in a sum spectrum in which the spectra of all six distances were added up. It was possible to identify nine transitions of ^{53}Ti (cf. Fig. 1) that are clearly visible after Doppler correction for the degraded component. These correspond to the decays of the first six

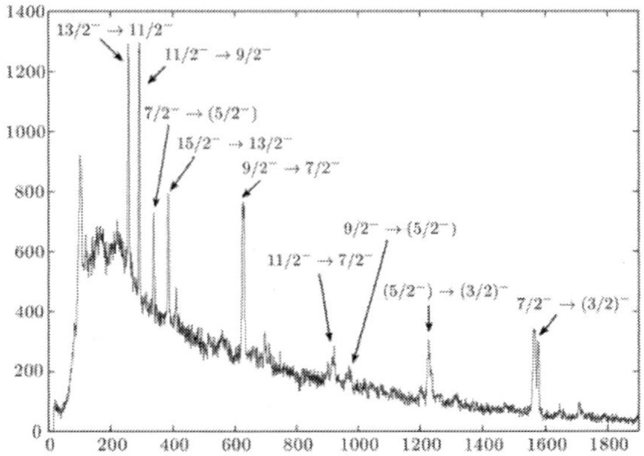

Figure 1. Doppler-corrected γ-ray spectrum in coincidence with ions identified as ^{53}Ti, summed over all six distances. Nine γ-ray decays are observed.

excited states up to the $15/2^-$ state of the yrast band in ^{53}Ti. The level scheme in Fig. 2 contains only the transitions and levels observed in this work.

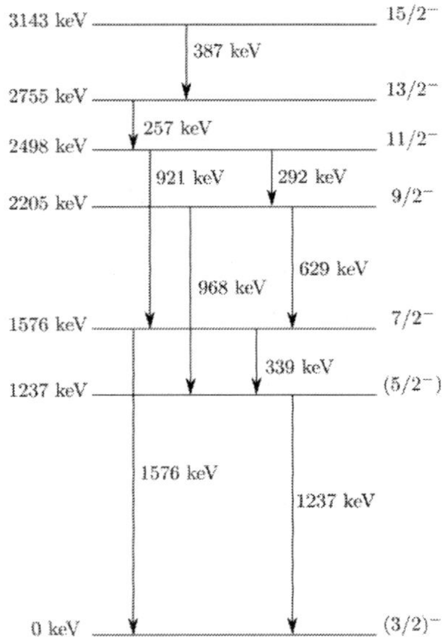

Figure 2. Partial level scheme of ^{53}Ti relevant for the present experiment.

For the evaluation of the $7/2^-$ state the decay into the $(3/2)^-$ state was used (1576 keV), on the one hand due to the clear separation of the fast and degraded components, respectively, and on the other hand because it is by far the strongest decay branch. Furthermore, the γ-ray energy of the transition $7/2^- \rightarrow (5/2^-)$ is significantly lower with 339 keV and the Doppler shift is therefore smaller, which leads to a worse separation of the two components. In addition, the latter decay has a relative intensity of only 33 [13] normalized to the strongest decay with a relative intensity of 100. Only the decay into the $7/2^-$ state

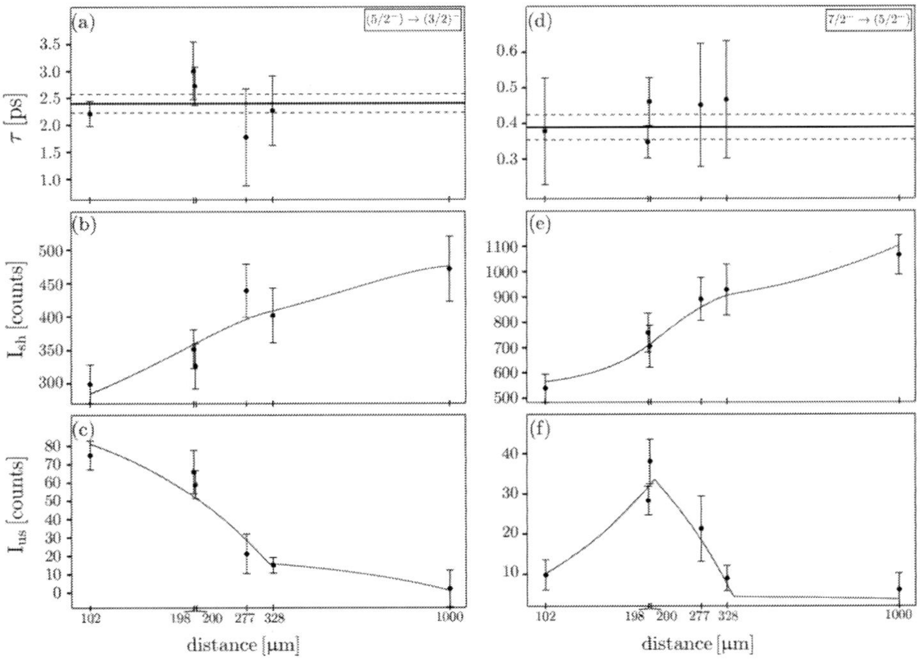

Figure 3. Plot of lifetime analysis for the $(5/2^-)$ and $7/2^-$ states in ^{53}Ti. In the upper panels (a), (d) the black solid lines represent the weighted mean value of the lifetimes; dashed lines mark the statistical uncertainty. The intensities of the fast and degraded components are shown in (b),(e) and (c),(f), respectively. For the intensity determination of the degraded component (c),(f) a feeding correction was considered. Note the logarithmic distance scale.

(629 keV) could be used to determine the lifetime of the $9/2^-$ state, since this decay was much stronger with a relative intensity of 100 than the decay into the $(5/2^-)$ state (968 keV, 10.5). However, a smaller Doppler-shift is present due to the smaller transition energy. Although two transitions depopulate the $11/2^-$ state, the transition to the $9/2^-$ state with only 292 keV is not suitable for the analysis, so the transition to the $7/2^-$ state (921 keV) was used even tough it only has a relative intensity of 17 as compared to the strongest decay branch with a relative intensity of 100. Due to the insufficient separation of the $15/2^- \rightarrow 13/2^-$ transition, it was not possible to evaluate the lifetime of the $15/2^-$ state.

The peak positions and widths of the degraded components were fixed with calibrations of the strongest decay of ^{50,52,53}Ti. The lifetime is extracted from intensities of the depopulating transition after subtraction of the observed feeding for each distance x according to:

$$\tau(x) = \frac{I^d_{us}(x) - \alpha \cdot I^p_{us}(x)}{\frac{d}{dx} I^d_{sh}(x)} \frac{1}{v_T} \qquad (2)$$

s where I^d_{us} and I^d_{sh} are the efficiency-corrected and normalized intensities of the degraded and fast components of the depopulating (d) and populating (p) state, α includes the branching ratios and $v_T = \beta_T \cdot c$ is the mean velocity of the recoils before the degrader. According to DDCM, lifetime values are derived at each target-to-degrader distance and a constant line of lifetime values as a function of the distance is expected. Deviations from this constant line are indicators of systematic effects that can be detected using this method. Preliminary lifetime fits for the $(5/2^-) \rightarrow (3/2)^-$ and $7/2^- \rightarrow (3/2)^-$ transitions performed with the program NAPATAU [14] are shown in Fig. 3. Similar fits could also be generated for the higher lying states $9/2^-$, $11/2^-$ and $13/2^-$.

The middle panels (b),(e) show the intensities of the fast components of the depopulating transition and the lower panels (c),(f) those of the degraded components taking into account a feeding correction for observed feeders of the respective level of interest. As a result of these two, the deduced lifetime is shown in the upper panels (a),(d). The data are still under analysis, thus experimental preliminary results of measured lifetime values for the $(5/2^-)$ to $13/2^-$ states in ^{53}Ti are summarized in Table 1.

Table 1. Results of the preliminary lifetime values of the first five yrast states in ^{53}Ti from this work.

I^π	E_{Level} (keV)	E_γ (keV)	τ (ps)
$(5/2^-)$	1237.1	1237.1	2.4 ± 0.5
$7/2^+$	1576.3	1576.3	0.4 ± 0.4
$9/2^-$	2205.8	629.6	1.1 ± 0.5
$11/2^-$	2498.1	921.8	3.5 ± 0.7
$13/2^-$	2755.9	257.8	2.4 ± 0.6

Since the slowing down of the recoils in the degrader does not happen instantaneously, but takes a finite time, slowing down effects occur here (Doppler Shift-Attenuation (DSA)). These lead to the fact that the degraded peak has no longer a Gaussian shape, but a tail in the direction of the flight component, which should be assigned to the slow component, because the degrader is already reached. This can be problematic for the analysis of RDDS measurements with insufficient separation of the components, since these decays in the spectrum cannot be distinguished from those in which a recoil decays in flight between target and degrader. This effect depends on the lifetime and increases strongly with decreasing lifetime. In general, a correction for lifetimes smaller than 2 ps is considered. This applies especially to the lifetimes of the $7/2^-$ and $9/2^-$ states for which such a correction must be made in a next step and is planned by means Monte-Carlo-Simulations using the G4LifeTimeG toolkit [15].

4 Summary

Preliminary lifetime values of five excited states of the yrast band in ^{53}Ti were determined for the first time. The next step is to finalize the results by considering DSA effects for very short lifetimes. The results can be used to test shell-model predictions for different interactions. In addition to the existing calculations in Ref. [3], further interactions should be used to evaluate their predictability. With the new experimental transition strengths the description of the nuclear structure in ^{53}Ti may be improved. Further, the predicted evolution of shell closures at $N = 32, 34$ can be tested.

Acknowledgements

The authors would like to thank the staff at GANIL involved in this work for their support. Special thanks go to the operators for the excellent focusing of the beam. This work was supported by the German Research Foundation "Bundesministerium für Bildung und Forschung" (BMBF) under contract number 05P18PKFN9.

References

[1] D.C. Dinca, R.V.F. Janssens, A. Gade, D. Bazin, R. Broda, B.A. Brown, C.M. Campbell, M.P. Carpenter, P. Chowdhury, J.M. Cook et al., Phys. Rev. C **71**, 041302 (2005)

[2] L.A. Parks, C.N. Davids, R.C. Pardo, Phys. Rev. C **15**, 730 (1977)

[3] B. Fornal, S. Zhu, R.V.F. Janssens, M. Honma, R. Broda, B.A. Brown, M.P. Carpenter, S.J. Freeman, N. Hammond, F.G. Kondev et al., Phys. Rev. C **72**, 044315 (2005)

[4] A. Dewald, O. Möller, P. Petkov, Progress in Particle and Nuclear Physics **67**, 786 (2012)

[5] S. Akkoyun, A. Algora, B. Alikhani, F. Ameil, G. de Angelis, L. Arnold, A. Astier, A. Ataç, Y. Aubert, C. Aufranc et al., Nuclear Instruments and Methods in Physics Research Section A: Accelerators, Spectrometers, Detectors and Associated Equipment **668**, 26 (2012)

[6] A. Dewald, O. Möller, P. Petkov, Progress in Particle and Nuclear Physics **67**, 786 (2012)

[7] S. Pullanhiotan, M. Rejmund, A. Navin, W. Mittig, S. Bhattacharyya, Nuclear Instruments and Methods in Physics Research Section A: Accelerators, Spectrometers, Detectors and Associated Equipment **593**, 343 (2008)

[8] M. Rejmund, B. Lecornu, A. Navin, C. Schmitt, S. Damoy, O. Delaune, J. Enguerrand, G. Fremont, P. Gangnant, L. Gaudefroy et al., Nuclear Instruments and Methods in Physics Research Section A: Accelerators, Spectrometers, Detectors and Associated Equipment **646**, 184 (2011)

[9] M. Vandebrouck, A. Lemasson, M. Rejmund, G. Frémont, J. Pancin, A. Navin, C. Michelagnoli, J. Goupil, C. Spitaels, B. Jacquot, Nuclear Instruments and Methods in Physics Research Section A: Accelerators, Spectrometers, Detectors and Associated Equipment **812**, 112 (2016)

[10] A. Goldkuhle, C. Fransen, A. Dewald, K. Arnswald, M. Bast, M. Beckers, A. Blazhev, T. Braunroth, G. Hackenberg, G. Häfner et al., The European Physical Journal A **55**, 53 (2019)

[11] A. Goldkuhle et al. (2019), submitted for Phys. Rev. C

[12] A. Dewald, S. Harissopulos, P. von Brentano, Zeitschrift für Physik A Atomic Nuclei **334**, 163 (1989)

[13] H. Junde, Nuclear Data Sheets **86**, 2689 (2000)

[14] B. Saha, *Computer code* NAPATAU, Institute for Nuclear Physics, University of Cologne, unpublished

[15] S. Agostinelli, J. Allison, K. Amako, J. Apostolakis, H. Araujo, P. Arce, M. Asai, D. Axen, S. Banerjee, G. Barrand et al., Nuclear Instruments and Methods in Physics Research Section A: Accelerators, Spectrometers, Detectors and Associated Equipment **506**, 250 (2003)

EPJ Web of Conferences **223**, 01023 (2019) https://doi.org/10.1051/epjconf/201922301023
NSD2019

Discovery of collective states in ^{208}Pb by complete spectroscopy

Andreas Heusler [1,*]

[1]Gustav-Kirchhoff-Strasse 7/1, Heidelberg, Germany

Abstract. Most neutron and proton bound states in ^{208}Pb are described in the shell model as one-particle one-hole configurations. Besides precise data obtained with the Q3D magnetic spectrograph of the MLL (Garching, Germany) an important reaction is the inelastic proton scattering via isobaric analog resonances in ^{209}Bi. It yields amplitudes of neutron one-particle one-hole configurations with relative signs in each state of ^{208}Pb. The orthogonality, normality, and sum rule relations allow to investigate the completeness of the transformation matrices of one-particle one-hole configurations describing the states in ^{208}Pb with spins from 0^- to 14^- and 0^+ to 12^+. By this method amplitudes of unobservable one-particle one-hole configurations can be determined. The comparison of spin, parity, and dominant particle-hole components thus derived in up to 30 states of a certain spin to shell model calculations allows to identify states described as collective excitations of the entire nucleus.

1 Introduction

^{208}Pb is the heaviest nucleus where more than 500 states are rather well known [1]. The excitation energies of most neutron bound states ($S(n) = 7368\,\text{keV}$) and part of the proton bound states ($S(n) = 8004\,\text{keV}$) are known with an uncertainty of less than 1 keV. Most states are one-particle one-hole (1p1h) configurations but three dozen states consist definitely of other configurations [2–5].

Knowledge of spin, parity, and dominant configuration up to a certain excitation energy defines complete spectroscopy. It is rather good for states at $E_x < 7.2\,\text{MeV}$. States at $6.2 < E_x < 7.2\,\text{MeV}$ [6, 7] are being analyzed from reconstructed data presented in [8]. States at $E_x < 6.2\,\text{MeV}$ are completely known with spin, parity, and dominant configuration [3]. Complete spectroscopy allows to find collective states in the heavy nucleus ^{208}Pb not described as 1p1h configurations.

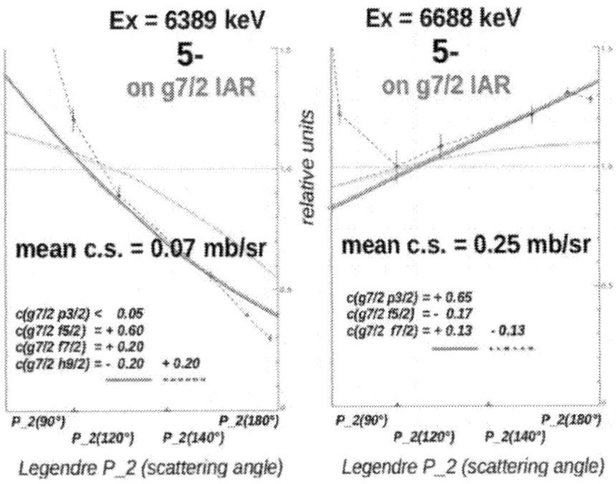

Figure 1. Angular distributions for ^{208}Pb(p,p') measured at the MPI-Heidelberg in 1969 using semiconductor detectors (12 keV resolution) at scattering angles $90° < \Theta < 170°$ on the $g_{7/2}$ IAR in ^{209}Bi [8]. Calculations of the differential cross section with known s.p. widths [7] yield up to four amplitudes for each state. The sensitivity is extremely high. The shown amplitudes $-1 < c(LJ^{+1}lj^{-1}) < +1$ for the configurations (left frame) $g_{7/2}p_{3/2}$, $g_{7/2}f_{5/2}$, $g_{7/2}f_{7/2}$, $g_{7/2}h_{9/2}$ and (right frame) $g_{7/2}p_{3/2}$, $g_{7/2}f_{5/2}$, $g_{7/2}f_{7/2}$ well reproduce the angular distribution at backward angles $140° \lesssim \Theta \leq 170°$. Reversing the sign of one amplitude as indicated does not yield a fit. Note that instead of the scattering angle Θ the Legendre polynomial $P_2(\Theta)$ is used as abscissa.

2 Experimental observations

Spectroscopy of ^{208}Pb started immediately after the discovery of radioactivity. Rather soon the high excitation energy of the lowest state in ^{208}Pb was recognized. The energy of two and a half MeV is the almost the highest across the whole nuclear chart. It is now known with a precision of 10 eV [1]. In 1954 the spin of the lowest state was determined as 3^- [9] which is again rather unique across the whole nuclear chart.

An important event was the discovery of isobaric analog resonances (IAR) in heavy nuclei [10]. From 1965 until 1969 important data were obtained for ^{208}Pb from the study of the proton decay of IARs in ^{209}Bi [1, 6–8, 11]. Yet only with the finishing of the Q3D magnetic spectrograph (Fig. 2) at the Maier-Leibnitz Laboratorium (MLL) at Garching (Germany) higher excited states ($4.0 < E_x < 8.0\,\text{MeV}$) could be studied.

*e-mail: A.Heusler@mpi-hd.mpg.de

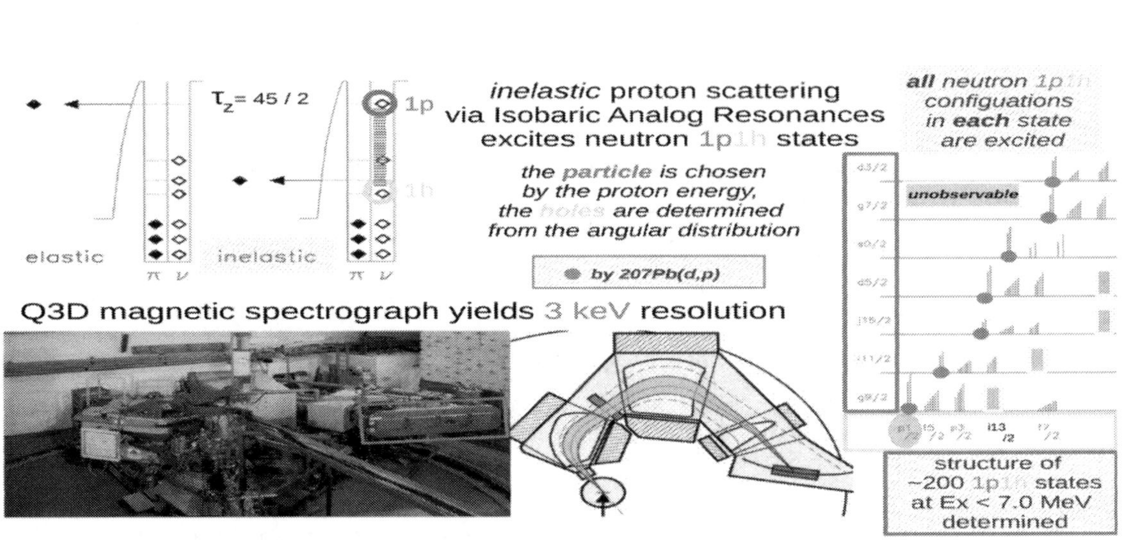

Figure 2. (upper and right frames) Schematic presentation of the reaction ^{208}Pb(p,p') via IARs in ^{209}Bi. (lower left frames) The quadrupole 3-dipole (Q3D) magnetic spectrograph at the MLL (Garching, Germany).

The resolution of 3 keV is barely sufficient to resolve the states in ^{208}Pb with a mean distance of 10 keV. The resolution is limited to 3 keV because the M-electrons in lead have such binding energies. L-electrons and K-electrons with binding energies of 15 and 88 keV produce annoying satellites to each peak.

Using the Q3D magnetic spectrograph from 2003 until now most negative parity states at $E_x < 7.2$ MeV and nearly all states at $E_x < 6.2$ MeV became known. Five states were described as neutron and proton vibrations [2]. Ten states were recognized as tetrahedral rotations and vibrations [4]. Two dozen states were described by the coupling of 1p1h configurations to the 3^- yrast state [5].

The question how to discern other collective states in ^{208}Pb from 1p1h states can be only solved by complete spectroscopy. It means that up to a certain limit in excitation energy all states must be found and for each of them spin, partity and dominant structure must be determined.

3 Complete spectroscopy

Complete spectroscopy is approached by studying several different reactions, especially ^{208}Pb(p,p'), ^{208}Pb(d,d'), ^{208}Pb(α,α'), ^{207}Pb(d,p) and ^{209}Bi($d,^3$He). However, the most important reaction is the inelastic proton scattering via IARs in ^{209}Bi. It allows to determine amplitudes with relative signs of all neutron one-particle one-hole configurations in each state.

In ^{208}Pb there are 44 excess neutrons. By adding one proton an IAR in ^{209}Bi is created with a typical energy of 15-20 MeV and a width of about 300 keV. In the subsequent proton decay all neutron particle-hole configurations in each state of ^{208}Pb are excited. The decay protons act in a coherent manner to create interference patterns in the angular distributions (Fig. 1).

The knowledge of some amplitudes with relative signs allows to investigate the orthogonality relations among the states. Together with the normality and sum rule relations amplitudes of unobservable configurations can be determined. They are unobservable either because the cross section is too low or because there is no target as is the case for proton particle-hole configurations built with a particle in a higher orbit.

Fig. 1 shows angular distributions of ^{208}Pb(p,p') for two states with spin 5^- taken on the $g_{7/2}$ IAR in ^{209}Bi. It demonstrates the high sensitivity of the inelastic proton scattering via IARs. Similar angular distributions for about hundred states were measured from which spin, parity, and several amplitudes were derived [1, 3, 4, 6, 7].

Fig. 3 shows transformation matrices from 1p1h configurations to states with spins 2^-, 3^-, 4^-, and 5^-. Transformation matrices for states with spins 0^-, 1^-, 6^-, 7^-, 8^-, and 1^+, 5^+, 6^+, 7^+, 8^+, 9^+, 10^+ look similar. Most large amplitudes are centered close to the diagonal, only sometimes the order of two configurations is exchanged. For states with spins 12^-, 13^-, 14^-, and 0^+, 2^+, 3^+, 4^+, 11^+, and 12^+ no amplitudes could be measured. No state with spin 9^-, 10^-, or 11^- is known [1].

Fig. 4 compares level schemes for the 3^- and 5^- states at $E_x < 7.2$ MeV to shell model calculations. One calculation (SDI) uses the surface delta interaction [12], another one (M3Y) the Michigan 3-Yukawa interaction [13]. The SDI calculates the multipole splitting and assumes no further residual interaction, the M3Y employs the mixing among 1p1h, 2p2h, and 3p3h configurations.

 https://doi.org/10.1051/epjconf/201922301023

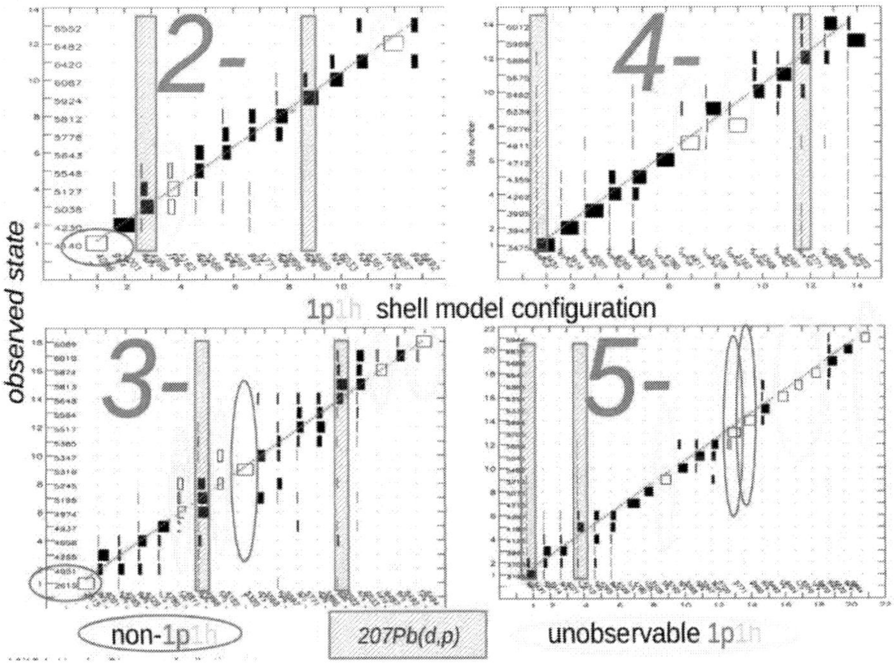

Figure 3. Transformation matrices from 1p1h configurations to states with spins 2^-, 3^-, 4^-, and 5^-. The normalized amplitudes are shown in the range $0 < 1$; black for measured values, white for values (unobservable configurations marked yellow) determined by using the orthogonality, normality, and sum rule relations. Amplitudes obtained in a redundant manner by ^{207}Pb(d,p) are marked by a frame. Amplitudes of states not described by 1p1h configurations are marked – the 2^- yrast, 3^- yrast states, another 3^- and two 5^- states at $E_x \approx$ 5.5 MeV.

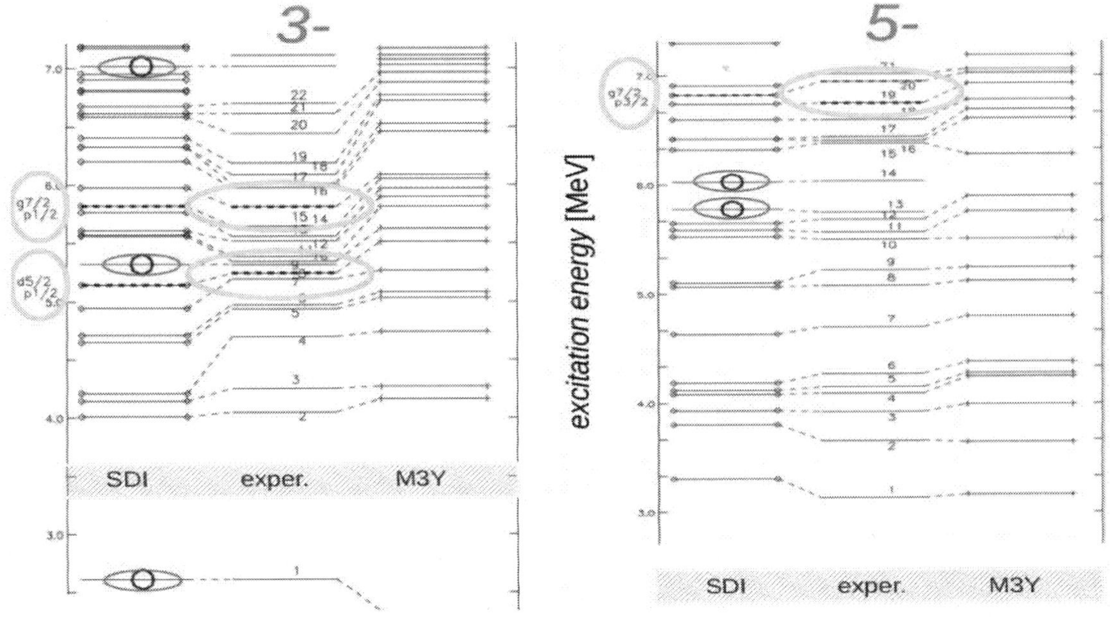

Figure 4. Comparison of level schemes for the 3^- and 5^- states at $E_x < 7.2$ MeV to shell model calculations with SDI [12] and M3Y [13]. Three 3^- and two 5^- states are not easily explained with M3Y by 1p1h configurations and with SDI not at all.

Both by SDI and M3Y describe the excitation energies for most states with spins 3^- and 5^- with a mean deviation of 30 keV, also for most states with spins 0^-, 1^-, 2^-, 4^-, 6^-, 7^-, 8^-, 12^-, 13^-, 14^-, and 3^+, 5^+, 7^+, 8^+, 9^+, 10^+, 11^+, and the 12^+ yrast state. Fig. 5 shows level schemes for the 6^+ states at $E_x < 6.2$ MeV and for the 12^+ states at $E_x < 10$ MeV (see also Figs. 5 and 9 in [3]). Fig. 3 in [5] shows the level scheme for all positive parity states at $E_x < 6.5$ MeV, see also [14]. The 6^+ yrast and 12^+ yrare states evidently appear in addition to 1p1h configurations.

 https://doi.org/10.1051/epjconf/201922301023

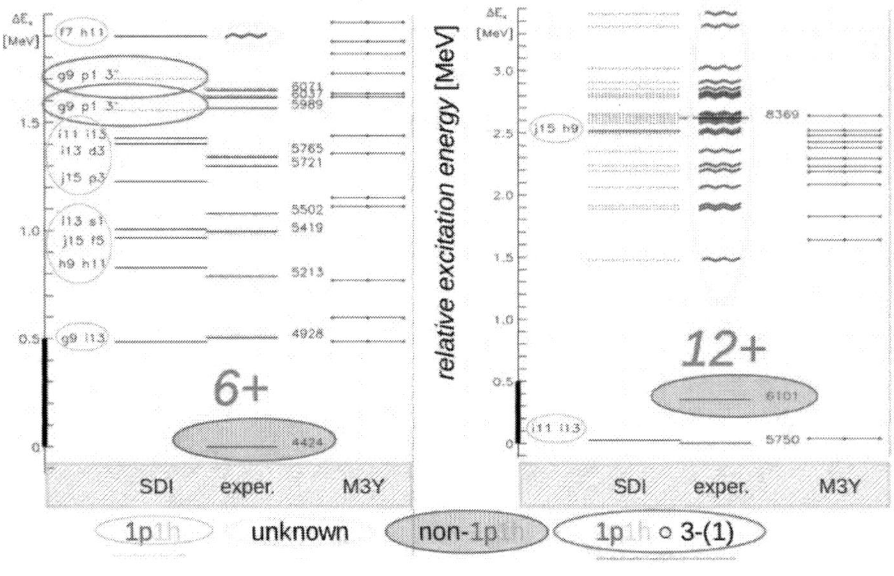

Figure 5. Comparison of level schemes for the 6^+ and 12^+ states at $E_x < 6.2$ MeV to shell model calculations with SDI [3] and M3Y [13]. (The energies are shown relative to the yrast state.) The 6^+ yrast and 12^+ yrare states are not explained by any existing model. Two 6^+ states at $E_x \approx 6.0$ MeV (1.6 MeV above the yrast state) are described by the coupling of $g_{9/2}p_{1/2}$ to the 3^- state. 2p2h configurations with spin 12^+ are predicted more than 1.5 MeV above the yrast state (wavy lines).

4 Summary

In the heavy nucleus ^{208}Pb several different classes of excitations are observed,
– overwhelmingly 1p1h configurations,
– two dozen states with 1p1h configurations coupled to the 3^- yrast state,
– five neutron and proton pairing vibrations,
– ten members of tetrahedral rotating and vibrating bands,
– the 6^+ yrast and 12^+ yrare states unexplained,
– one more 3^- state and two more 5^- states in the region at $5.2 < E_x < 6.05$ MeV not explained by any existing model.

Still a few positive parity states below $E_x = 6.2$ MeV are not yet identified and above $E_x = 6.2$ MeV most positive parity states are unknown. A mysterious question is about a 2^+ state near the ground state predicted as the coupling of both intruders $j_{15/2}$ and $i_{13/2}$ to the 3^- yrast state [5].

5 Outlook

Urgently needed is the theoretical description of dodecahedral and icosahedral configurations similar to the tetrahedral configurations by some algebraic cluster model.

On the experimental side the neutron capture on ^{207}Pb and the investigation of the subsequent γ-cascade is needed. Modern equipment exists but studying γ-transitions among the more than 500 states needs a great effort.

References

[1] M. J. Martin, Nucl. Data Sheets **108**, 1583-1808 (2007)

[2] R. V. Jolos, A. Heusler, P. von Brentano, Phys. Rev. C **92**, 011302 (2015)

[3] A. Heusler, R. V. Jolos, T. Faestermann, R. Hertenberger, H.-F. Wirth, P. von Brentano. Phys. Rev. C **93**, 054321 (2016)

[4] A. Heusler, Eur. Phys. J. A **50**, 92 (2017)

[5] A. Heusler, R. V. Jolos, P. von Brentano, Phys. Rev. C **99**, 034323 (2019)

[6] W. R. Wharton, P. von Brentano, W. K. Dawson, P. Richard, Phys. Rev. **176**, 1424 (1968)

[7] A.Heusler, H.-J. Glöckner, E.Grosse, C.F.Moore, J.Solf, P.von Brentano, Eur. Phys. J. A **50**, 92 (2014)

[8] H.-J. Glöckner, *Diploma thesis* (Universität Heidelberg, Germany, 1972)

[9] L. G. Elliott, R. L. Graham, J. Walker, J. L. Wolfson, Phys. Rev. **93**, 356 (1954)

[10] J. D. Fox, C. F. Moore, D. Robson, Phys. Rev. Lett. **12**, 198 (1964)

[11] C. F. Moore, L. J. Parish, P. von Brentano, S. A. A. Zaidi, J. L. Wolfson, Phys. Lett. **22**, 616 (1966)

[12] A. Heusler, R. V. Jolos, P. von Brentano, Yad. Fiz. **76**, 7 (2013); Phys. Atomic Nuclei, **76**, 807 (2013)]

[13] R. Broda *et al.*, Phys. Rev. C **95**, 064308 (2017)

[14] A. Heusler, *https://agenda.infn.it/event/16348/contributions/85701/attachments/63639/76670/Heusler_NSD2019.pdf* (Nuclear Structure and Dynamics - NSD2019, May 13-17, 2019)

EPJ Web of Conferences **223**, 01024 (2019) https://doi.org/10.1051/epjconf/201922301024
NSD2019

Shape transition in the neutron-rich Y nuclei and its evolution across the isotopic chain

Łukasz W. Iskra[1,2,*], *Bogdan* Fornal[2], *Silvia* Leoni[1,3], *Caterina* Michelagnoli[4], *Simone* Bottoni[1,3], *Natalia* Cieplicka-Oryńczak[2], *Michael* Jentschel[4], *Felix* Kandzia[4], *Yung Hee* Kim[4], *Ulli* Köster[4], and *Carlotta* Porzio[3]

[1]INFN sezione di Milano via Celoria 16, 20133, Milano, Italy
[2]Institute of Nuclear Physics, PAN, 31-342 Krakow, Poland
[3]Dipartimento di Fisica, Università degli Studi di Milano, I-20133 Milano,
[4]ILL, 71 Avenue des Martyrs, 38042 Grenoble CEDEX 9, France

Abstract. The evolution of the deformation across the Y isotopic chain, in the vicinity of N = 60 boundary, has been studied using gamma spectroscopy methods. The nuclei of interest have been produced by neutron induced fission of ^{235}U and ^{241}Pu targets, during two experimental campaigns named EXILL and FIPPS at the Institute Laue-Langevin in Grenoble. The emitted gamma rays have been collected by HPGe and $LaBr_3$ detectors providing the identification of the high-spin levels in the ^{94}Y and ^{96}Y isotopes up to 6 MeV excitation energy, as well as information about the half-lives of the in-band states in the ^{98}Y nucleus. The obtained results shed new light on the onset of deformation in the neutron-rich Y isotopes and, in general, in the A ≈ 100 region.

1 Introduction

The region around Z = 40 and N = 60 is especially important to test the mechanism of creation of deformed structures, as well as to study its evolution across specific isotopic chains. This is due to the sudden onset of deformation at the ground states in the nuclei with N ≥ 60, which is reported as the most dramatic shape change in the whole nuclear chart [1]. For lighter systems, the ground levels have shown spherical character, while rotational bands are observed at higher excitation energy. In the case of Zr (Z = 40) and Sr (Z = 38) isotopes, the coexistence of such structures has already been observed at N = 58,59 [2]. However, the parameters of deformation β, extracted from the lifetime of in-band states, in the ^{96,97}Sr and ^{98,99}Zr nuclei [2] are significantly smaller than for ground state rotational bands at N ≥ 60. In the case of the Y isotopic chain, the experimental information about the deformation size of the rotational band was far less complete. In a very recent study of T. W. Hagen *et al.*, [3], the half-lives of the several deformed states in ^{99}Y and ^{101}Y have been measured and a deformation parameter β ~ 0.4 has been obtained. A similar value was measured by the authors of this collaboration for the rotational band in ^{98}Y (N = 59) isotope, which was briefly reported in Ref. [4]. The observation of strongly deformed rotational band in the ^{98}Y nucleus raises the question if similar bands are present also in lighter isotopes, further away from the N = 60 boundary. The searching of such structures in the ^{94}Y and ^{96}Y isotopes will be presented in the current paper. Details about the lifetime measurements of the in-band states in the ^{98}Y nucleus will also be provided.

2 Experimental details

The data for the present study have been obtained from the two experimental campaigns conducted at the Institute Laue-Langevin in Grenoble. During the first one, named EXILL [5], the reactor neutron beam, with a flux of $10^8/(s \times cm^2)$, induced fission reactions (n,f) of the ^{235}U and ^{241}Pu targets. Gamma rays emitted from the fission product were detected by an array which consisted of eight HPGe clover detectors and 16 $LaBr_3$(Ce) FATIMA scintillators [6], mounted about 15 cm from the target position. The events were registered by using a fully digital acquisition system in a triggerless mode, with 100 MHz clock. For the gamma spectroscopy measurement, the $LaBr_3$ detectors where replaced by additional 6 large HPGe coaxial detectors and 2 clovers. The success of the EXILL campaign triggered the development of a permanent gamma-spectroscopy experimental setup at ILL. The new array, called FIPPS, (FIssion Product Prompt γ-ray Spectrometer) [7] consists of up to 16 HPGe clovers. For the new fission runs, ^{233}U and ^{235}U active targets have been used [8]. Scintillation light in the target cell provides precise information about the time when the

* Corresponding author: lukasz.iskra@mi.infn.it

(n,f) reaction occurs, giving a clean fission tag. Data from the FIPPS campaign were partially sorted and included into the results presented in the current work.

3 Data analysis and discussion

Part of the analysis concerning lifetime measurements based on the fast-timing technique has been described, for the EXILL-FATIMA campaign, by J.-M. Régis et al. [9] and employed in order to study lifetimes of selected states in the ^{90}Kr and ^{133}Sb isotopes (see Refs [10] and [11], respectively). The method assumes sorting the data into triple coincidence events, where one γ-ray is collected by HPGe detector (with good energy resolution) and two gamma-rays by $LaBr_3$(Ce) scintillators (with good time resolution). The half-lives can be extracted from the difference between the centroids of the time distributions which correspond to the gamma rays feeding and decaying from the state of interest, if visible in the $LaBr_3$ detectors [9]. The obtained results have to be corrected by using the prompt response difference (PRD) curve, as given in Ref. [9].

Further analysis has been performed for states in the ^{98}Y isotope [4]. The levels of interest are members of the rotational band built on the 4$^-$, 6.95-μs isomer [12, 13]. In order to measure the lifetime of the 5$^-$, 596-keV level, the 101- and 130-keV transitions have been selected for the TAC "start-stop" combination. Whereas, the 186-keV line has been used as gate in the HPGe array, providing a clean spectrum. During the analysis, various gating configurations have been used to avoid random coincidence event or contribution from other isotopes, when transitions with similar energy might occur. Figure 1 presents result of the analysis where the $T_{1/2}$ = 175(25) ps value for the 5$^-$ state has been established. The analysis was continued upwards, along the rotational band, giving values of 51(10)- and 45(15)-ps half-lives for the 6$^-$ and 7$^-$ states, respectively. For the 8$^-$, 1070-keV state the 15-ps upper limit has been obtained. The half-lives which have been obtained during the analysis were used to calculate the reduced transition probabilities B(M1) and B(E2). Branching ratios of in-band transitions have been adopted from [14] and corrected for electron conversion coefficients. Therefore, the final deformation parameter for this rotational band was calculated as an average from the partial results, leading to β = 0.40(14).

The presence of a strongly deformed band in the Y nucleus with N = 59 encourages the search for rotational bands in lighter Y isotopes. Such an investigation has been performed for ^{96}Y and ^{94}Y, using gamma spectroscopy methods. The data has been sorted into triple γ coincidence cubes with various time windows. During the analysis also prompt-delayed gamma coincidences were used, where the time between two gammas corresponds to the half-lives of the isomers occurring in the studied isotopes. A detailed description of the coincidences techniques can be found in Refs. [14, 15].

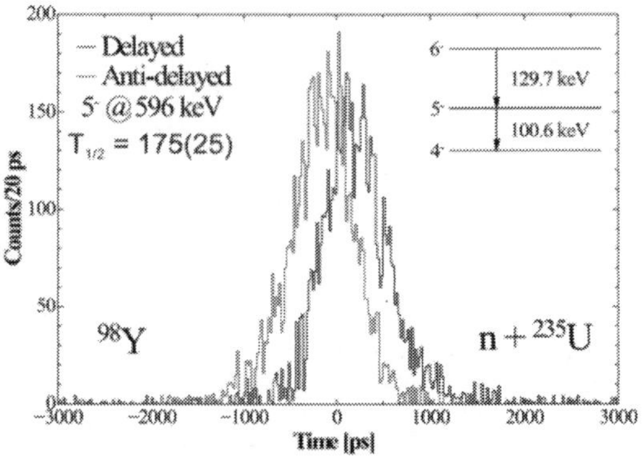

Fig. 1. TAC spectra of the corresponding distributions D_{stop} (black) and D_{start} (red) are shown

Laser spectroscopy studies of Ref. [17] provide information about the spherical character of the 0$^-$ ground state and the 8$^+$ isomer in the ^{96}Y isotope. The latter observation has been confirmed in gamma spectroscopy study, by identifying the structure located above the isomer [18]. On the other hand, we propose that the new 201-ns state at 1655-keV excitation energy is deformed [4], indicating the occurrence of shape coexistence in the ^{96}Y nucleus. This competition between shapes is even more pronounced in the heavier even-A Y isotopes, as it is shown in the pictorial view of Fig. 2. In the case of the ^{100}Y nucleus, a rotational band has already been observed on the 1$^+$, 11-keV state and there is no sign of a spherical structure at higher spin values. Whereas, for the ^{98}Y, N = 59 isotope, with 2 neutrons less, the presence of coexisting spherical and deformed shapes is very clear. In this case, the 0$^-$ ground state is spherical, as also confirmed by the laser spectroscopy study [17]. The deformation occurs at the higher excitation energy, around 0.5 MeV where a significantly developed rotational band above the 6.95-μs isomer is found. In Ref. [17], also the second highly deformed isomer has been measured, later established at 466-keV excitation energy by Urban *et al.* [13]. It is found that the overall feeding of the states with spherical and deformed configurations is more or less equal in the ^{98}Y isotope. This intensity balance is significantly unsettled in the ^{96}Y nucleus, with 2 neutrons less. Here the spherical configuration plays a dominant role, while the deformed structure is weak and moved even higher in excitation energy. The isomeric bandhead also has a shorter half-life in comparison to those in the ^{98}Y isotope. On the contrary, the spherical 8$^+$, 9.6-s isomer in ^{96}Y is strongly fed, while the population of the shorter 10$^-$, 0.8-μs counterpart in the heavier neighbor is relatively weak.

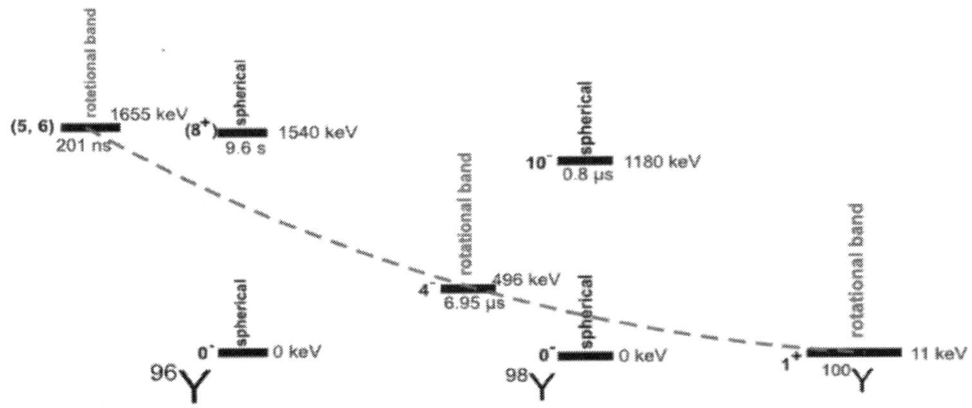

Fig. 2. Pictorial drawing of the evolution of deformed structures (in red) across the Y isotopic chain

Inspired by the new results in the ^{96}Y isotope, our collaboration made an effort to develop the level scheme of the ^{94}Y nucleus, located 5 neutrons away from the N = 60 boundary. Using data from EXILL (and crosschecked with partially sorted data from FIPPS), the level scheme was extended up to 5 MeV excitation energy [19], however, no evidence of deformed structure in this isotope was found. This is not surprising, taking into account the significant decrease in population of the rotational bands, in comparison with the spherical pattern when further away from N = 60. In addition, the fact that deformed structures move to higher excitation energy and isomeric bandheads disappear make even more difficult the identification of rotational bands in these nuclei.

4 Conclusion

The hypothesis that large deformation might occur also in yttrium isotopes with $N < 60$ is in line with recent results from the EXILL campaign, where, for the first time, a rotational band in N = 57 nucleus has been suggested [4]. According to this interpretation, deformed structures start to be present in lighter nuclei at high excitation energy and become dominant and yrast for isotopes with $N \geq 60$. This scenario calls for further studies of the high-spin structures in the neutron-rich Y isotopes. The issue is being investigated taking advantage of new high quality and high statistic data from the FIPPS campaign, where a fission trigger based on an active scintillating target is used.

Acknowledgments

The authors acknowledge the technical services of the ILL for supporting the EXILL and FIPPS campaigns. This work was supported by the Italian Istituto Nazionale di Fisica Nucleare, and by the Polish National Science Centre under Contract No. 2014/14/M/ST2/00738 and 2013/08/M/ST2/00257.

References

1. K. Heyde and J. L. Wood., Rev. Mod. Phys., **83** 1467 (2011).
2. W. Urban *et al.*, Nucl. Phys A **689**, 605 (2001).
3. T. W. Hagen *et al.*, Phys. Rev. C **95**, 034302 (2017).
4. Ł. W. Iskra *et al.*, Europhys. Lett. 117, 12001 (2017).
5. M. Jentschel *et al.*, J. Instrum. 12, P11003 (2017).
6. FATIMA Technical Design Report and O. Roberts et al., Nuc. Instr. Meth. 748, 91 (2014).
7. C. Michelagnoli *et al.*, EPJ Web Conf. 193, 04009 (2018).
8. F. Kandzia *et al.*, in preparation
9. J.-M. Régis *et al.*, Nucl. Instr. Meth. Phys. Res. A **763**, 210 (2014).
10. J.-M. Régis *et al.*, Phys. Rev. C 90, 067301 (2014).
11. G. Bocchi *et al.*, Phys. Lett. *B*, **760** 273 (2016).
12. S. Brant, G. Lhersonneau, and K. Sistemich, Phys. Rev. C 69, 034327 (2004).
13. W. Urban *et al.*, Phys. Rev. C 96, 044333 (2017).
14. Balraj Singh and Zhiqiang Hu, NDS 98, 335 (2003).
15. Ł. W. Iskra *et al.*, Phys. Rev. C89, 044324 (2014).
16. Ł. W. Iskra *et al.*, Phys. Lett. B 788, 396 (2019).
17. B. Cheal et al., Phys. Rev. Lett. B645, 133 (2007).
18. Ł. W. Iskra *et al.*, Acta Phys. Pol. B 48, 581 (2017).
19. Ł. W. Iskra *et al.*, Phys. Scr. 92, 104001 (2017).

Neutrinoless Double-Beta Decay and Realistic Shell Model

Nunzio Itaco[1,2,*], *Luigi* Coraggio[2], and *Riccardo* Mancino[1,2]

[1]Dipartimento di Matematica e Fisica, Università degli Studi della Campania "Luigi Vanvitelli", viale Abramo Lincoln 5 - I-81100 Caserta, Italy
[2]Istituto Nazionale di Fisica Nucleare,
Complesso Universitario di Monte S. Angelo, Via Cintia - I-80126 Napoli, Italy

Abstract. We report on the calculation of the neutrinoless double-β decay nuclear matrix element for ^{76}Ge within the framework of the realistic shell model. The effective shell-model Hamiltonian and the two-body transition operator describing the decay are derived by way of many-body perturbation theory. Particular attention is focused on the role played by the so-called Pauli blocking effect in the derivation of the effective operator.

1 Introduction

Neutrinoless double-β ($0\nu\beta\beta$) decay is an exotic nuclear process predicted by extensions of the Standard Model of particle physics. Observation of such a process would prove the non-conservation of lepton number, and show that neutrinos have a Majorana mass component (see Refs. [1, 2] and references therein).

In the framework of light Majorana neutrino exchange, the half life of the $0\nu\beta\beta$ decay is inversely proportional to the square of the effective Majorana neutrino mass $\langle m_\nu \rangle \equiv |\sum_i U_{ei}^2 m_{\nu_i}|$. Explicitly, it can be written as

$$\left[T_{1/2}^{0\nu}\right]^{-1} = G^{0\nu} \left|M^{0\nu}\right|^2 \langle m_\nu \rangle^2 \ , \tag{1}$$

where $G^{0\nu}$ is the so-called phase-space factor (or kinematic factor), and $M^{0\nu}$ is the nuclear matrix element (NME) directly related to the wave functions of the parent and grand-daughter nuclei. At present, the phase-space factors involved in the double β decays of nuclei of experimental interest are calculated with great accuracy [3, 4], therefore it is crucial to have precise values of the NME, both to improve the reliability of the $0\nu\beta\beta$ lifetime predictions, and to extract neutrino properties from the experimental results.

Several nuclear structure models have been employed to provide NME values as precise as possible, the most largely employed being, at present, the Interacting Boson Model (IBM) [5–7], the Quasiparticle Random-Phase Approximation (QRPA) [8–11], Energy Density Functional methods [12], and the Shell Model (SM) [13–19].

Each model can be more suitable than another for a certain class of nuclei, and, when comparing the calculated NMEs, it can be seen that, at present, the results obtained with different approaches can differ by a factor of two or three (see for instance the review [20]).

It is worth noting that, all the above models are based on the use of a truncated Hilbert space to overcome the computational complexity, but only within the realistic shell model an effective $0\nu\beta\beta$ operator, which takes into account the degrees of freedom which do not appear explicitly in the calculated wavefunctions, may be introduced. This approach has been pioneered by Kuo and coworkers [21, 22], and more recently pursued by Holt and Engel [23]. Actually, in the SM approach it is possible to consider the correlations involving single-particle orbitals outside the model space deriving an effective operator. This is the so-called realistic shell-model approach where both the effective Hamiltonian and the transition operators are derived from a realistic free nucleon-nucleon (NN) potential V_{NN} by way of the many-body theory [24, 25].

In two recent papers [26, 27], we have adopted the above approach to calculate properties related to the GT and $2\nu\beta\beta$ decays of nuclei $0\nu\beta\beta$-candidates, with mass ranging from $A = 48$ to $A = 136$. Our results and their comparison with the experimental data have shown that many-body perturbation theory (MBPT) allows to derive effective SM Hamiltonians and transition operators that reproduce quantitatively the observed spectroscopic and decay properties, without resorting to an empirical quenching of the axial coupling constant g_A.

On these grounds, we report in this paper on a preliminary calculation of the NME involved in the $0\nu\beta\beta$ decay of ^{76}Ge, that is currently investigated by GERDA experiment [28] at the Laboratori Nazionali del Gran Sasso (LNGS) of INFN, and by CDEX-1 experiment [29] at China Jinping Underground Laboratory.

Our theoretical framework is the many-body perturbation theory [25, 26, 30, 31]; we start from a realistic nuclear potential, and derive an effective SM Hamiltonian and a $0\nu\beta\beta$ decay operator that are employed to calculate the wavefunctions of the ground states of ^{76}Ge and ^{76}Se and the $0\nu\beta\beta$ NME. We focus our attention on the role of the so-called "blocking effect", that takes into account the Pauli exclusion principle in systems with more than two

*e-mail: nunzio.itaco@unicampania.it

valence nucleons [32, 33] in the derivation of the two-body effective $0\nu\beta\beta$ operator.

The paper is organized as follows. In the next section we give a brief description of how we derive the shell-model Hamiltonian and the two-body effective $0\nu\beta\beta$ decay operator. In Section 3, we report the results that we obtain for the ^{76}Ge NME, analyzing in detail the contribution arising from the core-polarization blocking diagrams. A brief summary is reported in the last section.

2 Outline of calculations

The starting point of our calculation is provided by the high-precision CD-Bonn NN potential [34], that is smoothed integrating out its repulsive high-momentum components by way of the so-called $V_{\text{low-k}}$ approach [35, 36]. In this way we get a softer NN potential defined up to a cutoff Λ, that preserves the physics of the original CD-Bonn interaction. The value of Λ is chosen, as in many of our recent papers [26, 27, 37–39] equal to 2.6 fm^{-1}, this value being a trade off between the need of minimizing the role of the missing three-nucleon force (3NF) [38] and that of ensuring the perturbative behavior of the potential. The Coulomb potential is explicitly taken into account in the proton-proton channel.

The harmonic oscillator (HO) potential U is introduced as an auxiliary one-body potential in order to break up the Hamiltonian for a system of A nucleons as the sum of a one-body term H_0, which describes the independent motion of the nucleons, and a residual interaction H_1:

$$\begin{aligned} H &= \sum_{i=1}^{A} \frac{p_i^2}{2m} + \sum_{i<j=1}^{A} V_{\text{low-k}}^{ij} = T + V_{\text{low-k}} = \\ &= (T+U) + (V_{\text{low-k}} - U) = H_0 + H_1 \ . \end{aligned} \tag{2}$$

It is now possible to define a truncated model space in terms of the eigenvectors of H_0. To study the $0\nu\beta\beta$ decay of ^{76}Ge we employ a model space spanned by the four $1p_{3/2}, 1p_{1/2}, 0f_{5/2}, 0g_{9/2}$ proton and neutron orbitals outside the doubly-closed ^{56}Ni core, and we derive an effective shell-model Hamiltonian H_{eff}, that takes into account the degrees of freedom that are not explicitly included in the shell-model framework.

We derive H_{eff} by resorting to the many-body perturbation theory, an approach that has been developed by Kuo and coworkers through the 1970s [24, 40]. More precisely, we use the well-known $\hat{Q}$ box-plus-folded-diagram method [41], where the $\hat{Q}$ box is defined as a function of the unperturbed energy ϵ of the valence particles:

$$\hat{Q}(\epsilon) = PH_1P + PH_1Q\frac{1}{\epsilon - QHQ}QH_1P \ , \tag{3}$$

where the operator P projects onto the model space and $Q = \mathbf{1} - P$. In the present calculations the $\hat{Q}$ box is expanded as a collection of one- and two-body irreducible valence-linked Goldstone diagrams up to third order in the perturbative expansion[31, 42].

Within this framework, it can be shown that the effective Hamiltonian H_{eff} can be written in terms of the $\hat{Q}$ box derivatives [43]

$$\hat{Q}_m = \frac{1}{m!}\frac{d^m\hat{Q}(\epsilon)}{d\epsilon^m}\bigg|_{\epsilon=\epsilon_0} \ , \tag{4}$$

$$H_{\text{eff}} = \sum_{i=0}^{\infty} F_i \ , \tag{5}$$

where

$$\begin{aligned} F_0 &= \hat{Q}(\epsilon_0) \\ F_1 &= \hat{Q}_1(\epsilon_0)\hat{Q}(\epsilon_0) \\ F_2 &= \left[\hat{Q}_2(\epsilon_0)\hat{Q}(\epsilon_0) + \hat{Q}_1(\epsilon_0)\hat{Q}_1(\epsilon_0)\right]\hat{Q}(\epsilon_0) \\ &\dots \end{aligned} \tag{6}$$

ϵ_0 being the model-space eigenvalue of the unperturbed Hamiltonian H_0.

The H_{eff} for one-valence-nucleon systems provides the single-particle (SP) energies for our SM calculations, while the two-body matrix elements (TBMEs) are obtained from H_{eff} derived for the nuclei with two valence nucleons, by subtracting the theoretical SP energies.

The so obtained SP energies and TBMEs can be found in [27], where it is also reported a detailed discussion of the perturbative properties of H_{eff}.

The NME involved in $0\nu\beta\beta$ decay, $M^{0\nu}$, may be expressed as

$$M^{0\nu} = M^{0\nu}_{GT} - \left(\frac{g_V}{g_A}\right)^2 M^{0\nu}_F \tag{7}$$

where the matrix elements are defined as follows:

$$M^{0\nu}_\alpha = \sum_{m,n}\left\langle 0^+_f \mid \tau^-_m\tau^-_n O^\alpha_{mn} \mid 0^+_i\right\rangle \ , \tag{8}$$

with $\alpha = (GT,\ F)$, having neglected the tensor term, whose effect has been estimated to be of the order of few percent [15]. The operators O^α_{mn} have the following expression:

$$\begin{aligned} O^{GT}_{mn} &= \vec{\sigma}_m \cdot \vec{\sigma}_n H_{GT}(r) \ , \\ O^F_{mn} &= H_F(r) \ , \end{aligned}$$

and the neutrino potentials H_α are defined using the closure approximation

$$H_\alpha(r) = \frac{2R}{\pi}\int_0^\infty f_\alpha(qr)\frac{h_\alpha(q^2)}{q + \langle E\rangle}G_\alpha(q^2)qdq, \tag{9}$$

where $f_{F,GT}(qr) = j_0(qr)$, $\langle E\rangle$ is the average energy of the virtual intermediate states used in the closure approximation, while the explicit expression of the form factors $h_\alpha(q^2)$ can be found for instance in Ref. [44].

As mentioned before, we want to derive the effective $0\nu\beta\beta$ decay operator tailored for the chosen model space. To this end, we resort to the formalism presented by Suzuki and Okamoto in Ref. [25]. In this approach, a non-Hermitian effective operator Θ_{eff} can be expressed as

$$\begin{aligned} \Theta_{\text{eff}} &= (P + \hat{Q}_1 + \hat{Q}_1\hat{Q}_1 + \hat{Q}_2\hat{Q} + \hat{Q}\hat{Q}_2 + \cdots) \\ &\quad \times(\chi_0 + \chi_1 + \chi_2 + \cdots) \ . \end{aligned} \tag{10}$$

https://doi.org/10.1051/epjconf/201922301025

where the operators χ_n are defined as:

$$\chi_0 = (\hat{\Theta}_0 + h.c.) + \hat{\Theta}_{00} \ , \quad (11)$$

$$\chi_1 = (\hat{\Theta}_1 \hat{Q} + h.c.) + (\hat{\Theta}_{01} \hat{Q} + h.c.) \ , \quad (12)$$

$$\chi_2 = (\hat{\Theta}_1 \hat{Q}_1 \hat{Q} + h.c.) + (\hat{\Theta}_2 \hat{Q} \hat{Q} + h.c.) + (\hat{\Theta}_{02} \hat{Q} \hat{Q} + h.c.) + \hat{Q} \hat{\Theta}_{11} \hat{Q} \ , \quad (13)$$

$$\ldots$$

where $\hat{\Theta}_m$, $\hat{\Theta}_{mn}$ have the following expressions:

$$\hat{\Theta}_m = \frac{1}{m!} \frac{d^m \hat{\Theta}(\epsilon)}{d\epsilon^m} \bigg|_{\epsilon=\epsilon_0} \ , \quad (14)$$

$$\hat{\Theta}_{mn} = \frac{1}{m!n!} \frac{d^m}{d\epsilon_1^m} \frac{d^n}{d\epsilon_2^n} \hat{\Theta}(\epsilon_1; \epsilon_2) \bigg|_{\epsilon_1=\epsilon_0, \epsilon_2=\epsilon_0} \ , \quad (15)$$

with

$$\hat{\Theta}(\epsilon) = P\Theta P + P\Theta Q \frac{1}{\epsilon - QHQ} Q H_1 P \ , \quad (16)$$

$$\hat{\Theta}(\epsilon_1; \epsilon_2) = P H_1 Q \frac{1}{\epsilon_1 - QHQ} Q \Theta Q \frac{1}{\epsilon_2 - QHQ} Q H_1 P \ , \quad (17)$$

Θ being the bare transition operator.

It is worth noting that using Eqs. 5, 6, Eq. 10 may be then rewritten as

$$\Theta_{\rm eff} = H_{\rm eff} \hat{Q}^{-1} (\chi_0 + \chi_1 + \chi_2 + \cdots) \ , \quad (18)$$

enlightening the connection existing between the effective Hamiltonian and the effective operators.

In present calculation for the effective two-body $0\nu\beta\beta$ decay operator, we arrest the χ_n series to the χ_2 term. Since χ_3 depends on the first, second, and third derivatives of $\hat{\Theta}_0$ and $\hat{\Theta}_{00}$, and on the first and second derivatives of the $\hat{Q}$ box (see Eq. 13), our estimation of these quantities leads to evaluate χ_3 being at least one order of magnitude smaller than χ_2. The calculation is performed starting from a perturbative expansion of $\hat{\Theta}_0$ and $\hat{\Theta}_{00}$, and in Fig. 1 we report all the two-body Θ_0 diagrams up to the first order in $V_{\rm low-k}$, the bare operator Θ being represented by a dashed line.

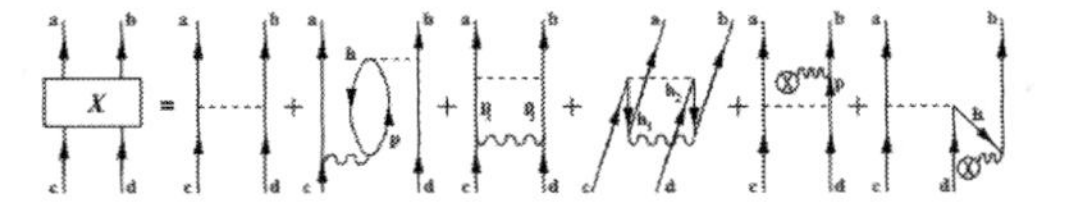

Figure 1. Two-body second-order diagrams included in the perturbative expansion of $\hat{\Theta}$. The dashed lines indicate the bare operator Θ, the wavy lines the two-body potential $V_{\rm low-k}$.

The circle with a cross inside represents a first-order ($V_{\rm low-k} - U$)-insertion, arising from the U term in the interaction Hamiltonian H_1 (see for example Ref. [31] for details).

The diagrams in Fig. 1 refer to the derivation of the effective operator for a system with two valence-nucleons. When dealing with nuclei with a larger number of valence nucleons, many-body diagrams come into play, accounting for the interaction via the two-body force of the many-valence nucleons with core excitations as well as with virtual intermediate nucleons scattered above the model space. The two topologies of second-order connected three-valence-nucleon diagrams for a two-body operator $\hat{\Theta}$ are reported in Fig. 2 (diagrams (a) and (b)). These diagrams correct the Pauli-principle violation introduced by diagram (a_1) and (b_1) when one of the intermediate particle states is equal to m.

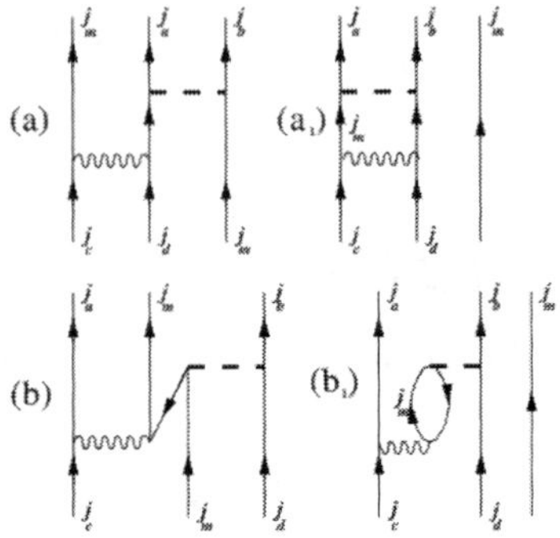

Figure 2. Three-body second-order diagrams that should appear in the perturbative expansion of $\hat{\Theta}$. As in Fig. 1, the dashed line indicates the bare operator Θ, the wavy line the two-body potential $V_{\rm low-k}$. For the sake of simplicity, for each topology only one of the diagrams which correspond to the permutation of the external lines is drawn.

Since there are no shell-model codes, at present, able to handle three-body transition operators, we derive a density-dependent two-body operator from the three-body one by summing and averaging over one incoming and outcoming particle of the connected diagrams (a) and (b) of Fig. 2. In this way we take into account the filling of the model-space orbitals when dealing with more than two valence nucleons.

More precisely, first we calculate exactly the three-body connected diagrams (k) $\langle [(j_a j_b)_J, j_m]_{J'} | (k) | [(j_c j_d)_J, j_n]_{J'} \rangle$, with k equal to a and b, and whose expressions can be found in Ref. [45], then we evaluate the corresponding density-dependent two-body diagram $\langle [(j_a j_b)_J | (\overline{k}) | [(j_c j_d)_J \rangle$, whose explicit expression is

$$\langle [(j_a j_b)_J | (\overline{k}) | [(j_c j_d)_J \rangle = \sum_{m,J'} \rho_m \frac{\hat{J}'^2}{\hat{J}^2} \langle [(j_a j_b)_J, j_m]_{J'} | (k) | [(j_c j_d)_J, j_m]_{J'} \rangle \quad (19)$$

where the summation over m-index runs in the model space, and ρ_m is the unperturbed occupation density of the orbital j_m accordingly to the number of valence nucleons.

The so obtained two-body contribution is finally added to the collection of the diagrams of the perturbative expansion of $\hat{\Theta}$.

3 Results

In this Section we present the results of our SM calculations for the NME of ^{76}Ge. As previously mentioned, all the calculations have been performed employing theoretical SP energies, TBMEs, and effective transition operators, describing ^{76}Ge as a core of ^{56}Ni plus 4 protons and 16 neutrons interacting in a reduced model space spanned by the four $1p_{3/2}, 1p_{1/2}, 0f_{5/2}, 0g_{9/2}$ orbitals.

In Ref. [27] we have compared the calculated low-energy spectrum, the electromagnetic properties, the GT$^-$ strength distributions and the calculated NMEs of the $2\nu\beta\beta$ decay with the available experimental data, showing a quantitative agreement.

In order to see how much the choice of the effective Hamiltonian can affect the value of the NME, in Table 1 we compare our calculated NME using the bare $0\nu\beta\beta$ operator without any renormalization, with the values obtained in two other SM calculations performed using the same model space but with different empirical SM hamiltonians [15, 46].

Table 1. Theoretical $0\nu\beta\beta$ NME calculated in the framework of the SM using different effective Hamiltonians. *I* and *II* are the results obtained by Menendez and coworkers [15] and by Sen'kov and coworkers [46], respectively. The $0\nu\beta\beta$ operator employed is not renormalized.

	Our	I	II
NME	3.40	2.96	3.25

As it can be seen, the theoretical results are in good agreement, the largest discrepancy non exceeding 15%.

In Table 2 we report the calculated $0\nu\beta\beta$ NME obtained using an effective operator derived by second order MBPT and taking into account all the intermediate states up to an excitation energy of 14 $\hbar\omega$, which are enough to provide convergent NME values. To clarify the role of Pauli blocking diagrams we have derived two different operators with and without taking into account diagrams (a) and (b) of Fig. 2.

Table 2. $M^{0\nu}$, $M^{0\nu}_{GT}$ and $M^{0\nu}_{F}$ for ^{76}Ge calculated using the $0\nu\beta\beta$ effective operator. See text for details.

	2nd	2nd + Pauli blocking
$M^{0\nu}_{GT}$	2.39	2.03
$M^{0\nu}_{F}$	-0.64	-0.66
$M^{0\nu}$	2.79	2.44

The inspection of Table 2 shows that including Pauli blocking diagrams provides a further reduction of $M^{0\nu}$ by an amount around 10%, and that this effect is mainly due to the reduction of the Gamow-Teller contribution.

To better understand the action of Pauli blocking diagrams, we report in Table 3 the values of the diagrams reported in Fig. 2 for GT-operator corresponding to the most relevant configurations involved in the ^{76}Ge $0\nu\beta\beta$ decay.

As expected, the density dependent diagrams (a) and (b) reduce the contribution of the Pauli violating diagrams (a_1) and (b_1).

Table 3. Values of the second-order diagrams reported in Fig. 2. The values of the diagrams (a) and (b) are calculated according to the expression in Eq. 19

$(j_a, j_b)_{J=0}$ $(j_c, j_d)_{J=0}$	(a)	(a_1)	(b)	(b_1)
$(f_{5/2}, f_{5/2})$ $(g_{9/2}, g_{9/2})$	0.157	-0.337	-1.096	0.335
$(p_{3/2}, p_{3/2})$ $(g_{9/2}, g_{9/2})$	0.189	-0.263	-0.219	0.058

4 Summary

In this paper we have reported on some preliminary results for the SM calculation of the nuclear matrix element involved in the ^{76}Ge $0\nu\beta\beta$ decay. The effective shell-model Hamiltonian and the two-body transition operator describing the decay are derived by way of many-body perturbation theory at third and second order, respectively. The role of the Pauli blocking three body diagrams that appear in the perturbative expansion of the effective operator when dealing with more than two valence nucleons, is taken into account in an approximate way by introducing two-body density-dependent diagrams.

References

[1] F.T. Avignone, S.R. Elliott, J. Engel, Rev. Mod. Phys. **80**, 481 (2008)
[2] J.D. Vergados, H. Ejiri, F. Šimkovic, Rep. Prog. Phys. **75**, 106301 (2012)
[3] J. Kotila, F. Iachello, Phys. Rev. C **85**, 034316 (2012)
[4] S. Stoica, M. Mirea, Phys. Rev. C **88**, 037303 (2013)
[5] J. Barea, F. Iachello, Phys. Rev. C **79**, 044301 (2009)
[6] J. Barea, J. Kotila, F. Iachello, Phys. Rev. Lett. **109**, 042501 (2012)
[7] J. Barea, J. Kotila, F. Iachello, Phys. Rev. C **87**, 014315 (2013)
[8] F. Šimkovic, A. Faessler, V. Rodin, P. Vogel, J. Engel, Phys. Rev. C **77**, 045503 (2008)
[9] F. Šimkovic, A. Faessler, H. Müther, V. Rodin, M. Stauf, Phys. Rev. C **79**, 055501 (2009)
[10] D.L. Fang, A. Faessler, V. Rodin, F. Šimkovic, Phys. Rev. C **83**, 034320 (2011)
[11] A. Faessler, V. Rodin, F. Simkovic, J. Phys. G **39**, 124006 (2012)
[12] T.R. Rodríguez, G. Martínez-Pinedo, Phys. Rev. Lett. **105**, 252503 (2010)
[13] E. Caurier, F. Nowacki, A. Poves, Eur. Phys. J. A **36**, 195 (2008)
[14] J. Menéndez, A. Poves, E. Caurier, F. Nowacki, Phys. Rev. C **80**, 048501 (2009)
[15] J. Menéndez, A. Poves, E. Caurier, F. Nowacki, Nucl. Phys. A **818**, 139 (2009)
[16] M. Horoi, Phys. Rev. C **87**, 014320 (2013)
[17] M. Horoi, B.A. Brown, Phys. Rev. Lett. **110**, 222502 (2013)
[18] A. Neacsu, M. Horoi, Phys. Rev. C **91**, 024309 (2015)
[19] B.A. Brown, D.L. Fang, M. Horoi, Phys. Rev. C **92**, 041301 (2015)

[20] J. Engel, J. Menéndez, Rep. Prog. Phys. **80**, 046301 (2017)

[21] H.Q. Song, H.F. Wu, T.T.S. Kuo, Phys. Rev. C **40**, 2260 (1989)

[22] A. Staudt, T.T.S. Kuo, H.V. Klapdor-Kleingrothaus, Phys. Rev. C **46**, 871 (1992)

[23] J.D. Holt, J. Menéndez, A. Schwenk, J. Phys. G **40**, 075105 (2013)

[24] T.T.S. Kuo, E. Osnes, *Lecture Notes in Physics, vol. 364* (Springer-Verlag, Berlin, 1990)

[25] K. Suzuki, R. Okamoto, Prog. Theor. Phys. **93**, 905 (1995)

[26] L. Coraggio, L. De Angelis, T. Fukui, A. Gargano, N. Itaco, Phys. Rev. C **95**, 064324 (2017)

[27] L. Coraggio, L. De Angelis, T. Fukui, A. Gargano, N. Itaco, F. Nowacki, Phys. Rev. C **100**, 014316 (2019)

[28] M. Agostini, M. Allardt, E. Andreotti, A.M. Bakalyarov, M. Balata, I. Barabanov, M. Barnabé Heider, N. Barros, L. Baudis, C. Bauer et al. (GERDA Collaboration), Phys. Rev. Lett. **111**, 122503 (2013)

[29] W. Zhao, Q. Yue, K.J. Kang, J.P. Cheng, Y.J. Li, S.T. Lin, Y. Bai, Y. Bi, J.P. Chang, N. Chen et al. (CDEX Collaboration), Phys. Rev. D **88**, 052004 (2013)

[30] T.T.S. Kuo, J. Shurpin, K.C. Tam, E. Osnes, P.J. Ellis, Ann. Phys. (NY) **132**, 237 (1981)

[31] L. Coraggio, A. Covello, A. Gargano, N.Itaco, T.T.S. Kuo, Ann. Phys. **327**, 2125 (2012)

[32] P.J. Ellis, E. Osnes, Rev. Mod. Phys. **49**, 777 (1977)

[33] I. Towner, Physics Reports **155**, 263 (1987)

[34] R. Machleidt, Phys. Rev. C **63**, 024001 (2001)

[35] S. Bogner, T.T.S. Kuo, L. Coraggio, Nucl. Phys. A **684**, 432c (2001)

[36] S. Bogner, T.T.S. Kuo, L. Coraggio, A. Covello, N. Itaco, Phys. Rev. C **65**, 051301(R) (2002)

[37] L. Coraggio, A. Covello, A. Gargano, N. Itaco, T.T.S. Kuo, Phys. Rev. C **91**, 041301 (2015)

[38] L. Coraggio, A. Gargano, N. Itaco, JPS Conference Proceedings **6**, 020046 (2015)

[39] L. Coraggio, A. Gargano, N. Itaco, Phys. Rev. C **93**, 064328 (2016)

[40] T.T.S. Kuo, F. Krmpotić, K. Suzuki, R. Okamoto, Nucl. Phys. A **582**, 205 (1995)

[41] T.T.S. Kuo, S.Y. Lee, K.F. Ratcliff, Nucl. Phys. A **176**, 65 (1971)

[42] L. Coraggio, A. Covello, A. Gargano, N. Itaco, Phys. Rev. C **81**, 064303 (2010)

[43] E.M. Krenciglowa, T.T.S. Kuo, Nucl. Phys. A **235**, 171 (1974)

[44] R.A. Sen'kov, M. Horoi, Phys. Rev. C **88**, 064312 (2013)

[45] A. Polls, H. Müther, A. Faessler, T.T.S. Kuo, E. Osnes, Nucl. Phys. A **401**, 124 (1983)

[46] R.A. Sen'kov, M. Horoi, Phys. Rev. C **90**, 051301 (2014)

Enhanced monopole and dipole transitions in nuclei induced by α cluster structures

Makoto Ito[1,*], *Makoto* Nakao[1], and *Tomoyuki* Okuno[1] *Syuuichiro* Ebata[2]

[1]Department of Pure and Applied Physics, Kansai University, Yamatecho, 3-3-35, Suita 564-8680, Japan
[2]Laboratory for Advanced Nuclear Energy, Institute of Innovative Research, Tokyo Institute of Technology, 2-12-1-N1-16 Ookayama, Meguro-ku 152-8550, Japan

Abstract. Enhancement of the monopole and dipole transitions in the low-lying state is discussed on the basis of the microscopic and macroscopic α cluster models. Theoretical calculation clearly demonstrates that the strength of the monopole and dipole transitions are strongly enhanced by the excitation in the relative motion of the α cluster and the residual nucleus. The transition strength induced by the α excitation appears as the discrete distribution, and its excitation energy is much lower than the excitation energy of the single nucleon excitation, which is expected from the naive mean field picture.

1 Introduction

Basic properties of ground state in nuclei can be described by mean field picture, in which individual nucleons perform single particle motions in a self-consistent mean field [1]. Coherent excitations of such the single particle motion generate various collective excitations of the nuclei [1]. On the contrary, the so-called α cluster structures are known to be realized in the excited states of the nuclei [2]. The α particle is quite stable and inert and hence, it becomes a subunit, which is a building block in constructing the intrinsic structures of the nuclear systems. The typical examples of the α cluster state are ^{8}Be = 2α, ^{12}C = 3α, ^{16}O = α + ^{12}C and ^{20}Ne = α + ^{16}O in the lighter mass region.

The picture of the α cluster structures is extended to much heavier systems beyond $A = 40$, such as ^{44}Ti = α + ^{40}Ca [2, 4, 5], neighboring nuclei of ^{94}Mo = α + ^{90}Zr [4, 6], and ^{212}Po = α + ^{208}Pb [4, 7]. In the studies of the heavy systems, the macroscopic α cluster model, in which the local potential for the system of α – residual nucleus is an initial ingredient, are mainly applied. Amongst, the ^{44}Ti nucleus is the most deeply analyzed not only by the local potential approaches [3, 4] but also by the full microscopic approaches [5].

The α cluster structure mainly appears as the first excited 0^+ state or the yrast 1^- state. The latter state corresponds to the negative parity partner of the parity doublet with the ground 0^+ state. Recently, the isoscalar (or electric) monopole and dipole transitions have been proposed as effective probes to identify the cluster excitation [8, 9]. The monopole transition is strongly increased for the radial excitation in the relative wave function of the α cluster and a resitudal nucleus, while the dipole operator strongly responds to the dipole excitation in asymmetric cluster systems with an α particle plus a heavy core. When the α cluster structures are well developed, the excited 0^+ state and the yrast 1^- state appear at much lower excitation energy than the excitation energy based on an independent nucleon motion in a mean field picture. Therefore, the low-lying enhancement in the monopole and dipole transition is considered to be a clear sign of the excitation of α cluster degrees of freedom.

In this report, we show our analyses for the light nuclei, such as ^{12}Be [10] and ^{18}O [11], which are based on the microscopic and semi-microscrpic α cluster models, and point out that the strong monopole transition is induced by the α cluster excitation. Furthermore, we analyze the monopole and dipole transitions in the heavier systems, ^{44}Ti and $^{104-110}$Te, by applying the macroscopic α cluster model, which is based on the double folding potential of the α particle and residual core nuclei. From the calculation of the macroscopic α cluster model, we will demonstrate that the monopole and dipole transition is prominently enhanced at anomalously low excitation energy.

2 Theoretical framework

In this section, we briefly explain theoretical framework for the calculation of the light and heavy mass systems. The (semi) microscopic cluster model is applied to the light systems, while we use the macroscopic α cluster model in the calculation of the heavy systems.

2.1 Micorscopic and semi-microscopic cluster models for ^{12}Be and ^{18}O

In the structure calculations of the ^{12}Be and ^{18}O nuclei, we apply the microscopic and semi-microscopic α cluster models, respectively, in which the anti-symmetrization effect among the nucleons are taken into account. As for

*e-mail: itomk@kansai-u.ac.jp

^{12}Be, we assume the six-body systems of $\alpha + \alpha + 4N$ and employ the generalized two-center cluster model (GTCM) [10], which is possible to handle the atomic structure with the binary He-clusters and the covalent structure of excess four neutrons around two α cores. On the contrary, the orthogonality condition model (OCM) with the α + ^{14}C cluster configuration is used for the structure calculation of ^{18}O [11]. Both calculations nicely reproduce the level sequence of the 0^+ states. The details of these cluster models are explained in Refs.[10] and [11].

2.2 Macroscopic α cluster model in ^{44}Ti and Te isotopes

In the heavy systems, the explicit treatment of the antisymmetrization among the nucleon becomes complicated. Thus, we employ the macroscopic α cluster model for ^{44}Ti with α + ^{40}Ca and $^{104-110}$Te with α + $^{100-106}$Sn. The basic framework is same as those shown in Refs. [3, 4]. In the macroscopic α cluster model, the nuclear potential of α and heavy cores (^{40}Ca or Sn) is calculated from the double folding (DF) model [12], which is symbolically written as a function of the α – core relative coordinate **R**,

$$U_{\rm DF}(\mathbf{R}) = \iint \rho_\alpha(\mathbf{r}_\alpha)\rho_C(\mathbf{r}_C) \times v_{\rm NN}^{\rm DDM3Y}(s,\rho)d\mathbf{r}_\alpha d\mathbf{r}_C \quad (1)$$

with $s = |\mathbf{r}_C - \mathbf{r}_\alpha - \mathbf{R}|$. Here $\mathbf{r}_\alpha$ ($\mathbf{r}_C$) denotes a coordinate measured from the center of mass in α (core nucleus). $\rho_\alpha(\mathbf{r}_\alpha)$ is the density of α particle, which reproduces the charge form factor, while $\rho_C(\mathbf{r}_C)$ represents the density of the core nucleus, which is calculated by the mean-filed model [13].

In Eq.(1), $v_{\rm NN}^{\rm DDM3Y}$ represents the effective nucleon-nucleon (NN) interaction which acts between a pair of nucleons contained in the α particle and the core nucleus. In the present calculation, we adopt the DDM3Y (density dependent Michigan three range Yukawa) interaction [12]. The DF potential with DDM3Y is well known to reproduce the differential cross section of the α + ^{40}Ca elastic scattering [3]. We calculate the energy levels with the DF potential in Eq. (1). In the calculation of the energy levels, we take into account the pseudo potential to exclude the Pauli's forbidden state and the absorbing boundary condition to handle the unbound states [14]. The Pauli's forbidden states are prepared by assuming the lowest shell model configuration for the core nucleus and considering the condition about the number of the radial nodes in the α – Core relative wave function.

The continuum strength function for the transition from the ground state to the unbound continua, which is induced by the external field, is calculated from the ABC calculation [14]. As for the operator of the external field, we consider the monopole ($\lambda = 0$) and dipole ($\lambda = 1$) operators, which are basically written in the nucleon degree of freedom like [8, 9]

$$\hat{O}_{\lambda=0} = \sum_{i=1}^{A \text{ or } Z} r_i^2 \quad (2)$$

$$\hat{O}_{\lambda=1} = \sum_{i=1}^{A} r_i^3 Y_{1,0}(\hat{\mathbf{r}}_i) \ . \quad (3)$$

In Eq. (2), the summation is taken over a whole nucleons (A) in the case of isoscalar monopole (IS0) transition, while it is restricted to protons (Z) for the electric monopole (E0) transition. In the calculations of semi-microscopic and macroscopic α cluster models, we extract a part of the α–core relative coordinate (**R**) from Eqs. (2) and (3) according to the definitions in Refs. [8, 9].

3 Resuts

3.1 Monopole transition in ^{12}Be and ^{18}O

The results of the 0^+ energies and the isoscalar monopole (IS0) matrix elements for ^{12}Be are summarized in table 1. In this table, the strengths are shown in units of the single particle strength for $0p \to 1p$. All of the excited states, which appear below the excitation energy of $E_x \leq$ 15 MeV, have the strength comparable to or larger than the single particle excitation. In particular, the excitation to 0_3^+ around $E_x \sim 10$ MeV is the most enhanced of all the excited 0^+ state because this state corresponds to the direct excitation in the α – ^{8}He relative motion from the ground ^{12}Be nucleus [10]. The theoretical strength of $0_1^+ \to 0_3^+$ nicely reproduce the observed strength of 2.0 [15].

Table 1. 0^+ energies and strength of IS0 matrix elements in ^{12}Be. The theoretical energies are shown in units of MeV, while the energies in parentheses mean the experimental data. The IS0 strength of theory (Th.) and experment (Exp.) are shown in unit of the single particle strength [10]. Experimental data of IS0 is taken from Ref. [15]

State	Energy	IS0 (Th.)	IS0 (Exp.)
0_2^+	2.2 (2.3)	2.6	–
0_3^+	9.8 (10.5)	3.5	2.0
0_4^+	11.9	0.9	–
0_5^+	12.6	1.5	–
0_6^+	14.8	1.6	–

In table 2, the calculated electric monopole (E0) matrix elements in ^{18}O are compared with the experimental data. In the α + ^{14}C cluster model, the 0_3^+ state is missing because this state is considered to have the dominant configuration of $^{16}{\rm O} \otimes \nu(1s_{1/})^2$, which is difficult to describe by the α cluster model. The 0_2^+ state appears at the low excitation energy of $E_x \sim 4$ MeV but the observed E0 strength to this state is 1.5 times larger than the strength for the single particle excitation [16]. Theoretical calculation is 1.1 in a unit of the single particle strength, and this result nicely reproduces the observed strength [11].

Table 2. Same as table 1 but for ^{18}O with the E0 strength [11]. Experimental data of E0 and energy of 0_4^+ are taken from Refs. [16] and [17], respectively.

State	Energy	E0 (Th.)	(Exp.)
0_2^+	4.3 (3.6)	1.1	1.5
0_4^+	8.7 (9.9)	0.5	–
0_5^+	15.6	1.6	–

From the results shown in tables 1 and 2, we can clearly understand that the α cluster model is possible to reproduce the enhanced and discrete monopole strength at the low excitation energy below E_x = 15 MeV. The reproduction of the low-lying discrete strength is difficult in the naive mean field picture because the monopole excitation of the single nucleon motion requires the $2\hbar\omega$ jump in energy, which amounts to about 30 MeV in this mass region. For instance, in the mean field model, the monopole strength appears at much higher energy with the broad width structure [8].

Since the sophisticated shell model calculations have been done in recent studies, the comparison of such the modern shell model calculation with the α cluster model is very interesting. For instance, no-core shell model (NCSM) was applied to ^{10,12}Be, and the energy levels in ^{10}Be, which included the intruder 0_2^+ state, were beautifully reproduced [18]. The IS0 strength was not shown in the NCSM calculation but the comparison of the α cluster model and NCSM is very interesting in future studies.

3.2 IS0 and IS1 transition in ^{44}Ti

The macroscopic α cluster model is applied to the α + ^{40}Ca system, and we have found that the excited 0^+ state appears at the excitation energy of E_x = 12.5 MeV with the decay width of Γ = 0.7 MeV. The α – ^{40}Ca wave function in this excited 0^+ state have the one higher node in comparison to the wave function in the ground 0^+ state. Thus, the excitation from the ground 0^+ state to the excited 0^+ state corresponds to the nodal excitation in the α – ^{40}Ca relative motion. In the previous experiment of ^{40}Ca(^{6}Li, d)^{44}Ti at E_α = 37 and 50 MeV, the higher nodal 0^+ state is identified at E_x = 10.9 MeV from the analysis of the differential cross section [19]. The higher nodal 0^+ state at E_x = 12.5 MeV obtained in theory seems to correspond to the observed 0^+ state at E_x = 10.9 MeV.

The continuum strength function for the IS0 transition is calculated, and we have confirmed that the prominent peak appears at the excitation energy of about 12 MeV, which corresponds to the higher nodal 0^+ state. Since the IS0 operator strongly induces the nodal excitation in the cluster relative motion, a strong enhancement occurs in the transition from the ground state to the higher nodal state.

We calculate the ratio of the IS0 matrix element of the resonant 0^+ state to the single particle strength. Up to the resonance energy, we sum up the total strength of the IS0 matrix element, and we have found that the total strength amounts to about 1.5 times larger than the respective single particle strength. Moreover, the enhancement factor reaches about 3 around the energy region of E_x = 20 MeV. In a simple mean field picture, the monopole excitation requires $2\hbar\omega$ excitation of a nucleon motion in the pf-shell, which reaches about 25 MeV. In the cluster excitation, however, the IS0 strength exceeds the single particle strength at much lower excitation energy ($E_x \lesssim 12$ MeV).

We have also performed the similar calculation for the dipole 1^- state. In the this calculation, two 1^- resonances are obtained; the 1_1^- state at E_x = 5.3 MeV with extremely sharp width and the 1_2^- state at E_x = 12.7 MeV with Γ = 4.4 MeV. The former state corresponds to the negative parity partner of the ground 0^+ state, while the latter state has the higher nodal state of the yrast 1_1^- state at E_x = 5.3 MeV. In the ^{40}Ca(^{6}Li,d)^{44}Ti reaction, the 1^- state is observed as the fragmented levels (E_x ~ 6.2 MeV and 8.2 MeV) with the centroid energy of E_x = 7.2 MeV [19]. Since this fragmented 1^- state has the large spectroscopic factor, S_α ~ 0.3, the 1^- state observed at the centroid of E_x = 7.2 MeV seems to correspond to the 1_1^- state at E_x = 5.3 MeV, which is obtained by the present α + ^{40}Ca cluster model, although the theoretical energy is a little lower than the experimental energy. The similar assignment was also done in previous works [19].

As for the 1_2^- state, there is no experimental identification but a considerable mixture of the 1^- strength is confirmed in the 0^+ resonance at E_x = 10.9 MeV [19]. Since the 0^+ (E_x = 10.9 MeV and Γ = 0.7 MeV) and 1_2^- (E_x = 12.7 MeV and Γ = 4.4) resonances in the calculation overlap each other in energy, the considerable mixture of the 0^+ and 1^- strengths in the experiment is consistent to the theoretical calculation.

In the calculation of the continuum IS1 strength, the strong and sharp enhancement is obtained at the resonance position of E_x ~ 5 MeV. The magnitude of the IS1 matrix element from the ground 0^+ state to the 1_1^- resonant state is about 4.5 times larger than the single particle strength. In the simple mean field picture, the excitation to the 1^- state requires the $1\hbar\omega$ excitation, which corresponds to about 12 MeV. In the case of the α – ^{40}Ca relative excitation, the enhanced strength occurs at a half of the energy in the single particle excitation, E_x ~ 5 MeV. This is the similar feature to the IS0 transition. The contribution from 1_2^- at E_x ~ 15 MeV further increases the total strength of the matrix elements, and the ratio of the total strength to the single particle strength finally reaches about a factor of about 6 in the energy region up to 15 MeV. The theoretical calculation predicts the strong enhancement in the IS1 strength and hence, the experimental measurements are strongly desired in future studies.

3.3 IS1strength in $^{104--110}$Te

We have applied the macroscopic α cluster model to even Te isotopes, such as $^{104-110}$Te with α + $^{100-106}$Sn. The strength of the DF potential is slightly tuned (1.03) so as to reproduce the experimental Q-value of the α decays in $^{104-110}$Te isotopes. The continuum strength function for the IS1 transition are calculated by employing the renormalized DF potentials.

In all of the Te isotopes, we have confirmed that the sharp and broad peaks appear in the IS1 strength around the excitation energy of E_x ~ 5 MeV and E_x ~ 13 MeV, respectively. The IS1 matrix elements for the E_x = 5 MeV states are comparable to the single particle unit, and the strength in ^{104}Te is the largest of all isotopes. The IS1 matrix element in ^{104}Te is about 0.9 in a unit of the single particle strength. Since $1\hbar\omega$ for the single particle excitation is about 9 MeV in this mass region, we can easily understand that the dipole excitation of the α particle occurs at much lower excitation energy (E_x ~ 5 MeV) in comparison to the single particle excitation.

In even Te isotopes, the experimental observations of the low-lying electric dipole (E1) strength were reported in the mass region around the β stability line, $^{122-130}$Te [20]. Some enhanced structures were observed in $E_x \leq$ 5 MeV, which were analyzed by two phonon model [20] and quasi-particle random phase approximation with deformed single particle basis [21]. It is very interesting to extend the measurement of the low-lying E1 strength to the lighter Te isotopes, $^{104-110}$Te, while the application of α cluster model to the heavier Te isotope is important subject in future work.

4 summary and discussion

In summary, we have applied the microscopic and semi-microscopic α cluster models to the lighter systems, such as ^{12}Be and ^{18}O, and clearly demonstrated that the prominent enhancements occurs in the isoscalar or electric monopole matrix elements going to the low-lying discrete 0^+ levels. This enhancement is induced by the excitation in the relative motion between the α cluster and the residual nucleus. Nevertheless the strength of the monopole transition is comparable to or exceeds the single particle strength, the excitation energy of such the α monopole excitation is much lower than the energy of the single nucleon excitation in the mean field picture. Therefore, the low-lying and discrete strength in the monopole transition is characteristic phenomena in the α cluster structure.

We have also extended the analysis about the isoscalar monopole (IS0) transition to the heavier system of ^{44}Ti, in which the macroscopic cluster model with α + ^{40}Ca known to work successfully. The continuum strength of the IS0 transition is calculated, and we have found that the strong IS0 transition, which exceeds the respective single particle strength, appear at the low excitation energy region, which is about a half of $2\hbar\omega$ for the single nucleon excitation. Thus, the low-lying enhancement in the monopole matrix element has also been confirmed in the heavy system.

In addition, we have also calculated the continuum strength of the isoscalar dipole (IS1) transition for the α + ^{40}Ca system, which has been proposed as a effective probe for the asymmetric cluster system [9]. The matrix element of the IS1 transition is prominently increased in the low excitation region, and this result is similar to the result in the IS0 transition. Furthermore, the calculations of the continuum IS1 strength is also performed for $^{104-110}$Te isotopes by applying the α + $^{100-106}$Sn cluster model. The calculation of Te isotopes has also supported the low-lying enhancement of the IS strength generated by the α cluster excitation.

From the results of a series of our calculations, the low-lying enhancement in the monopole and dipole transitions, which is induced by the α cluster excitation, seems to generally occur over a wide range of nuclear systems. The excitation in the relative motion of the α cluster and the residual nucleus can naturally describe the low-lying enhancement in the strength. In the analysis of the heavy systems, however, we must consider the breaking effect of α cluster generated by coupling to the mean field configuration more explicitly. In particular, the coupling of the α configuration and the shell model configuration is quite important in the initial ground state in the heavy systems. This coupling will reduce the total strength of the monopole and dipole transition. Moreover, the decays of the final α cluster state into the more complicated compound states will be important because the excited monopole and dipole states are embedded in continuum. This decay effect will spread the discrete IS strength.

The revised theoretical calculation to take into account the breaking and spreading effects on the α cluster state are now underway.

References

[1] P. Ring and P. Schuck, 2004 Nuclear Many-Body Problem1st edn (Berlin: Springer), and references therein.

[2] H. Horiuchi *et al.*, Suppl. Prog. Theor. Phys. **192**, 1 (2012), and references therein.

[3] U. Atzrott, P. Mohr, H. Abele, C. Hillenmayer, and G. Stadt, Phys. Rev. C**53**, 1336 (1996).

[4] F. Michel, S. Ohkubo and G. Reidmeister, Suppl. Prog. Theor. Phys. bf132, 7 (1998), and references therein.

[5] M. Kimura and H. Horiuchi, Nucl. Phys. A**767**, 58 (2006).

[6] M. A. Souza and H. Miyake, Phys. Rev. C**91**, 034320 (2015), and references therein.

[7] Y. Suzuki and S. Ohkubo, Phys. Rev. C**82**, 041303(R) (2010), and references therein.

[8] T. Yamada, Y. Funaki, T. Myo, H. Horiuchi, K. Ikeda, G Röpke, PŚchuck, and A. Thosaki, Phys. Rev. C**85**, 034315 (2012).

[9] Y. Chiba, Y. Taniguchi and M. Kimura, Phys. Rev. C**95**, 044328 (2017).

[10] M. Ito and K. Ikeda, Rep. Prog. Phys. **77** 096301, (2014).

[11] M. Nakao, H. Umehara, S. Ebata and M. Ito, Phys. Rev. C**98**, 054318 (2018).

[12] M. E. Brandan and G. R. Satchler, Nucl. Phys. **A487**, 477 (1988).

[13] S. Ebata, T. Nakatsukasa, T. Inakura, K. Yoshida, Y. Hashimoto, and K. Yabana, Phys. Rev. C**82**, 034306 (2010).

[14] M. Iwasaki, R. Otani, Y. Takenaka and M. Ito, PTEP **2015**, 023D01, and references therein.

[15] Z. H. Yand *et al.*, Phys. Rev. C**91** 024304 (2015).

[16] K. H. Souw *et al.*, Phys. Rev. C**11**, 1899 (1975).

[17] M. L. Avila et al., Phys. Rev. C**90**, 024327 (2014).

[18] L. Liu et al., Phys. Rev. Cbf96, 014302 (2012).

[19] T. Yamaya, K. Katori, M. Fujiwara, S. Kato and S. Ohkubo, Suppl. Prog. Theor. Phys. **132**, 73 (1998), and references therein.

[20] R. Schwengner et al., Nucl. Phys. A**620**, 277 (1997).

[21] E. Guliyev et al., Phys. Lett. B**532**, 173 (202).

Interesting states in A = 10 mass region, populated in ^{10}B + ^{10}B nuclear reactions

Deša Jelavić Malenica[1,*], *Matko* Milin[2], *Alessia* Di Pietro[3], *Pierpaolo* Figuera[3], *Marcello* Lattuada[3], *Đuro* Miljanić[1], *Agatino* Musumarra[3], *Maria Grazia* Pellegriti[3], *Valentina* Scuderi[3], *Natko* Skukan[1], *Neven* Soić[1], *Suzana* Szilner[1], *Domenico* Torresi[3], and *Milivoj* Uroić[1]

[1]Ruđer Bošković Institute, Zagreb, Croatia
[2]Faculty of Science, University of Zagreb, Zagreb, Croatia
[3]INFN - Laboratori Nazionali del Sud and Sezione di Catania, Catania, Italy

Abstract. The ^{10}B+^{10}B reactions are measured at beam energies of 50 and 72.2 MeV. The large spin of ^{10}B nucleus (J^π= 3$^+$) makes this reaction particularly suitable to populate high spin states in the exit channels. Population and decay of different states in $A\approx$10 mass region is studied, and the results are discussed from the structure point of view. In particular, a new state in ^{12}C at E_x= 24.4 MeV is observed to be strongly populated in the triple α-particle coincidences.

1 Introduction

States of different structures occur in light nuclei around A=10, at similar excitation energies: from shell-model to very exotic ones. Experimental data in this region are incomplete and often controversial.

The most popular nucleus in this mass region is ^{12}C, because of its essential role in stellar nucleosynthesis. The so-called Hoyle state [1] - the second 0$^+$ state of ^{12}C, at excitation energy E_x= 7.65 MeV, is responsible for dramatic speed-up of helium burning [2] and crucial for production of all heavier elements. Modeling the Hoyle state is a non-trivial task because it seems to have a unique structure and can not be described within *e.g.* shell model calculations, even when $4\hbar\omega$ excitations are included [3]. Many recent experimental studies are focused on Hoyle state and its possible excitations. Recent work by Marin Lámbarri *et al.* [4] made an important step in completing and systematizing the ^{12}C spectroscopy in general, with a new state (5$^-$ at E_x= 22.4 MeV) reported, and most of the low-lying ^{12}C states explained as corresponding to the triangular oblate spinning top with a $\mathcal{D}_{3h}$ symmetry [5]. More details on theoretical progress made in explaining the ^{12}C structure are given in a very recent review of microscopic clustering in nuclei [6].

From the structure point of view, there are many other interesting states in $A\approx$10 mass region. One such example are molecular states, with the intense experimental research in the last decades (*e.g.* [7]). Molecular states were finally clearly established in ^{10}Be [8, 9], forming a rotational band starting with the 0^+_2 state at E_x= 6.18 MeV. Such states are predicted to exist in a number of other nuclei [10], but all of them have in common that the valence nucleons are in fact *neutrons*. *Protons* as valence particles are expected to make the molecules less stable, due to the Coulomb repulsion, especially while filling the σ-orbitals.

Possible molecular states with protons as valence particles are predicted to exist in ^{11}B and ^{11}C nuclei [11, 12], making their theoretical description in a framework of a single model rather demanding. While ground and low-lying states of these nuclei are well described as shell-model structures, in the highly excited states there are two types of cluster structures: three-body $2\alpha + t$ (or ^{3}He) cluster structures (corresponding to the 3α cluster structures in ^{12}C) and molecular structures with three valence nucleons (one or two of them being protons), moving in the mean field of 2α particles (which can be associated with molecular structures in ^{10}Be). These kind of coexistence of three-body cluster and molecular orbital structures inside similar energy region is one of the unique features of ^{11}B and ^{11}C nuclei [12].

The present paper reports on results of ^{10}B+^{10}B nuclear reactions, measured at beam energies of 50 and 72.2 MeV. ^{10}B nucleus is very special by itself - it is one of five stable nuclei with odd number of protons and neutrons and it is the one nucleus with the largest ground state spin J$^\pi$ = 3$^+$. Although ^{10}B ground state can be simply described as shell model stretched configuration with two nucleons in $p_{3/2}$ subshell coupled to maximum possible spin, fundamental no-core shell model calculations correctly reproduce its spin only if the three-nucleon force is taken into account [13]. This peculiar structure of ^{10}B, which can be described as a mixture of shell model and cluster configurations of the type ^{6}Li$_{gs}$+α or ^{6}Li(0^+_2,1)+α [11], together with a high spin of the ground state J^π= 3$^+$, is expected to enable the population of a range of different

*e-mail: djelavic@irb.hr

high-spin states in a variety of nuclei around $A\approx10$, at high excitation energies.

Only few selected results will be presented in next sections, while the complete results for ^{12}C states can be found in Ref. [15].

2 Experiment

The ^{10}B+^{10}B experiment was performed at INFN-LNS Catania, using the SMP Tandem accelerator. The selectivity in populating different single particle and cluster states of neighbouring nuclei was studied, together with a sequential decay of states in question. This is the first time that such a reaction was employed for nuclear structure studies.

Two beam energies of 50.0 and 72.2 MeV were used, together with targets enriched in ^{10}B up to 99.8%. Highly segmented detector setup consisted of four ΔE-E silicon telescopes, each composed of a thin ΔE detector (57-67 μm), divided into 4 quadrants and a thick DSSSD detector (500 or 1000 μm), divided into 16 strips in both front and back sides. Three detector setups were used, with different azimuthal angles of the detector centers in a range from 20° to 53°.

Segmentation of DSSSD detectors makes the analysis of obtained experimental data rather demanding. Besides the calibration of a large number of apparently independent detector channels, there are complications associated with dead layers [16] and interstrip gaps [17, 18] that had to be taken in the account properly. Several novel procedure improvements [19] for easy and accurate calibration of DSSSD detectors were used in the analysis - details can be found in Ref. [20].

The detection efficiencies for a given setup was obtained from a Monte Carlo (MC) code that simulates two-body ^{10}B+^{10}B reaction, assuming an isotropic c.m. distribution of products, followed by sequential decay into fragments, again assuming an isotropic c.m. distribution.

At both beam energies the number of detected α particles was remarkably higher than the number of any other detected nuclei - even triple α-particle coincidences were detected with rather large statistics. Fig. 1 gives an example of ΔE-E spectrum for one quadrant of the telescope at 22.4°. Lines correspond to different isotopes of H, He, Li, Be, B and C, from bottom to top, respectively, being the most abundant ^{4}He and heaviest detected ^{13}C.

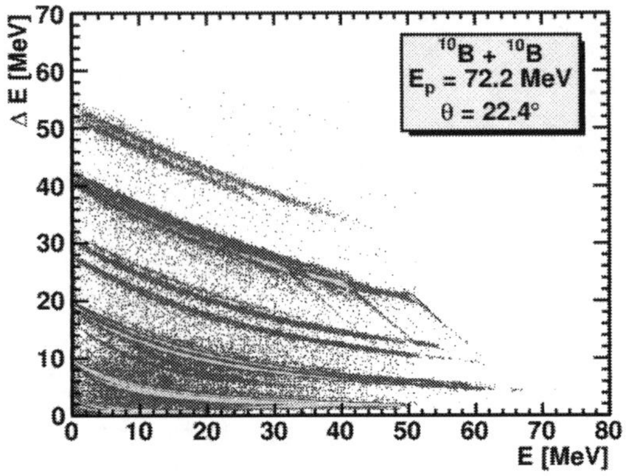

Figure 1: ΔE-E spectrum for the ^{10}B+^{10}B nuclear reaction measured at 72.2 MeV.

3 Results

3.1 Interesting states in ^{12}C

In the case where two α particles are detected in coincidence, by assuming the reaction to be a three body, one can reconstruct the excitation energy spectra of undetected ^{12}C, using the procedure suggested in [21]. An example of such spectra is shown in the inset of Fig. 2a. This spectra was obtained with the constrain that two detected α particles are coming from the ground state of the intermediate nucleus, ^{8}Be. After subtracting the background, one obtains the spectrum in the main panel of Fig. 2a.

Several strong peaks (states in ^{12}C) can be easily identified in Fig. 2a: 0^+ ground state, 2^+ state at 4.44 MeV, 3^- at 9.64 MeV and 5^- at 22.4 MeV (as proposed in Ref. [4]). The population of the g.s. is suppressed due to the large Q-value mismatch (Q= 19.25 MeV, $Q_{opt}\approx -12$ MeV). Furthermore, there is a small, but visible peak corresponding to the known ^{12}C state at 7.65 MeV (Hoyle state), and also peaks at E_x= 24.3 and 30.3 MeV, the last one being very pronounced in other reaction channel: ^{10}B(^{10}B, d^{10}B), reported in Ref. [15].

Even if rather restricted in angular range, angular distributions for the events given in Fig. 2a are very forward peaked, suggesting that the corresponding reaction mechanism is deuteron transfer - all the populated states have therefore non-negligible ^{10}B+d strength.

In a case where three α particles are detected in coincidence, the undetected part of the exit channel corresponds to the ^{8}Be nucleus, which is particle unstable, and decays exclusively (for E_x(^{8}Be) up to $\approx$17.3 MeV) through the α+α channel. So, by detecting three α particles in coincidence, one can reconstruct the details of ^{10}B+^{10}B$\rightarrow$5α reaction and its different sequential paths, which is a rarely seen and measured reaction channel.

Fig. 2b gives the ^{12}C excitation energy spectrum, reconstructed from the triple α-particle coincidences, with two of them coming from the ^{8}Be g.s. decay (with an energy of only 92 keV this decay has a clear signature in the detectors and is easily recognized and extracted), and two undetected α particles coming from the ^{8}Be first excited state at the E_x(^{8}Be)= 3.03 MeV. The contribution from different two-body intermediate steps, which would proceed through the ^{16}O states, is dismissed throughout plotting Dalitz plots [21] and introducing additional cuts to the data. Spectrum in Fig. 2b shows only a fraction of the collected data - only events where 2α particles are detected in

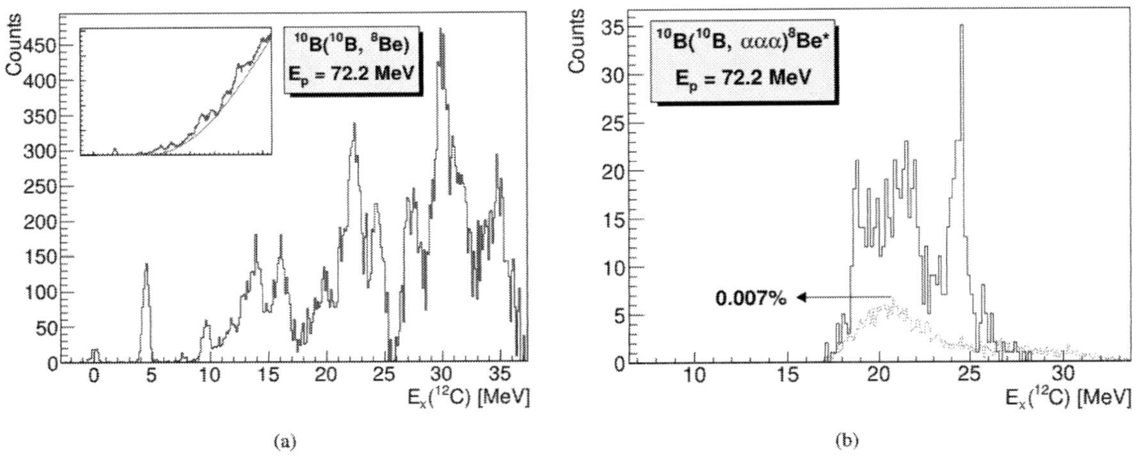

Figure 2: The results for ^{12}C measured in (a) double and (b) triple α coincidences (see text for details). The (red) dotted line in (b) shows the corresponding Monte Carlo efficiency curve.

telescope at 20°, and third α detected at 30° are included. Spectra for different detector combinations and setups can be seen in Ref. [15, 20]. All this spectra show same states in ^{12}C, but since the detection efficiency depend on excitation energy (being the consequence of geometry issues), not all states are seen in different detector combinations.

For many spectra obtained with different combinations of detectors and setups the dominant state at higher excitation energy (as clearly seen in Fig. 2b) is the state at E_x= 24.4 MeV. The peak is seen always at the same excitation energy (E_x= 24.4 MeV) and with the same width (Γ= 0.50 MeV); considering the smooth behavior of the efficiency in this energy region, we can be sure that this state at E_x= 24.4 MeV is a genuine state in ^{12}C.

Besides this state, for detector combination and setups with higher efficiency in low energy region, two additional states are measured in 3α particles coincidences: the states at E_x= 7.65 and 9.64 MeV. These states, together with the state at 24.4 MeV, were less populated in deuteron transfer channel shown in Fig. 2a. Considering that, and the reaction channel measured, we can suggest that the state at 24.4 MeV has a structure that is closer to the Hoyle state, than to the compact ^{12}C ground state. This is probably also a reason why this state is not seen in the recent precise measurements in which the ^{12}C states are populated with the inelastic scattering of ^{3}He or ^{4}He [4].

By comparing the state at E_x= 24.4 MeV (with $\Gamma\approx$ 500±40 keV), measured in this experiment, with energy and widths of states listed in last compilation for ^{12}C [22], we conclude that it is a new state, not listed before. The spin and parity assignments for this new state are essential to find its place in the spectroscopy systematics of ^{12}C.

3.2 Interesting states in ^{11}C

Excitation spectra for ^{11}C, measured in reaction channel ^{10}B(^{10}B, α^7Be)^{9}Be*, in which undetected ^{9}Be is left in its second excited state at E_x= 2.43 MeV, shows four strong peaks, as seen in Figures 3a and 3b. Two of them, at 8.11 and 8.43 MeV (Fig. 3a), are well known states that predominantly decay by alpha emission [22]. The state at 8.11 MeV, populated only through this channel in present experiment, is a cluster state, with very dilute structure of the type α+α+^{3}He. Such a state is a famous Hoyle state analog, as discussed in Ref. [11], which can be considered also as a dilute Bose-Einstein condensate. On the other hand, neighbouring state at 8.43 MeV, is a shell model state, with very compact structure. That state is also populated trough one nucleon transfer reaction channel, unlike the state at 8.11 MeV. Third state in low energy region, at 8.67 MeV, is not listed as alpha-decaying state, but is seen to decay by alpha emission in Ref. [23].

The fourth state seen in this channel lies in the higher excitation energy region, shown in Fig. 3b, at $E_x\approx$14MeV. Since the efficiency in this region is not smooth, and background is rather high, it is difficult to say weather this is a real state in ^{11}C. The fact that it is seen also in other detector combinations and for other excited states of undetected ^{9}Be, adds in favour that it is a real physical state. It is important to mention that a possible analog of this state is seen in the mirror reaction channel ^{10}B(^{10}B, α^7Li)^{9}B, where a broad state in ^{11}B is measured at $E_x\approx$15MeV.

3.3 Other interesting results

Besides shown results, there are many other channels measured in this experiment, which led to some interesting and rarely seen states.

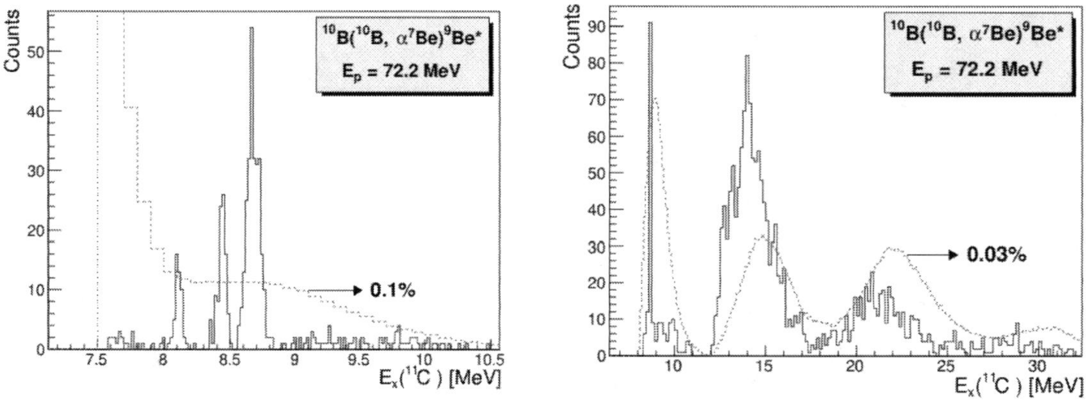

(a) (b)

Figure 3: The ^{11}C excitation energy spectra obtained from measured ^{7}Be and α-particle coincidences: (a) in detectors that lie on the same side of the beam (events in which both particles are detected in the same detector are included); (b) in all detector combinations, except the ones where both particles are detected in same detector. The (red) dotted lines show the corresponding Monte Carlo efficiency curves.

Excitation energy spectra of ^{13}C show a very strong α-decaying state at E_x = 18.9 MeV, just above the energy region reported in Ref. [24]. This state has a pronounced $^{9}Be+\alpha$ structure, and is a good candidate for molecular state with one valence nucleon orbiting around 3α centers.

Two very important states in ^{14}N at E_x = 13.2 and 15.39 MeV are populated in α-transfer reaction channel. Both of them fit nicely to a recent antisymmetrized molecular dynamics calculations as the head and the 5^+ state of the $^{10}B(3^+) + \alpha$ rotational band ($K^\pi = 3^+$) [25].

Acknowledgments

The authors acknowledge financial support from the Croatian Science Foundation under Project No. 7194 and in part under project no. IP-2018-01-1257. This work was also partly supported by the European Regional Development Fund for the Center of Excellence for Advanced Materials and Sensing Devices (Grant No. KK.01.1.1.01.0001).

References

[1] F. Hoyle, Astrophy. J. Suppl. Ser. **1**, 121 (1954).
[2] M. Freer and H.O.U. Fynbo, Prog. Part. Nucl. Phys. **78**, 1 (2014).
[3] P. Navrátil, J.P. Vary, and B.R. Barrett, Phys. Rev. Lett. **84**, 5728 (2000).
[4] D.J. Marín-Lámbarri *et al.*, Phys. Rev. Lett **113**, 012502 (2014).
[5] R. Bijker and F. Iachello, Ann. Phys. (Amsterdam) **298**, 334 (2002).
[6] M. Freer *et al.*, Rev. Mod. Phys. **90**, 035004 (2018).
[7] N. Soić *et al.*, Europhys. Lett. **34**, 7 (1996).
[8] M. Milin *et al.*, Nucl. Phys. A **753**, 263 (2005).
[9] M. Freer *et al.*, Phys. Rev. Lett. **96**, 042501 (2006).
[10] W. von Oertzen, M. Freer, Y. Kanada-En'yo, Phys. Rep. **432**, 43 (2006).
[11] Y. Kanada-En'yo, Phys. Rev. C **75**, 024302 (2007).
[12] T. Suhara, Y. Kanada-En'yo, Phys. Rev. C **85**, 054320 (2012).
[13] S.C. Pieper, Riv. Nuovo. Cim. **31**, 709 (2008).
[14] Y. Kanada-En'yo, H. Morita and F. Kobayashi, Phys. Rev. C **91**, 054323 (2015).
[15] D. Jelavić Malenica *et al.*, Phys. Rev. C **99**, 064318 (2019).
[16] O. Tengblad *et al.*, Nucl. Instrum. Meth. Phys. Res. A **525**, 458 (2004).
[17] D. Torresi *et al.*, Nucl. Instrum. Meth. Phys. Res. A **713**, 11 (2013).
[18] L. Grassi *et al.*, Nucl. Instrum. Meth. Phys. Res. A **767**, 99 (2014).
[19] M. Uroić *et al.*, Eur. Phys. J. A **51**, 93 (2015).
[20] D. Jelavić Malenica, Ph.D. thesis, University of Zagreb, 2015.
[21] E. Costanzo, M. Lattuada, S. Romano, D. Vinciguerra, and M. Zadro, Nucl. Instrum. Methods Phys. Res. A **295**, 373 (1990).
[22] J.H. Kelley, J.E. Purcell, and C.G. Sheu, Nucl. Phys. A **968**, 71 (2017).
[23] N. Soić *et al.*, Nucl. Phys. A **742**, 271 (2004).
[24] I. Lombardo, Phys. Rev. C **97**, 034320 (2018)
[25] Y. Kanada-En'yo, Phys. Rev. C **92**, 064326 (2015).

Masses and Beta-decay Studies of Neutron-rich Nuclei using the X-array and Gammasphere

F.G. Kondev[1,*], D.J. Hartley[2], R. Orford[1,3], J.A. Clark[1,4], G. Savard[1,5], K. Auranen[1], A.D. Ayangeakaa[1,2], S. Bottoni[1,6], M.P. Carpenter[1], P. Copp[1], K. Hicks[2], C.R. Hoffman[1], R.V.F. Janssens[7], B.P. Kay[1], T. Lauritsen[1], J. Li[1], S.T. Marley[8], G.E. Morgan[8], G. Mukherjee[9], S. Nandi[9], W. Reviol[1,10], J. Sethi[1,11], D. Seweryniak[1], S. Stolze[1], J. Wu[1], R. Yadav[12], and S. Zhu[1]

[1]Physics Division, Argonne National Laboratory, Lemont, Illinois 60439, USA
[2]Department of Physics, U.S. Naval Academy, Annapolis, Maryland 21402, USA
[3]Nuclear Science Division, Lawrence Berkeley National Laboratory, Berkeley, California 94720, USA
[4]Department of Physics and Astronomy, University of Manitoba, Winnipeg, Manitoba R3T 2N2, Canada
[5]Department of Physics, University of Chicago, Chicago, Illinois 60637, USA
[6]Universitet degli Studi di Milano and INFN, Via Celoria 16, 20133 Milano, Italy
[7]Department of Physics and Astronomy, University of North Carolina, Chapel Hill, North Carolina 27599, USA and Triangle Universities Nuclear Laboratory, Duke University, Durham, North Carolina 27708, USA
[8]Department of Physics and Astronomy, Louisiana State University, Baton Rouge, Louisiana 70803, USA
[9]Variable Energy Cyclotron Centre, 1/AF Bidhan Nagar, Kolkata 700064, India and Homi Bhabha National Institute, Training School Complex, Anushaktinagar, Mumbai-400094, India
[10]Department of Chemistry, Washington University, St. Louis, Missouri 63130, USA
[11]Department of Chemistry and Biochemistry, University of Maryland, College Park, Maryland 20742, USA
[12]Department of Biological and Physical Sciences, South Carolina State University, Orangeburg, South Carolina 29117, USA

Abstract.
Properties of neutron-rich nuclei in the A∼160 region are important for achieving a better understanding of the nuclear structure in this region where little is known owing to difficulties in the production of these nuclei at the present nuclear physics facilities. These properties are essential ingredients in the interpretation of the rare-earth peak at A∼160 in the *r* process abundance distribution, since theoretical models are sensitive to nuclear structure input. Predicated on these ideas, we have initiated a new experimental program at Argonne National Laboratory. During the first experiment, beams from the Californium Rare Isotope Breeder Upgrade radioactive beam facility were used in conjunction with the SATURN decay station and the X-array. We focused initially on several odd-odd nuclei, where β decays of both the ground state and an excited isomer were investigated. Because of the spin difference, a variety of structures in the daughter nuclei were selectively populated and characterized based on their decay properties. Mass measurements using the Canadian Penning Trap aimed at establishing the excitation energy of the β-decaying isomers were also carried out. Evidence was found for a change in the single-particle structure, which in turn results in the formation of a sizable N=98 sub-shell gap at large deformation. Results from the first experimental campaign using the newly-commissioned β-decay station at Gammasphere are also presented.

1 Introduction

The structure of deformed, neutron-rich nuclei in the rare-earth region is of significant interest for both the nuclear-structure and astrophysics fields. Although much progress is being made in our understanding of the *r* process, a satisfactory explanation for the elemental peak in abundance near A=160 is still elusive [1]. The quest to understand the origin of this peak may be key to correctly identifying the astrophysical conditions for the *r* process. Theoretical models of element production are dependent on masses and lifetimes of neutron-rich, deformed rare-earth nuclei in this region where little or no information is available. The available nuclear-structure information is also scarce, owing to difficulties in the production of these nuclei. In order to address these issues, an experimental program has been initiated at Argonne National Laboratory using high-purity radioactive beams produced by the CAlifornium Rare Isotope Breeder Upgrade (CARIBU) facility. Mass measurements using the Canadian Penning Trap (CPT) and $\beta-\gamma-\gamma$ coincidence studies using the Scintillator And Tape Using Radioactive Nuclei (SATURN) moving tape system and the X-array spectrometer were carried out. During the first experimental campaign, several long-lived, β-decaying states were identified in the neutron-rich ^{160}Eu and ^{162}Eu nuclei [2]. Evidence was found for a change in the single-particle structure, which in turn re-

*e-mail: kondev@anl.gov

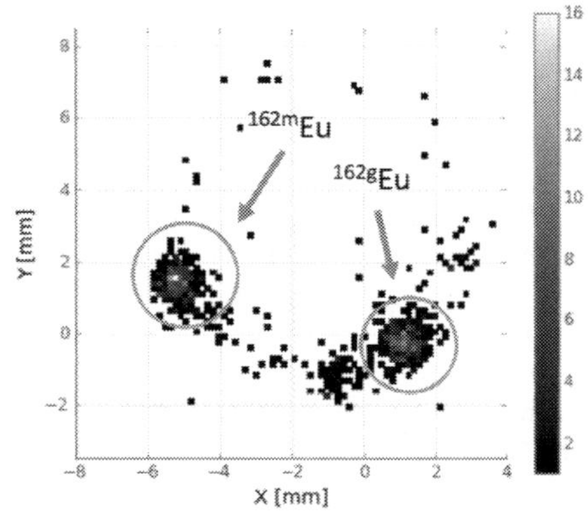

Figure 1. A histogram of a phase-imaging measurement showing the positions of ^{162m,g}Eu ions on the detector plane.

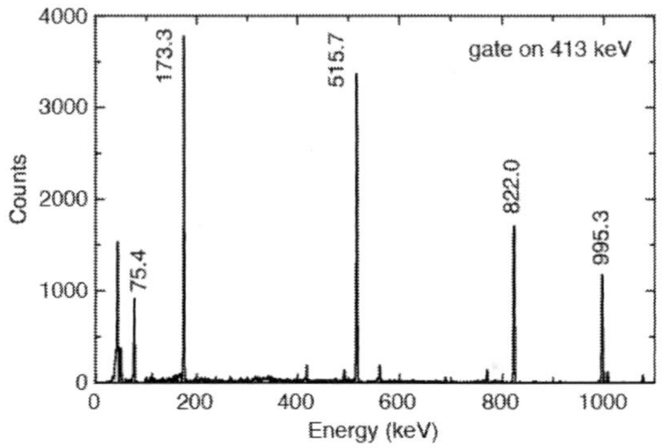

Figure 2. Gamma rays in coincidence with the 413-keV γ ray following the decay of the high-spin, β-decaying state in ^{160}Eu.

sults in the formation of a sizable N=98 sub-shell gap at large deformation [2].

The nuclear applications community is also interested in the decay properties of nuclei in this fission-product region. Past studies using the high-resolution γ-ray spectroscopy technique usually involved a single, or a few, germanium detectors, which resulted in incomplete decay schemes. Recently, measurements using the calorimetric Total Absorption Gamma-ray Spectroscopy (TAGS) method (see for example Refs. [3–5] and references therein) provide useful information on average β-decay properties. However, the technique utilizes low-resolution NaI detectors and depends sensitively on detailed knowledge of the decay schemes, which are usually not well known. As a consequence, TAGS relies on extensive theoretical considerations and Monte-Carlo simulations, which contribute to results that are model dependent and uncertainties are not well determined. We report here on the first $\beta - \gamma - \gamma$ spectroscopy measurements using the newly-commissioned β-decay station at Gammasphere, where both the discrete and calorimetric approaches can simultaneously be used.

2 Mass measurements with the CPT and decay spectroscopy studies using the X-array

The mass and β-decay spectroscopy experiments were both conducted in the low-energy experimental area of the CARIBU facility [6] located at the ATLAS facility at Argonne National Laboratory. The CARIBU beams were further purified by a Multi-Reflection Time-Of-Flight (MR-TOF) mass separator [7] before being delivered to the CPT mass spectrometer [8], where the Phase-Imaging Ion-Cyclotron-Resonance (PI-ICR) technique [9, 10] was used. From the new results, isomers in both ^{160}Eu and ^{162}Eu were established for the first time at excitation energies of 93.0 (12) keV and 160.2 (24) keV, respectively. It is worth noting that previous studies of ^{160}Eu and ^{162}Eu postulated the existence of a single β-decaying state [11–13]. An example of the final phase measurement showing the well-resolved $^{162m,g}Eu^{2+}$ ions is given in Figure 1.

The nuclear-structure properties of the ground and isomeric states in ^{160}Eu and ^{162}Eu were subsequently investigated through β-decay measurements. The CARIBU beams were directed onto a β-decay counting station comprised of the SATURN moving-tape system, four scintillator detectors for β-particle detection, and the X-array spectrometer with four germanium clover detectors, and one Low Energy Photon Spectrometer (LEPS), for γ-ray detection [14].

Combining the newly obtained experimental data with predictions of multi-quasiparticle calculations, the structure of the ground- and the isomeric states in ^{160}Eu and ^{162}Eu were assigned the $K^\pi=5^-$ (g.s.) and 0^-(isomer) states of the $\pi5/2[413]\otimes\nu5/2[523]$ configuration and the $K^\pi=1^+$ (g.s.) and 6^+ (isomer) states of the $\pi5/2[413]\otimes\nu7/2[633]$ configuration, respectively [2]. In ^{160}Eu, the $K^\pi=5^-$ ground state was found [2, 15] to preferentially β decay via an allowed transition to the $K^\pi=5^-$, $\pi^2(5/2[413],5/2[532])$ state at 1999 keV in the daughter nucleus ^{160}Gd. Such a transition involves the $\nu5/2[523]\rightarrow\pi5/2[532]$ configuration change and it is very fast. The $K^\pi=5^-$ state in ^{160}Gd has multiple γ-ray decay branches, the strongest one being via the 516-keV transition to the newly observed $K^\pi=4^+$, $\pi^2(5/2[413],3/2[411])$ state at 1483 keV [2, 15]. The latter subsequently decays via the 413-keV transition to the previously known $K^\pi=4^+$ state at 1071 keV which has a dominant $\nu^2(3/2[521],5/2[523])$ configuration [15]. A sample γ-ray spectrum produced by gating on the 413-keV transition is shown in Figure 2.

In ^{162}Eu, only β decay of the high-spin isomer was observed. It predominately feeds the newly observed $K^\pi=(6^-)$, $\nu^2(5/2[523],7/2[633])$ state in ^{162}Gd at 1453 keV that depopulates via the 206- and 331-keV γ rays to the $I^\pi=(6^+)$ and (5^+) members of the $K^\pi=2^+$ γ-vibrational band, respectively [2].

Figure 3. The new β-decay station at Gammsphere showing the tape-transport system and the array of six plastic scintillator detectors for β-particle triggering.

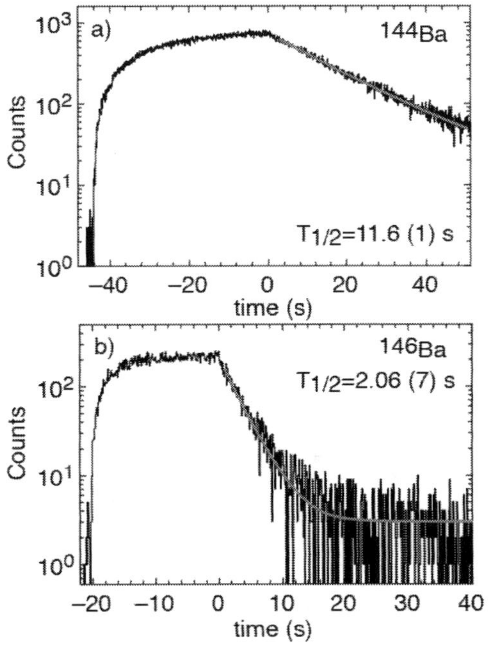

Figure 4. Time spectra produced by gating on: (a) - sum of 104-, 157-, 173- and 388-keV γ rays following the ^{144}Ba decay; (b) - sum of 121-, 141-, 197-, 232- and 251-keV γ rays following the ^{146}Ba decay.

3 Decay spectroscopy studies using Gammasphere

Gammasphere is one of most powerful γ-ray spectrometers in the world, owing to its good energy resolution, high efficiency and exceptional resolving power. The calorimetric potential of Gammasphere has only been used in the past for measuring entry regions in heavy nuclei [16] and never for β-decay studies. In principle, Gammasphere should be far superior compared to any TAGS device, as it retains all the information from many decay paths, instead of adding them up inside the detector. These data can be combined in many ways in the off-line analysis, including the performance of calorimetric studies for the high-energy part of the decay energy spectrum simultaneously with the discrete γ-ray spectroscopy data. In addition, the high coincidence power of Gamasphere allows dedicated $\gamma-\gamma$ angular correlations to be studied, which can provide information on the γ-ray transition multipolarities. Such data allow spins and parities of the excited states to be determined and this information is needed in the analysis of both the discrete and calorimetric β-decay data.

One reason that Gammasphere has never been used in β-decay spectroscopy studies in the past is the so-called "parent-daughter" problem: any interesting and exotic nucleus produced far from stability in a reaction will quickly decay to a series of much longer-lived daughter and granddaughter isobars that rapidly build up to secular equilibrium with much higher activity. As a consequence, the long-lived activities quickly swamp the exotic decay of interest. A solution to this problem is a tape-transport system which can be used to implant the CARIBU beams into the center of the spectrometer for an optimal collection time (usually 1-5 half-lives of the nuclide of interest) and then to remove the accumulated activity after a suitable counting time (usually 5-10 half-lives).

A new β-decay counting station for optimal use with Gammasphere was recently completed and commissioned. It comprises of a tape-transport system and an array of 6 plastic scintillators for β-particle triggering [17], as shown in Figure 3. During the commissioning experiment, data was collected using CARIBU beams of several A=144 (^{144}Ba and ^{144}La) and A=146 (^{146}Ba, ^{146m,g}La and ^{146}Ce) nuclei. Figure 4 shows time spectra produced by gating on selected γ-ray transitions following β decays of ^{144}Ba and ^{146}Ba. The measured half-lives of 11.6 (1) s (^{144}Ba) and 2.06 (7) s (^{146}Ba) are in good agreement with the presently evaluated values [18, 19].

The neutron-rich ^{146}La nuclide is of interest to several decay-data applications, including decay heat from nuclear reactors and antineutrino spectra reconstructions [20]. This nucleus is known to have β-decaying ground and isomeric states that were assigned $I^{\pi}=(2^-)$ and (6^-), respectively, based on a shell-model predictions [19]. However, the excitation energy of the isomer was not known and the decays of the two states were not well separated, thus resulting in incomplete and misleading level schemes. Controversially, while the decay data for the high-spin, β-decaying state in ^{146}La are included in the ENSDF database [19], they are absent in the NUDAT (claimed to be a "friendly face" of ENSDF) and ENDF/BVIII.0 [21], and JEFF3.3 [22] libraries, which are commonly used by the nuclear applications community. Using the newly-collected data, we were able to separate and significantly improve the decay schemes for the β-decaying ground and isomeric states in ^{146}La. A γ-ray spectrum showing transitions in coincidence with the 503-keV γ ray, the 6^+ to 4^+ transition within the ground-state band of ^{146}Ce, is shown in Figure 5. The excitation en-

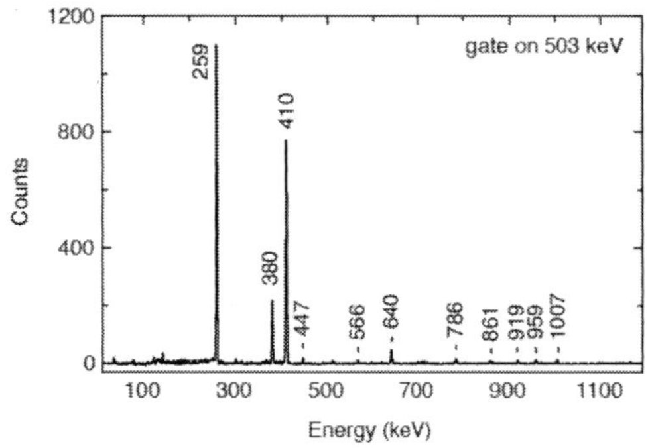

Figure 5. Gamma-ray spectrum from the high-spin β-decay of ^{146}La showing the main transitions in prompt coincidence with the 503-keV γ ray, depopulating the $I^{\pi}=6^{+}$ level of the ground-state band in ^{146}Ce daughter nucleus.

ergy of the isomer was recently established as 141.4(25) keV using direct mass measurements at the CPT and the phase-imaging technique [10].

The decay scheme of ^{144}La was also significantly improved and comparisons with the previous TAGS data [3] shows some notable differences, although it is worth pointing out that no uncertainties were given in Ref. [3].

This work is funded by the U.S. Department of Energy, Office of Nuclear Physics, under Contract No. DE-AC02-06CH11357 (ANL) and by the National Science Foundation under Grant Nos. PHY-1502092 (USNA). The work was also sponsored in part by the U.S. Department of Energy, National Nuclear Security Administration, Office of Defense Nuclear Nonproliferation R&D (NA-22). This research used resources of Argonne National Laboratory's ATLAS facility, which is a DOE Office of Science User Facility.

References

[1] M.R. Mumpower, G.C. McLaughlin, R. Surman, and A.W. Steiner, J. Phys. G: Nucl. Part. Phys. **44**, 034003 (2017).

[2] D.J. Hartley *et al.*, Phys. Rev. Lett. **120**, 182502 (2018).

[3] R.C. Greenwood *et al.*, Nucl. Instr. and Meth. Phys. Res. A**390**, 95 (1997).

[4] B.C. Rasco *et al.*, Phys. Rev. C**95**, 054328 (2017).

[5] E. Valencia *et al.*, Phys. Rev. C**95**, 024320 (2017).

[6] G. Savard, S. Baker, C. Davids, A. Levand, E. Moore, R. Pardo, R. Vondrasek, B. Zabransky, and G. Zinkann, Nucl. Instr. and Meth. Phys. Res. B**266**, 4086 (2008).

[7] T. Hirsh *et al.*, Nucl. Instr. and Meth. Phys. Res. B**376**, 229 (2016).

[8] J. Van Schelt *et al.*, Phys. Rev. Lett. **111**, 061102 (2013).

[9] S. Eliseev, K. Blaum, M. Block, C. Droese, M. Goncharov, E. Minaya Ramirez, D. A. Nesterenko, Yu. N. Novikov, and L. Schweikhard, Phys. Rev. Lett. **110**, 082501 (2013).

[10] R. Orford *et al.*, Nucl. Instr. and Meth. Phys. Res. B, in press.

[11] C.W. Reich, Nucl. Data Sheets **105**, 557 (2005).

[12] C.W. Reich, Nucl. Data Sheets **108**, 1807 (2007).

[13] J. Wu *et al.*, Phys. Rev. Lett. **118**, 072701 (2017).

[14] A.J. Mitchell *et al.*, Nucl. Instr. and Meth. Phys. Res. A**763**, 232 (2014).

[15] D.J. Hartley *et al.*, to be published.

[16] P. Reiter *et. al.*, Phys. Rev. Lett. **84**, 3542 (2000).

[17] P. Copp *et al.*, to be published

[18] A.A. Sonzogni, Nucl. Data Sheets **93**, 599 (2001).

[19] Yu. Khazov, A. Rodionov, and G. Shulyak, Nucl. Data Sheets **136**, 163 (2016)

[20] P. Dimitriou and A.L. Nichols, *Total Absorption Gamma-ray Spectroscopy for Decay Heat and Other Applications* (INDC(NDS)-**0676**, IAEA, Vienna, February 2015) 1-36.

[21] NUDAT and ENDF databases at www.nndc.bnl.gov.

[22] JEFF3.3 database at www.oecd-nea.org/dbdata.

Spin-tensor decomposition:

A useful tool for shell model effective interaction

Pawan Kumar*, *Shahariar* Sarkar, *Pushpendra P.* Singh and *P. K.* Raina

Department of Physics, Indian Institute of Technology Ropar, Punjab, India - 140001

Abstract. The spin-tensor decomposition is employed to construct a new interaction, named CKHeN, for $0p$-shell. This new interaction is used to calculate the effective single-particle energies of $\pi 0p_{3/2}$ and $\pi 0p_{1/2}$ orbitals in Li isotopes, and the level structures of ^{7,8,9}Li isotopes. The calculated level structures are found in good agreement with experimental data.

1 Introduction

Spin-tensor decomposition (STD) is a useful tool to decompose the model-space dependent shell model effective two-nucleon interaction into its central, spin-orbit and tensor force structure [1]. For last one and half decades, it has been used with the aim to understand the role of different components of two-nucleon interaction in the shell evolution in neutron-rich nuclei [2–4]. It has also been used to show why microscopic shell model interactions fail to describe the shell evolution in neutron-rich nuclei [3]. In this study, we use STD for the CK(8-16) interaction derived for $0p$-shell nuclei, and examine the properties of its total spin (J) averaged proton-neutron central and tensor force matrix elements;

$$\bar{V}^{\pi\nu}_{jj'} = \frac{\sum_J (2J+1) V^{\pi\nu}_{jj'}}{(2j+1)(2j'+1)}, \qquad (1)$$

where, sum runs only over the Pauli principal allowed J values.

For bare-tensor force, Otsuka *et. al.*, [6], showed that proton-neutron interaction $\bar{V}^{\pi\nu}_{jj'}$ corresponding to proton spin-up orbital ($j_> = l + 1/2$) and neutron spin-down orbital ($j'_< = l' - 1/2$) (or vice-versa) is attractive, whereas, if both proton and neutron orbitals are spin-up (or spin-down), the interaction is repulsive. It is also demonstrated that bare-tensor force matrix elements barely change and hold its nature after dealing with short-range repulsion part of two-nucleon problem and in-medium effects [7]. Furthermore, the numerical analysis shows that tensor interaction of well-established shell model effective interaction, *e.g.*, USDB, has same nature as for bare-tensor force.

The proton-neutron central component $\bar{V}^{\pi\nu}$ of shell model effective interaction is found to possess strong-orbital node *(nl)* and weak-spin *(j)* dependency [8]. It means that proton-neutron central force matrix elements $\bar{V}^{\pi\nu}_{j_> j'_>}$, $\bar{V}^{\pi\nu}_{j_< j'_<}$, and $\bar{V}^{\pi\nu}_{j_> j'_<}$ or ($\bar{V}^{\pi\nu}_{j_< j'_>}$) corresponding to proton l-orbital and neutron l'-orbital are nearly same. The strong-orbital node property of central interaction has been analytically demonstrated by Smirnova *et. al.*, [9] using the spin-exchange zero-range δ potential.

*e-mail: pawan.kumar@iitrpr.ac.in

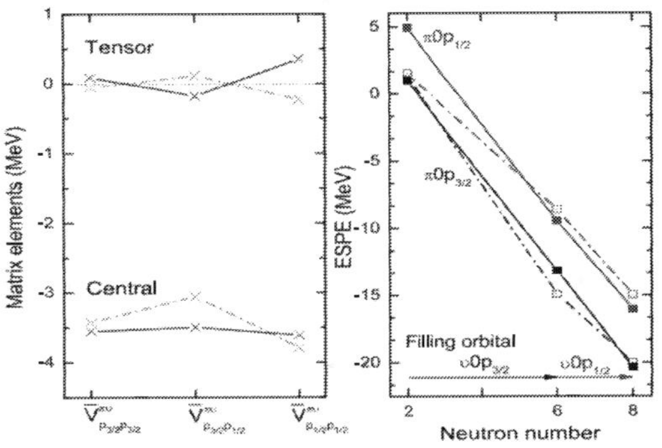

Figure 1. Left: J-averaged proton-neutron central and tensor force matrix elements in $0p$−shell. Right: ESPEs of $\pi - 0p$ orbitals in Li isotopes. The dot-dash and solid line are used for CK(8-16) and CKHeN interactions, respectively

In Fig. 1, we show proton-neutron central and tensor matrix elements of the CK(8-16) interaction. Here, central force matrix elements do not have similar strength which manifest that central component of CK(8-16) lacks weak-spin dependency. Further, tensor force matrix elements are present with opposite nature than its regular nature. In present work, we construct a new interaction in which these discrepancies are not present.

2 Spin-tensor decomposition and New effective interaction - CKHeN

Spin-tensor decomposition: Nucleons are intrinsic spin 1/2 fermions; therefore, the interaction between two-nucleon can be written as the linear sum of scalar product of configuration space operator Q and spin space operator S of rank k [1];

$$V = \sum_{k=0}^{2} V(k) = \sum_{k=0}^{2} Q^k . S^k \quad (2)$$

where rank $k = 0$, 1 and 2 represent central, spin-orbit and tensor force, respectively. Using the LS-coupled two-nucleon wave functions, the matrix element for each V^k can be calculated from matrix element V [10].

CKHeN interaction: To construct a new effective interaction for $0p$-shell, we have considered the single-particle matrix elements and isospin T = 1 two-body matrix elements of the interaction developed in Ref. [11], and T = 0 two-body matrix elements of the CK (8-16) interaction. The two-body matrix elements of this integrated interaction, mainly diagonal matrix elements, have been modified to gain common features for central and tensor matrix elements. Note that the spin-tensor decomposition has been performed at each step of modification to check the properties of central and tensor force matrix elements. The final interaction is named CKHeN, and its proton-neutron central and tensor force matrix elements are shown in Fig. 1. In this new interaction, central force matrix elements are present with good weak-spin dependency, and tensor force matrix elements are present with its characteristic properties.

3 Shell model calculations

The CKHeN interaction consists of $p_{3/2}$ and $p_{1/2}$ orbitals for protons and neutrons above ^{4}He core, and is tested for the effective single-particle energies (ESPEs) of $\pi 0p_{3/2}$ and $\pi 0p_{1/2}$ orbitals in Li isotopes and the level structures of ^{7,8,9}Li isotopes. The ESPE (ε') of π-j orbital in Li isotopes is given by [8];

$$\varepsilon'^{\pi}_{j} = \varepsilon^{\pi}_{j} + \sum_{j'} n^{\nu}_{j'} \bar{V}^{\nu\pi}_{j'j} \quad (3)$$

where, ε_j is unperturbed single-particle energy of orbital j, and $n_{j'}$ is number of neutrons in orbital j'. The level structures of ^{7,8,9}Li isotopes are calculated using shell model code-NUSHELLX@MSU [12]. The calculated ESPEs and level structures are shown in Fig. 1 and 2, respectively. The results obtained using the CK(8-16) interaction are also shown in these figures.

With the CKHeN interaction, the energy gap between spin-orbital partners, $\pi 0p_{3/2}$ and $\pi 0p_{1/2}$, in Li isotopes is found to remain nearly constant (see Fig. 1). This is similar to the observation seen in F and Sc isotopes where the energy gap between $\pi 0d_{5/2}$ and $\pi 0d_{3/2}$ orbitals, and $\pi 0f_{7/2}$ and $\pi 0f_{5/2}$ orbitals, respectively, remains nearly constant [8, 13]. However, with the CK(8-16) interaction, the energy gap between spin-orbital partners in Li isotopes increases when neutrons occupy $\nu 0p_{3/2}$ orbital, which is not consistent with the systematic.

In Fig. 2, the CKHeN interaction is shown to reasonably predict the experimental low-lying states of ^{7,8,9}Li relative to the CK(8-16) interaction.

4 Summary

In this work, spin-tensor decomposition is employed to discuss the discrepancies of CK(8-16) interaction and to construct a new effective interaction, named CKHeN, for $0p-$shell. The new interaction is tested for Li isotopes and found reasonably predicting their spectroscopic properties.

Acknowledgments

PK acknowledges SERB, India for the travel support to attend NSD-2019 conference.

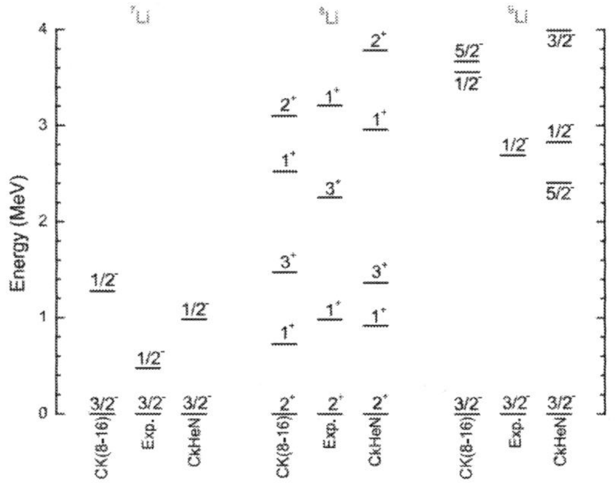

Figure 2. Level structure of ^{7,8,9}Li. Experimental data is taken from Ref. [14].

References

[1] M. W. Krison, Phys. Lett. B **47**, 110 (1973).
[2] N. A. Smirnova *et. al.*, Phys. Lett. B **686**, 109 (2010).
[3] N. A. Smirnova *et. al.*, Phys. Rev. C **86**, 034314 (2012).
[4] A. Umeya *et. al.*, Phys. Rev. C **69**, 024306 (2004).
[5] S. Cohen *et. al.*, Nucl. Phys. **73**, 1 (1965)
[6] T. Otsuka *et. al.*, Phys. Rev. Lett. **95**, 232502 (2005).
[7] N. Tsunoda *et. al.*, Phys. Rev. C **84**, 044322 (2011).
[8] P. Kumar *et. al.*, Phys. Rev. C **100**, 024328 (2019).
[9] N. A. Smirnova *et. al.*, Phys. Rev. C **69**, 044306 (2004).
[10] J. P. Elliott *et. al.*, Nucl. Phys. A **121**, 241 (1968).
[11] J. Stevenson *et. al.*, Phys. Rev. C **37**, 2220 (1988).
[12] B. A. Brown *et. al.*, Nucl. Data Sheets **120**, 115 (2014).
[13] P. Kumar *et. al.*, Nucl. Phys. A **983**, 210 (2019).
[14] https://www.nndc.bnl.gov/

EPJ Web of Conferences **223**, 01030 (2019) https://doi.org/10.1051/epjconf/201922301030
NSD2019

Application of the THM to the investigation of reactions induced by unstable nuclei: the ^{18}F(p,α)^{15}O case

Marco La Cognata[1,*], *Rosario G.* Pizzone[1], *Jordi* José[2,3], *Margarita* Hernanz[3,4], *Silvio* Cherubini[1,5], *Marisa* Gulino[1,6], *Giuseppe G.* Rapisarda[1,5], and *Claudio* Spitaleri[1,5]

[1]INFN - Laboratori Nazionali del Sud, Catania, Italy
[2]Departament de Física, EEBE, Universitat Politècnica de Catalunya, 08019 Barcelona, Spain
[3]Institut d'Estudis Espacials de Catalunya, 08034 Barcelona, Spain
[4]Institut de Ciencies de l'Espai (ICE-CSIC), Campus UAB, c/ Can Magrans s/n, 08193 Bellaterra, Spain
[5]Dipartimento di Fisica e Astronomia, Università degli Studi di Catania, Catania, Italy
[6]Facoltà di Ingegneria ed Architettura, Kore University, Viale delle Olimpiadi 1, 94100 Enna, Italy

Abstract. The Trojan Horse Method is applied to the investigation of the ^{18}F(p,α)^{15}O reaction, by extracting the quasi free contribution to the ^{2}H(^{18}F,α^{15}O)n process. For the first time the method is applied to a reaction of astrophysical importance involving a radioactive nucleus. After investigating the reaction mechanism populating the $\alpha + {}^{15}$O + n exit channel, we could extract the ^{18}F(p,α)^{15}O cross section and calculate the astrophysical factor over the 0 – 1 MeV energy interval. The possibility of exploring the cross section with no need of extrapolation allowed us to to point out the possible occurrence of a $7/2^+$ state at 126 keV, which would strongly influence the trend of the astrophysical factor at the energies of astrophysical interest. However, the low energy resolution prevents us to draw definite conclusions. Possible astrophysical consequences are also discussed, motivating further work on this reaction.

1 ^{18}F in astrophysics

Classical novae are among the most energetic thermonuclear explosions in the Cosmos, releasing about 10^{44} erg. They occur in a binary system made up of a white dwarf and a less evolved companion. If matter is transferred from the less evolved star to the compact companion, it accumulates under degenerate conditions until the thermonuclear runaway is ignited. In the envelope temperatures in excess of 10^8 K are reached and nuclei undergo a sequence of proton captures producing β-unstable nuclei. Moreover, during these events about 10^{-3} – 10^{-7} M$_\odot$ are ejected into the interstellar medium, enriched in CNO and intermediate-mass elements such as Ne, Na, Mg, Al (see [1–4] for recent reviews on classical novae).

The production of β-unstable nuclei suggests the possibility to observe both γ-ray emission due to the disintegration of short-lived radioactive species such as ^{13}N and ^{18}F and a late emission, due to medium-lived unstable nuclei such as ^{7}Be and ^{22}Na, determining the production of definite γ-lines from the de-excitation of excited states of their daughter nuclei. However, while classical novae have been observed in all wavelengths, from radio-waves to γ-rays with energies in excess of 100 MeV, they have been quite elusive in the ~ 0.1 – 10 MeV range, where the contribution of γ-rays from radioactive nuclei is expected.

In detail, the γ-ray spectrum should be dominated by an emission at energies ≤511 keV, originated by the annihilation of positrons from ^{13}N and ^{18}F β^+ decay. The major contribution is predicted to come from ^{18}F, owing to its lifetime $T_{1/2}$ ~ 110 s. This would allows ^{18}F to survive until the outer layers of the nova becomes transparent to γ-rays [3]. For this reason, the detection of the γ-lines from ^{18}F would help constraining models. However, only upper limits on the ^{18}F annihilation line have been set to date, leading to the establishment of a detectability distance, namely, the maximum distance of a classical nova for which the γ-line from ^{18}F can be observed. At present, the most stringent detectability distance of the 511 keV line has been set by the SPI spectrometer on board the space-borne γ-ray observatory INTEGRAL, ~ 3 kpc [5, 6].

Such constraint is strongly dependent on the predicted ^{18}F abundances produced during novae outbursts. In turn, accurate modeling of the elemental yields are critically dependent on the cross sections of the production and destruction nuclear reactions entering ^{18}F nucleosynthesis.

2 Present status of the ^{18}F(p,α)^{15}O cross section measurements

Production of ^{18}F in classical novae starts with the ^{16}O(p,γ)^{17}F reaction. Then, ^{18}F is either produced by proton capture on ^{17}F, through the ^{18}Ne(β^+)^{18}F chain, or following ^{17}F β^+ decay by proton capture on the daughter ^{17}O nucleus. Since the ^{18}F half-life is quite large in comparison with the outburst timescale, it is primarily destroyed

*e-mail: lacognata@lns.infn.it

through proton-captures, the $^{18}F(p,\gamma)^{19}Ne$ and, chiefly, the $^{18}F(p,\alpha)^{15}O$ reaction, ~ 1000 times more efficient than the former [4]. However, the $^{18}F(p,\alpha)^{15}O$ reaction is the most uncertain process in ^{18}F production (and destruction) network, even if many direct and indirect measurements as well as theoretical studies (see, e.g. Ref.[7]) have focused on its investigation at astrophysical energies.

Direct measurements are very challenging because of the energy range of interest, below about 400 keV, where cross sections are vanishingly small, and because of the need of a radioactive beam, whose produced intensities seldom exceed 10^6 ions per second. Direct measurements include the results of Ref.[8–13], focusing on the exploration of a broad $3/2^+$ resonance at about 700 keV in the center-of-mass system. A few measurements could reach astrophysical energies, measuring few points around 300 keV, where an additional resonance ($J^\pi = 3/2^-$) was observed [14, 15]. However, owing to poor energy resolution and large uncertainties, such measurements, though extremely complicated, could not set strong constraint on the trend of the cross section or, equivalently, of the astrophysical factor. The trend of the cross section is indeed very important to calculate the reaction rate, that is the input parameter of astrophysical calculations. Therefore, R-matrix was often used to extrapolate and interpolate the trend of the astrophysical factor; however, the occurrence of interfering resonances made it impossible to establish the behaviour of the astrophysical factor with the accuracy requested by astrophysical models.

With this respect, indirect techniques proved to be very helpful in the search of resonances at astrophysical energies, by performing the spectroscopy of ^{19}Ne intermediate compound system or of its mirror nucleus ^{19}F. Among others, the (d, p) reaction on ^{18}F and the $^{15}N - \alpha$ elastic scattering were used to explore the ^{19}F spectrum at energies corresponding to those of astrophysical interest (see [16] and [17], respectively). Similarly, p inelastic scattering off ^{19}Ne was used to perform its spectroscopy [18], as well as the (d, n) transfer reaction on ^{18}F [19] and the (p, d) neutron pickup on ^{20}Ne [20, 21]. These studies and charge symmetry considerations show that low-energy s-wave resonances are extremely influential, in particular the $3/2^+$ levels observed and/or predicted above and below the proton emission threshold (about 6.410 MeV) in ^{19}Ne. This is due to the occurrence of a strong $3/2^+$ resonance at $E_{c.m.} = 665$ keV in the $^{18}F(p,\alpha)^{15}O$ astrophysical factor, possibly interfering with lower-lying $3/2^+$ state. The consequent interference pattern would strongly influence the trend of the astrophysical factor well within the energy window of astrophysical importance. Since this is presently the largest source of uncertainty affecting the $^{18}F(p,\alpha)^{15}O$ astrophysical factor, recently the $^{19}F(^3He, t\gamma)^{19}Ne$ and the $^{19}F(^3He, t)^{19}Ne$ reactions were studied aiming at determining the energies and proton widths of the predicted $3/2^+$ ^{19}Ne states (see [22] and [23], respectively).

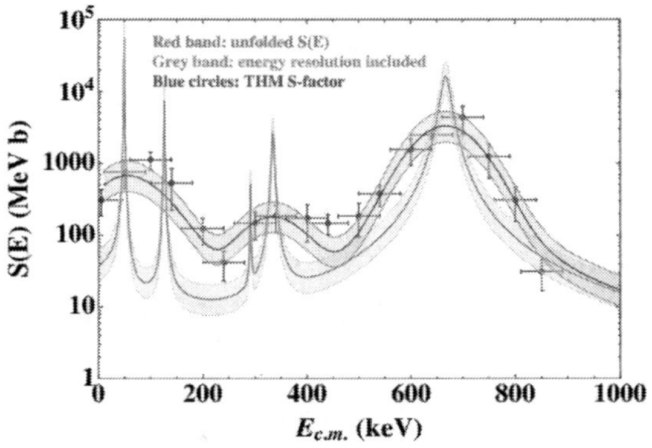

Figure 1. R-matrix analysis of the THM astrophysical factor (blu points), under the assumption of $J^\pi = 3/2^+$ for the 6460 keV ^{19}Ne state as discussed in [25, 26]. The black line is the smoothed R-matrix calculation, accounting for a 53 keV energy spread (standard deviation), with parameters given in table 1. The red line is the corresponding deconvoluted astrophysical factor. The grey and the red bands show the uncertainty on the R-matrix analysis (standard deviation) due to the experimental errors affecting the THM S-factor.

3 Indirect measurement of the $^{18}F(p,\alpha)^{15}O$ cross section using the THM

Because of the ambiguities in the trend of the $^{18}F(p,\alpha)^{15}O$ astrophysical factor at ultra-low energies, we decided the apply the Trojan Horse Method (THM) to its determination. Indeed, THM is an alternative and valuable approach to obtain the bare-nucleus cross section of a reaction induced by charged particles at energies lower than the Coulomb barrier, thanks to the use of suitable reactions with three particles in the exit channel. A recent review of the method is given in Ref.[24]. Here we recall that THM is based on the selection of the quasi-free reaction mechanism, namely, we single out the condition under which a cluster system (for instance, a deuteron), usually referred to as Trojan Horse nucleus, is used to transfer a participant cluster while the remaining cluster is emitted without influencing the participant-projectile interaction. In the case of the $^{18}F(p,\alpha)^{15}O$ reaction, we studied the $^2H(^{18}F,\alpha^{15}O)n$ three-body reaction, the neutron being the spectator to the nuclear reaction of astrophysical interest. Since the energy of the incident ^{18}F is chosen large enough to overcome the Coulomb barrier of the ^{18}F–d interaction, the deuteron breakup takes place inside the nuclear field, so that Coulomb repulsion is greatly suppressed. Such energies, of the order of several AMeV, are usually much larger than those corresponding to the atomic degrees of freedom, so electron screening effects are also negligible. On the other hand, both the energy spent to break the Trojan Horse nucleus and intercluster motion make it possible to span the astrophysical energy region with a single beam energy.

Two experiments were carried out, one at CNS-RIKEN (see [25] for more details) and one at the Texas A&M University [26]. Results are in agreement with each other within uncertainties and their weighed average is shown as blue circles in figure 1. In the figure, vertical error bars account for statistical and normalization errors, while the horizontal error bar indicates the calculated energy resolution, strictly linked to the bin width used in the data analysis. More details can be found in the original works [25, 26]. We then performed an R-matrix analysis of the THM data, starting from the work [20] and modifying the parameters in order to best reproduce the THM S-factor. The adopted procedure is discussed at length in [27]. Here we remind that the best fit curve (shown as a black line in figure 1) is achieved by assuming the (++)(++) interference pattern, following to the notation in Fig. 3 of [20], excluding the 7 keV resonance corresponding to the 6.417 MeV state in ^{19}Ne (see table 1) and inserting a $7/2^+$ state of ^{19}Ne at 6.537 MeV, whose occurrence was already pointed out in [25, 26]. Also, to compare the calculated astrophysical factor with the experimental THM one, the effect of the energy resolution is taken into account, by smearing the theoretical curve to match the 53 keV (standard deviation) calculated energy resolution. A very good agreement is found between the fitted curve and the THM S-factor, as for 18 degrees of freedom, a reduced $\chi^2 = 1.5$ is obtained. The R-matrix astrophysical factor, devoid of effects due to energy resolution, is shown as a red line in figure 1. Finally grey and red bands are used to highlight the uncertainty affecting the R-matrix calculation, for the smeared and unfolded astrophysical factors respectively, taking into account the experimental uncertainties affecting THM data. Table 1 contains the resonance parameters leading to the best fit of the THM data.

To test the effect of the inclusion of the 6417 keV state, R-matrix calculations comprising and excluding such state were implemented. Inserting such a state with the resonance parameters of the literature [20] leads to a reduced $\chi^2 = 3.1$ for 18 degrees of freedom after smearing, two times larger than the calculation completed without its inclusion. This suggests that THM data tend to exclude it from the astrophysical factor. The reason of the increase of the reduced χ^2 is also apparent from the inspection of figure 1. The inclusion of the 6417 keV state would cause an increase of the R-matrix S-factor by a factor of more than 3 at 5 keV (from about 300 MeVb up to about 1000 MeVb), leading to a discrepancy of 5.5σ with respect to the THM astrophysical factor at the same energy. However, more work is necessary to definitely exclude such contribution as our conclusions are presently based on a single point. Yet, it is worth nothing that the THM result agrees with the lack of observation of the mirror state in the ^{15}N$(\alpha,\alpha)^{15}$N cross section [28].

The second important feature is the occurrence of the 6537 keV state reported in [25, 26]. This level, corresponding to a 126 keV resonance, was not included in the R-matrix calculation by [20]. However, its omission in the present R-matrix calculation would determine an increase in the reduced χ^2 to 1.8. Owing to the THM data energy resolution, the deviation of the calculated S-factor not including the 126 keV resonance from the experimental THM one is well below 5σ, so a new measurement with improved energy resolution is necessary to clarify its occurrence. This is a very important point since this resonance occurs at astrophysical energies, right in the position where the largest influence from interference between s-wave resonances is expected. Therefore, the presence of the 6537 keV is of significant astrophysical relevance, calling for further indirect high-resolution studies of the ^{18}F(p,α)^{15}O reaction using the THM. A more extended discussion on the R-matrix analysis can be found in [27].

Table 1. Parameters of the R-matrix calculation (red line) in figure 1. Resonance energies, corresponding levels in ^{19}Ne, spin-parities, Γ_p and Γ_α are reported, respectively. For the sub threshold state, the ANC is given instead of Γ_p. A complete discussion on the parameters can be found in [27] and in [20] and references therein.

E_{res} (keV)	E_x (keV)	J^π	Γ_p (keV)	Γ_α (keV)
-124	6286	$1/2^+$	83.5 fm$^{-1/2}$	11.6
7	6417	$3/2^-$	1.6 10^{-41}	0.5
29	6440	$1/2^-$	3.8 10^{-19}	220
49	6460	$3/2^+$	2.3 10^{-13}	0.9
126	6537	$7/2^+$	7.1 10^{-8}	1.5
291	6702	$5/2^+$	2.4 10^{-5}	1.2
334	6745	$3/2^-$	2.2 10^{-3}	5.2
665	7075	$3/2^+$	15.2	23.8
1461	7872	$1/2^+$	55	347

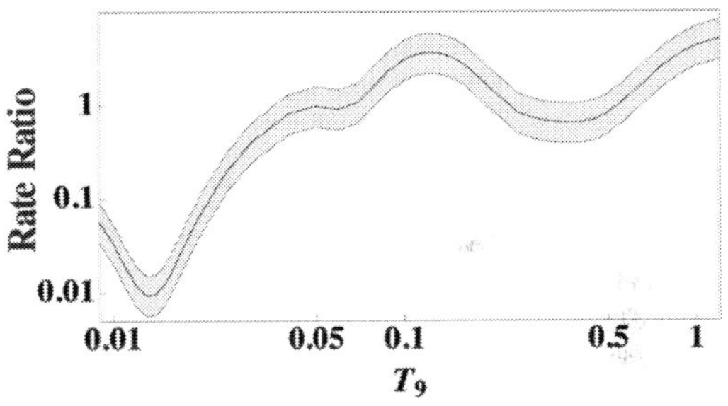

Figure 2. Ratio of the ^{18}F$(p,\alpha)^{15}$O reaction rate calculated using the deconvoluted THM S-factor (red band of figure 1) to the one reported in the JINA REACLIB database [30]. Uncertainties of the reaction rate are represented as a grey band.

4 Astrophysical consequences

Using standard equations (see, for instance, [29]), the reaction rate for the ^{18}F(p,α)^{15}O reaction was calculated using the best-fit R-matrix astrophysical factor, devoid of energy resolution effects (the red band in figure 1). The resulting reaction rate was then divided by the one listed in the JINA REACLIB database [30], for ease of comparison. In fact, as it is shown in [27], the reaction rate is exponentially decreasing with decreasing temperature. The ratio of the THM reaction rate to the REACLIB one is shown in figure 2 as a black line. The REACLIB rate is used for comparison since it is routinely used in novae modeling. The shaded area displays the uncertainty interval corresponding to the red band in figure 1. Temperature on the horizontal axis is given in units of 10^9 K, namely $T_9 = T/10^9$ K. For $0.1 \lesssim T_9 \lesssim 0.5$, typical temperatures of novae nucleosynthesis, the THM reaction rate is about a factor of 2 on the average larger than the one in the REACLIB database [30], in agreement with the results by [20]. The effect of the missing 7 keV resonance causes a decrease of the reaction rate ratio below about $T_9 \sim 0.1$, outside the temperature range of astrophysical energies.

The THM reaction rate was then implemented into one-dimensional, hydrodynamic simulations of novae thermonuclear runaways using the `SHIVA` code [4, 31]. In particular, we tested the influence of the reaction rate for two models of 1.25 $M_\odot$ oxygen-neon white dwarfs, accreting H-rich material from the stellar companion at a rate of 2×10^{-10} $M_\odot$ yr^{-1}. By changing the $^{18}F(p,\alpha)^{15}O$ reaction rate only, using in one case the THM one and in the other the one listed in REACLIB, we found no change in the dynamical properties of the explosion. Conversely, important differences in the chemical composition of the ejected matter was found, revealing a decrease of the ^{18}F supply for the THM case by a factor of ~ 2. This would imply a reduction of the detectability distance of the 511 keV annihilation line by γ-ray satellites by a factor $\sim \sqrt{2}$. This may help explaining the present lack of observations of this γ-ray line.

References

[1] S. Starrfield, C. Iliadis and W.R. Hix, *Classical Novae, 2nd Edition* (Cambridge Astrophysics Series. Cambridge, 2006) 77

[2] S. Starrfield, C. Iliadis and W.R. Hix, Publ. Astron. Soc. Pac., **128**, 051001 (2016)

[3] J. José and S.N. Shore, *Classical Novae, 2nd Edition* (Cambridge Astrophysics Series. Cambridge, 2'008) 121

[4] J. José, *Stellar Explosions: Hydrodynamics and Nucleosynthesis* (CRC/Taylor and Francis, Boca Raton, 2016)

[5] M. Hernanz et al., Astrophys. J. **526**, L97 (1999)

[6] M. Hernanz and J. José, *5th INTEGRAL Workshop "the INTEGRAL Universe"* (ESA SP-552, 2004) 95

[7] M.Dufour, P.Descouvemont, Nucl.Phys. A **785**, 381 (2007)

[8] K.E.Rehm et al., Phys.Rev. C **52**, R460 (1995)

[9] K.E.Rehm et al., Phys.Rev. C **53**, 1950 (1996)

[10] J.-S.Graulich et al., Phys.Rev. C **63**, 011302 (2001)

[11] D.W.Bardayan et al., Phys.Rev. C **63**, 065802 (2001)

[12] A. St. J. Murphy et al., Phys.Rev. C **79**, 058801 (2009)

[13] N. de Sereville et al., Phys.Rev. C **79**, 015801 (2009)

[14] D.W.Bardayan et al., Phys.Rev.Lett. **89**, 262501 (2002)

[15] C.E.Beer et al., Phys.Rev. C **83**, 042801 (2011)

[16] R.L.Kozub et al., Phys.Rev. C **71**, 032801 (2005)

[17] M La Cognata et al., Phys.Rev. C **99**, 034301 (2019)

[18] J.C.Dalouzy et al., Phys.Rev.Lett. **102**, 162503 (2009)

[19] A.S.Adekola et al., Phys.Rev. C **83**, 052801 (2011)

[20] D.W.Bardayan et al., Phys.Lett. B **751**, 311 (2015)

[21] D.W.Bardayan et al., Phys.Rev. C **96**, 055806 (2017)

[22] M. R. Hall et al., Phys. Rev. Lett. **122**, 052701 (2019)

[23] D. Kahl et al., Eur. Phys. J. A **55**, 4 (2019)

[24] R.E. Tribble et al., Rep. Prog. Phys. **77**, 106901 (2014)

[25] S. Cherubini et al., Phys. Rev. C **92**, 015805 (2015)

[26] R.G. Pizzone et al., Eur. Phys. J. A **52**, 24 (2016)

[27] M. La Cognata et al., Astrophys. J. **846**, 65, (2017)

[28] D.W. Bardayan, R.L. Kozub and M.S. Smith, Phys. Rev. C **71**, 018801 (2005)

[29] C. Iliadis, *Nuclear Physics of Stars* (Wiley, New York, 2007)

[30] R.H. Cyburt et al., Astrophys. J. Suppl. Ser. *189*, 240 (2010)

[31] J. José and M. Hernanz, Astrophys. J. **494**, 680 (1998)

Heavy Ion Charge Exchange Reactions as Probes for Beta-Decay

Horst Lenske[1,2,*]

[1]Institut für Theoretische Physik, Justus-Liebig-Universität Gießen, D-35392 Gießen
[2]NUMEN Collaboration, LNS Catania

Abstract. Peripheral heavy ion single and double charge reactions are described by fully quantum mechanical distorted wave methods. A special class of nuclear double charge exchange (DCE) reactions proceeding as a one-step reaction through a two-body process are shown to proceed by nuclear matrix elements of a diagrammatic structure as found also in $0\nu2\beta$ decay. These hadronic Majorana-type DCE reactions (MDCE) have to be distinguished from second order DCE reactions, given by double single charge exchange (DSCE) processes, resembling $2\nu2\beta$ decay. The theoretical concepts of MDCE are discussed. First results show that ion-ion DCE reactions are the ideal testing grounds for investigations of rare second order nuclear processes, giving insight into nuclear in-medium two-body correlation.

1 Introduction

There is broad consensus that $0\nu2\beta$ decay will be a highly promising gateway to physics beyond the standard model of elementary particle physics. Once observed, it will give direct evidence on the Majorana-nature of neutrinos with far reaching implications for neutrino masses, neutrino-matter interactions and flavour mixing up to the question of the matter-antimatter asymmetry in the universe [1–5]. Such a signal has to be distinguished from the two-neutrino beta-decay ($2\nu2\beta$) [6, 7] which is allowed in the standard model. A few nuclei are known to decay by this already rather rare process, as discussed e.g. in Ref. [8]. While the matrix elements are accessible by the observed $2\nu2\beta$ transitions such a check against data does not yet exist for $0\nu2\beta$ processes. Thus, estimates of life times and transition probabilities are relying on theoretical investigations, notoriously showing an uncomfortably large spread of values. Independent tests of the nuclear structure input under controllable dynamical conditions are highly necessary, allowing to evaluate and gauge the theoretical results by an independent process. The field will profit tremendously if a surrogate process could be identified which is technically and physically easily accessible. While single charge exchange (SCE) reactions with light and heavy ions have been studied intensively, see e.g. [9, 10], close to nothing is known about double charge exchange reactions. Only very recently, the NUMEN project has been initiated [11], using heavy ion reactions to explore that unknown territory, also aiming at establishing the relation to double beta decay. Clearly, to establish that connection requires additional efforts in our theoretical understanding of nuclear multi-step reactions. Hence, in section 2 we start with a brief introduction into the theoretical background for single and double single charge exchange reactions. In section 3 the physical concept of Majorana DCE reactions is briefly introduced. First results are discussed in section 4 and in section 5 an outlook will be given.

2 Single Charge Exchange and Double Single Charge Exchange Reactions

Single charge exchange reactions (SCE) have become a widely used tool for studying the spin-isospin response of nuclei. The discovery of the giant Gamow-Teller resonance (GTR) by the pioneering experiments at IUCF [12] initiated widespread experimental and theoretical research activities, continuing with even increasing intensity until today. Over the years, a wealth of data has been accumulated and was used to derive the charge exchange response of nuclei. Comprehensive reviews of the experimental and theoretical methods and achieved results are found in [9] for light ion reactions and in [10] for heavy ion-induced reactions. Peripheral heavy ion collisions, corresponding to direct reactions, are as useful for spectral studies as light ion scattering. An especially appealing aspects is the broad range of projectile-target combinations which, for example, allow to project out selectively specific features, e.g. spin flip and non-spin flip transitions.

As discussed in detail in Ref. [13], the SCE reaction amplitudes are expressed as DWBA matrix elements of the nucleon-nucleon T-matrix with spin-isospin elements of tensorial rank 0 and form factors $V^{(C)}_{ST}$, and rank 2 with corresponding form factors $V^{(Tn)}_{ST}$. They are connecting the initial channel $\alpha = a + A$ and the final channel $\beta = b + B$. The SCE reaction kernel is given by products of nuclear

*e-mail: horst.lenske@physik.uni-giessen.de

form factors

$$K^{(ST)}_{\alpha\beta}(\mathbf{p}) = (4\pi)^2 (V^{(C)}_{ST}(p^2) F^{(ab)\dagger}_{ST}(\mathbf{p}) \cdot F^{(AB)}_{ST}(\mathbf{p}) + \delta_{S1}\sqrt{\frac{24\pi}{5}}\, V^{(Tn)}_{ST}(p^2) Y^*_2(\hat{\mathbf{p}}) \cdot \left[F^{(ab)\dagger}_{ST}(\mathbf{p}) \otimes F^{(AB)}_{ST}(\mathbf{p})\right]_2) \tag{1}$$

where the rank-2 tensorial coupling relates to the spin degrees of freedom only. Through the form factors $F^{(ab),(AB)}_{ST}$, the kernels contain the spectroscopic information on the nuclear transitions, and the dynamics by the interaction form factors $V^{(C),(Tn)}_{ST}$. In the central interaction part, the scalar product indicates the contraction of the projectile and target form factor with respect to the spin and isospin degrees of freedom. The isospin degrees of freedom are of course projected by the nuclear transitions to the proper combination of $\tau_\pm$ operators. The one-step SCE reaction amplitude is obtained as

$$M_{\alpha\beta}(\mathbf{k}_\alpha, \mathbf{k}_\beta) = \int d^3p N_{\alpha\beta}(\mathbf{p}) \mathcal{U}_{\alpha\beta}(\mathbf{p}). \tag{2}$$

where the transition potential is defined by summation over the reaction kernels, Eq.(1). Initial and final state interactions are contained in the distortion coefficient [13, 14]

$$N_{\alpha\beta}(\mathbf{p}) = \frac{1}{(2\pi)^3} \langle \chi^{(-)}_\beta | e^{-i\mathbf{p}\cdot\mathbf{r}} | \chi^{(+)}_\alpha \rangle. \tag{3}$$

The distortion coefficient $N_{\alpha\beta}$ is closely related to the elastic scattering amplitude: For $p \to 0$ and $k_\beta \to k_\alpha$ the definition of the elastic S-matrix is recovered. Thus, in leading order, the above equation corresponds to the folding of the nuclear transition form factors with the ion-ion elastic scattering amplitude. Because of the strong absorption, the distortion coefficient acts mainly as a scaling factor, typically reducing the forward cross section by several orders of magnitudes compared to the plane wave limit. Only at momentum transfers exceeding 100 MeV/c $N_{\alpha\beta}$ leads to modifications of the momentum structure of cross sections.

If we consider, on the other hand, the effective operator underlying the conventional double-SCE two-step reaction mechanism, we find

$$V^{(DSCE)}(\mathbf{13}, \mathbf{24}) \sim \sum_{cC} T_{NN}(\mathbf{3}, \mathbf{4}) \mathcal{G}_{cC}(\mathbf{2}-\mathbf{4}, \mathbf{1}-\mathbf{3}) T_{NN}(\mathbf{2}, \mathbf{1}) \tag{4}$$

where T_{NN} is the isovector nucleon-nucleon T-matrix and $\mathcal{G}_{cC}$ denotes the (full many-body) propagator of the intermediate nuclei reached in the first SCE reaction step. The DSCE reaction amplitude is given effectively by a second order distorted wave expression

$$M^{(DSCE)} = \langle \chi^{(-)}_\beta bB | V^{(DSCE)} | aA \chi^{(+)}_\alpha \rangle. \tag{5}$$

3 Majorana Double Charge Exchange Reactions

Second order quantal processes like heavy ion double charge exchange reactions are of genuine reaction theoretical interest. First of all, until now heavy ion DCE reactions have not been studied, neither experimentally nor theoretically. Some attempts were made on (π^+, π^-) reactions but the notoriously bad energy definition of the incoming pion beams is unfavorable for spectroscopic work. Thus, double charge exchange reactions with heavy ions are much better suited for explorations of weakly populated transitions. Here, we consider collisional charge exchange processes given by elementary interactions between target and projectile nucleons. In accordance with explicit calculations, the mean-field driven transfer contributions are neglected because they are at least of 4th order for DCE reactions considered here [15]. Thus, only processes with changes of the charge partitions but leaving the projectile-target mass partition unaltered will be discussed.

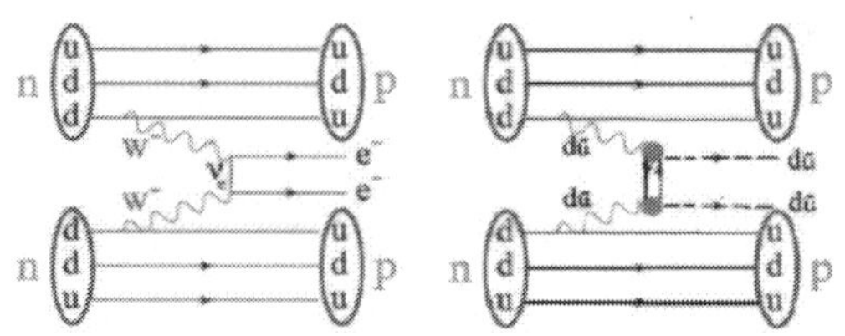

Figure 1. The elementary weak interaction process mediating nuclear $0\nu2\beta$ decay (left) and a corresponding strong interaction process (right) are depicted schematically. The QCD counterpart is given by the simultaneous emission of two $[d\bar{u}]$ pairs in an isovector-vector, e.g. 1^-, configuration (wavy lines), decaying into a pionic $[d\bar{u}]$ configuration and a charge-neutral $q\bar{q}$ pair.

A central question is whether we can identify on the elementary level a correspondences between strong and weak interaction processes. The answer is yes, as illustrated in Fig. 1. Under nuclear structure aspects, the $0\nu2\beta$ decay of a nucleus is nothing but special class of two-body correlation, sustained by the exchange of a (pair of) Majorana neutrino(s) between two nucleons where the interaction vertices are given by the emission of virtual $W^\pm$ gauge bosons. The strong interaction counterpart is a two-nucleon correlation built up by the exchange of a virtual charge-neutral quark-antiquark ($q\bar{q}$) pair accompanied by the emission of a charged $q\bar{q}$ component, thus changing at the same time the nucleonic charges. Similar to the weak process, the strong vertices are originating from gauge bosons, here given by the initial emission of gluons which materialize into two $q\bar{q}$ pairs. At the end, the highly off-shell $q\bar{q}$ compounds will decay into mesons, preferentially into pions but also multi-pion configurations like the scalar and vector mesons.

Both processes become of interest if they reveal their existence and nature in observable signals. In this respect, we encounter a fundamental difference between $0\nu2\beta$ decay and the hadronic process: Only the former may occur in an isolated nucleus while the latter one is inhibited by energy conservation. Thus, in order to observe the double-meson emission by a nucleon pair a partner nucleus is required which takes care of the virtuality of the process by absorbing the two charged virtual mesons. For that pur-

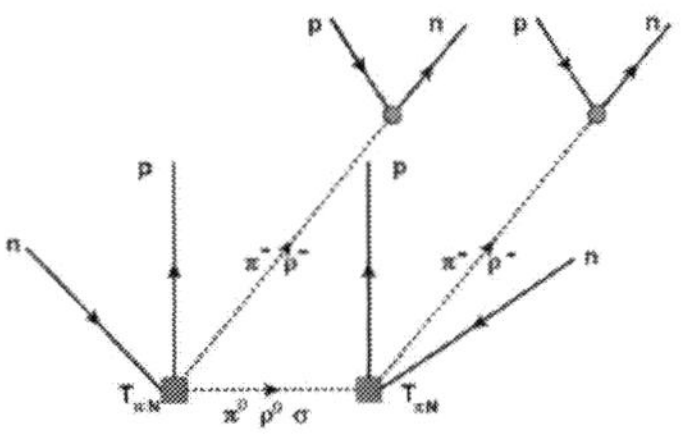

Figure 2. Generic diagram illustrating the hadronic surrogate process for $0\nu 2\beta$ decay. A virtual $nn \rightarrow pp\pi^-\pi^-$ scattering process, causing the $\Delta Z = +2$ target transition $A \rightarrow B$, is accompanied by $nnp^{-1}p^{-1}$ double-CC excitation in the projectile. Crossed diagrams are not displayed.

pose, heavy ion double charge exchange reactions are the ideal tool. The diagrammatic structure of such a reaction is indicated in Fig. 2. The target undergoes a correlated double-meson pair decay $nn \rightarrow pp\pi^-\pi^-$ (in fact a $2p2n^{-1}$ 2p-2h transition with coherent emission of a pion pair) and the projectile absorbs the pions by the simultaneous excitation of two np^{-1} -type configurations. Other mesons will contribute as well.

The whole reaction will proceed as a one-step reaction via a special kind of two-body interaction generated by the correlation diagram. Denoting the (in-medium) pion-nucleon T-matrix by $T_{\pi N,\pi' N'}$, the target-part of the interaction is in a somewhat symbolic notation

$$V^{(MDCE)}(\mathbf{13},\mathbf{24}) \sim T_{\pi^- p,\pi^0 n}(\mathbf{1},\mathbf{3})D_{\pi^0}(\mathbf{1}-\mathbf{2})T_{\pi^0 n,\pi^- p}(\mathbf{2},\mathbf{4}) \tag{6}$$

where the $n \rightarrow p$ target transitions are denoted by **1** and **2**, respectively. The coordinates **3** and **4** indicate the outgoing charged pions, inducing the complementary transitions in the projectile. The correlation built up by the neutral pion is described by the propagator D_{π^0}. A decomposition into irreducible tensors gives rise in particular to an effective rank-2 iso-tensor projectile-target interaction of operator structure $[\tau_{1A} \otimes \tau_{2A}]_2 \cdot [\tau_{3a} \otimes \tau_{4a}]_2$. We recognize immediately the similarity to the nuclear matrix element of $0\nu 2\beta$ decay, justifying the name Majorana-DCE. At present, the strengths of the nucleon-nucleon and the pion-nucleon T-matrices are taken from data. More refined description will be scrutinized in future work by an effective field theoretical description and by referring to the data base available for free space $nn \rightarrow pp\pi^-\pi^-$ reactions. Previously, the charge-conjugated reaction $pp \rightarrow nn\pi^+\pi^+$ reaction and other double-pion production channels were investigated at CELSIUS and COSY [16–20] and more recent also at HADES [21]. Theoretical studies of the on-shell reaction combining meson exchange and resonance excitation are found in [22, 23].

4 Heavy Ion DCE Reactions and Data

The full DCE reaction amplitude is given by the coherent sum of the MDCE and the DSCE amplitudes:

$$\begin{aligned} M_{\alpha\beta} &\sim \langle \chi_\beta^{(-)\dagger}, bB | V^{(MDCE)} + V^{(DSCE)} | aA, \chi_\alpha^{(+)} \rangle \\ &= M_{\alpha\beta}^{(MDCE)} + M_{\alpha\beta}^{(DSCE)} \end{aligned} \tag{7}$$

The results discussed below were obtained fully quantum mechanically by one-step DWBA calculations for the MDCE amplitudes and second order DWBA calculations for the DSCE amplitudes, respectively. Thus, ion-ion interactions are treated to all orders. As discussed in Ref. [13] the strong absorption in grazing ion-ion collisions allows to evaluate the distortion coefficient in black disk approximation. As mentioned before, under such conditions the ISI/FSI effects are resulting effectively in a scaling factor, allowing to relate at forward angles the cross sections to the corresponding plane wave cross sections. Thus, in principle spectroscopic information can be extracted from the data, provided the elastic interactions are known to the necessary precision. In the present calculations, double folding potentials have been used.

First results for a DCE reaction along the line discussed above were measured recently by the NUMEN collaboration [24] for the reaction $^{18}O + ^{40}Ca \rightarrow ^{18}Ne + ^{40}Ar$ at $T_{lab} = 15$ AMeV. The reaction leads from the 0^+ ground states of the initial to the 0^+ ground states of the final nuclei, constraining the total angular momentum transfer to $J^P = 0^+$. Second-order DWBA calculations with QRPA transition strengths have been performed along the line of Ref. [13]. For these exploratory investigations only the pionic contributions were included. This leaves open an overall scaling factor which was fixed by normalizing the MDCE cross section to the data point at the smallest scattering angle. The forward peak of the angular distribution is dominated, in fact, by the $(L = 0, S = 1)$ MDCE component, but the $(L = 2, S = 2)$ components are of comparable size at larger scattering angles. Moreover, they are essential for the description of the data. Overall, the shape of the measured angular distribution is described decently well in view of the exploratory character of the calculations. In Fig. 3, DSCE and MDCE cross sections are displayed and compared to data as a function of the momentum transfer. Remarkably, the measured angular range ($\theta \lesssim 12^\circ$) covers a momentum range of more than 400 MeV/c. The DSCE cross section was normalized to the large angle region because higher order reactions typically prevail at larger momentum transfers. That conjecture is confirmed by the DSCE angular distribution: Aside from the typical L=0 forward structure the main body of the angular distribution oscillates around a mean-value of a few times $10^{-4} mb/sr$. Cross sections of higher multipolarities are carrying less strengths and are of flatter shape. With all caution, we may conclude that the data are in favor of the one-step MDCE angular distribution. Even if the DSCE cross section would be scaled to the measured forward angle cross section, its shape would not match the observed angular distribution, an observation giving further evidence to the dominance of the MDCE one-step reaction mechanism.

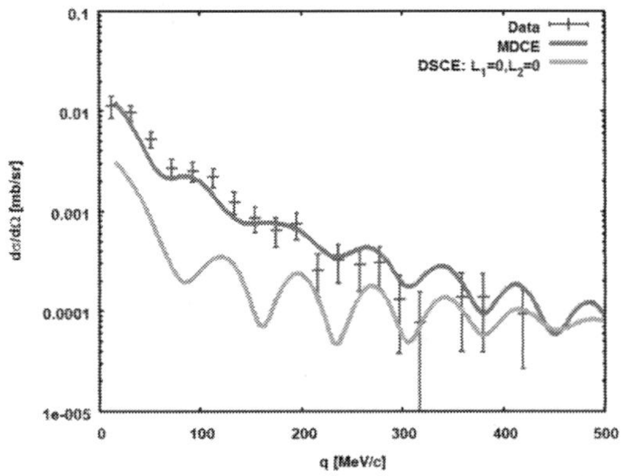

Figure 3. Angular distribution of the DCE reaction $^{18}O+^{40}Ca \rightarrow$ $^{18}Ne+^{40}Ar$ at T_{lab} = 270 MeV. The one-step MDCE and the two-step DSCE cross sections are shown separately in comparison to the data of Ref. [24].

5 Outlook

A new theoretical scenario for heavy ion double charge exchange reactions was introduced. At the diagrammatic level, structures similar to $0\nu2\beta$ matrix elements have been identified. The hadronic Majorana-DCE process is accessible only by reactions of composite nuclei. We have discussed explicitly the case of a DCE reaction with medium mass ions at relatively low incident energy. ISI and FSI ion-ion interactions were taken into account and the quantum mechanical coherence of the MDCE and the DSCE reaction mechanism was treated properly. The strongly forward peaked measured angular distributions indicate a direct mechanism which indeed is confirmed by the calculations. These first results are very promising by indicating a new way of accessing second order nuclear matrix elements of charge changing interactions. Together with the much better studied SCE reactions and their established usefulness for spectroscopic work, heavy ion DCE reactions are opening a new window to high-precision spectroscopy. Although it will not be possible to insert the extracted matrix elements directly into a $0\nu2\beta$ analysis, DCE reactions provide an unique way to validate nuclear structure models under controllable laboratory conditions by comparison to data on processes of comparable physical content. New impact on theoretical investigations in both reaction and nuclear structure theory is demanded for a quantitative understanding of these special reactions. Although the present calculations do not yet include the full spectrum of contributions, they are establishing the hadronic Majorana-DCE reaction mechanism. The refinements may lead to changes in detail but will not alter the overall picture. An exciting and encouraging result is that the MDCE process is clearly visible, even dominating the cross section at extreme forward angles.

References

[1] W.G. Haxton, G.J. Stephenson, Prog. Part. Nucl. Phys. **12**, 409 (1984)
[2] T. Tomoda, Rept. Prog. Phys. **54**, 53 (1991)
[3] J. Suhonen, O. Civitarese, Phys. Rept. **300**, 123 (1998)
[4] J.D. Vergados, H. Ejiri, F. Šimkovic, Rep. Prog. Phys. **75**, 106301 (2012)
[5] O. Cremonesi, M. Pavan, Adv. High Energy Phys. **2014**, 951432 (2014), 1310.4692
[6] M. Goeppert-Mayer, Phys. Rev. **48**, 512 (1935)
[7] D. Stefanik, F. Simkovic, A. Faessler, Phys. Rev. C **91**, 064311 (2015), arXiv:1506.00835 [nucl-th]
[8] A.S. Barabash, Nucl. Phys. **A935**, 52 (2015), 1501.05133
[9] D. Frekers, A. M., Eur. Phys. J. **A 54**, 177 (2018)
[10] M. Lenske, F. Cappuzzello, M. Cavallaro, M. Colonna, Prog. Part. Nucl. Phys. **NN**, in print (2019)
[11] F. Cappuzzello et al., Eur. Phys. J. **A54**, 72 (2018)
[12] C.D. Goodman, C.A. Goulding, M.B. Greenfield, J. Rapaport, D.E. Bainum, C.C. Foster, W.G. Love, F. Petrovich, Phys. Rev. Lett. **44**, 1755 (1980)
[13] H. Lenske, J.I. Bellone, M. Colonna, J.A. Lay, Phys. Rev. **C98**, 044620 (2018), 1803.06290
[14] H. Lenske, J. Phys. Conf. Ser. **1056**, 012030 (2018)
[15] J.A. Lay, S. Burrello, J.I. Bellone, M. Colonna, H. Lenske, Lenske (within the NUMEN project), J. Phys. Conf. Ser. **1056**, 012029 (2018)
[16] J. Johanson et al., Nucl. Phys. A **712**, 75 (2002)
[17] H. Clement et al., Int. J. Mod. Phys. A **20**, 1747 (2005)
[18] S.A. El-Bary et al., Eur. Phys. J. A **37**, 267 (2008)
[19] T. Tsuboyama, F. Sai, N. Katayama, T. Kishida, S.S. Yamamoto, Phys. Rev. C **62**, 034001 (2000)
[20] V.V. Sarantsev et al., Phys. At. Nucl. **70**, 1885 (2007)
[21] G. Agakishiev et al., Phys. Lett. B **750**, 184 (2015), arXiv:1503.04013 [nucl-ex]
[22] L. Alvarez-Ruso, E. Oset, E. Hernandez, Nucl. Phys. A **633**, 519 (199)
[23] X. Cao, B.S. Zou, H.S. Xu, Phys. Rev. C **81**, 065201 (2010), [arXiv:1004.0140 [nucl-th]]
[24] F. Cappuzzello et al., Eur. Phys. J. **A51**, 145 (2015), 1511.03858

Revealing microscopic origins of shape coexistence in the Ni isotopic chain

Silvia Leoni[1,*], *Bogdan* Fornal[2], *Nicolae* Marginean[3], *Michele* Sferrazza[4], *Yusuke* Tsunoda[5] and *Takaharu* Otsuka[5,6,7,8,9]

[1] Department of Physics, University of Milano and INFN, Via Celoria 16, 20133, Milano, Italy
[2] Institute of Nuclear Physics, PAN, 31-342 Kraków, Poland
[3] Horia Hulubei National Institute of Physics and Nuclear Engineering - IFIN HH, Bucharest 077125, Romania
[4] Département de Physique, Université libre de Bruxelles, B-1050 Bruxelles, Belgium
[5] Center for Nuclear Study, University of Tokyo, Hongo, Bunkyo-ku, Tokyo 113-0033, Japan
[6] Department of Physics, University of Tokyo, Hongo, Bunkyo-ku, Tokyo 113-0033, Japan
[7] RIKEN Nishina Center, 2-1 Hirosawa, Wako, Saitama 351-0198, Japan
[8] Instituut voor Kern- en Stralingsfysica, KU Leuven, B-3001 Leuven, Belgium
[9] National Superconducting Cyclotron Laboratory, Michigan State University, East Lansing, Michigan 48824, USA

Abstract. In a two-neutron transfer experiment, performed in Bucharest in July 2016 at sub-Coulomb barrier energy, a photon decay hindered - solely - by a nuclear shape change was identified in the ^{66}Ni nucleus. Such a rare process, at spin zero, was clearly observed before only in actinide nuclei in the 1970's, where fission isomers were found. The experimental findings on ^{66}Ni have been well reproduced by the Monte Carlo Shell Model Calculations of the Tokyo group, which predict a multifaceted scenario of coexistence of spherical, oblate and prolate shapes in neutron-rich Ni isotopes. The results on ^{66}Ni encouraged a comprehensive gamma-spectroscopy investigation of neutron-rich Ni isotopes, in particular ^{62}Ni and ^{64}Ni, at IFIN-HH (Bucharest), IPN Orsay and ILL (Grenoble), employing different reaction mechanisms to pin down the wave function composition of selected excited states. The aim is to shed light on the microscopic origin of deformation in neutron-rich Ni nuclei, possibly locating other examples of "shape-isomer-like" structures in this region.

1 Introduction

The shape is one of the most fundamental properties of the atomic nucleus. Spherical shapes naturally appear in the vicinities of magic nuclei. On the contrary, when moving away from shell closures, different shapes may compete, and often coexist in the very same nucleus at low spins, in a restricted interval of low excitation energy [1,2]. Shapes of quadrupole symmetric forms are the most abundant, a special class being represented by superdeformed states: they have been experimentally observed next to spherical/oblate excitations at high spin and high energy, even in the proximity of magic numbers [3-5].

Low-lying 0^+ excited states, characterized by deformations different from the ground state, represent one of the most striking fingerprints of shape coexistence, in even-even systems. They have been discovered in different regions of the nuclear chart, from the "light" Si/Mg and Ni/Zn/Ge regions, to the much heavier Po/Pb/Hg and Rn/Ra systems. Indeed, it is in the U region that, back in the 60's, the first and most striking examples of shape coexistence, namely shape isomers, were discovered. In ^{236}U and ^{238}U, E2 γ branches from excited 0^+ states showed a retardation of the order of 10^7, in competition with fission [6-8]. Such states were interpreted as deformed structures, residing in secondary minima of the nuclear potential energy surface (PES), separated by a high barrier from the main ground state configuration [9]. Shape isomerism has been predicted to occur also in other regions of the nuclear chart, in particular in the neutron-rich Ni region, by various mean-field models [10-12]. However, it is only very recently that an example of "shape-isomer-like" structure was found by our collaboration in the ^{66}Ni nucleus [13], while coexistence of spherical, oblate and prolate shapes along the Ni isotopic chain was observed in experiments performed at ISOLDE/CERN, MSU and RIKEN [14-16]. The phenomenon of shape coexistence in neutron-rich Ni isotopes can nowadays be also described by fully microscopic approaches, taking advantage of the most powerful supercomputing systems. One of the most prominent examples is the Monte Carlo Shell Model (MCSM) of the Tokyo group [17], which correctly predicts the occurrence of shape isomerism in ^{66}Ni, together with other possible cases in the lighter ^{62,64}Ni stable isotopes, although with reduced magnitude.

In this proceeding we report on the research program of our collaboration, aiming at a detailed investigation, with different probes, of the microscopic structure of neutron-rich Ni isotopes, from mass A=62 to A=66. After recalling the observation of a "shape-isomer-like" structure in ^{66}Ni [13], obtained in a 2016 experiment performed at the Tandem Laboratory of the Horia Hulubei National Institute of Physics and Nuclear Engineering (IFIN-HH), in Bucharest, we briefly describe the following series of measurements on the Ni chain, currently ongoing in Bucharest, IPN-Orsay and ILL (Grenoble). The aim of this research program is to shed light on the microscopic origin of nuclear deformation, not least shape isomerism, using different reaction mechanisms, to probe MCSM predictions.

* Corresponding author: silvia.leoni@mi.infn.it

2 Experimental Investigation

The research program of our collaboration concerns the detailed spectroscopic investigation of neutron-rich Ni nuclei, from ^{62}Ni to ^{66}Ni, by employing sub-Coulomb barrier transfer reactions with heavy-ion stable beams and neutron capture reactions with intense neutron beams (see Fig. 1). Particular emphasis is given to lifetime measurements, as well as to comparative studies of the population of specific states by different reaction mechanisms (e.g., proton and neutron transfer reactions with ^{11}B and ^{18}O heavy ions). Such reactions are found to selectively populate specific excited states, thus suggesting a sensitivity to the state wave function composition.

We like to note that similar spectroscopy measurements in heavier Ni isotopes (see Fig. 1) would necessarily require intense radioactive beams from new generation facilities, such as SPES, HIE-ISOLDE or SPIRAL2.

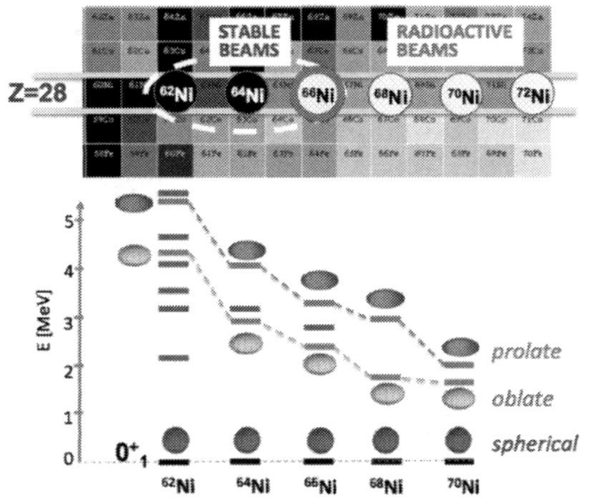

Fig. 1. Top: region of the nuclear chart focusing on neutron-rich Ni isotopes in the proximity of the valley of stability. Even-even A=62, 64, 66 (and the even-odd 65), are currently investigated by the present collaboration, employing intense stable ion beams and neutron beams. Heavier Ni isotopes require, instead, intense radioactive beams. Bottom: Monte Carlo Shell Model predictions for 0^+ states of spherical (blue), oblate (green) and prolate (red) shape, for Ni isotopes between A = 62 and 70.

2.1. Spectroscopy of ^{62}Ni, ^{64}Ni and ^{66}Ni by two-neutron transfer reactions

The first experiment of this research program, on the Ni chain, was performed by our collaboration in July 2016, at the Tandem Laboratory of IFIN-HH, in Bucharest. It focused on the spectroscopy of ^{66}Ni populated by the two-neutron transfer reaction induced by a ^{18}O beam on a ^{64}Ni target, at the sub-Coulomb barrier energy of 39 MeV [13]. This low energy severely hindered the fusion-evaporation channel. The experiment made use of a steady beam current of about 30 pnA, for more than 40 days and the γ-decay from excited states in ^{66}Ni was measured with the ROSPHERE HPGe array [18], consisting at that time of 14 HPGe detectors and 11 LaBr$_3$(Ce) scintillators. The first part of the experiment was performed with a 5 mg/cm^2 thick target, in order to obtain a very clean spectrum by gating on the 1425-keV $2^+ \rightarrow 0^+$ ground state transition. All observed transitions were found to belong to ^{66}Ni and depopulate states reported by Broda et al. [19] at excitation energy below ≈ 4.1 MeV. After a closer inspection of the gamma-ray spectra, several transitions clearly showed tails, pointing to emission in flight during the slowing down process of ^{66}Ni inside the target. By using a Doppler Shift Attenuation Method (DSAM) lineshape analysis [20-21], half lives values in the range from 1 to 2 ps were obtained for a total of six states, i.e., the 7-, 3+, 4+, 2+, 4+ and 3- states at 4089, 2971, 3185, 3229, 3614 and 3687 keV depopulated by the 490-, 1546-, 1760-, 1804-, 2189- and 2262-keV lines, respectively. Extracted B(E/Mλ) values resulted in fair agreement with MCSM predictions, as shown in Table 1 (for the 3- state at 3687 keV, no predictions are available). This analysis gives strong support to the predictive power of the model, which turns out to reproduce extremely well also the beta-decay properties of ^{66}Co into ^{66}Ni [22].

In the second part of the experiment, the lifetimes of all three 0^+ excited states (located at 2443-, 2671- and 2974-keV excitation energies, respectively) were extracted via the plunger technique, using the same ^{18}O+^{64}Ni reaction and a 5 mg/cm^2 thick Ta stopper, placed between 10 and 3000 μm from a 1 mg/cm^2 ^{64}Ni target (12 distances were considered). Half lives values of 7.6(8), 134(9) and 20(7) ps were obtained, corresponding to B(E2) decay probabilities of 4.3±0.5, 0.09±0.01, and 0.21±0.07 W.u., for the 0^+_2, 0^+_3 and 0^+_4 states, respectively. In the case of 0^+_3, the independent measurement of Olaizola et al. [23] confirmed our value. Table 1 summarizes the results obtained from the lifetime analysis of ^{66}Ni.

A detailed analysis of the wave functions composition of the excited 0^+ states, calculated by the MCSM, was then performed to interpret the experimental results. It turned out that in the case of 0^+_3, the retardation of the E2 decay is arising from cancellation effects among E2 matrix elements, while for the 0^+_4 state located at 2974 keV, the hindrance is of different nature. In this case, it is caused by a sizable barrier in the potential energy surface (PES), separating a deep prolate secondary minimum, in which the 0^+_4 resides, from the spherical ground state minimum. Further, it was found that this prolate local minimum of 0^+_4 arises from a promotion of neutrons up to the $g_{9/2}$ orbital, what reduces the proton shell gap at Z=28 and, in consequence, favors the promotion of protons across this gap [17]. Such a result makes ^{66}Ni a unique example of nuclear system, apart from the much heavier actinides, in which a photon decay is hindered - solely - by a nuclear shape change (a "shape-isomer-like" structure). Figure 2, on the right, shows a comparison between a partial level scheme of ^{66}Ni, focusing on the decay from the 0^+ states, and predictions from the MCSM.

As shown in the bottom of Fig. 1, MCSM calculations predict a scenario of coexisting shapes also in ^{62}Ni and ^{64}Ni, with the possible appearance of "shape-isomer-like" structures similarly to the case of ^{66}Ni, although occurring at higher excitation energy and with reduced magnitude. In ^{64}Ni, as in ^{66}Ni, the first three excited 0^+ states are calculated to be oblate, spherical and prolate, respectively, while in ^{62}Ni up to 9 excited 0^+

	E [keV]	$T_{1/2}$ [ps]	Eγ [keV]	$I_i \rightarrow I_f$ (E/Mλ)	B(E/Mλ)$_{exp.}$ [W.u.]	B(E/Mλ)$_{Th.}$ [W.u.]
PLUNGER	2443	7.6(8)	1018	$0^+ \rightarrow 2^+$ (E2)	4.3(5)	4.1
PLUNGER	2671	134(9)	1245	$0^+ \rightarrow 2^+$ (E2)	0.09(1)	0.01
PLUNGER	2974	20(7)	1549	$0^+ \rightarrow 2^+$ (E2)	0.21(7)	0.006
PLUNGER	3541	1456(478)	356	$5^- \rightarrow 4^+$ (E1)	6.0×10^{-6}	-
DSAM	2971	0.9(1)	1546	$3^+ \rightarrow 2^+$ (M1)	$6.3(7)\times10^{-3}$	2.8×10^{-3}
DSAM	3185	1.4(1)	1760	$4^+ \rightarrow 2^+$ (E2)	1.5(1)	5.9
DSAM	3229	1.5(2)	1804	$2^+ \rightarrow 2^+$ (M1)	$2.5(4)\times10^{-3}$	2.4×10^{-3}
DSAM	3614	0.6(1)	2189	$4^+ \rightarrow 2^+$ (E2)	1.1(1)	0.74
DSAM	3687	0.9(1)	2262	$3^- \rightarrow 2^+$ (E1)	$3.9(5)\times10^{-5}$	-
DSAM	4089	1.3(2)	490	$7^- \rightarrow 6^-$ (M1)	0.15(2)	0.073

Table 1. Summary of the ^{66}Ni lifetime analysis results obtained with Plunger (top) and DSAM (bottom) techniques, in comparison with theory predictions from MCSM [17].

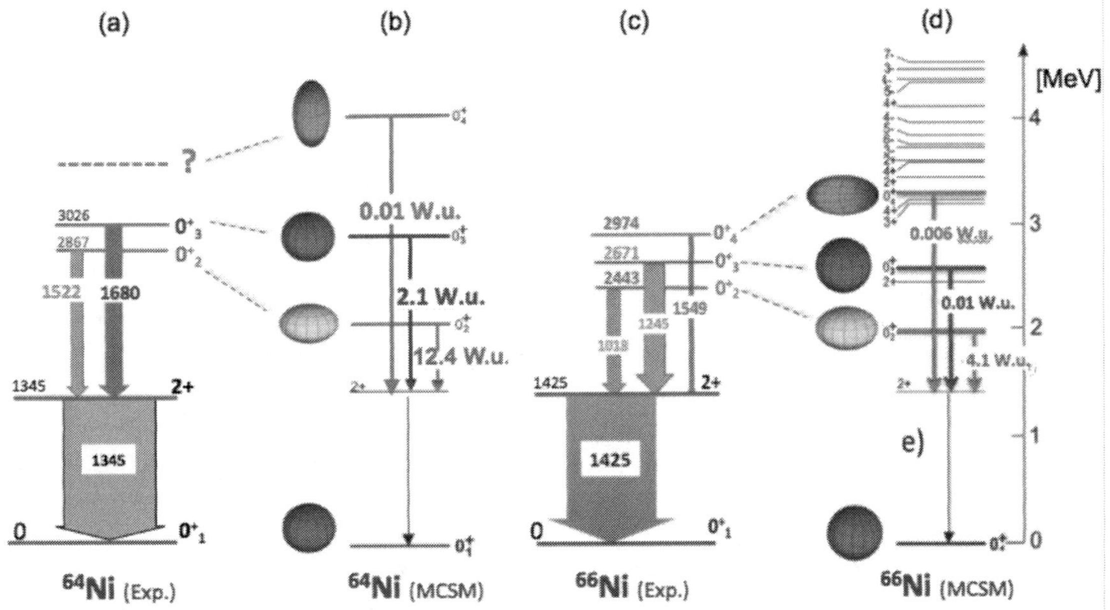

Figure 2. Partial level schemes of ^{64}Ni (a) and ^{66}Ni (c), focusing on the decay from 0^+ states, as observed experimentally. Corresponding calculations from Monte Carlo Shell Model are shown in (b) and (c), with cartoons indicating the predicted intrinsic shape for the 0^+ states (i.e., spherical (blue), oblate (green) and prolate (red)). Calculated B(E2) values from each 0^+ excitation to the first excited 2^+ state are also given in Weisskopf units (W.u.).

states are expected. They are all predicted to be spherical, apart from 0^+_6 and 0^+_8, which should be characterized by oblate and prolate shapes, respectively. According to calculations, ^{62}Ni should mark the appearance of coexisting deformations in the Ni isotopes, in the neutron rich side of the nuclear chart.

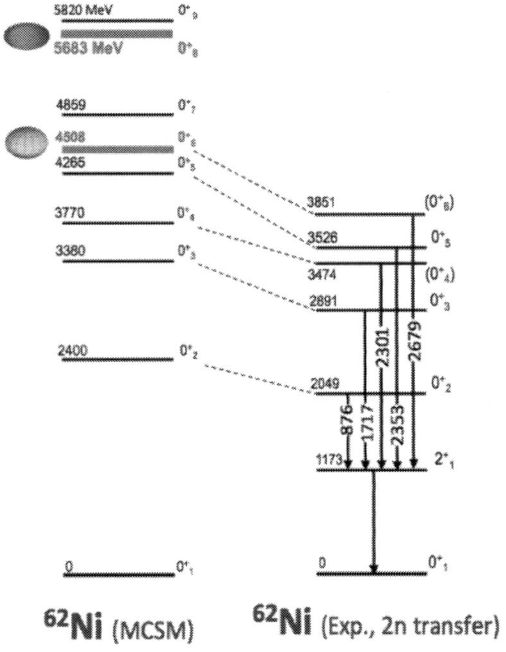

Figure 3. Preliminary comparison between MCSM predictions (left) and experimental data (right) for excited 0^+ states in ^{62}Ni. The experiment was performed with the Nu-Ball array at ALTO IPN-Orsay, employing the 2n-transfer reaction ^{60}Ni(^{18}O,^{16}O)^{62}Ni, at sub-Coulomb barrier energy.

Encouraged by the above predictions, our collaboration has started a detailed investigation of 0^+ excitations in ^{62}Ni and ^{64}Ni, by employing two-neutron transfer reactions, at sub-Coulomb barrier energy, with a ^{18}O beam an ^{60}Ni and ^{62}Ni targets. The ^{64}Ni experiment was performed at IFIN-HH, with the ROSPHERE array [18], while the ^{62}Ni investigation was carried out at ALTO IPN-Orsay, with the Nu-Ball array, consisting of 24 Compton-suppressed HPGe Clover detectors (from EUROBALL), 10 Compton-suppressed HPGe coaxial detectors and 20 $LaBr_3$ scintillator detectors, for lifetime measurements by fast-timing techniques.

Figures 2 (left) and 3 show preliminary results from the analysis of ^{64}Ni and ^{62}Ni two neutron-transfer data, in comparison with MCSM predictions. In the case of ^{64}Ni, the decay from the second and third excited 0^+ states, located at 2867 and 3026 keV, is clearly observed (see also top panel of Fig. 5), while no firm conclusions can be drawn, thus far, on the location of the prolate deformed 0^+ state, predicted by theory around 4 MeV. Preliminary results from plunger measurements provide lifetime values in the ps range for both 0^+_2 and 0^+_3, in fair agreement with MCSM calculations, thus giving further support to their predictive power.

In the case of ^{62}Ni, a preliminary spectroscopic investigation allowed to identify the γ decay from the first 5 excited 0^+ states, i.e., $0^+_2$to 0^+_6 (see Fig. 3 and Fig. 4) - the spin assignment of the proposed 0^+_4 is being confirmed by angular correlation studies. According to MCSM calculations, the highest observed 0^+ excitation, i.e., the 0^+_6 state, should be of oblate nature, while the prolate 0^+ state, i.e., 0^+_8, should be more than 1 MeV higher in excitation energy. It is interesting to note that one of the two 0^+ states, predicted to be deformed by the MCSM, could be the band head of the positive parity rotational band recently identified by Albers et al. in a high-spin experiment performed with GAMMASPHERE and FMA [24]. If this scenario is experimentally confirmed, the connection between deformation, developing at high spins, and the occurrence of coexisting shapes, at spin 0, could be established for the first time.

2.2. Selectivity of transfer reactions

In order to gain further insight into the structure of the excited 0^+ states, observed in ^{62,64}Ni in the 2n transfer studies discussed above, proton-pick-up measurements induced by a ^{11}B beam on ^{63}Cu and ^{65}Cu targets, at sub-Coulomb barrier energy, were performed in 2018 and 2019 at IFIN-HH. As shown in Figs. 4 and 5, a different selectivity in the population of excited states in ^{62}Ni and ^{64}Ni is clearly observed when comparing the spectra from the two-neutron transfer and one-proton pick-up reactions. As a general feature, in both nuclei the population of negative parity states is found to be strongly suppressed in the proton-pick-up reaction (see, for example, the 1825-keV 5- → 4+ decay in ^{62}Ni, in Fig. 4, and the 1475-keV 5- → 4+ or the 2213-keV 3- → 2+ decays in ^{64}Ni, in Fig. 5).

What concerns the 0^+ states, in both ^{62}Ni and ^{64}Ni a clear change in their relative population is observed, when comparing the two-neutron-transfer and one-proton-pick-up reactions. For both nuclei, the proton-pick-up reaction strongly suppresses the population of certain 0^+ states, i.e., 0^+_3 in ^{62}Ni, and both 0^+_2 and 0^+_3 in ^{64}Ni (only the 1521-keV γ ray from 0^+_2 is weakly seen), as shown in Figs. 4 and 5.

Finally, the most striking feature, when comparing the two different transfer reactions, is the appearance of new transitions in the proton-pick-up case. In ^{64}Ni, for example, two γ transitions are observed at 2138 and 2502 keV, which are not seen in the two-neutron-transfer reaction (see Fig. 5). They would depopulate states at 3484 and 3848 keV, respectively. The 2502-keV line is particularly strong and shows pronounced tails, what points to a lifetime comparable to the stopping time of the ^{64}Ni product in the target (i.e., ≤ 1 ps).

We like to note that Monte Carlo Shell Model Calculations may provide a first interpretation of the selective population observed in the proton-pick-up reactions, both for ^{62}Ni and ^{64}Ni. According to MCSM predictions, the structure of ^{63}Cu and ^{65}Cu – our target nuclei - is different: while the wave function of the $3/2^-$ ground state of ^{63}Cu shows fragmented contributions distributed over spherical and (slightly deformed) oblate and prolate configurations, the ground state of ^{65}Cu is expected to exhibit a more defined spherical-oblate structure. This may play a role in determining the relative population of the different observed states (including 0^+

states) in the final ^{62}Ni and ^{64}Ni nuclei, after a proton removal from the ground state of ^{63}Cu and ^{65}Cu, respectively. In particular, it is found that the observed relative population of specific states in ^{62}Ni and in ^{64}Ni, could be, at least partially, explained, by looking at the Spectroscopic Factors (SF) predicted by the MCSM – they define the overlap between the initial state (i.e., the ground state of the target nucleus) and the final state wave functions. For example, in ^{62}Ni, the direct population of 0^+_3 and 0^+_4, with respect to 0^+_2, is about 0.2(1) and 1.1(1), to be compared with the corresponding 0.2 and 2.2 SF's ratios. Similarly, in ^{64}Ni, the experimental population of 0^+_3 with respect to 0^+_2 is about 0.1(1), while the SF's ratio gives 0.03. Such qualitative observations support the need for a detailed comparison between experiment and theory, in which both the microscopic structure of the state of interest and the specific features of the reaction mechanism are carefully taken into account.

To shed more light on the complex structure of both ^{62}Ni and ^{64}Ni, further experimental investigations are planned, in order to fully characterize the newly located excitations – they include neutron capture reactions with intense neutron beams at Institut Laue-Langevin (ILL, Grenoble), where the FIPPS HPGe array [25] can be used for high-precision γ-spectroscopy measurements.

3 Conclusion

A photon decay from a 0^+ state in ^{66}Ni, hindered, solely, by a nuclear shape change, has been recently discovered by our collaboration [17]. This is a clear-cut evidence of a "shape-isomer-like structure", an extremely rare case of totally different microscopic configurations (prolate and spherical in this case) coexisting at similar excitation energy. This experimental finding is well described by large scale shell model calculations based on Monte Carlo computational schemes, which provide a microscopic description of neutron-rich Ni isotopes and predict a coexistence of spherical, oblate and prolate structures along the entire chain. According to calculations, a deep prolate (secondary) minimum could also appear in lighter even-even neutron-rich Ni isotopes, such as ^{62}Ni and ^{64}Ni, although at higher excitation energies and with possibly reduced magnitude.

Inspired by the above results, our collaboration has started en extended research program aiming at a high-precision γ-spectroscopy investigation of neutron rich Ni isotopes, from A = 62 to 66. Experimental campaigns are currently ongoing in Bucharest, ALTO IPN-Orsay and ILL (Grenoble), employing different reaction mechanisms to probe the state wave function composition. Preliminary results clearly show a different/strong selectivity in the population of specific states, when using reactions involving transfer of neutrons or protons, thus pointing to the need of a theoretical interpretation in which both structural properties and reaction dynamics are properly taken into account.

It is quite certain that such a detailed experimental investigation, performed with a strong support from theory, will significantly contribute to our understanding of the microscopic origin of nuclear deformation, a key issue in nuclear structure physics.

This work was supported by the Italian Istituto Nazionale di Fisica Nucleare, by the Polish National Science Centre under Contracts No. 2014/14/M/ST2/00738 and No. 2013/08/M/ST2/00257, by the Romanian MCI Project Nucleu PN 19 06 01 02 and by the Interuniversity Attraction Poles Programme (IAP) via the Belgian Science Policy Office (BriX network P7/12). It was supported by the Grants No. PN-II-RU-TE-2014-4-2003 and No. PNIII-IFA-CERN-RO-03-ISOLDE. It was also partially supported by Grants-in-Aid for Scientific Research (23244049), by HPCI Strategic Program (hp150224), by MEXT and JICFuS, by Priority Issue (Elucidation of the fundamental laws and evolution of the universe) to be Tackled by Using Post "K" Computer (hp160211, hp170230), and by CNS-RIKEN joint project for largescale nuclear structure calculations.

References

1. J.L. Wood et al., Phys. Rep. 215, 101 (1992).
2. K. Heyde and J. L. Wood, Rev. Mod. Phys. **83**, 1467 (2011).
3. B. Singh, R. Zywina, and R.B. Firestone, Nuclear Data Sheets **97**, 241 (2002).
4. A. Lopez-Martens et al., Prog. Part. Nuc. Phys. **89**, 137 (2016).
5. S. Leoni and A.Lopez-Martens, Phys. Scr. **91**, 063009 (2016).
6. S. M. Polikanov, Sov. Phys. Usp. **15**, 486 (1973).
7. J. Kantele et al., Phys. Rev. Lett. **51**, 91 (1983).
8. P. Butler et al., J. Phys. G **6**, 1165 (1980).
9. P. Walker and J. Dracoulis, Nature **399**, 35 (1999).
10. P. Bonche et al., Nucl. Phys. A 500, 308 (1989).
11. M. Girod et al., Phys. Rev. Lett. **62**, 2452 (1989).
12. P. Möller et. al., Phys. Rev. Lett. **103**, 212501 (2009).
13. S. Leoni, B. Fornal, N. Marginean et al., Phys. Rev. Lett. **118**, 162502 (2017).
14. B. P. Crider et al., Phys. Lett. B **763**, 108 (2016).
15. A.I. Morales et al., Phys. Rev. C **93**, 034328 (2016).
16. A.I. Morales et al., Phys. Lett. B **765**, 328 (2017).
17. Y. Tsunoda et al., Phys. Rev. C **89**, 031301 (2014).
18. D. Bucurescu et al., Nuc. Inst. Meth. Phys. Res. A **837**, 1 (2016).
19. R. Broda et al., Phys. Rev. C **86**, 064312 (2012).
20. P. Petkov et al., Nuc. Inst. Meth. Phys. Res. A **488**, 555 (2002).
21. P. Petkov et al., Nuc. Inst. Meth. Phys. Res. A **783**, 6 (2015).
22. M. Stryjczyk et al., Phys. Rev. C **98**, 064326 (2018).
23. B. Olaizola et al., Phys. Rev. C **95**, 061303(R) (2017).
24. M. Albers et al., Phys. Rev. C **94**, 034301 (2016)
25. C. Michelagnoli et al., EPJ Web Conf. **193**, 04009 (2018).

 https://doi.org/10.1051/epjconf/201922301032

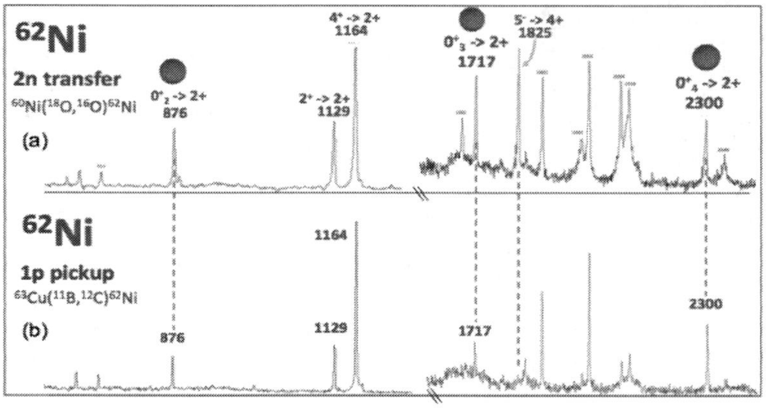

Figure 4. Portions of γ spectra of ^{62}Ni, as obtained from the two-neutron transfer reaction ^{60}Ni(^{18}O,^{16}O)^{62}Ni (a) and the proton pick-up reaction ^{63}Cu(^{11}B,^{12}C)^{62}Ni (b) (both performed at beam energies ~1 MeV below the Coulomb barrier energy), focusing on the transition energy region where the decay from the first three excited 0^+ states is expected. A clear difference in population pattern is generally observed, with strong suppression of transitions from negative parity states, in the case of the proton pick-up reaction (see for example the 1825-keV 5-→ 4+ decay).

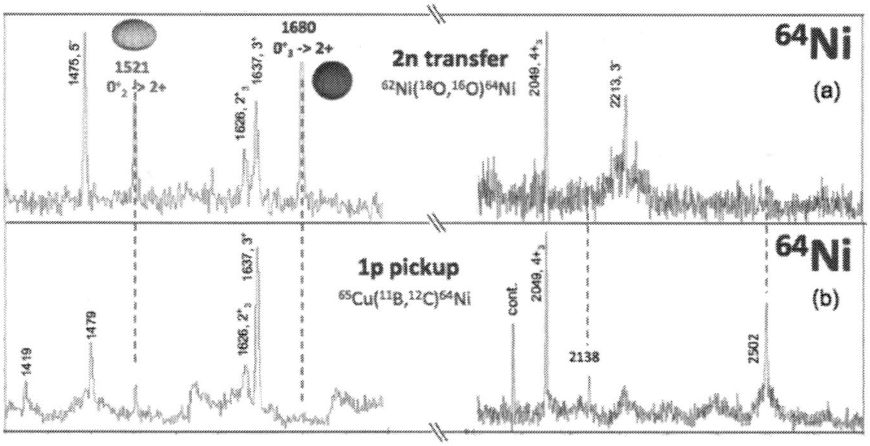

Figure 5. Portions of γ spectra of ^{64}Ni, as obtained from the two-neutron transfer reaction ^{62}Ni(^{18}O,^{16}O)^{64}Ni (a) and the proton-pick-up reaction ^{65}Cu(^{11}B,^{12}C)^{64}Ni (b) (both performed at beam energies ~1 MeV below the Coulomb barrier energy), focusing on the transition energy region where the decay from the second and third excited 0^+ states is expected. As in similar studies on ^{62}Ni (see Fig. 4), a clear difference in population pattern is generally observed. In the case of the proton pick-up reaction, a strong suppression in the population of both negative parity states and 0^+ states is observed (see text for details).

Nuclear response at zero and finite temperature

Elena Litvinova[1,2,*], *Peter* Schuck[3,4,**], and *Herlik* Wibowo[1,***]

[1]Department of Physics, Western Michigan University, Kalamazoo, MI 49008, USA
[2]National Superconducting Cyclotron Laboratory, Michigan State University, East Lansing, MI 48824, USA
[3]Institut de Physique Nucléaire, IN2P3-CNRS, Université Paris-Sud, F-91406 Orsay Cedex, France
[4]Université Grenoble Alpes, CNRS, LPMMC, 38000 Grenoble, France

Abstract. We present some recent developments on the nuclear many-body problem, such as the treatment of high-order correlations and finite temperature in the description of in-medium two-nucleon propagators. In this work we discuss two-time propagators of the particle-hole type, which describe the response of finite nuclei to external probes without nucleon transfer. The general theory is formulated in terms of the equation of motion method for these propagators with the only input from the bare nucleon-nucleon interaction. The numerical implementation was performed on the basis of the effective mason-nucleon Lagrangian in order to study the energy-dependent kernels of different complexity. The finite-temperature extension of the theory with $ph \otimes phonon$ configurations is applied to a study of the multipole response of medium-mass nuclei.

Introduction — Response of many-body quantum systems to external perturbations comprises a large class of problems studied in many subfields of quantum physics. The observed spectral distributions are uniquely associated with quantum correlation functions, which encrypt the entire information about strongly-correlated complex media. Response of atomic nuclei to various experimental probes represents a very rich playground to study correlation functions, which was further extended with the advent of rare beam facilities [1, 2]. Nuclear correlation functions have a long history of theoretical studies based on the quantum field theory (QFT) techniques. For instance, the idea of coupling between single-particle and emergent collective degrees of freedom in nuclei [3–5], which explained successfully many of the observed phenomena, can be related to the non-perturbative versions of QFT-based and, in principle, exact, equations of motion (EOM) for correlation functions in nuclear medium [6–11]. In this work we discuss how a hierarchy of non-perturbative approximations to the dynamical kernels of the EOM's for one-fermion and two-time two-fermion propagators can be mapped to the kernels of the phenomenological nuclear field theories (NFT), where such kernels are commonly referred to as particle-vibration coupling (PVC) or quasiparticle-phonon (QPM) models. This mapping provides an understanding of the emergent collective phenomena, explaining the mechanism of their formation from the underlying strongly-interacting degrees of freedom. Last decade, the latter type of approaches was linked to the contemporary density functional theories [12–15, 15–24], advancing the PVC models to self-consistent frameworks and providing a better understanding of the low-energy nuclear structure [25–29]. However, very little progress has been made on the conceptual advancements of the many-body aspects of the non-perturbative NFT's, except for some specific rare topics, such as PVC with charge-exchange phonons [18, 20] or PVC-induced ground state correlations [30, 31]. More specifically, a relatively little effort was made on developments and numerical implementations of NFT's beyond the two-particle-two-hole $2p-2h$ level [32, 33], although the phenomenological multiphonon approach [5, 34, 35] indicates the possibility to meet the shell-model standards in large model spaces. Also, the problem of consistent linking those approaches with the underlying bare interactions remains unsolved. In the present work we revise the past and recent developments of the EOM method and compare them to the PVC approach in the time blocking approximation (TBA)[13, 36]. As an outcome, we formulate an extended strongly-coupled theory, which includes higher-order configurations with emergent long-range correlations corresponding to collective degrees of freedom. Moreover, we set a prospect for a consistent and systematically improvable approach to fermionic correlation functions, which is directly linked to the bare two-fermion interaction. The presented numerical implementations are, however, still based on an effective interaction, which allows us to test the long-range correlation sector of the theory at zero and finite temperatures.

Formalism and calculations — We consider a system of fermions with the Hamiltonian[1]:

*e-mail: elena.litvinova@wmich.edu
**e-mail: schuck@ipno.in2p3.fr
***e-mail: herlik.wibowo@wmich.edu

[1]We elaborate here on the detailed equation of motion based on this general Hamiltonian, which is most commonly used in nuclear physics. The generalizations to multinucleon Hamiltonians, relativity and adding explicit bosonic degrees of freedom are straightforward.

 https://doi.org/10.1051/epjconf/201922301033

$$H = \sum_{12} t_{12}\psi^{\dagger}{}_1\psi_2 + \frac{1}{4}\sum_{1234} \bar{v}_{1234}\psi^{\dagger}{}_1\psi^{\dagger}{}_2\psi_4\psi_3 = T + U, \quad (1)$$

where the matrix elements of kinetic energy $t_{12} = \delta_{12}\varepsilon_1$ and antisymmetrized interaction $\bar{v}_{1234} = v_{1234} - v_{1243}$ are given in the basis which diagonalizes t_{12}. The number subscripts '1' represent the full set of the one-fermion quantum numbers in this basis. The operators $\psi(1), \psi^{\dagger}(1)$ are fermionic (for instance, nucleonic) fields in the Heisenberg picture:

$$\psi(1) = e^{iHt_1}\psi_1 e^{-iHt_1}, \qquad \psi^{\dagger}(1) = e^{iHt_1}\psi^{\dagger}{}_1 e^{-iHt_1}, \quad (2)$$

The excitation spectrum of such a system is characterized by the two-time particle-hole propagator (response function):

$$R(12, 1'2') \equiv R_{12,1'2'}(t-t') = -i\langle T(\psi_1^{\dagger}\psi_2)(t)(\psi_{2'}^{\dagger}\psi_{1'})(t')\rangle =$$
$$= -i\langle T\psi^{\dagger}(1)\psi(2)\psi^{\dagger}(2')\psi(1')\rangle, \quad (3)$$

assuming that $t_1 = t_2 = t, t_{1'} = t_{2'} = t'$ and T is the chronological ordering operator, while the averaging in Eq. (3) $\langle ... \rangle$ is performed over the ground state in the case of $T = 0$ and over the grand canonical ensemble at $T > 0$.

First we discuss the case of zero temperature. Differentiation of Eq. (3) with respect to the time arguments generates the equation of motion for $R_{12,1'2'}(t-t')$, which takes the form of the Dyson equation after the Fourier transformation to the energy domain:

$$R(\omega) = R^{(0)}(\omega) + R^{(0)}(\omega)K(\omega)R(\omega). \quad (4)$$

The $R^{(0)}(\omega)$ denotes the amplitude of particle-hole propagation without interaction (uncorrelated propagator):

$$R^{(0)}_{12,1'2'}(\omega) = \frac{\mathcal{N}_{121'2'}}{\omega - \varepsilon_2 + \varepsilon_1}, \qquad \mathcal{N}_{121'2'} = \delta_{11'}\delta_{22'}\mathcal{N}_{12}, \quad (5)$$

where $\mathcal{N}_{12} = n_1 - n_2$ with $n_1 = \langle\psi^{\dagger}{}_1\psi_1\rangle$ being the occupation number of the single-particle state 1. The interaction kernel $K(\omega)$ of Eq. (4) is the amplitude

$$K_{12,1'2'}(t-t') = \Big[-\delta(t-t')\langle[[V, \psi^{\dagger}{}_1\psi_2], \psi^{\dagger}{}_{2'}\psi_{1'}]\rangle +$$
$$+ i\langle T[V, \psi^{\dagger}{}_1\psi_2](t)[V, \psi^{\dagger}{}_{2'}\psi_{1'}](t')\rangle^{irr}\Big]\mathcal{N}_{12}^{-1}\mathcal{N}_{1'2'}^{-1}, \quad (6)$$

irreducible in the particle-hole channel and Fourier-transformed to the energy domain. It is clearly split into the instantaneous $K^{(0)}$ and the time-dependent $K^{(r)}$ parts: $K(t-t') = K^{(0)}\delta(t-t') + K^{(r)}(t-t')$. The instantaneous term represents the self-consistent phonon-mean field which is, thereby, derived from the bare interaction and contains two-body fermionic density being the equal-times limit of the particle-hole response function (3) [37, 38]. The time-dependent part $K^{(r)}$ of the kernel K is given in the diagrammatic form in Fig. 1, where $K^{(r)} = \sum_{ij} K^{(r;ij)}$, the subscripts i, j take the values of 1 and 2, and we omitted constant factors $\pm i/4$ in front of each term [39]. All of them include the two-time two-particle-two-hole (4-fermion) Green function $G^{(4)}$ defined as:

$$G^{(4)}(543'1', 5'4'31) =$$
$$= \langle T(\psi^{\dagger}{}_1\psi^{\dagger}{}_3\psi_5\psi_4)(t)(\psi^{\dagger}{}_{4'}\psi^{\dagger}{}_{5'}\psi_{3'}\psi_{1'})(t')\rangle. \quad (7)$$

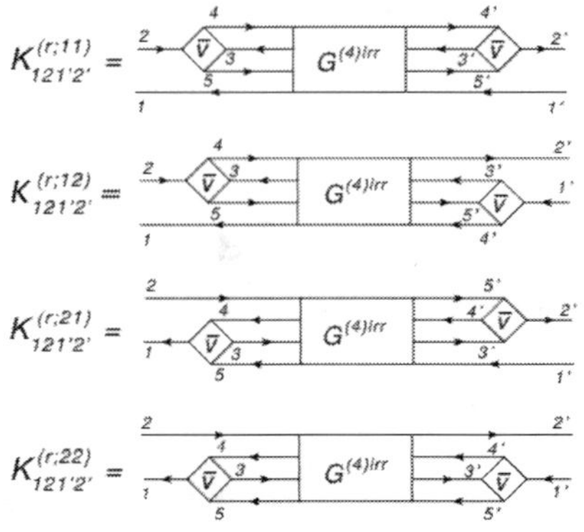

Figure 1. The components of the dynamical kernel $K^{(r)}_{12,1'2'}(t-t')$. Straight solid lines stand for fermionic propagators, the square blocks denote the antisymmetrized nucleon-nucleon interaction $\bar{v}$, and the rectangular blocks $G^{(4)}$ correspond to the two-particle-two-hole Green function (7).

Evaluating this propagator is the central problem for the description of the dynamical kernel beyond the lowest-order with respect to the interaction $\bar{v}$, that is essential for strongy-coupled systems. The direct EOM for this propagator generates higher-rank propagators, resulting in an infinite sequence of EOM's. Alternatively, efficient truncation schemes have been suggested, for instance, in Refs. [11, 37, 38], where $G^{(4)}$ is treated as a superposition of products of two-fermion particle-hole (ph) correlation functions (3) and particle-particle (pp) ones. As discussed in detail in Ref. [39], in this non-perturbative approximation the problem reduces to the closed system of equations:

$$\hat{R}(\omega) = \hat{R}^{(0)}(\omega) + \hat{R}^{(0)}(\omega)K[\hat{R}(\omega)]\hat{R}(\omega), \quad (8)$$

where $\hat{R} = \left\{R^{(ph)}, R^{(hp)}, R^{(pp)}, R^{(hh)}\right\}$ and the same structure is implied for the uncorrelated propagators $\hat{R}^{(0)}$.

To solve Eq. (8) in practice, one needs a reasonable initial approximation for $\hat{R}$ to generate a convergent iterative procedure. This turns out to be possible in approaches based on the nuclear energy density functionals, which are adjusted to approximate the static part $K^{(0)}$ of the kernel K by the effective interaction. In this case the initial $\hat{R}$ can be computed with only the static part of the integral kernel of Eq. (6), i.e. in the random phase approximation (RPA). Thus, reasonably good results were obtained in the relativistic quasiparticle time blocking approximation (RQTBA), where this static part is described by the exchange of effective mesons whose coupling constants and masses are adjusted to the nuclear ground state properties on the Hartree level [40]. The dynamical part of the kernel was modeled by the PVC in the lowest order with respect to the phonon coupling vertex, following the non-relativistic (quasiparticle) time blocking approximation ((Q)TBA) [36, 41, 42]. As we show in Ref. [39], the (RQ)TBA kernel is topologically equivalent to the part of the EOM kernel with a single two-fermion correlation function while another pair of fermions remains uncorrelated. This equivalence was found by performing the map-

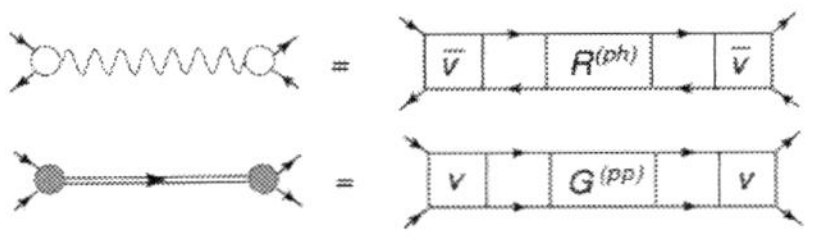

Figure 2. Diagrammatic mapping of the EOM to the PVC-TBA. Empty and filled circles denote the coupling vertices of the normal and pairing phonons, respectively, and the wiggly and double lines their propagators. $R^{(ph)}$ and $G^{(pp)}$ are the particle-hole response function and the particle-particle Green function, respectively.

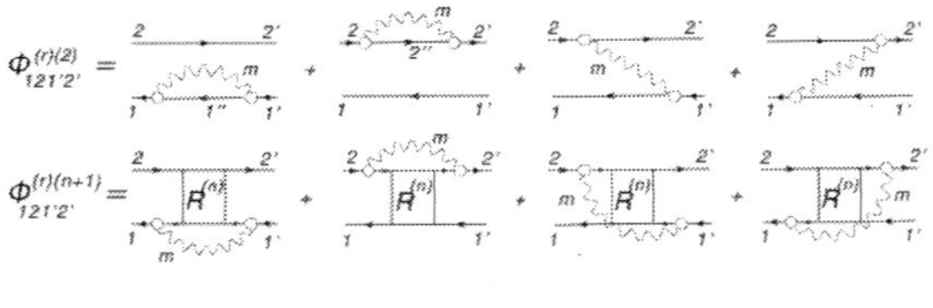

Figure 3. Dynamical kernel of the conventional PVC-TBA (top) and generalized EOM-TBA (bottom).

ping illustrated diagrammatically in Fig. 2 and the resulting (RQ)TBA kernel is displayed in the upper part of Fig. 3. The topological equivalence means that in the kernel of the (RQ)TBA approach based on the effective interaction this interaction replaces the bare interaction of the EOM while the one-fermion propagators become the mean-field propagators. Notice that the use of the effective interaction provides a good convergence of the iterations - in practice, one iteration is usually sufficient.

In this work we follow the EOM method and combine it with the relativistic nuclear field theory (RNFT). The corresponding dynamical kernel of the new approach named EOM/RQTBA^3 is given in the lower part of Fig. 3, where the index "(n)" denotes the iteration order. The numerical implementation has been performed for $n = 2$ based on the effective NL3 meson-nucleon interaction [43] of quantum hadrodynamics (QHD), i.e. the static part of the interaction kernel is still approximated by the effective meson-exchange interaction. The second correlation function introduced into the dynamical kernel leads to an additional fragmentation of the strength distributions. The results of calculations for the giant dipole resonance (GDR) in ^{42,48}Ca are displayed in Fig. 4 in comparison to data. These cross sections were studied in Ref. [44] for the role of the $2q \otimes phonon$ (two quasiparticles coupled to a phonon) configurations in the formation of the spreading width, which was found quite significant as compared to relativistic quasiparticle random phase approximation (RQRPA) with only $2q$ (two-quasiparticle) configurations. However, the total width and the high-energy strength obtained in RQTBA are still insufficient to reproduce the data. Now we can see that EOM/RQTBA^3 with more complex $2q \otimes 2phonon$ configurations shows the potential of resolving those problems by inducing a stronger fragmentation of the GDR and its additional spreading toward both higher and lower energies.

Another recent extension of the RNFT was performed for the case of finite temperature [47]. For this purpose, the conventional TBA for the time-dependent part of the in-medium nucleon-nucleon interaction amplitude was adopted for the thermal (imaginary-time) Green's function formalism. We found, in particular, that introducing a soft blocking, instead of a sharp blocking at zero temperature, brings the temperature-dependent Bethe-Salpeter equation to a single frequency variable equation similar to Eq. (8). The method was implemented self-consistently in the framework of QHD and investigated the temperature dependence of spectra of even-even medium-heavy nuclei. An illustration of these studies is given in Fig. 5 for the dipole response of ^{68}Ni. Of particular interest was the spreading of the giant dipole resonance induced by the dynamical kernel of FT-RTBA, which remained strong in the high-temperature regimes, and the appearance of significant low-energy strength caused by the thermal unblocking, see Refs. [47, 48] for further details. The temperature dependence of the spin-isospin response was investigated as well and found to be even more sensitive to low and moderate temperatures [49], that has important implications for astrophysics.

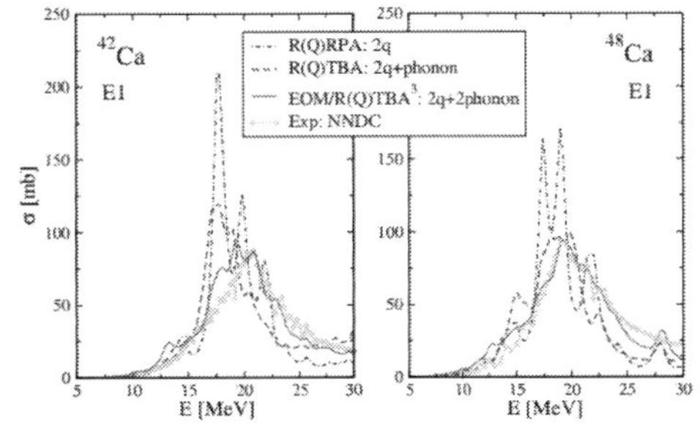

Figure 4. Theoretical dipole photoabsorption cross sections in ^{42,48}Ca compared to the evaluated experimental data of Refs. [45, 46].

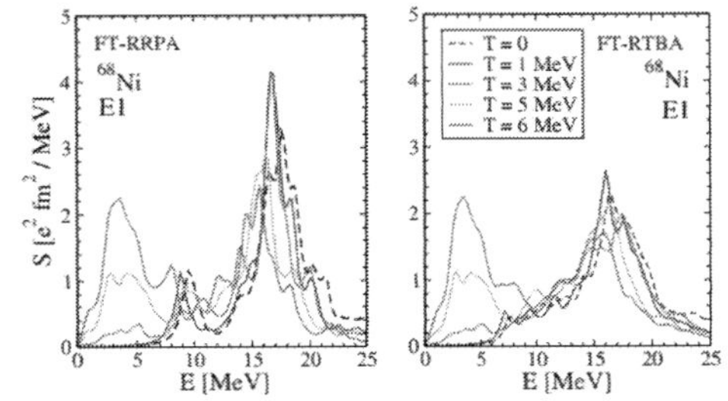

Figure 5. Dipole strength distribution in ^{68}Ni at various temperatures in the finite-temperature relativistic RPA (FT-RRPA) and in the finite-temperature relativistic time blocking approximation (FT-RTBA).

Summary — A novel microscopic approach to the nuclear response developed by synthesis of the equation of motion method and quantum hadrodynamics is presented. It allows for a consistent description of nuclear excited states with the explicit treatment of beyond-mean-field correlations between up to six fermions, such as $2q \otimes 2phonon$. The first calculations of the dipole response in medium-heavy nuclei demonstrate important improvements of this description in both low- and high-energy sectors. We also present a finite-temperature version of the theory with explicit $2q \otimes phonon$ correlations. The obtained results are consistent with the existing experimental data and predict the evolution of the nuclear strength functions with temperature taking into account spreading effects. In particular, a noticeable enhancement of the low-lying dipole as well as of the spin-isospin strength distributions has been found at the astrophysical conditions, that may affect the astrophysical modeling of the r-process nucleosynthesis and core-collapse supernovae.

Acknowledgement — This work was partly supported by the US-NSF Career Grant PHY-1654379.

References

[1] I. Tanihata, Nuclear Physics News **8**, 4 (1998)
[2] T. Glasmacher, B. Sherrill, W. Nazarewicz, *et al.*, Nuclear Physics News **27**, 28 (2017)
[3] A. Bohr, B.R. Mottelson, *Nuclear structure*, Vol. 1 (World Scientific, 1969)
[4] G. Bertsch, P. Bortignon, R. Broglia, Reviews of Modern Physics **55**, 287 (1983)
[5] V. Soloviev, *Theory of Atomic Nuclei: Quasiparticles and Phonons* (Institute of Physics Publishing, 1992)
[6] D.J. Rowe, Review of Modern Physics **40**, 153 (1968)
[7] P. Schuck, Zeitschrift für Physik A **279**, 31 (1976)
[8] S. Adachi, P. Schuck, Nuclear Physics **A496**, 485 (1989)
[9] J. Dukelsky, G. Röpke, P. Schuck, Nuclear Physics **A628**, 17 (1998)
[10] A. Storozhenko, P. Schuck, J. Dukelsky, G. Ropke, A. Vdovin, Annals of Physics **307**, 308 (2003)
[11] P. Schuck, M. Tohyama, European Physical Journal A **52**, 307 (2016)
[12] E. Litvinova, P. Ring, Physical Review C **73**, 044328 (2006)
[13] E. Litvinova, P. Ring, V. Tselyaev, Physical Review C **75**, 064308 (2007)
[14] E. Litvinova, P. Ring, V. Tselyaev, Physical Review C **78**, 014312 (2008)
[15] E. Litvinova, P. Ring, V. Tselyaev, Physical Review Letters **105**, 022502 (2010)
[16] E. Litvinova, P. Ring, V. Tselyaev, Physical Review C **88**, 044320 (2013)
[17] A.V. Afanasjev, E. Litvinova, Physical Review C **92**, 044317 (2015)
[18] E. Litvinova, Physics Letters B **755**, 138 (2016)
[19] C. Robin, E. Litvinova, European Physical Journal A **52**, 205 (2016)
[20] C. Robin, E. Litvinova, Physical Review C **98**, 051301 (2018)
[21] V.I. Tselyaev, Physical Review C **88**, 054301 (2013)
[22] V. Tselyaev, N. Lyutorovich, J. Speth, *et al.*, Physical Review C **94**, 034306 (2016)
[23] Y. Niu, Z. Niu, G. Colò, E. Vigezzi, Physical Review Letters **114**, 142501 (2015)
[24] Y. Niu, Z. Niu, G. Colò, E. Vigezzi, Physics Letters B **780**, 325 (2018)
[25] J. Endres, E. Litvinova, D. Savran, *et al.*, Physical Review Letters **105**, 212503 (2010)
[26] R. Massarczyk, R. Schwengner, F. Dönau, *et al.*, Physical Review C **86**, 014319 (2012)
[27] I. Poltoratska, R. Fearick, A. Krumbholz, *et al.*, Physical Review C **89**, 054322 (2014)
[28] E. Lanza, A. Vitturi, E. Litvinova, D. Savran, Physical Review C **89**, 041601 (2014)
[29] B. Özel-Tashenov, J. Enders, H. Lenske, *et al.*, Physical Review C **90**, 024304 (2014)
[30] S.P. Kamerdzhiev, G.Ya. Tertychny, V.I. Tselyaev, Physics Letters **B267**, 12 (1991)
[31] C. Robin, E. Litvinova, arXiv:1903.09182 (2019)
[32] V. Tselyaev, N. Lyutorovich, J. Speth, P.G. Reinhard, Physical Review C **97**, 044308 (2018)
[33] E. Litvinova, Physical Review C **91**, 034332 (2015)
[34] V.Yu. Ponomarev, P.F. Bortignon, R.A. Broglia, V.V. Voronov, Nuclear Physics **A687**, 170 (2001)
[35] F. Andreozzi, F. Knapp, N. Lo Iudice, A. Porrino, J. Kvasil, Physical Review C **78**, 054308 (2008)
[36] V. Tselyaev, Soviet Journal of Nuclear Physics **50**, 780 (1989)
[37] V. Olevano, J. Toulouse, P. Schuck, Journal of Chemical Physics **150**, 084112 (2018)
[38] P. Schuck, M. Tohyama, Physical Review B **93**, 165117 (2016)
[39] E. Litvinova, P. Schuck, in preparation (2019)
[40] D. Vretenar, A.V. Afanasjev, G.A. Lalazissis, P. Ring, Physics Reports **409**, 101 (2005)
[41] E. Litvinova, V. Tselyaev, Physical Review C **75**, 054318 (2007)
[42] V.I. Tselyaev, Physical Review C **75**, 024306 (2007)
[43] G.A. Lalazissis, J. König, P. Ring, Physical Review C **55**, 540 (1997)
[44] I.A. Egorova, E. Litvinova, Physical Review C **94**, 034322 (2016)
[45] V.A. Erokhova, M.A. Yolkin, A.V. Izotova, *et al.*, Bull. Rus. Acad. Sci. Physics **67**, 1636 (2003)
[46] National Nuclear Data Center, https://www.nndc.bnl.gov (NNDC)
[47] E. Litvinova, H. Wibowo, Physical Review Letters **121**, 082501 (2018)
[48] H. Wibowo, E. Litvinova, Physical Review C **100**, 024307 (2019)
[49] E. Litvinova, C. Robin, H. Wibowo, arXiv:1808.07223 (2018)

EPJ Web of Conferences **223**, 01034 (2019) https://doi.org/10.1051/epjconf/201922301034
NSD2019

Analysis of excited states in ^{13}C and their cluster structure

Ivano Lombardo[1,*]

[1]INFN-Sezione di Catania, Via Santa Sofia, 64 - 95123 Catania, Italy

Abstract. An accurate understanding of the spectroscopy of ^{13}C is key to understand clustering phenomena in light nuclear systems. In this nucleus, many theoretical models predict the existence of α-cluster configurations, stabilized by the presence of an extra neutron. Despite of its importance, the spectroscopy of ^{13}C is still affected by ambiguities, demanding for new data against which benchmarking the models. We improve the knowledge on the ^{13}C level scheme by means of a comprehensive R-matrix analysis of several experimental data sets. Our new spectroscopic information contributes to solve existing ambiguities in ^{13}C structure and will serve a valid benchmark for theoretical models attempting to describe clustering in neutron rich systems.

1 Introduction

Carbon isotopes represent a unique opportunity to investigate clustering phenomena in light nuclear systems as a function of the neutron excess. The clustered structure of carbon isotopes is a long standing problem. For a long time, the self-conjugate ^{12}C was associated to a linear chain configuration formed by 3α-particles [1]. Recently, high-precision experiments pointed out instead the occurrence of D_{3h} triangular symmetry in such nucleus [2, 3], triggering a renovated interest for three-centers cluster structures in nuclei. In this framework, it is fundamental to fully understand the spectroscopy of the carbon isotopes that are obtained by removing or adding a neutron to the self-conjugate ^{12}C, namely ^{11}C and ^{13}C. The first has been the subject of recent investigations [4–6] that pointed out excited states associated to cluster-like structures and discussed their relevance in astrophysics [7]. For the latter, the neutron rich ^{13}C, the situation is much more complicated. Recently, theoretical investigations predicted the possible manifestation of 3α cluster structures in this nucleus with the formation of highly-deformed shapes and rotational bands extending well above the α-emission threshold [8–10]. However, the presence of several ambiguities affecting the spectroscopy of such nucleus [11], in particular above the α-emission threshold (i.e. E_x = 10.648 MeV), prevent a fully unambiguous identification of the proposed bands, demanding for more firm experimental findings [12].

Because of the clustered structure of ^{9}Be, α+^{9}Be reactions have been proposed to be very powerful tools to explore the spectroscopy of ^{13}C above the α-threshold [13]. When an α-particle strikes a ^{9}Be, dominant reactions are the elastic or inelastic resonant scatterings, respectively ^{9}Be(α,α) and ^{9}Be(α,α') [13–19], and the ^{9}Be(α,n) reactions [20, 21]. To help to clarify the spectroscopy of ^{13}C above the α-threshold, we performed, for the first time, a completely unified and comprehensive analysis of the above mentioned reactions in terms of the R-matrix theory.

In this paper, we report on the results of our R-matrix analysis that strongly contributes to better constrain the spectroscopy of ^{13}C. In Section 3, we describe the data sets used for the present analysis. Section 4 contains details on the R-matrix fit of the data sets. In conclusion, Section 4 contains interpretations of such results in terms of the clustered structure of ^{13}C. More details on this analysis and related interpretations can be found in Ref. [22].

2 Construction of data sets

Data regarding the elastic, ^{9}Be(α,α), and inelastic, ^{9}Be(α,α'), scatterings have been obtained from a dedicated experiment carried out at the TTT3 tandem accelerator [23, 24] of the University of Naples "Federico II", Italy. A ^{4}He beam with excellent qualities of collimation and energy resolution was delivered to a scattering chamber, where several detectors where installed, centered at 160°, 150°, 135°, 110° and 70° in the laboratory frame. Reactions were induced by a thin ($\approx$ $122 \mu g/cm^2$) self-supporting ^{9}Be target. The beam energy was varied in 60 keV steps, covering the energy range E_α = 3.5-10 MeV. More details regarding the experimental setup and results of the experiment are discussed in a separate publication [25, 26]. Differential cross section (DCS) data of the elastic scattering have been measured at the backward angles 160°, 150°, 135°, 110°, while the inelastic scatterings ^{9}Be(α,α_1), ^{9}Be(α,α_2) and ^{9}Be(α,α_4), leaving the residual ^{9}Be in the E_x = 1.684 MeV, E_x = 2.429 MeV and E_x = 3.049 MeV states (respectively), have been obtained at 70° by using the procedure described in [22]. In the present data set, we include the measured excitation functions of the elastic scattering and the ^{9}Be(α,α_1) inelastic data. The ^{9}Be(α,α_2) and ^{9}Be(α,α_4) data have

*e-mail: ivano.lombardo@ct.infn.it

been excluded from our data set because of possible contaminations arising from the broad peak associated to the ^{9}Be(α,α_3) process as described in Ref. [22]. To better characterize the ^{13}C excitation energy region close to the α emission threshold, $E_\alpha < 2$ MeV, we added to our data sets also the differential cross section data of Ref. [15] in the bombarding energy region E_α = 1.0-1.7 MeV at θ_{lab} = 160°, 150°. We allowed for the presence of normalization factors for such data (within the range 0.85-1.15), to account for possible absolute normalization errors affecting the analysis done in Ref. [15].

We also included in our data set the ^{9}Be(α,n) reactions previously published in the literature. We used absolute integrated cross section data for the n_0 (i.e. leaving the ^{12}C in its ground state) and n_1 (when the ^{12}C residual nucleus is populated in its E_x = 4.44 MeV state) channels extracted from Refs. [27–29]. The n_1 data was normalized by a factor of 0.52 as discussed in Ref. [22].

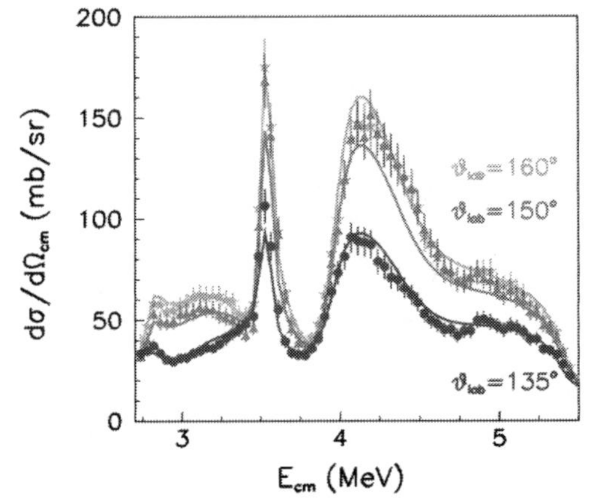

Figure 1. ^{9}Be(α,α) excitation functions for θ_{lab} = 135°, 150°, 160° backward angles. The solid lines represent the result of the simultaneous R-matrix fit of ^{9}Be(α,α), ^{9}Be(α,α_1) and ^{9}Be(α,$n_{0,1}$) as discussed in the text.

3 R-matrix analysis of data

Data sets described in the previous section were interpreted in terms of the R-matrix theory to probe the spectroscopy of the ^{13}C nucleus formed as the intermediate state of the collision. We performed a simultaneous fit of the ^{9}Be(α,α) excitation functions at backward angles, the ^{9}Be(α,α_1) data at 70° and the ^{9}Be(α,n_0) and ^{9}Be(α,n_1) integrated cross section data. The latter, in particular, is a fundamental ingredient to adequately constrain the partial widths of ^{13}C states involved in the analysis. Additionally, the presence of several detection angles in the ^{9}Be(α,α) excitation functions enables to give firm J^π assignments to many states for which ambiguities were previously reported in the literature. As the starting parameters for the fit, we considered the spectroscopy of ^{13}C as reported in the compilation of Ref. [11], and we carefully took into account the findings of previous calculations published in Refs. [13, 15]. We used $\ell = 8$ as the maximum order for the partial waves contributing in the reaction or scattering events. We included in our calculation also the effects of the finite target thickness.

Results of the simultaneous R-matrix fit, performed with AZURE2 [30], are shown in Fig. 1 for the elastic scattering ^{9}Be(α,α) at three backward angles, θ_{lab} = 135°, 150°, 160°. With our refined spectroscopy, the overall agreement of the R-matrix calculation and the data is excellent at all energies, as clearly visible in figure. The reduced χ^2 values for our fit range between ≈ 0.6 to ≈ 2 depending on the reaction channel. We find a satisfactory agreement with the findings of Refs. [13, 15] at low value of excitation energy $E_x \leq 16$ MeV. In this energy region, some of the quoted J^π values of ^{13}C where ambiguous in the literature. When necessary, and taking advantage of the presence of a big body of data, J^π values previously quoted in the literature have been changed to reproduce our data. The resulting parameters are summarized in Ref. [22] and improve the overall spectroscopy of ^{13}C in this energy region.

At higher energies, 16MeV < E_x < 19MeV, the situation is much more complicated as previous R-matrix analyses of the ^{9}Be(α,α) scattering don't extend to this high energies [13]. Furthermore, in the adopted ^{13}C level scheme of Ref. [11], one can observe a significant lack of information in such excitation energy region. Our ^{9}Be(α,α) extends to a higher energy region enabling us to improve the spectroscopy of ^{13}C in a wider energy domain.

More details regarding this analysis can be found in Ref. [22], the complete spectroscopy of ^{13}C above the α-threshold is reported in Table 1 of such publication.

4 Conclusions and perspectives

The spectroscopy of ^{13}C is key to identify three-centers molecular states and their rotational bands in light nuclear systems. To contribute to solve ambiguities affecting the spectroscopy of this nucleus above the α disintegration threshold, we have performed a new comprehensive R-matrix fit of a large body of data. We have carefully reviewed previous experiments reported in the literature and involving α +^{9} Be reactions. The latter were chosen because of their high selectivity to cluster states in ^{13}C. Results of our R-matrix fit improve significantly the spectroscopy of ^{13}C, enabling to more firm J^π assignments thanks to the presence of several detection angles.

The improved spectroscopy of ^{13}C allows us to some speculations regarding the possible appearance of rotational bands characterized by cluster nature. Our results seem to support the theoretical suggestions proposed in the recent study of Ref. [31] and will be the subject of a successive more refined analysis in the effort to identify rotational bands in the neutron rich ^{13}C nucleus.

In the near future, we will exploit detection arrays with high granularity and/or large solid angle acceptance [32–36] to carry out a new experimental campaign to probe the spectroscopy of light nuclear systems with high precision. This, together with the opportunity to carry out new experiments by using large acceptance detectors as γ-ray arrays [37], will help to improve the understanding on clustering in neutron-rich nuclei.

References

[1] H. Morinaga et al., Phys. Rev. **101**, 254 (1956)
[2] M. Itoh et al., Phys. Rev. C **84**, 054308 (2011)
[3] M. Freer et al., Phys. Rev. C **86**, 034320 (2012)
[4] M. Freer et al., Phys. Rev. C **85**, 014304 (2012)
[5] H. Yamaguchi et al., Phys. Rev. C **87**, 034303 (2013)
[6] I. Lombardo et al., J. Phys. G: Nucl. Part. Phys. **43**, 045109 (2016)
[7] C. Spitaleri et al., Phys. Rev. C **95**, 035801 (2017)
[8] Y. Chiba, M. Kimura, J. Phys. Conf. Ser. **569**, 012047 (2014)
[9] N. Furutachi, M. Kimura, Phys. Rev. C **83**, 021303(R) (2011)
[10] T. Yamada, Y. Funaki, Phys. Rev. C **92**, 034326 (2015)
[11] F. Ajzenberg-Selove, Nucl. Phys. A **523**, 1 (1991)
[12] M. Milin, W. von Oertzen, Eur. Phys. J. A **14**, 295 (2002)
[13] M. Freer et al., Phys. Rev. C **84**, 034317 (2011)
[14] R. Taylor, N. Fletcher, R. Davis, Nucl. Phys. **65**, 318 (1965)
[15] J. Goss et al., Phys. Rev. C **7**, 1837 (1973)
[16] Z. Saleh et al., Ann. der Phys. **7**, 76 (1974)
[17] J. Leavitt et al., Nucl. Instrum. Meth. Phys. Res. B **85**, 37 (1994)
[18] J. Liu, Z. Zheng, W. Chu, Nucl. Instrum. Meth. Phys. Res. B **108**, 247 (1996)
[19] M. Zadro et al., Nucl. Instrum. Meth. Phys. Res. B **259**, 836 (2007)
[20] A. Obst, T. Grandy, J. Weil, Phys. Rev. C **5**, 738 (1972)
[21] D.D. Martini, C. Soltesz, T. Donoghue, Phys. Rev. C **7**, 1824 (1973)
[22] I. Lombardo et al., Phys. Rev. C **97**, 034320 (2018)
[23] I. Lombardo et al., J. Phys. G: Nucl. Part. Phys. **40**, 125102 (2013)
[24] I. Lombardo et al., Bulletin of the Russian Academy of Sciences: Physics **78**, 1093 (2014)
[25] I. Lombardo et al., Nucl. Instrum. Meth. Phys. Res. B **302**, 19 (2013)
[26] I. Lombardo et al., J. Phys. Conf. Ser. **569**, 012068 (2014)
[27] R. Kunz et al., Phys. Rev. C **53**, 2486 (1996)
[28] L. van der Zwan, K. Geiger, Nucl. Phys. A **152**, 481 (1970)
[29] K. Geiger, L. van der Zwan, Nucl. Instr. Meth. **131**, 315 (1975)
[30] R. Azuma et al., Phys. Rev. C **81**, 045805 (2010)
[31] R. Bijker, F. Iachello, Phys. Rev. Lett **122**, 162501 (2019)
[32] D. Dell'Aquila et al., Nucl. Instrum. Meth. Phys. Res. A **877**, 227 (2018)
[33] D. Dell'Aquila et al., Phys. Rev. Lett. **119**, 132501 (2017)
[34] A. Pagano, M. Alderighi, F. Amorini, A. Anzalone, L. Arena, L. Auditore, V. Baran, M. Bartolucci, I. Berceanu, J. Blicharska et al., Nucl. Phys. A **734**, 504 (2004)
[35] E.D. Filippo et al., Acta Phys. Pol. B **40**, 1199 (2009)
[36] G. Pastore et al., Nucl. Instrum. Meth. Phys. Res. A **860**, 42 (2017)
[37] G. Cardella et al., Nucl. Instr. and Meth. A **799**, 64 (2015)

Role of correlations in two-neutron transfer reactions

Jesus Lubian[1,*], *Jonas* L. Ferreira, *Roberto* Linares, *Erica* N. Cardozo, *Barbara* Paes, *Francesco* Cappuzzello[2], *Diana* Carbone[2], *Manuela* Cavallaro[2], *Uiran* U. da Silva[3], *Ruben* Lichtenthäler[3]

[1]Instituto de Física, Universidade Federal Fluminense, Niterói, Rio de Janeiro 24210-340, Brazil[2]INFN, Laboratori Nazionali del Sud, I-95125, Catania, Italy
[3]Departamento de Física Nuclear, Universidade de São Paulo, C.P. 66318, 05389-970, São Paulo, Brazil

Abstract. This work presents theoretical results compared with the experimental data for the two-neutron transfer angular distribution in which a beam of ^{18}O nucleus, at 84 MeV incident energy, has collided on the ^{13}C, ^{28}Si, and ^{64}Ni targets. The two-neutron transfer in the ^{9}Be(^{7}Be,^{9}Be)^{7}Be reaction, at 23.1 MeV incident energy, was also analyzed. The main goal was to verify the relevance of the pairing correlation of the two transferred neutrons on the cross sections and to show its role when both neutrons are transferred to states with a low and high degree of collectivity.

1 Introduction

The study of pairing correlation in two-nucleon transfer reaction has intensively been performed in the last decades. In the past, this kind of direct reactions was usually probed by the *(t,p)* and *(p,t)* reactions. Due to restrictive radioprotection rules, these reactions were prohibited. Another alternative reaction was proposed in the Istituto Nazionale di Fisica Nucleare – Laboratori Nazionali del Sud (Italy) by considering the (^{18}O,^{16}O) reaction, once the ^{18}O nucleus can be considered as composed by an inert core of ^{16}O and two valence neutrons. In this sense, some reactions were measured considering the ^{18}O nucleus as projectile at 84 MeV incident energy, and ^{12,13}C [1, 2], ^{16}O [3, 4], ^{28}Si [5] and ^{64}Ni [6] as target to analyze the feasibility of the (^{18}O,^{16}O) reaction as substitute of the *(t,p)* one, and then to investigate the role of the pairing correlation of both transferred neutrons in the transfer reaction cross sections obtained performing coupled reaction channel calculations [7]. The ^{12}C(^{18}O,^{16}O)^{14}C reaction has experimentally shown, similarly to *(t,p)*, the feasibility in populating the two-neutron configuration of the ^{14}C nucleus.

In particular, the two-neutron transfer in the ^{18}O+^{16}O reaction includes the complexity of dealing with identical ingoing and outgoing partitions, as well as the recently studied two-neutron transfer in the ^{9}Be+^{7}Be reaction [8] at 23 MeV, and the alpha-particle transfer in the ^{16}O+^{12}C at 80 and 84 MeV incident energy [9]. The complexity appears because the elastic scattering and elastic transfer are indistinguishable processes, and they can compete. In these cases, the quantum interference effect between elastic scattering and elastic transfer should be explicitly taken into account in the calculations.

Another important aspect of the transfer reactions is their high selectivity, which is very important to study the nuclear structure of the involved nuclei. By means of these reactions, it is possible to obtain spectroscopic information as the spectroscopic amplitudes that are connected to the strength of each single- or two-particle component of the nuclear wave functions (or even to the alpha-particle component). As it is well known, the two valence particles may be simultaneously or sequentially transferred from the projectile to the target in stripping reactions, and these processes may compete with each other. While the simultaneous two-nucleon transfer reactions (or one-step) reveal the two-particle component of the nuclear wave functions, the sequential two-nucleon transfer (or two-step) raises their one-particle nature. In this work, we discuss the competition between the collectivity and the pairing correlation in the states populated by the two transferred particles. The two-neutron transfer angular distribution has been quite well described by the theoretical calculation for some specific measured channels. The theoretical results show that the states with high collectivity are preferably accessed by the sequential two-neutron transfer process suggesting that the correlation of the two transferred neutrons was broken by the collectivity of the populated final state.

This work is organized as ^{18}O follows. The next section is devoted to the results and discussions while in the last section, we give some remarks and show some perspectives on our work guidelines.

2 Results

The more straightforward theoretical treatment involving the rearrangement reaction can be reached by considering only the ground state of the partner nuclei in the initial and final partition, as well as, assuming the coupling between the partitions as weak. In this context, the distorted wave Born approximation (DWBA) approach might be applied, once the coupling interaction is regarded in the first order. On the other hand, if the coupling interaction between the partitions is extended to higher-orders, the coupled reaction channel (CRC) method will be invoked. Moreover, in two-neutron stripping transfer reaction like, for instance, $a(b+2n)+A \rightarrow b+B(A+2n)$ the transfer amplitude can be determined by the equation

$$T_{\alpha\rightarrow\gamma} = \langle \chi_{\gamma}{}^{(-)} | W_{\alpha\gamma} | \chi_{\alpha}{}^{(+)} \rangle, \qquad (1)$$

* Corresponding author: lubian@if.uff.br

so that the residual interaction responsible to the transfer process is written as $W_\alpha = U(R) + v(\boldsymbol{r}_1) + v(\boldsymbol{r}_2) - U(R_\alpha)$. In addition, the matrix elements defined by $W_{\alpha\gamma} \equiv \langle \phi_\gamma | W_\alpha | \phi_\alpha \rangle$ contains all the structure information of the collision partners. $\chi_\alpha^{(+)}$ and $\chi_\gamma^{*(-)}$ represent the wave functions of the relative motion in the entrance and final partitions, respectively, with the α and γ being all the quantum numbers needed to specify the scattering state. In the collision, the target interacts with the core (in which the two valence neutrons are bound) by mean of the $U(R)$ interaction, and with each valence particles by $v(\boldsymbol{r}_1)$ and $v(\boldsymbol{r}_2)$. $U(R_\alpha)$ represents the interaction between projectile and target nuclei concerning the center of mass coordinate R_α. Moreover, while $U(R)$ and $U(R_\alpha)$ are complex potentials, the $v(\boldsymbol{r}_1)$ and $v(\boldsymbol{r}_2)$ interactions are assumed to be real. From the transfer amplitudes we can determine the transfer angular distribution given by

$$\frac{d\sigma_{\alpha\to\gamma}}{d\Omega} = \frac{\mu_\alpha \mu_\gamma}{(2\pi\hbar^2)^2} \frac{k_\gamma}{k_\alpha} \frac{1}{(2I_a+1)(2I_A+1)} \left| T_{\alpha\to\gamma} \right|^2, \quad (2)$$

where μ_i and k_i are the reduced masses and the wave number in partition i. In the two-particle transfer, the T amplitude given in equation (1) is explicitly represented by two terms. The first, corresponds to the direct two-particle transfer, in which the two particles are transferred from the initial partition to the final one, while, the second term, is related to the sequential two-particle transfer, where both valence particles are transferred one by one passing through an intermediate partition, let say β. In this case, the equation (1) is represented as $T_{\alpha\to\gamma} = T^{(1)}{}_{\alpha\to\gamma} + T^{(2)}{}_{\alpha\to\beta\to\gamma}$[7]. The direct and sequential two-particle transfer can compete with each other and both terms should be considered in the same calculation. However, it is essential to consider each transfer mechanism to study the degree of correlation between the transferred particles.

In the present work, the two-neutron transfer angular distributions were calculated considering the finite-range coupled reaction channel method using the Fresco code [10].

To perform microscopic transfer reaction calculations, it is important to know the one and two particle spectroscopic amplitudes, as well as, to use a parameter-free optical potential (like a double folding potential). In this work, the theoretical transfer calculations were performed using the São Paulo double folding potential [11] in the real and imaginary parts of the optical potential ($U(R) = (N_r + iN_i)V^{SP}(R)$). In the ingoing partition, the imaginary part was normalized by a factor 0.6 [12], which consider that the more relevant couplings to the bound states were explicitly included, but some dissipation to continuum occurs. On the other hand, in the outgoing partition, the imaginary part was multiplied by a factor 0.78 [13], once no couplings were considered. The spectroscopic amplitudes concerning the one- and two-neutron wave function overlaps were derived from the shell model calculation by using the NuShellX code [14]. The single-particle wave functions were generated considering a Woods-Saxon potential whose depth was varied to fit the experimental biding energies of one and two neutrons.

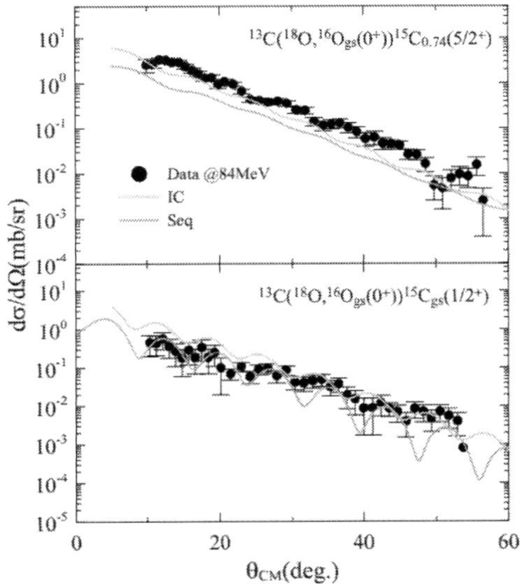

Fig. 1. Comparison of the experimental data and theoretical results corresponding to the $^{13}C(^{18}O,^{16}O)^{15}C$. IC means independent coordinates CRC calculations. Seq stands for the sequential CCBA calculations.

To probe the pairing correlations in the $^{12,13}C(^{18}O,^{16}O)^{14,15}C$, $^{16}O(^{18}O,^{16}O)^{18}O$, $^{28}Si(^{18}O,^{16}O)^{30}Si$, and $^{64}Ni(^{18}O,^{16}O)^{66}Ni$ reactions, we have separately analyzed the direct and sequential two-neutron transfer reactions. In particular, the microscopic independent coordinates scheme was considered to analyze the direct two-neutron transfer in finite range coupled reaction channel calculations. Here, both valence particles are treated as particles spatially separated and simultaneously transferred from the initial to the final state. In this case, the spectroscopic amplitudes correspond to two-particle configurations. Details about this method can be found in Ref. [10]. On the other hand, we also considered the sequential process in which the two neutrons are transferred one by one passing through an intermediate partition. In these transfer calculations, high-order couplings between states inside the partition and to first-order between different partitions were considered. This means that the coupled channel Born approximation was employed for the sequential two-neutron transfer calculations. In a full quantum treatment, the direct and sequential two-neutron transfer should be considered together in the same calculation.

In the two-neutron transfer reaction studies mentioned above, the cross section for the direct and sequential processes that populated the ground state of the residuals $^{14,15}C$, ^{18}O, ^{30}Si, and ^{66}Ni nuclei had the same order of magnitude. The result of this study is shown in Figures 1, 2, and 3, in which the ^{18}O was the projectile impinging on light- and medium-mass targets. Conversely, for the two-neutron transfer in the $^{206}Pb(^{18}O,^{16}O)^{208}Pb$ reaction [16] the cross section of the sequential process was larger than the direct one.

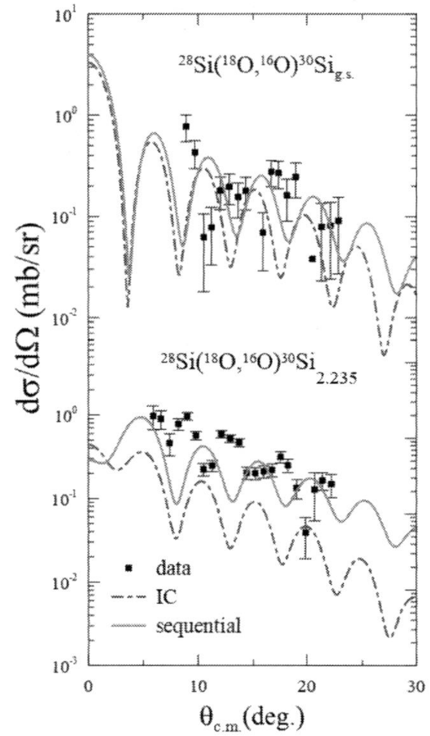

Fig. 2. Comparison of the experimental data and theoretical results corresponding to the ^{28}Si(^{18}O,^{16}O)^{30}Si. IC means independent coordinates CRC calculations. Sequential stands for the sequential CCBA calculations.

The scenario is a little bit different when the two neutrons populate some excited states of the residual nuclei with strong collectivity. In this case, the correlation amount a considerable number of nucleons is more relevant, and the pairing correlation does not appear to be very important. In this situation, the sequential process becomes the most suitable reaction mechanism. On the other hand, the excited states with lower collectivity were preferably populated by the direct two-neutron transfer in the light residual nuclei ^{14,15}C and ^{18}O. In Figure 1, we just show the results for the angular distribution populating the ground ($1/2^+$) and first excited ($5/2^+$) states of the ^{15}C nucleus. For the light-medium ^{30}Si and medium ^{66}Ni mass residual nuclei, the sequential two-neutron transfer was more likely to occur, as can be seen in Figures 2 and 3. This behavior has been associated with the competition between the collective and pairing residual interactions. In the ^{13}C(^{18}O,^{16}O)^{15}C some evidence of the existence of giant pairing vibration [15] were found which are produced by the coherent particle-particle excitation during the removal or addition of two nucleons with relative motion characterized by the s-wave.

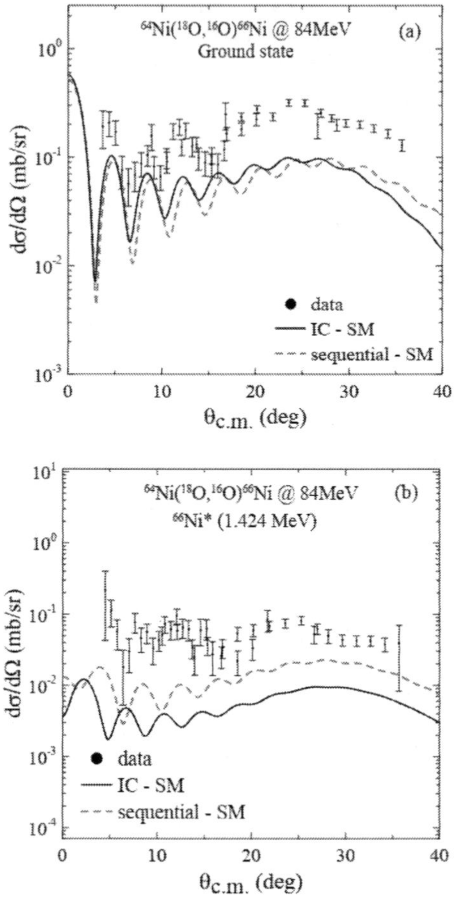

Fig. 3. Comparison of the experimental data and theoretical results corresponding to the ^{64}Ni(^{18}O,^{16}O)^{66}Ni. IC means independent coordinates CRC calculations. Sequential stands for the sequential CCBA calculations. SM means that the spectroscopic amplitudes are derived from shell-model calculations.

The ^{9}Be(^{7}Be,^{9}Be)^{7}Be [8] elastic transfer reaction shows features similar to the (^{18}O,^{16}O) [3,4] elastic transfer mentioned above, in the sense that the direct two-neutron mechanism is very important (in fact, in this case, it is the preferred mechanism). In this reaction, the elastic scattering and elastic transfer are indistinguishable processes, once the ingoing and outgoing partitions are the same. Therefore, the quantum effect of interference between both processes have been taken into account. It was shown that, while the elastic scattering is the predominant process at forward angles, the elastic-transfer is more likely to occur at backward angles. On the other hand, in the alpha-transfer reaction ^{12}C(^{16}O,^{12}C)^{16}O, it is observed a strong competition between the elastic scattering and elastic transfer processes at backward angles. In Figure 4, we show the comparison between experimental data and theoretical predictions for the angular distribution concerning the quasi-elastic collision ^{7}Be+^{9}Be. One can observe that the direct two-neutron transfer is the most important process at backward angles when we compared with the sequential one.

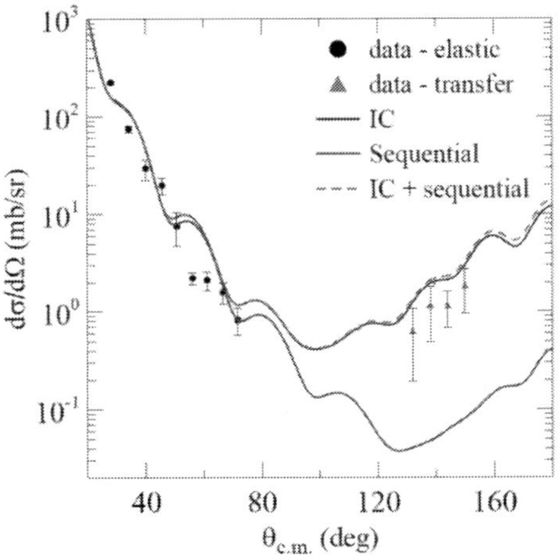

Fig. 4. Comparison of the experimental data and theoretical predictions corresponding to the elastic scattering and elastic-transfer angular distributions in the $^{7}Be+^{9}Be$ collision at 23.1 MeV.

3 Conclusions

In this work, we present some measured angular distributions corresponding to the two-neutron transfer reaction in comparison with the theoretical calculation by using the finite range coupled reaction channel method. The data for the $^{12,13}C(^{18}O,^{16}O)^{14,15}C$, $^{16}O(^{18}O,^{16}O)^{18}O$, $^{28}Si(^{18}O,^{16}O)^{30}Si$ and $^{64}Ni(^{18}O,^{16}O)^{66}Ni$ reactions, at 84 MeV incident energy, were measured at Istituto Nazionale di Fisica Nucleare – Laboratori Nazionali del Sud (Italy). Besides, the experimental data for the $^{9}Be(^{7}Be,^{9}Be)^{7}Be$ reaction were showed and compared with the theoretical results. This data have been measured at São Paulo University using the RIBRAS (Radioactive Ion Beams in Brazil) facility. The results show the importance of the pairing correlation between the two valence neutrons in the transfer reactions.

In final states with high collectivity, the sequential two-neutron transfer is favored. On the other hand, nuclear states with low collectivity have been populated through the direct two-neutron transfer.

The Brazilian authors acknowledge the partial financial support from CNPq, FAPERJ, and CAPES and fromINCT-FNA (Instituto Nacional de Ciência e Tecnologia- Física Nuclear e Aplicações).

References

1. M. Cavallaro, F. Cappuzzello, M. Bondi, D. Carbone, V.N. Garcia, A. Gargano, S. M. Lenzi, J. Lubian, C. Agodi,F.Azaiez, et al., Phys. Rev. C **88**, 054601, (2013).
2. D. Carbone, J.L. Ferreira, F. Cappuzzello, J. Lubian, C. Agodi, M. Cavallaro, A. Foti, A. Gargano, S.M. Lenzi, R. Linares, G. Santagati, Phys. Rev. C **95**, 034603, (2017).
3. M.J. Ermamatov, F. Cappuzzello, J. Lubian, M. Cubero, C. Agodi, D. Carbone, M. Cavallaro, J.L. Ferreira, A. Foti, V.N. Garcia, A. Gargano, J. A. Lay, S.M. Lenzi, R. Linares, G. Santagati, and A. Vitturi, Phys. Rev. C **94**, 024610, (2016).
4. M.J. Ermamatov, R. Linares, J. Lubian, J.L. Ferreira, F. Cappuzzello, D. Carbone, M. Cavallaro, P.N. de Faria, A. Foti, G. Santagati, and V.A.B. Zagatto, Phys. Rev. C **96**, 044603, (2017).
5. E.N. Cardozo, J. Lubian, R. Linares, F. Cappuzzello, D. Carbone, M. Cavallaro, J. L. Ferreira, A. Gargano, B.Paes, and G. Santagati, Phys. Rev. C **97**, 06461, (2018).
6. B. Paes, G. Santagati, R.M. Vsevolodovna, F. Cappuzzello, D. Carbone, E.N. Cardozo, M. Cavallaro, H.Garcia-Tecocoatzi, A. Gargano, J.L. Ferreira, S.M.Lenzi, R. Linares, E. Santopinto, A. Vitturi, and J. Lubian, Phys. Rev. C **96**, 044612, (2017).
7. G. R. Satchler, *Direct Nuclear Reactions* (Oxford University Press, Oxford, 1983).
8. U. Umbelino, K.C.C. Pires, R. Lichtenthaler, V. Scarduelli, G.A. Scotton, A. L´epine-Szily, V. Guimarães, J. Lubian, B. Paes, J.L. Ferreira, M.A. G. Alvarez, J.M.B. Shorto, S. Appannababu, M. Assunção, R.P. Condori, V. Morcelle, Phys. Rev. C **99**, 064617, (2019).
9. J.L. Ferreira, J. Lubian, R. Linares, M.J. Ermamatov, H. Yépez-Martínez and P.O.Hess, Eur. Phys. J. A **55**, 94, (2019).
10. I.J. Thompson, Comput. Phys. Rep. **7**, 167, (1988).
11. L.C. Chamon, D. Pereira, M.S. Hussein, M.A. Cândido Ribeiro, and D. Galetti, Phys. Rev. Lett. **79**, 5218, (1997).
12. D. Pereira, J. Lubian, J. Oliveira, D. Sousa, and L. Chamon, Phys. Lett. B **670**, 330, (2009).
13. L.R. Gasques, L.C. Chamon, P.R.S. Gomes, and J. Lubian, Nucl. Phys. A **764**, 135, (2006).
14. NuShellX for Windows and Linux http://www.garsington.eclipse.co.uk/.
15. F. Cappuzzello, D. Carbone, M. Cavallaro, M. Bondì,C.Agodi, F. Azaiez, A.Bonaccorso, A. Cunsolo, L. Fortunato, A. Foti, *et al.*, Nat. Comm.**6**, 6743, (2015).
16. A. Parmar, Sonika, B.J. Roy, V. Jha, U.K. Pal, T. Sinha, S.K. Pandit, V.V. Parkar, K. Ramachandran, K. Mahata, S.Santra, A.K. Mohanty, Nucl. Phys. A **940**, 167, (2015).

Four α-particles as a final state of $^{16}O^*$ Quasi Projectile decay

Maria Vittoria Managlia[1], *Mauro* Bruno[1,2], *Michela* D'Agostino[1,2], *Luca* Morelli[1,2], *Giorgio* Baiocco[3], *Francesca* Gulminelli[4], *Catalin* Frosin[5,6], *Sandro* Barlini[5,6], *Antonio* Buccola[5,6], *Alberto* Camaiani[5,6], *Giovanni* Casini[6], *Magda* Cicerchia[7], *Marco* Cinausero[7], *Daniela* Fabris[8], *Fabiana* Gramegna[7], *Giorgia* Mantovani[7], *Tommaso* Marchi[7], *Pietro* Ottanelli[5,6], *Gabriele* Pasquali[5,6], *Silvia* Piantelli[6], and *Simone* Valdré[6]

[1]Dipartimento di Fisica e Astronomia dell'Università di Bologna, Italy.
[2]INFN Sezione di Bologna, Bologna, Italy.
[3]Dipartimento di Fisica dell'Università, Pavia, Italy.
[4]Normandie Univ. ENSICAEN, UNICAEN, CNRS/IN2P3, LPC Caen, 14000 Caen, France.
[5]Dipartimento di Fisica e Astronomia dell'Università di Firenze, Italy.
[6]INFN Sezione di Firenze, Firenze, Italy.
[7]INFN, Laboratori Nazionali di Legnaro, Legnaro, Italy.
[8]INFN Sezione di Padova, Padova, Italy.

Abstract. Four α-particles as a final state of $^{16}O^*$ quasi-projectile decay produced in peripheral $^{16}O+^{12}C$ reactions at 130 MeV is thoroughly studied. The different decay channels leading to the four α-particles final state are reconstructed by carrying out an event-by-event analysis of α correlations in the population of intermediate ^{8}Be and ^{12}C. Although small, a non negligible contribution due to $^{8}Be_{gs}$ evaporation is found. A comparison between predictions of an accurate Hauser-Feshbach decay code and branching ratios of the different decay channels is performed. Significant deviations are observed, among these the Hoyle state population which is considerably lower than the one predicted according to the statistical model, thus suggesting possible structure effects in the Coulomb barrier and/or in the transmission coefficients.

1 Introduction

The aim of this experimental campaign carried out by the NUCL-EX collaboration is investigating the properties of light nuclei [1]. In this contribution we will present results of peripheral $^{16}O+^{12}C$ collisions with the decay of the excited $^{16}O^*$ Quasi Projectile (QP) in a final state of four α-particles. A comparison with the predictions of a statistical decay code (HFℓ) based on Hauser-Feshbach formalism and including all the discrete levels of light nuclei, taken from the NUDAT2 database [2], was made. Agreement or disagreement with statistical decay code could be a hint of the importance of non statistical effects. The analysis explores the two channels of the decay [3]: the first foresees the formation of two $^{8}Be_{gs}$ which both decay in two α particles

$$^{16}O^* \rightarrow ^{8}Be_{gs} + ^{8}Be_{gs} \rightarrow \alpha + \alpha + \alpha + \alpha, \quad (1)$$

the second consists in the emission of an α-particle leaving a $^{12}C^*$ as a residue, which in turn decays in three α-particles, through a ^{8}Be intermediate state.

$$^{16}O^* \rightarrow ^{12}C^* + \alpha \rightarrow ^{8}Be_{gs} + \alpha + \alpha \rightarrow \alpha + \alpha + \alpha + \alpha \quad (2)$$

where $^{12}C^*$ can be in the Hoyle state (0_2^+, 7.65 MeV) [4] or in a 3^- excited state at 9.64 MeV.

2 The experiment and the event selection

Measurements were carried out at the INFN laboratories in Legnaro using a pulsed beam of ^{16}O at 130 MeV [5]. The apparatus consists of GARFIELD, a multi-detector located inside a large scattering chamber, coupled with the Ring Counter (RCo), a three-stage annular detector with a truncated cone geometry aimed for the detection at forward angles [6]. ΔE-E correlation and fast-slow PSA techniques were used for particle identification. The selection of events of interest has been performed by considering only events which feature four α-particles, in the final state, detected at forward angles (by the RCo in $5° \div 17°$ angular range) and no product detected by GARFIELD ($30° \div 150°$).

3 Analysis

From energy and momentum conservation one obtains the target and projectile excitation energy:

$$E_{tot_i} = E_{beam} = E_{tot_f} = \sum_{i=1}^{4} E_{\alpha_i} + E_{rec} + E_{exc_t} - Q \quad (3)$$

where E_{α_i} is kinetic energy of the ith α-particle, E_{exc_t} the QT excitation energy, Q is the reaction Q-value and E_{rec} is the QT recoil energy.

Taking into account also the energy loss in the target, E_{rec} turns out to be always lower than the threshold for ^{12}C detection.

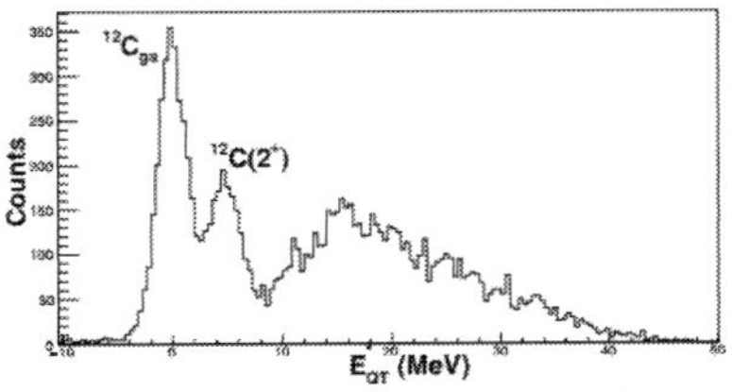

Figure 1. QT excitation energy as calculated in eq. (3).

In the energy spectrum of the QT, shown in Fig. 1, it is possible to observe the corresponding peaks at the ground state and first γ excited state at 4.4 MeV of the QT (^{12}C) nucleus. A tail to higher excitation energies can be seen as well. The analysis presented here includes only the events in which the QT remains in its ground state. However, other analyses that consider the other QT excitation energies have been carried out [7]. In order to determine the decay channel in two $^8Be_{gs}$, the relative energy of the pair of α-particles providing the lowest value out of the four is considered together with the relative energy of the two complementary α-particles. The vertical red lines in Fig. 2

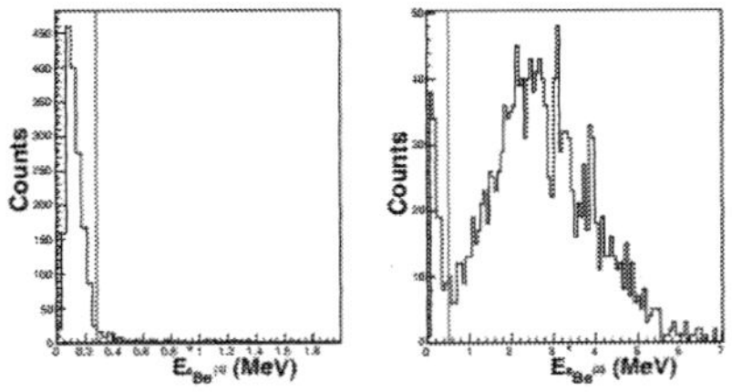

Figure 2. Left panel: minimum relative energy of couples of α-particles. Right panel: relative energy of the complementary couple.

represent the selection performed in order to consider the two relative energies corresponding to the ground state of the ^8Be. In this way, only events with the production of two contemporary $^8Be_{gs}$ are taken into account. This accounts for 7.5 % of the total number of events with the QT in the ground state.

To consider the events corresponding to the second decay channel, from the excitation energy of three α-particles out of the four, the excitation energy of carbon is calculated as the minimum energy of the four combinations. This amount is added to the energy of the fourth α particle to obtain the excitation energy of the ^{16}O.

In Fig. 3 we show the excitation energy of the $^{12}C^*$. The two excited carbon states (Hoyle 0_2^+ and 3^- state, labelled as C_1^* and C_2^*) can be distinguished. The selection of the data that fall within the 7.65 MeV or 9.64 MeV peaks enables a separate analysis of the two decay channels in which an excited carbon nucleus is formed.

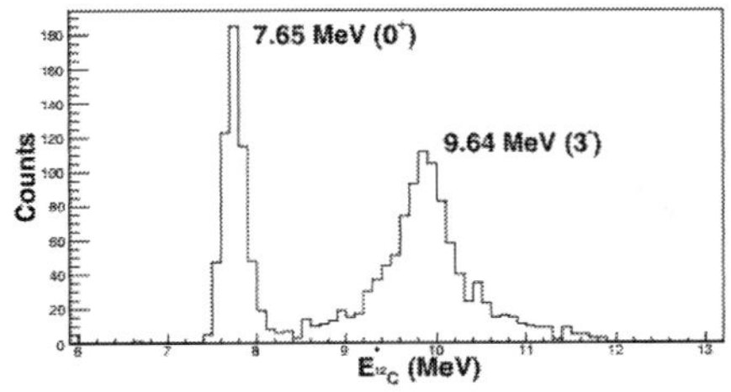

Figure 3. Excitation energy spectrum of $^{12}C^*$ in which $^{16}O^*$ decays with the emission of an α particle.

The decay process with the formation of ^{12}C in the Hoyle state accounts for 32 % of the occurrences, whereas the one with carbon in 3^- state represents 49 %. This results in the preferential decay of $^{16}O^*$ through these decay channels.

Finally, we have performed the comparison with the HFℓ predictions where we considered the decay of a mixing of excited levels with characteristics extracted from experimental data. Table 1 shows the results of this comparison: the observed branching ratios differ significantly

Table 1. Experimental branching ratios of different reactions compared to HFℓ predictions. C_1^* corresponds to the Hoyle state, C_2^* to 9.64 MeV.

reaction	EXP.(%)	HFℓ (%)
$^{12}C(^{16}O,^8Be_{gs}+^8Be_{gs})^{12}C_{gs}$	7.5	20.8
$^{12}C(^{16}O,^{12}C_1^* + \alpha)^{12}C_{gs}$	32.0	53.0
$^{12}C(^{16}O,^{12}C_2^* + \alpha)^{12}C_{gs}$	49.0	19.3
background	11.5	6.8

from the theoretical ones, especially as far as Hoyle state is concerned. A possible explanation might be the influence of the well-known α-cluster structure of the Hoyle state on the Coulomb barrier and the associated transmission coefficients of the corresponding evaporation channel.

References

[1] L. Morelli *et al.*, J. Phys. G 41, 075107 (2014). L. Morelli *et al.*, J. Phys. G 41, 075108 (2014). L. Morelli *et al.*, J. Phys. G 43, 045110 (2016), F. Gulminelli *et al.* in **Nuclear Particle Correlations and Cluster Physics** ed. U. Schroeder, p 257, L. Morelli *et al. ibid* p.283.

[2] *www.nndc.bnl.gov/nudat2/.*

[3] M. Freer *et al.* Phys. Rev C **51**, 1682 (1995).

[4] F. Hoyle *The Astrophysical Journal, Supplement Series* **1**, 12 (1954).

[5] C. Frosin, **Thesis: The ^{16}O+^{12}C reaction at 90.5, 110 and 130 MeV beam energy**, University of Bologna, 2016.

[6] M. Bruno *et al.*, Eur. J. Phys. A 49, 128 (2013).

[7] M. Bruno *et al.* submitted to J. Physics G.

Study of High-Energy Fission in Inverse Kinematics

G. Mantovani[1,2,3], D. Ramos[4,5], M. Caamaño[3], A. Lemasson[5], M. Rejmund[5], L. Audoin[4], H. Álvarez Pol[3], J. Frankland[5], B. Fernández-Domínguez[3], E. Galiana-Baldó[3,6], J. Piot[5], F. Gramegna[1], T. Marchi[1], M. Cicerchia[2,1], D. Ackermann[5], S. Biswas[5], E. Clement[5], D. Durand[6], F. Farget[5,6], M. O. Fregeau[5], D. Galaviz[6], A. Heinz[8], A. I. Henriques[9], B. Jacquot[5], B. Jurado[9], Y. Kim[5], P. Morfouace[5], D. Ralet[10], T. Roger[5], C. Schmitt[11], P. Teubig[6], I. Tsekhanovich[9]

[1]INFN, Laboratori Nazionali di Legnaro, Legnaro (Padova), Italy.
[2]Universitá degli Studi di Padova, Padova, Italy.
[3]Universidade de Santiago de Compostela, Santiago de Compostela, Spain.
[4]IPN Orsay, Université de Paris-Saclay, CNRS/IN2P3, Orsay Cedex, France.
[5]GANIL, CEA/DRF-CNRS/IN2P3, Caen Cedex, France.
[6]Universidade de Lisboa, Lisboa, Portugal.
[7]LPC Caen, Université de Caen Basse-Normandie, ENSICAEN-CNRS/IN2P3, Caen Cedex, France.
[8]Chalmers University of Technology, Göteborg, Sweden.
[9]CENBG, IN2P3/CNRS-Université de Bordeaux, Gradignan Cedex, France.
[10]CSNSM, CNRS/IN2P3, Université de Paris-Saclay, Orsay, France.
[11]IPHC Strasbourg, Université de Strasbourg-CNRS/IN2P3, Strasbourg Cedex, France.

Abstract. Fission at low excitation energy, is a process in which both macroscopic and microscopic aspects are involved. Some features in the total kinetic energy and in the N/Z distributions of the fragments, commonly associated with shell effects, came out in a series of recent experiments with high excitation energy fusion-fission reactions in inverse kinematics. In the latest experiment of this campaign, a study of high-energy fission and quasi-fission between a ^{238}U beam and a series of light targets was carried out by using the aforementioned technique, in order to probe the role of the shell structure in these processes.

1 Introduction

Since its discovery, fission appeared as a complex process where different nuclear properties interplay and shape the characteristics of the emerging fission fragment distributions [1]. Indeed, nuclear fission at low excitation energy, is a phenomenon in which both macroscopic and microscopic aspects are involved.

According to a pure liquid drop description fission should be a symmetric process. However, the large amount of experimental data on the fragment distributions reveal fission of actinides to be an asymmetric process that produces an heavy fragment with an average mass number rather constant, A ~ 140, regardless of the identity of the fissioning system, and a light partner that complements the total mass of the system. This behaviour is understood as a consequence of structure effects on the fragments.

Therefore, fission-fragment distributions are crucial observables for modeling the fission process. Indeed, their evolution depending on the fissioning actinide and excitation energy can give hints on the role that collective and intrinsic degrees of freedom play in fission reaction dynamics. However, the access to such observables was, until recently, limited to the mass information, where the contribution of proton and neutron numbers are mixed. In fact, in experiments performed in direct kinematics, such as thermal neutron-induced fission [2–4], due to the low velocity of the fission fragments, the nuclear charge information is limited to the light fragment when is possible. So then, is not possible to ascribe the heavy-partner stabilization to a given N or Z shell-closure. Later on, it has been demonstrated that this problem can be bypassed with the use of inverse kinematics [5]. Using both mass and nuclear charge observables, measured together as isotopic fragment yields, the role of neutrons and protons in the fission process can be probed. A new experimental approach, which exploits the inverse kinematics technique to have access to the nuclear charge information and the VAMOS magnetic spectrometer in GANIL for high-resolution measurements of the fragment mass, is currently being used in a fission campaign started in 2008 [6–8]. In this approach, in-flight fission of heavy actinides, induced by transfer and fusion reactions, is used. The work presented in the following is focused on the preliminary results of the third experiment within this fission campaign.

2 Experimental Details

The experiment involves a $^{238}U^{+31}$ beam at 5.9 MeV/u, with a final intensity of 10^9 pps, impinging on a set of light targets: ^{9}Be (500 μg/cm^2); ^{Natural}B (100 μg/cm^2); ^{24}Mg (500 μg/cm^2) and ^{27}Al (200 μg/cm^2). A series of fissioning system from ^{238}U up to ^{265}Db was populated by means of transfer and fusion reactions.

The experimental setup, described in Fig. 1, composed by the VAMOS large-solid angle magnetic spectrometer coupled with SPIDER, a ΔE-E telescope which consists of two annular DSSSD and covers polar angles between 30° and 47° [10, 11].

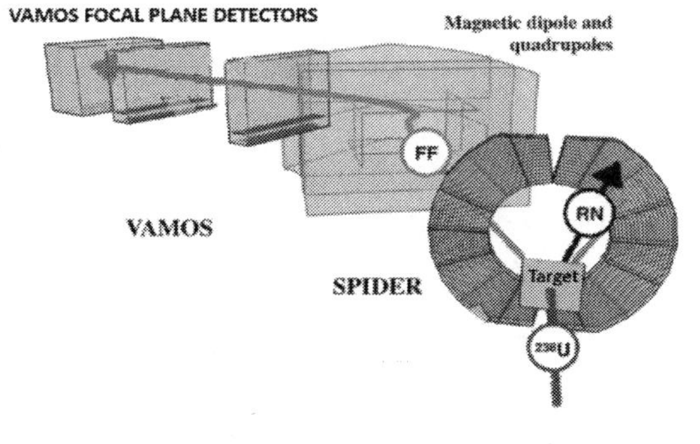

Figure 1. Sketch of the VAMOS + SPIDER experimental setup.

According to the inverse kinematics, both fission fragments were emitted at forward angles in the laboratory reference frame within a cone of 28° around the beam axis which pass through the central hole of SPIDER, whereas the target-like recoil from transfer reactions was more likely emitted at higher angles and could reach the SPIDER telescope, where was isotopically identified. In such a way, the binary reaction and the excitation energy of the fissioning system were reconstructed. A fusion reaction, more frequent than any transfer channel [9], is assumed for every fission event that is registered in coincidence without any target-like recoil in the telescope. In the next section only fusion-fission event data will be discussed. When the excited compound nucleus decays by fission, one of the fragments, if the acceptance allows it, enters the VAMOS spectrometer, is deflected by the magnetic field and finally identified at the focal plane setup.

3 Analysis and Preliminary Results

Once the fusion-fission events have been selected using SPIDER as a veto, the identification of the fission fragments at the VAMOS focal plane setup can be performed in terms of atomic number (Z), mass number (A) and ionic charge state (q) [8]. The isotopic fission-fragment identification allows to calculate the isotopic fission-fragment yields, Y(Z,A), defined as the normalized production of one particular fragment per fission. They are calculated as the number of events of each isotope N(Z,A) measured during the experiment divided by the total number of produced isotopes, corrected by the geometric (ϵ_g^{FF}) and intrinsic (ϵ_i^{FF}) efficiencies for fission fragment detection.

ϵ_g^{FF} is determined by the angular and magnetic rigidity acceptance of VAMOS while ϵ_i^{FF} is the intrinsic efficiency of the detectors, both of them depending on the kinematics of the fragment.

At the VAMOS focal plane setup the detection is limited to a single fragment, and the large acceptance of VAMOS, ±30 % in momentum and ±5.7° in polar angle [10] is not sufficient to cover the entire fragment distribution.

For this reason, two different VAMOS settings were used during the experiment in order to increase the acceptance both in magnetic rigidity and in angle (Bρ = 1.24 Tm, θ = 14°; Bρ = 1.1 Tm, θ = 21.5°). This is translated in a wider covered range of fragment masses.

The geometric efficiency is given by the product of the acceptance in polar angle $f_{\theta_{FF}^{CM}}(A, \theta_{FS}, \beta_{FS})$, for the acceptance in azimuthal angle, $f_\phi(\theta_{FF}^{lab}, \delta)$. The latter must be calculated for each (θ_{FF}^{lab}, δ) selected window as the ratio between the range of ϕ_{lab} accepted in VAMOS and the total physical ϕ_{lab} range covered by fission fragments:

$$\frac{f_\phi(\theta_{FF}^{lab}, \delta) = |\phi_{max}^{lab} - \phi_{min}^{lab}|}{2\pi} \tag{1}$$

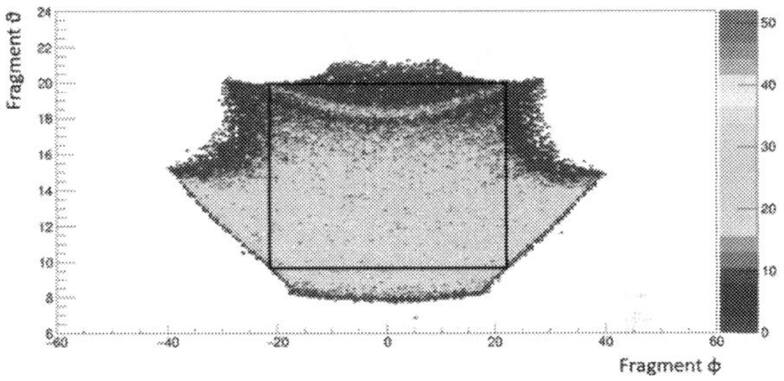

Figure 2. Selected window in θ and ϕ for a central Bρ range (δ = 0.9975 - 1.0025) for VAMOS setting Bρ = 1.24 Tm, θ = 14° to calculate the ϕ-angular acceptance, $f_\phi(\theta_{FF}^{lab}, \delta)$.

In Fig. 2 is reported an example of a selected window in (θ_{FF}^{lab}, δ) for the f_ϕ acceptance calculation.

4 Conclusions and Perspectives

The inverse kinematics technique allows to identify the fission fragments both in mass and nuclear charge. ϕ- and θ-angular acceptance calculations, needed to obtain the fission fragment yields, are still in progress. A further step is necessary to characterize the scission point in terms of fission fragment velocities and masses, in order to calculate the total kinetic energy and the total excitation energy.

References

[1] N. Bohr and J. A. Wheeler, Phys. Rev. **56**, 426 (1939)
[2] W. Lang et al., Nucl. Phys. A **345**, 34 (1980)
[3] U. Quade et al., Nucl. Phys. A **487**, 1 (1988)
[4] J. P. Bocquet et al., Z. Phys. A **335**, 41 (1990)
[5] K.-H. Schmidt et al., Nucl. Phys. A **665**, 221 (2000)
[6] M. Caamaño et al., Phys. Rev. C **88**, 024605 (2013)
[7] M. Caamaño et al., Phys. Rev. C **92**, 034606 (2015)
[8] D. Ramos et al., Phys. Rev. C **97**, 054612 (2018)
[9] D. C. Biswas et al., Phys. Rev. C **56**, 1926 (1997)
[10] M. Rejmund et al., Nucl. Instr. and Meth. in Physics Research Section A **646**, 184 (2011)
[11] M. Vandebrouck et al., Nucl. Instr. and Meth. in Physics Research Section A **812**, 112 (2016)

Shape coexistence in ^{94}Zr studied via Coulomb excitation

Naomi Marchini[1,2,*], *Marco* Rocchini[2], *Adriana* Nannini[2], *Daniel T.* Doherty[3], *Magdalena* Zielińska[4], *Paul E.* Garrett[5], *Katarzyna* Hadyńska-Klęk[6], *Dmitry* Testov[7,8], *Alain* Goasduff[7,8], *Giovanna* Benzoni[9], *Franco* Camera[9,10], *Samuel D.* Bakes[3], *Dino* Bazzacco[8], *Andreas* Bergmaier[11], *Thomas* Berry[3], *Harris* Bidaman[5], *Vinzenz* Bildstein[5], *Daniele* Brugnara[7,12], *Vincent H.* Brunet[3], *Wilton N.* Catford[3], *Matteo* De Rizzo[7], *Alejandra* Diaz Varela[5], *Thomas* Fäestermann[13], *Franco* Galtarossa[4], *Nicla* Gelli[2], *Andrea* Gottardo[12], *Andrea* Gozzellino[12], *Ralf* Hertenberger[14], *Andres* Illana[12], *James* Keatings[15], *Adam R.L.* Kennington[3], *Daniele* Mengoni[7,12], *Lisa* Morrison[3], *Daniel R.* Napoli[12], *Marco* Ottanelli[2], *Pietro* Ottanelli[2], *Giorgia* Pasqualato[7,12], *Francesco* Recchia[7,8], *Serena* Riccetto[16], *Marcus* Scheck[15], *Marco* Siciliano[4], *Giovanni* Sighinolfi[7], *Jacqueline* Sinclair[3], *Pietro* Spagnoletti[15], *José J.* Valiente Dobón[12], *Marine* Vandebrouck[4], *Katarzyna* Wrzosek-Lipska[6], and *Irene* Zanon[7,12]

[1] Università degli Studi di Camerino, Camerino, Italy.
[2]INFN Sezione di Firenze, Firenze, Italy.
[3]University of Surrey, Guildford, UK.
[4]IRFU, CEA, Université Paris-Saclay, France.
[5]University of Guelph, Guelph, Canada.
[6]Heavy Ion Laboratory, University of Warsaw, Poland
[7]Università degli Studi di Padova, Padova, Italy.
[8]INFN Sezione di Padova, Padova, Italy.
[9]INFN Sezione di Milano, Milano, Italy.
[10]Università degli Studi di Milano, Milano, Italy.
[11]Universität der Bundeswehr München, Germany;
[12]INFN Laboratori Nazionali di Legnaro, Legnaro (Padova), Italy.
[13]Technische Universität München, Germany;
[14]Ludwig-Maximilians Universität München, Germany;
[15]University of West of Scotland, Paisley, UK.
[16] Università degli Studi di Perugia and INFN Sezione di Perugia, Perugia, Italy.

Abstract. In recent years, a number of both theoretical and experimental investigations have been performed focusing on the zirconium isotopic chain. In particular, state-of-the-art Monte Carlo shell-model calculations predict shape coexistence in these isotopes. In this context, the ^{94}Zr nucleus, which is believed to possess a nearly spherical ground state, is particularly interesting since the purported deformed structure is based on the low-lying 0^+_2 state, making it amenable for detailed study. In order to provide definitive conclusions on the shapes of the low-lying states, two complementary experiments to study ^{94}Zr by means of low-energy Coulomb excitation were performed. This data will allow the quadrupole moments of the $2^+_{1,2}$ levels to be extracted as well as for the deformation parameters of the $0^+_{1,2}$ states to be determined and, thus, definitive conclusions to be drawn on the role of shape coexistence in this nucleus for the first time.
The first experiment was performed at the INFN Legnaro National Laboratory with the GALILEO-SPIDER setup, which, for the first time, was coupled with 6 lanthanum bromide scintillators ($LaBr_3$:Ce) in order to maximize the γ-ray detection efficiency. The second experiment was performed at the Maier-Leibnitz Laboratory (MLL) in Munich and used a Q3D magnetic spectrograph to detect the scattered ^{12}C ions following Coulomb excitation of ^{94}Zr targets.

1 Introduction

The shape of the nucleus is one of its fundamental properties and is governed by the correlations present among the nucleons and evolves with their number. Some of the most dramatic cases of shape evolution in the ground states of nuclei are encountered along the Sr and Zr isotopic chains. ^{90}Zr, with a closed proton sub-shell at $Z = 40$ and a closed neutron shell at $N = 50$, behaves to a large extent like a doubly-magic nucleus. The neutron number $N = 56$ becomes an effective sub-shell closure at $Z = 40$, and ^{96}Zr also exhibits properties of a doubly-magic nucleus. With the addition of only four neutrons, however, the dramatic lowering of the first excited 2^+ state in ^{100}Zr is indicative of a sudden onset of deformation. Sudden shape changes may be interpreted as a result of an inversion of two distinct configurations associated with different nu-

*e-mail: naomi.marchini@fi.infn.it

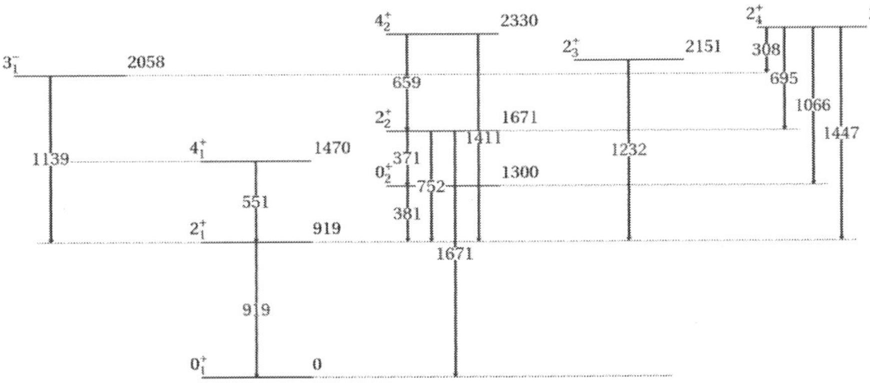

Figure 1. Level scheme for ^{94}Zr displaying the γ-ray transitions observed in the analysis performed to date for the ^{94}Zr+^{208}Pb Coulomb excitation experiment. The energies in keV for each transition are reported.

Figure 2. The experimental setup used for the ^{94}Zr Coulomb-excitation experiment at the INFN Legnaro National Laboratory. The GALILEO array, the SPIDER array and 6 $Labr_3$:Ce detectors are shown.

clear shapes, and the implication is that the distinct configurations should be present also in neighboring nuclei, although they may exist at high excitation energy. Thus, we may expect that the deformed configuration that forms the ground state for $N \geq 60$ persists in the lighter Zr isotopes. The first strong evidence for this was obtained for ^{94}Zr [1] where the observation of a $2^+_3 \to 0^+_2$ γ-ray branch, combined with a 2^+_3 level lifetime, enabled the determination of $B(E2; 2^+_3 \to 0^+_2) = 19 \pm 2$ W.u. indicative of a deformed collective structure. Recently, the even zirconium isotopes have been studied extensively in the framework of the Monte-Carlo Shell Model (MCSM) [2] that reproduces well the trends of the excitation energies and $B(E2)$ values. These calculations predict a spherical ground state and an oblate deformed structure for the 0^+_2 state for the ^{94}Zr isotope (see Fig. 1 of Ref. [2]).

The goal of the present study is to provide a definitive determination of the shapes involved for the 0^+_1 and 0^+_2 states, and for this purpose we employed Coulomb excitation which is the most direct method to study nuclear collectivity and shapes through the measurement of dynamic and static electromagnetic moments. Coulomb excitation selectively excites low-lying collective states with cross sections that are a direct measure of the electromagnetic matrix elements. When the beam energy is chosen to ensure that the nuclear surfaces are separated by a minimum of 5 fm [3], there is negligible contribution to the reaction process from nuclear forces and the matrix elements can be obtained in a model independent way.

2 Experimental details

A low-energy Coulomb excitation experiment was performed in 2018 at the INFN Legnaro National Laboratory using a 370-MeV beam of ^{94}Zr ions impinging on a 1 mg/cm^2 thick ^{208}Pb target. This beam energy was chosen as a compromise between compliance of the "safe-energy" criterion [3] and the necessity of maximising the population of high-lying states, allowing the excitation up to the 2^+_4 state. The level scheme of the ^{94}Zr states populated in this experiment is reported in Fig. 1. The γ rays emitted in the de-excitation of the excited states were detected by the GALILEO γ-ray array [4] and 6 large-volume $3'' \times 3''$ $LaBr_3$:Ce detectors [5], while the backscattered ^{94}Zr particles were detected with the SPIDER array [6] as shown in Fig. 2). The GALILEO γ-ray spectrometer (photo-peak efficiency ~2.1% and FWHM ~0.2% both at 1332.5 keV) consisted of 25 Compton-suppressed HPGe detectors arranged into 4 rings (at 152°, 129°, 119° and 90° with respect to the beam direction) while the SPIDER array for heavy-ion detection was composed of 7 segmented Si detectors each divided into 8 independent strips assembled in a cone-like shape. In order to enhance the sensitivity to the second-order effects, such as multi-step excitation or reorientation, SPIDER was positioned at backward angles in the GALILEO scattering chamber, covering the polar angular range from 123° to 161° in the laboratory frame. For the first time, $LaBr_3$:Ce detectors were coupled to the GALILEO-SPIDER installation in order to increase the

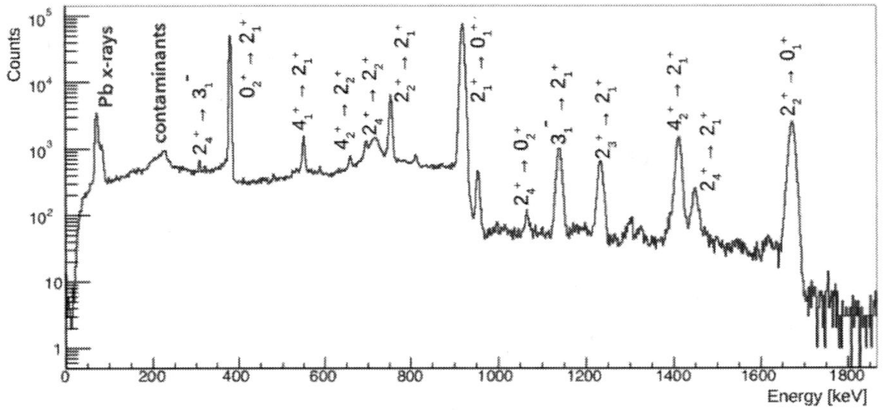

Figure 3. Spectrum of Doppler-corrected γ-ray energies measured by the GALILEO array with events in coincidence with backscattered ^{94}Zr ions detected by SPIDER for the ^{94}Zr+^{208}Pb Coulomb excitation reaction.

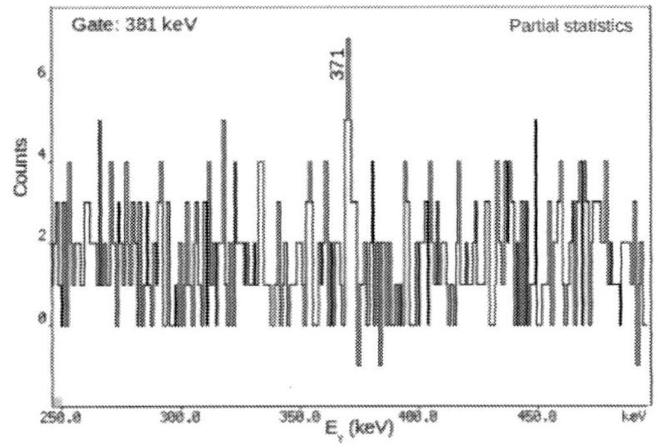

Figure 4. Confirmation for the existence of the de-exciting 371 keV transition, obtained by gating on the $0_2^+ \rightarrow 2_1^+$ transition at 381 keV in $\gamma - \gamma$–particle coincidence matrix.

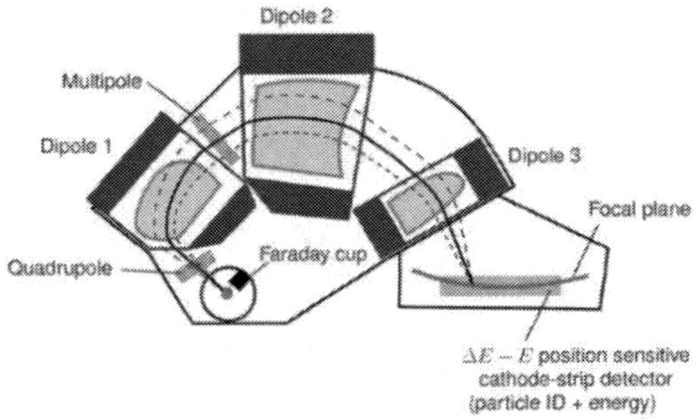

Figure 5. Schematic view of the Q3D spectrometer at the MLL.

efficiency for high-energy γ-ray detection [7]. The coincidence between γ rays and particles allows for the reconstruction of the event kinematics and to perform an event-by-event Doppler correction of detected γ-ray energies. A preliminary Doppler-corrected γ-ray energy spectrum, obtained by selecting the coincidence of events registered in GALILEO and SPIDER, is reported in Fig. 3.

3 On-going analysis

As indicated in Fig. 1, γ rays depopulating the purported shape-coexisting 2_2^+ level are observed. Shown in Fig. 4 is a partial γ-ray spectrum obtained with a coincidence condition set on the 381 keV $0_2^+ \rightarrow 2_1^+$ γ ray. A peak at 371 keV is clearly visible in this spectrum, confirming the result of Ref. [1].

The electromagnetic matrix elements will be extracted from the experimental data using the GOSIA code [8]. This code performs a fitting procedure to the experimental γ-yields (76 in this experiment) with the matrix elements as free parameters. Additional experimental spectroscopic data such as lifetimes, branching ratios and $E2/M1$ multipole mixing ratios, taken from Ref. [9], are also included in the analysis. The transitions needed to obtain the spectroscopic quadrupole moments of the $2_{1,2}^+$ states are clearly observed and, applying the quadrupole sum rules method [10], we will extract the deformation parameters of the ground state and the 0_2^+ state. As shown in Fig. 3, the transitions that involve the 3^- state are also observed. These transitions are necessary to reliably determine the $B(E3; 3_1^- \rightarrow 0_1^+)$ value.

In order to improve the precision of the analysis, we have extended our studies by performing single-step Coulomb-excitation studies at the Maier-Leibnitz Laboratory (MLL) in Munich [11]. Beams of 45 MeV of ^{12}C ions

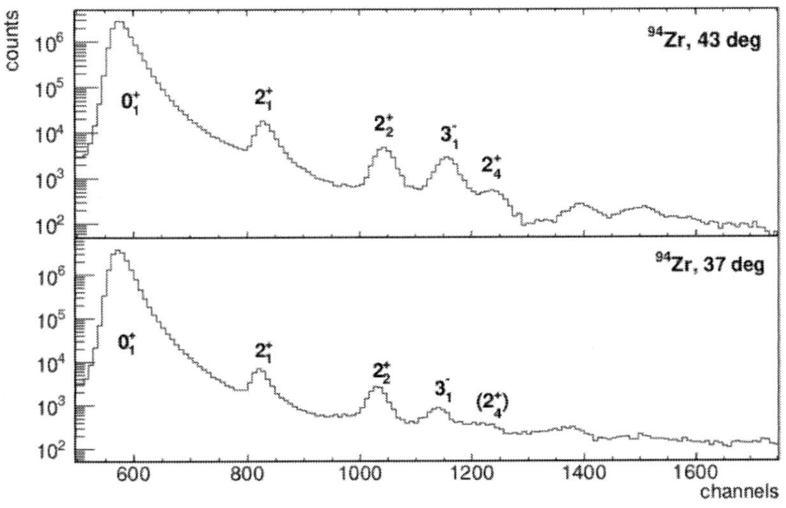

Figure 6. Spectrum of momentum-analysed ^{12}C ions detected at the focal plane of the Q3D spectrograph at angles of 43° (upper part) and 37° (lower part).

up to 100 pnA impinged on a ^{94}Zr target approximately 50 μb/cm^2 thickness. The scattered ^{12}C ions were momentum analysed using a Q3D magnetic spectrograph (a schematic of the apparatus is shown in Fig. 5), at angles of 34°, 37°, 41°, and 43°. Shown in Fig.6 are the ^{12}C spectra observed at angles of 37° and 43°, and clearly show peaks associated with the 2_1^+, 2_2^+, 3_1^-, and 2_4^+ states. Also present in each spectrum is the peak due to the elastically scattered ^{12}C ions. The ratio of the inelastic-to-elastic peak areas yield the excitation probabilities that are directly related to the $B(E\lambda; 0_1^+ \to I^\pi)$ values that serve as valuable additional constraints for the detailed GOSIA analysis of the γ-ray data, significantly aiding in the extraction of spectroscopic quadrupole moments.

4 Conclusion and further perspectives

We have performed complementary experiments using the Coulomb excitation of ^{94}Zr with ^{12}C and ^{208}Pb using a magnetic spectrograph for the former and a large-scale γ-ray detector array and particle detectors for the latter. From these experiments, we anticipate we can extract precise values for the quadrupole moments of the $2_{1,2}^+$, and through an invariant analysis the deformations parameters for the 0_1^+ and 0_2^+ states that can be directly compared with theoretical calculations.

We aim to extend this investigation to the closed neutron sub-shell nucleus ^{96}Zr. For this isotope, the MCSM calculations predicted spherical shapes for both the ground and 2_1^+ states, coexisting with triaxial shapes for both the 0_2^+ and 2_2^+ states. We have recently proposed a Coulomb-excitation experiment employing a ^{96}Zr target using the GALILEO + SPIDER + LaBr$_3$:Ce experimental setup at the Legnaro National Laboratory, while complementary data from ^{12}C scattering using the Q3D at the MLL have already been obtained.

References

[1] A. Chakraborty *et al.*, Phys. Rev. Lett. **110**, 022504 (2013)
[2] T. Togashi *et al.*, Phys. Rev. Lett. **117**, 172502 (2016)
[3] D. Cline Annu. Rev. Nucl. Part. Sci. **36**, 683 (1986)
[4] J.J. Valiente-Dobon *et al.*, LNL-INFN Annual Report 2014, 95 (2015)
[5] A. Giaz *et al.*, Nuclear Instruments and Methods in Physics Research A **729**, 910 (2013)
[6] M. Rocchini *et al.*, Physica Scripta **92**, 074001 (2017)
[7] N. Marchini *et al.*, LNL-INFN Annual Report 2017, 88 (2018)
[8] T. Czosnyka *et al*, Bull. Amer. Phys. Soc. **28**, 745 (1983) http://www.pas.rochester.edu/ cline/Gosia/
[9] B. Singhand and J.A Cameron, Nuclear Data Sheets **92**, 1 Brookhaven ENDSF database (2001) http://www.nndc.bnl.gov/ensdf
[10] K. Kumar, Phys. Rev. Lett. **28**, 249 (1972)
[11] M. Zielińska and P. Garrett. Private Communication

Multinucleon transfer reactions and proton transfer channels

Tea Mijatović[1,*], *Suzana* Szilner[1], *Lorenzo* Corradi[2], *Franco* Galtarossa[2], *Samuel* Bakes[3,4], *Daniele* Brugnara[2], *Gabriele* Carozzi[3], *Giulia* Colucci[3], *Petra* Čolović[1], *Enrico* Fioretto[2], *Andres* Gadea[5], *Alain* Goasduff[3], *Andrea* Gottardo[2], *Grzegorz* Jaworski[6], *Deša* Jelavić Malenica[1], *Tommaso* Marchi[2], *Daniele* Mengoni[3], *Matko* Milin[7], *Giovanna* Montagnoli[3], *Deni* Nurkić[7], *Giovanni* Pollarolo[8], *Romana* Popočovski[1], *Fernando* Scarlassara[3], *Marco* Siciliano[2], *Neven* Soić[1], *Alberto* Stefanini[2], *Dmitry* Testov[2,3], *Calin* Ur[9], *Jose Javier* Valiente - Dobón[2], *Nikola* Vukman[1], and *Irene* Zanon[3]

[1]Ruđer Bošković Institute, Zagreb, Croatia
[2]INFN - Laboratori Nazionali di Legnaro, Legnaro, Italy
[3]Dipartimento di Fisica, Università di Padova and INFN, Padova, Italy
[4]University of Surrey, Guildford, United Kingdom
[5]IFIC CSIC, Valencia University, Spain
[6]Heavy Ion Laboratory, University of Warsaw, Warsaw, Poland
[7]University of Zagreb, Zagreb, Croatia
[8]Università di Torino, and INFN, Torino, Italy
[9]ELI-NP, Bucharest, Romania

Abstract. Transfer reactions have always been of great importance for nuclear structure and reaction mechanism studies. So far, in multinucleon transfer studies, proton pickup channels have been completely identified in atomic and mass numbers at energies close to the Coulomb barrier only in few cases. We measured the multinucleon transfer reactions in the ^{40}Ar+^{208}Pb system near the Coulomb barrier, by employing the PRISMA magnetic spectrometer. By using the most neutron-rich stable ^{40}Ar beam we could populate, besides neutron pickup and proton stripping channels, also neutron stripping and proton pickup channels. Comparison of cross sections between different systems with the ^{208}Pb target and with projectiles going from neutron-poor to neutron-rich nuclei, as well as between the data and GRAZING calculations, was carried out. Finally, recent results concerning the measurement of the excitation function from the Coulomb barrier to far below for the ^{92}Mo+^{54}Fe system, where both proton stripping and pickup channels were populated with similar strength, will be discussed.

1 Introduction

Among the quasi-elastic processes of particular relevance are the ones related to the transfer of nucleon pairs. This type of reactions is perceived as a very promising tool for the study of correlations via simultaneous comparison of ±nn, ±pp and ±np pairs. They are especially advantageous since one could envisage the exchange of multiple nucleon pairs among the reactants for both neutrons and protons.

In studies of pair correlations in heavy-ion collisions several experiments have been performed in the past. In these reactions the signature of correlations is an enhancement of the cross section that, experimentally, may be estimated from the comparison of the measured two-particle transfer probability (or cross section) with the prediction of models using uncorrelated states. Unfortunately the experimental evidence was marred by the fact that almost all existing studies involved inclusive cross sections at energies higher than the Coulomb barrier and at angles forward of the grazing where the reaction mechanism is more complex due to the interplay between nuclear and Coulomb trajectories. Therefore, it is still an open question how the transfer of a nucleon pairs and the correlations affect the quasi-elastic processes.

*e-mail: tea.mijatovic@irb.hr

2 Multinucleon transfer reactions above the barrier

In the quasi-elastic regime the mass and charge distributions of transfer products are governed by optimum Q-value considerations and transfer form factors. For nuclei close to the stability line, these optimum Q-value arguments favor the neutron pickup and the proton stripping channels (pickup and stripping are referring to the lighter reaction fragment) [1]. This is the reason why multinucleon transfer reactions have been used as a competitive tool for the production of neutron-rich nuclei in the vicinity of the light partner [2–8]. This is presented in the Fig. 1 that shows the measured cross sections for the stable open shell ^{58}Ni and closed shell ^{40}Ca+^{208}Pb systems where the reaction mechanism strongly favors the proton stripping and neutron pickup channels. It is also evident how the transfer flux changes (shifts) with the use of the neutron-rich (stable) projectiles like ^{40}Ar [9]. In particular one sees how proton pickup channels open up. Even if these proton

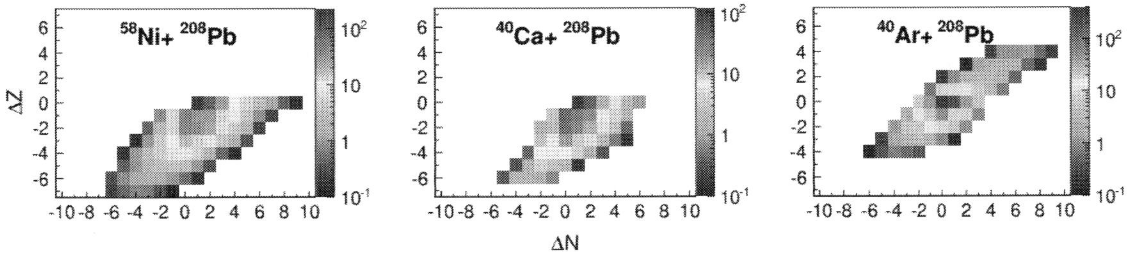

Figure 1. Angle and energy integrated total cross section for ^{40}Ar, ^{40}Ca, and ^{58}Ni projectiles on the ^{208}Pb target, at energies E_{lab} = 6.4, 6.2, and 6 MeV/A, respectively. The cross section values of the elastic (+inelastic) channel have been scaled down by a factor 100 to better display the behavior of the transfer channels [9].

pickup channels have lower cross sections than the proton stripping ones, the observed *Z-N* distribution turns out to be more symmetric.

Figure 2 shows the data compared with the calculations performed with the GRAZING code [10] for ^{40}Ar, ^{40}Ca, and ^{58}Ni projectiles on the ^{208}Pb target [9]. In all systems GRAZING well describes the one nucleon transfer channels. For other pure neutron transfer channels the total cross section is also very well reproduced, which is particularly true for neutron pickup channels. When neutron "stripping" channels are involved, we have to keep in mind that the evaporation of neutrons plays a very important role. Deviations between experimental data and calculations are more marked for channels involving the transfer of many protons. This fact has been discussed in previous publications where experimental cross sections were compared with different semi-classical models in order to see if the addition of new modes, in particular the transfer of a pair of nucleons (both neutrons and protons), may be justified [11]. The inclusion of these pair-transfer modes may be essential. However, we cannot rule out the contribution from deep-inelastic processes. The deviations between experimental data and calculations are more pronounced in the proton pickup sector, in particular when neutron pickup channels are involved, where the contributions from the deep-inelastic collision may be very substantial.

The pure neutron transfer channels have major contribution close to the optimum *Q*-value (~0 MeV), typical of the quasi-elastic regime, with an increasing strength for large energy losses as more neutrons are transferred. For the transfers of more particles, one sees a widening of the experimental *Q*-value and the appearance of the large energy losses. This contribution of the large energy loss components is more pronounced for the proton pickup channels, and when more nucleons are transferred. Such behaviour may indicate that processes more complex than the direct transfer dominate the observed yields. These large energy losses may lead to neutron evaporation which plays a major role in the final mass distribution. Therefore, these additional processes complicate the theoretical calculations even when not considering correlations. Much more work is needed to properly understand, both theoretically and experimentally, the behavior of proton transfer channels, in particular the proton pickup ones.

A definite dominance of the proton pickup and neutron stripping channels in the distribution of the transfer flux is predicted to occur with an additional increase of the neutron excess in projectile. Such situation leads to the population of neutron-rich nuclei in the corresponding heavy partner. However, one has to keep in mind that the primary yield can be influenced by secondary processes that generally shift the mass distributions toward lower values.

3 Multinucleon transfer reactions below the barrier

At energies closer and well below the Coulomb barrier the contribution of different channels and secondary effects is much lower, which is beneficial to better understanding of proton transfer channels. The interacting nuclei are only slightly influenced by the nuclear potential and *Q* values are restricted to few MeV for the open transfer channels. These conditions diminish the complexity of coupled channel calculations, since one needs to take into account only few populated states, and information may be extracted on the nucleon-nucleon correlations.

Measurement itself is more complicated, and significant progress has been made by performing the experiments in inverse kinematics [12]. In this way ions are forward focused which results in high detection efficiency and they have high kinetic energy which is important for energy and therefore mass resolution.

Recently, several systems have been measured with the large solid angle spectrometer PRISMA [13] for the study of nucleon-nucleon correlations [14–16]. PRISMA has a position-sensitive micro-channel plate detector placed at the entrance of the spectrometer, providing a start signal for time-of-flight measurements and bi-dimensional position signals. Ions pass through the optical elements of the spectrometer (a quadrupole and a dipole) and enter a focal plane detector which is made of a parallel plate detector of multi-wire type, providing timing and bidimensional position signals with resolutions similar to the entrance detector. Located at the end of the focal plane is an array of a transverse field multi-parametric ionization chambers (IC), providing nuclear charge via energy loss (ΔE) and

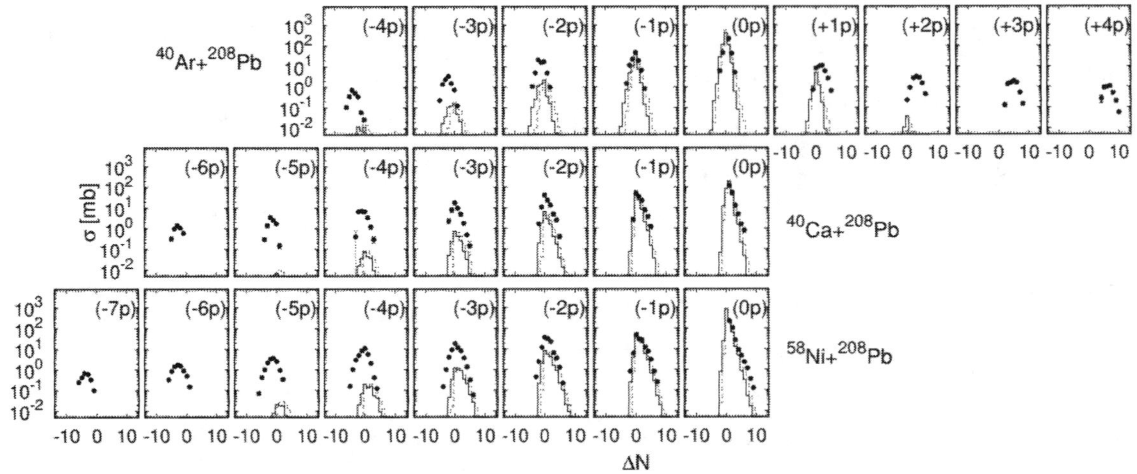

Figure 2. Total experimental cross section for ^{40}Ar, ^{40}Ca, and ^{58}Ni induced reactions on the ^{208}Pb target, at beam energies E_{lab} = 6.4, 6.2, and 6 MeV/A, respectively (points), and the GRAZING calculations with (solid line) and without (dashed line) neutron evaporation [9].

total energy (E) measurements. The detector system gives all the necessary information for the complete ion identification, which is performed via an event-by-event reconstruction of the trajectory inside the magnetic elements.

Making use of inverse kinematics, target recoils have been detected in multinucleon transfer reactions for the ^{96}Zr+^{40}Ca (closed shell) [14] and ^{116}Sn+^{60}Ni (super-fluid nuclei) [15, 16] systems. An excitation function has been obtained from the Coulomb barrier to 20-25% below. This type of experiments is best represented with transfer probabilities, defined as the ratio of the transfer yield over the quasi-elastic one, plotted as a function of distance of closest approach for a Coulomb trajectory, that takes into account both energy and angle.

In the proton sector the experimental data are very scarce. Proton transfer channels are in fact more difficult to measure below the barrier since their cross sections drop off more rapidly than those of neutron transfer channels. Therefore, the proton transfer processes in heavy-ion collisions are much less understood than the ones of neutrons, since large modification in the trajectories of entrance and exit channels are involved due to the modification of the Coulomb field. The single particle level density for protons is less studied than the one of neutrons and the corresponding single-particle form factors are less known (even the one-proton transfer cross sections are not very well described in the DWBA, in fact this theory predicts angular distributions that are shifted in comparison with the data). Moreover, transfer probabilities for protons are generally smaller at the same distances of closest approach. For this reason most of previous works were limited to near-barrier energies and to small values of distances of closest approach, where a non negligible overlap with the absorption region can significantly affect the interpretation of the transfer probabilities and therefore the extracted enhancement coefficients.

We recently made an experiment dedicated to the study of proton transfer channels below the Coulomb barrier where we expect lower contribution of the secondary processes. We chose the ^{92}Mo+^{54}Fe system involving proton-rich nuclei, where proton pickup and neutron stripping channels on ^{54}Fe become available, together with proton stripping and neutron pickup, following optimum Q-value considerations. Open channels can be seen in the Fig. 3, that shows GRAZING calculations for the system.

The ^{92}Mo+^{54}Fe system was measured from above to below the barrier in inverse kinematics using the PRISMA spectrometer. The high efficiency allowed to achieve sufficient yield to get an excitation function encompassing the Coulomb barrier (not far below Coulomb barrier but significantly below the strong absorption radius). We found that the yields of the pure two proton transfer channels are similar to those of the "alpha" transfer channels whose probability turns out to be much stronger than predicted by a simple phenomenological analysis of the probabilities, assuming an independent nucleon transfer process. However, one has to keep in mind that the cross sections are derived by integrating the whole total kinetic energy distributions, therefore, it is important to investigate how the transfer strength is distributed. This information is crucial to understand to what extent the strong population of the $(\pm 2p)$ and "$(\pm\alpha)$" channels can be due to simple reaction mechanism effects or whether nucleon-nucleon correlations play a role.

Therefore, we measured, for the same system, fragment-gamma coincidences. We used a ^{54}Fe beam at the bombarding energy E_{lab}=230 MeV (close to the Coulomb barrier) onto a ^{92}Mo target with a thickness of 200 μg/cm^2. Beam-like products were detected with the PRISMA spectrometer placed at θ_{lab} = 62° and 70°, where Fe-like scattered ions have kinetic energies ~2 MeV/A, sufficient to obtain A and Z separation. These angles, close to the grazing, are chosen in order to have a sufficient

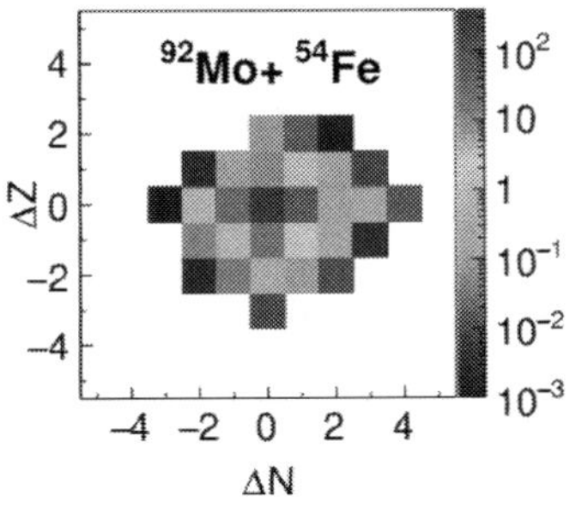

Figure 3. GRAZING calculated cross section for the target-like products in the collision of ^{92}Mo on ^{54}Fe at 370 MeV. Relatively symmetric population of neutron and proton pickup and stripping channels is obtained.

transfer yield under the condition that the relative motion trajectories are weakly affected by the nuclear components of the interaction. In this condition the transfer yields are coming from impact parameters leading to large distance of closest approach (where the transfer form factors are of exponential character with their tail governed by the binding energy of the involved single particle states) and are not affected by the absorption.

To measure electromagnetic transitions we used the new array of six 2" × 2" $LaBr_3$ γ detectors, installed on the top platform of the PRISMA scattering chamber. Each detector was located at 7 cm from the target and at 45° with respect to the reaction plane. In this way the intrinsic efficiency for 1.3 MeV γ-ray was ~1.3%. Tests have been performed with different γ sources, getting a timing resolution ~450 ps and an energy resolution ~ 25-30 keV (for ~ 1 MeV γ-ray). This energy resolution is sufficient to separate various transitions, which should allow us to measure the strength distributions.

Data analysis is under-way. Doppler corrected γ-spectrum for ^{54}Fe shows the strong population of 2^+ and 4^+ states. Similar approach will be used to extract the strength distributions for different isotopes.

PRISMA also allows to measure differential cross sections by integrating events in steps of ~ 1.5° in the reaction plane. This will be used to complement the transfer probabilities obtained via the excitation function previously measured in inverse kinematics. The probabilities obtained via excitation function and angular distribution should be very similar provided that large energy loss components correspond to a small fraction of the cross sections.

4 Summary

Transfer reactions with heavy ions are a powerful tool to investigate reaction mechanism and structure properties of nuclei. The comparison of systems going from the neutron-poor ^{40}Ca and ^{58}Ni to the neutron-rich ^{40}Ar projectiles on the ^{208}Pb target shows how the population trend evolves from proton stripping and neutron pickup to the opposite direction. Such processes are relevant for the production of neutron-rich target-like nuclei and high resolution fragment-γ studies are needed to improve the predictability of theoretical models.

Microscopic calculations are still challenging for proton transfer channels. Even if it is expected that at the energies below the Coulomb barrier the energy distribution (TKEL) will become narrower and will concentrate in the single state, the fragment-γ coincidences will be crucial to determine which states are dominantly populated, and thus to define the correct phase space used in the calculations.

Acknowledgments

This work has been supported in part by the Croatian Science Foundation under project no. 7194 and in part under project no. IP-2018-01-1257.

References

[1] L. Corradi, G. Pollarolo, and S. Szilner, J. Phys. G: Nucl. Part. Phys. **36**, 113101 (2009).
[2] S. Lunardi *et al.*, Phys. Rev. C **76**, 034303 (2007).
[3] J. J. Valiente-Dobón *et al.*, Phys. Rev. Lett. **102**, 242502 (2009).
[4] S. Szilner *et al.*, Phys. Rev. C **84**, 014325 (2011).
[5] S. Szilner *et al.*, Phys. Rev. C **87**, 054322 (2013).
[6] R. Chapman *et al.*, Phys. Rev. C **93**, 044318 (2016).
[7] A. Vogt *et al.*, Phys. Rev. C **93**, 054325 (2016).
[8] J. Litzinger *et al.*, Phys. Rev. C **97**, 044323 (2018).
[9] T. Mijatović *et al.*, Phys. Rev. C **94**, 064616 (2016).
[10] A. Winther, Nucl. Phys. A **572**, 191(1994); A. Winther, Nucl. Phys. A **594**, 203 (1995); program GRAZING http://www.to.infn.it/~nanni/grazing
[11] S. Szilner *et al.*, Phys. Rev. C **71**, 044610 (2005).
[12] F. Galtarossa *et al.*. Phys. Rev. C **97**, 054606 (2018).
[13] S. Szilner *et al.*, Phys. Rev. C **76**, 024604 (2007).
[14] L. Corradi *et al.*, Phys. Rev. C **84**, 034603 (2011).
[15] D. Montanari *et al.*, Phys. Rev. Lett. **113**, 052501 (2014).
[16] D. Montanari *et al.*, Phys. Rev. C **93**, 054623 (2016).

Alpha correlations in terms of the hyper-Pfaffian

Takahiro Mizusaki[1,*] and *Peter* Schuck[2,3]

[1]Institute of Natural Sciences, Senshu University, Tokyo 101-8425, Japan
[2]Institut de Physique Nucléaire, IN2P3-CNRS, Université Paris-Sud, 91406 Orsay, France
[3]Univ. Grenoble Alpes, CNRS, LPMMC, 38000 Grenoble, France

Abstract. We find that the overlap between alpha-like four-body condensation wave function and m-scheme basis state can be represented by the hyper-Pfaffian, which is a natural extension of the Pfaffian for the pair-condensation wave function. This overlap is useful for nuclear structure calculations with the variational Monte Carlo method. As the hyper-Pfaffian has, however, only a few mathematical relations, it is quite difficult to use for numerical computations. We also find that a specific case of hyper-Pfaffian can be expressed by the sum of standard Pfaffians.

1 Introduction

The alpha states for C and O have been studied over a half-century. On the other hand, alpha-like correlation or quartet structure in the shell model calculations has also been discussed, for example, Ref. [1]. The alpha-like four-nucleon correlations have been discussed in Refs. [2, 3] with the closure approximation. Recently renewed interests have been paid to such quartet structures in the nuclear shell model by N. Sandulescu and his collaborators [4–6] with the use of recursion relations.

Recently the variational Monte Carlo (VMC) method with the pair-condensate trial state has been applied to fermion many-body systems [7, 8], including the nuclear shell model [9, 10]. The VMC trial wave function with the pair-condensation state is given as

$$|\Psi^{2n}\rangle = \left(\frac{1}{2}\sum f_{ij}c_i^\dagger c_j^\dagger\right)^n |0\rangle \tag{1}$$

where $c^\dagger$'s are nucleon creation operators, n is the number of pairs and $|0\rangle$ is the fermion vacuum. The overlap between this pair-condensate wave function and the m-scheme basis $|\boldsymbol{m}\rangle$, which is defined as $|\boldsymbol{m}\rangle = c_{m_1}^\dagger \cdots c_{m_{2n}}^\dagger|0\rangle$, is shown by

$$\langle \boldsymbol{m}|\Psi^{2n}\rangle = n!Pf(\tilde{f}) \tag{2}$$

where Pf is the Pfaffian. (For its detail, see [9, 10] for example.) Note that the Pfaffian has been recently studied in the overlap between the HFB wave functions, for examples, [11–13].

The energy by the VMC is formulated as

$$E = \sum_m \rho(m)E_L(m), \tag{3}$$

where the sum regarding m is estimated by the Monte Carlo sampling. The local energy $E_L(m)$ is defined as

$$E_L(m) = \sum_{m'} h_{m,m'}\frac{\langle m'|\phi\rangle}{\langle m|\phi\rangle}. \tag{4}$$

The Monte Carlo sampling is carried out with the probability density $\rho(m)$, defined by

$$\rho(m) = \frac{|\langle m|\phi\rangle|^2}{\sum_m |\langle m|\phi\rangle|^2}, \tag{5}$$

where $\rho(m) \geq 0$ and $\sum_m \rho(m) = 1$. Thus, what we need to evaluate the energy, is the overlaps $\langle m|\phi\rangle$. If we obtain the gradient vector regarding the parameters contained in the trial state, we can perform the energy minimization with the Monte Carlo method. In this formulation, the complicated matrix element of the Hamiltonian between the two trial states is not explicitly needed. Therefore we focus on the overlap matrix element hereafter.

2 Quartet-condensate wave function and hyper-Pfaffian

In this work, we extend the pair-condensate wave function to the quartet-condensate wave function as

$$|\Psi^{4n}\rangle = \left(\frac{1}{4!}\sum f_{pqrs}c_p^\dagger c_q^\dagger c_r^\dagger c_s^\dagger\right)^n |0\rangle \tag{6}$$

where n is the number of quartets. We found [14] that the overlap between this quartet–condensate wave function and m-scheme basis, $|\boldsymbol{m}\rangle = c_{m_1}^\dagger \cdots c_{m_{4n}}^\dagger|0\rangle$, can be expressed by the hyper-Pfaffian as

$$\langle \boldsymbol{m}|\Psi^{4n}\rangle = \langle \boldsymbol{m}|\left(\sum_{p<q<r<s} f_{pqrs}c_p^\dagger c_q^\dagger c_r^\dagger c_s^\dagger\right)^n|0\rangle = n!Pf^{[4]}(f) \tag{7}$$

*e-mail: mizusaki@isc.senshu-u.ac.jp

where the $Pf^{[k]}$ is the hyper-Pfaffian by Jean-Gabriel Luque and Jean-Yves Thibon [15]. It is defined for fully anti-symmetric tensor $M_{i_1,\cdots,i_k}$ ($M_{i_{\sigma(1)},\cdots,i_{\sigma(k)}} = \mathrm{sgn}(\sigma) M_{i_1,\cdots,i_k}$ for any permutation σ) as,

$$Pf^{[k]}(M) \equiv \sum_{\sigma \in \mathfrak{S}_{kn,k}} \mathrm{sgn}(\sigma) \prod_{i=1}^{n} M_{\sigma(k(i-1)+1),\sigma(k(i-1)+2),\cdots,\sigma(ki)} \tag{8}$$

where permutation σ is restricted with the following condition as,

$$\begin{aligned} \mathfrak{S}_{kn,k} \equiv & \{\sigma \in \mathfrak{S}_{kn} | \sigma(k(i-1)+1) < \cdots < \sigma(ki)), \\ & \sigma(k(p-1)+1) < \sigma(kp+1)), \\ & 1 \le i \le n, 1 \le p \le n-1\}. \end{aligned} \tag{9}$$

In the Appendix, we show two examples. The hyper-Pfaffian has been very recently studied by several mathematicians [15–18]. Eq.(7) with this hyper-Pfaffian is a natural extension of the usual Pfaffian form. In general k-body condensation corresponds to the k-th order hyper-Pfaffian. However, the hyper-Pfaffian has, unfortunately, only a few mathematical relations, unlike the Pfaffian.

Next, we try to find useful relations by assuming the following form as

$$\langle \boldsymbol{m} | \Psi^{4n} \rangle = \langle \boldsymbol{m}^{\pi} \boldsymbol{m}^{\nu} | \left(\frac{1}{4} \sum f_{pqrs} c_p^{\pi\dagger} c_q^{\pi\dagger} c_r^{\nu\dagger} c_s^{\nu\dagger} \right)^n |0\rangle \tag{10}$$

where π and ν are proton and neutron labels. The m-scheme state is also a product of the proton and neutron m-scheme states $|\boldsymbol{m}^{\pi}\rangle$ and $|\boldsymbol{m}^{\nu}\rangle$, respectively. This form means alpha-like condensation.

For $n = 2$ and 3, we easily find that the corresponding hyper-Pfaffians can be written with the sum of standard Pfaffians. For $k = 4, n = 2$, the hyper-Pfaffian is reduced into the ordinary Pfaffian as

$$Pf^{[4]}(f) = \sum_{\sigma \in \mathfrak{S}_4} \mathrm{sgn}(\sigma) Pf\left(f_{\sigma(1)\sigma(2),\sigma(3)\sigma(4)}\right) \tag{11}$$

where $\sigma(1) < \sigma(2)$ and $\sigma(3) < \sigma(4)$. For $k = 4, n = 3$, the hyper-Pfaffian is also reduced into the ordinary Pfaffian as

$$\begin{aligned} Pf^{[4]}(f) = & \sum_{\sigma \in \mathfrak{S}_6} \mathrm{sgn}(\sigma) [Pf\left(f^3_{\sigma(1),\sigma(2),\sigma(3),\sigma(4),\sigma(5),\sigma(6)}\right) \\ & + z_{\sigma(1),\sigma(2),9,10} Pf\left(f^2_{\sigma(3),\sigma(4),\sigma(5),\sigma(6)}\right)] \end{aligned} \tag{12}$$

where $\sigma(1) < \sigma(2)$, $\sigma(3) < \sigma(4)$ and $\sigma(5) < \sigma(6)$. These relations are derived in [14], while for $n > 3$, there is no simple relation. To find a more general formula beyond this limitation and the gradient formula is currently under our investigation.

This work was supported by the research grant of the Senshu Research Abroad Program (2018) for one of the authors (T.M.).

Appendix: Example of the hyper-Pfaffian

Here we show two examples of the hyper-Pfaffian. For the $k = 4, n = 1$,

$$Pf^{[4]}[f] = f_{1234}, \tag{13}$$

For $k = 4, n = 2$,

$$\begin{aligned} Pf^{[4]}(f) = & f_{1234}f_{5678} - f_{1235}f_{4678} + f_{1236}f_{4578} \\ - & f_{1237}f_{4568} + f_{1238}f_{4567} + f_{1245}f_{3678} \\ - & f_{1246}f_{3578} + f_{1247}f_{3568} - f_{1248}f_{3567} \\ + & f_{1256}f_{3478} - f_{1257}f_{3468} + f_{1258}f_{3467} \\ + & f_{1267}f_{3458} - f_{1268}f_{3457} + f_{1278}f_{3456} \\ - & f_{1345}f_{2678} + f_{1346}f_{2578} - f_{1347}f_{2568} \\ + & f_{1348}f_{2567} - f_{1356}f_{2478} + f_{1357}f_{2468} \\ - & f_{1358}f_{2467} - f_{1367}f_{2458} + f_{1368}f_{2457} \\ - & f_{1378}f_{2456} + f_{1456}f_{2378} - f_{1457}f_{2368} \\ + & f_{1458}f_{2367} + f_{1467}f_{2358} - f_{1468}f_{2357} \\ + & f_{1478}f_{2356} - f_{1567}f_{2348} + f_{1568}f_{2347} \\ - & f_{1578}f_{2346} + f_{1678}f_{2345}. \end{aligned} \tag{14}$$

As n increases, the number of terms explosively increases.

References

[1] A. Arima and V. Gillet, Annal of Phys. **66**, 117 (1971).
[2] M. Hasegawa, S. Tazaki, R. Okamoto, Nucl. Phys. **A592** 45 (1995).
[3] M. Hasegawa, S. Tazaki, Nucl. Phys. **A633** 266 (1998).
[4] N. Sandulescu and D. Negrea, J. Dukelsky, and C. W. Johnson, Phys. Rev. **C 85** 061303(R) (2012).
[5] N. Sandulescu, D. Negrea, and C. W. Johnson, Phys. Rev. **C86**, 041302(R) (2012).
[6] M. Sambataro and N. Sandulescu, Phys. Rev. **C93**, 054320 (2016).
[7] M. Bajdich, L. Mitas, G. Drobny, L. K. Wagner, and K. E. Schmidt, Phys. Rev. Lett. **96**, 130201 (2006).
[8] D. Tahara and M. Imada, J. Phys. Soc. Jpn. **77**, 114701 (2008).
[9] T. Mizusaki and N. Shimizu, Phys. Rev. **C85**, 021301(R) (2012).
[10] N. Shimizu and T. Mizusaki, Phys. Rev. **C98**, 054309 (2018).
[11] L. M. Robledo, Phys. Rev. **C79**, 021302(R) (2009).
[12] B. Avez and M. Bender, Phys. Rev. **C85**, 034325 (2012).
[13] T. Mizusaki and M. Oi, Phys. Lett. **B715**, 219 (2012).
[14] T. Mizusaki and P. Schuck, Bulletin of the Institute of Natural Sciences, Senshu University **50**, 19, (2019).
[15] J.-G. Luque, J.-Y. Thibon, Adv. in Appl. Math. **29** 620 (2002).
[16] A.I. Barvinok, Mathematical Programming, **69** 449 (1995).
[17] D. Redelmeier, Ph.D. University of Waterloo (2006).
[18] Sho Matsumoto, Journal of Algebra **320** 612 (2008).

EPJ Web of Conferences **223**, 01041 (2019) https://doi.org/10.1051/epjconf/201922301041
NSD2019

Fusion hindrance in light and heavy systems

Giovanna Montagnoli[1,2,*]

[1]University of Padova
[2]INFN Sezione di Padova

Abstract.
The phenomenon of hindrance in sub-barrier heavy-ion fusion has been confirmed by several experimental evidences and it is now recognised as a general phenomenon of heavy-ion fusion process. In many cases the signature of fusion hindrance lies in the trend of the logarithmic slope of the excitation function and of the S factor at low energies. The comparison with standard Coupled-Channels calculations is a more quantitative evidence for its existence. In many medium-heavy systems the hindrance effect has been recognised with different features depending on the various couplings to the inelastic and transfer channels. Different theoretical approaches have been proposed to explain the hindrance but the underlying physics is still a matter of debate. Hindrance is observed in light systems, independent of the sign of the fusion Q-value, with different features. In the case of the ^{12}C + ^{30}Si system the effect is small but it is clearly observed. Near-by cases show evidence for systematic behaviours. A very recent experiment has concerned the lighter case ^{12}C + ^{24}Mg where hindrance shows up clearly, because a maximum of the S factor appears already at a relatively high cross section σ=1.6 mb. The consequences for the dynamics of stellar evolution have to be clarified by further experimental and theoretical work.

1 Introduction

Heavy-ion fusion reactions near and below the barrier are typical phenomena where the interplay of reaction dynamics and nuclear structure leads to a large enhancement of the fusion probability with respect to the one-dimensional quantum tunnelling limit. An important step forward in the fusion studies has been the introduction of the concept of barrier distribution [1] generated by the channel couplings and the possibility to obtain it from the experimental excitation functions. The barrier distribution has been exploited to disentangle the influence of the various coupled channels and its sensitivity to the static nuclear deformation [2], to the collective multi-phonon excitations [3] and to the transfer channels with positive Q-value [4], is remarkable.

Extending down in energy it has been observed that the fusion excitation functions of many systems decrease more steeply than the predictions of standard Coupled-Channels (CC) calculations based on a Wood-Saxon potential. This phenomenon known as fusion hindrance has been recognised as a general feature forcing us to rethink the shape of the inner part of the potential barrier as some theoretical models indicate.

In this talk the cases of ^{12}C + ^{30}Si, ^{24}Mg, recently studied at LNL, will be presented and discussed, showing the sub-barrier trend of their excitation function and of the S factor. Indeed, if the hindrance phenomenon exists in light systems, the fusion cross sections (viz. the S factors) near the Gamow peak might be substantially smaller than expected previously by simple extrapolations of the high-energy trends, where sets of measurements already exist. Fusion reactions between light heavy ions have a prominent role in the dynamics of stellar evolution.

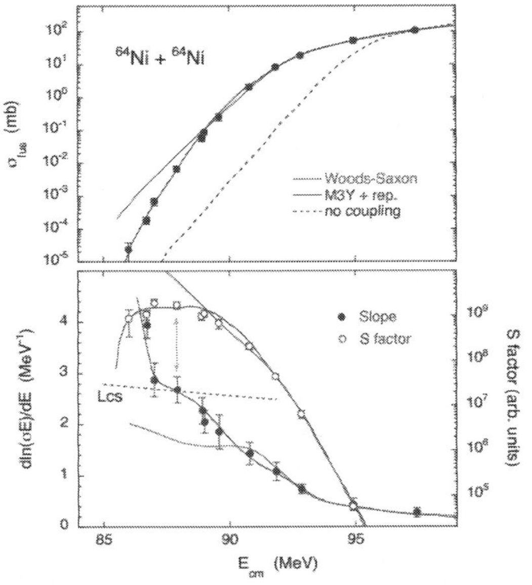

Figure 1. Fusion excitation function (upper panel), logarithmic slope and astrophysical S factor (lower panel) of ^{64}Ni + ^{64}Ni [10].

*e-mail: montagnoli@pd.infn.it

It was pointed out [5] that reactions such as $^{12}C + ^{12}C$, $^{12}C + ^{16}O$ and $^{16}O + ^{16}O$ (all of them have positive Q values for fusion) are important for the evolution of massive stars, beyond the helium burning, and the associated nucleosynthesis [6]. The existence of hindrance in these cases would lead to significant changes of the abundances of many isotopes. The two medium-light systems $^{12}C + ^{30}Si$, ^{24}Mg may be helpful to establish a reliable systematic behaviour to guide the extrapolation to lighter cases of astrophysical interest.

2 The phenomenon of fusion hindrance at deep sub-barrier energies

Hindrance of heavy-ion fusion at extreme sub-barrier energies, characterized by a steep fall off of the fusion cross section with decreasing energy, was discovered 15 years ago [6]. By plotting the cross section in terms of the S factor, $S(E) = \sigma E \exp(2\pi\eta)$, where η is the Sommerfeld parameter and E is the center of mass energy, fusion hindrance is easily recognized by a maximum of $S(E)$ at an energy E_s [7–9].

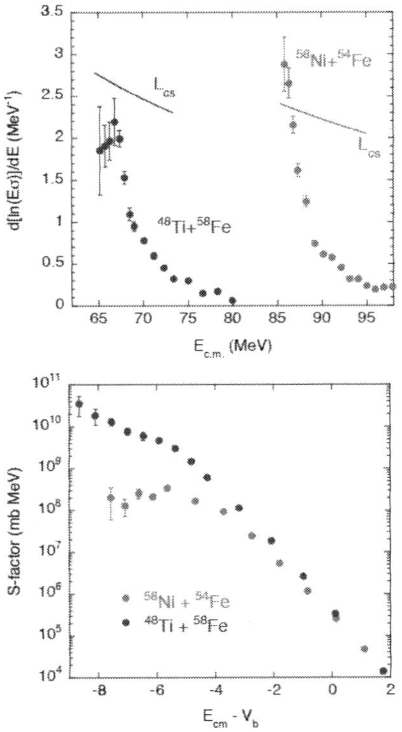

Figure 2. Logarithmic slopes (upper panel) and the S factor (lower panel) of $^{48}Ti+^{58}Fe$ and $^{58}Ni+^{54}Fe$ [11, 12].

This phenomenon was first studied in medium-heavy systems. In this mass region, the fusion Q value is always negative, so that the S factor is 0, when the incident energy approaches $E = -Q$. Thus, under such conditions an S factor maximum is unavoidable [7–9]. The classical example is $^{64}Ni + ^{64}Ni$ where the fusion cross section was measured down to about 20 nb. This is shown in Fig. 1. The excitation function decreases more steeply than the prediction of a standard CC calculation. This is better evidenced by the representation in terms of logarithmic derivative $L(E)$ of the excitation function which is reported in the lower panel of Fig. 1. It is known [7] that when the slope reaches the value $L_{CS} = \pi\eta/E$ the S factor develops a maximum and the corresponding energy is usually taken as the threshold for the hindrance effect.

Even for the hindrance phenomenon the role of nuclear structure is very important as it has been observed by comparing the two systems $^{48}Ti + ^{58}Fe$ and $^{58}Ni + ^{54}Fe$ [11, 12] (see Fig. 2). We notice that the two behaviours are remarkably different; indeed, the slope of $^{48}Ti + ^{58}Fe$ saturates below the barrier and remains much lower than L_{CS}. No maximum of the S factor develops, so that, in other words, no fusion hindrance seems to show up in this case in the measured energy range. On the other hand a clear maximum of the S factor can be observed for $^{58}Ni + ^{54}Fe$. The different trends originate from the dissimilar low energy nuclear structure of the involved nuclei. Indeed, ^{48}Ti and ^{58}Fe are soft and have a low-lying quadrupole excitation lying at 800-900 keV only. Instead, ^{58}Ni and ^{54}Fe have a closed shell (protons and neutrons, respectively) and are rather stiff.

3 The two systems $^{12}C + ^{30}Si$, ^{24}Mg: sub-barrier trends of the excitation functions and of the S-factors

A different situation might be found in medium-light and light systems where the Q-value for fusion may be positive. Indeed in such cases there is no energy threshold for fusion and consequently it is not necessary that the S factor develops a maximum. Whether there is also an S factor maximum at very low energies for such systems is an experimentally challenging question and various experiments have been recently performed in this sense (see e.g. Ref. [13]). However the existence of an S factor maximum has not yet been clearly established. Recently a detailed study on the system $^{12}C + ^{30}Si$ (Q_{fus}= +14.1 MeV) in a wide energy range showed an empirical evidence of the hindrance effect because the slope reaches and overcomes the L_{CS} value and the S factor appears to develop a maximum with decreasing energy. This is shown in Fig. 3 and is supported by CC calculations within the adiabatic model [14], using both the WS and the YPE potentials [15]. A damping of the coupling strengths improves the data fit at low energies and, overall, the YPE potential gives better results. The hindrance effect is weak but clearly observable from the comparison with the calculations.

The near-by system $^{12}C + ^{24}Mg$ (Q_{fus}= + 16.3 MeV) is even closer to the cases of astrophysical interest. Preliminary LNL data extending much below the barrier are reported in Fig. 4 that combines the trend of both excitation function and of the S factor. The S factor develops a

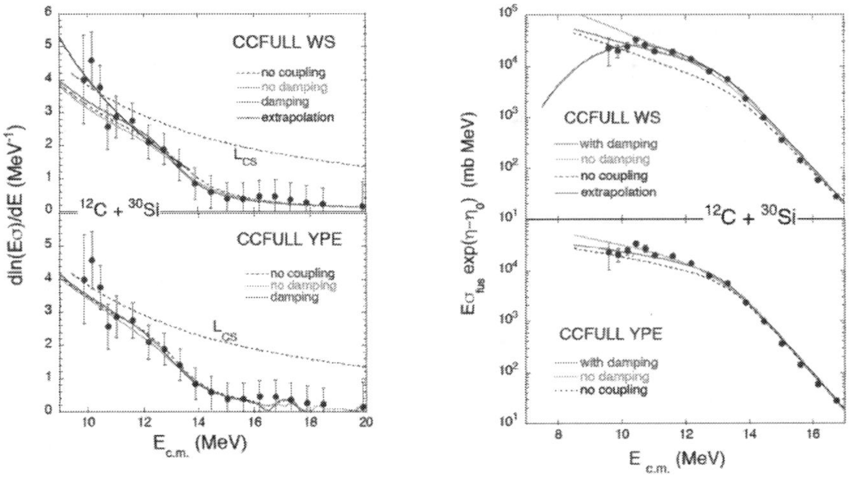

Figure 3. The logarithmic slope (left) crosses the L_{CS} value at the lowest energies where the S factor (right) appears to develop a maximum. In both cases a phenomenological extrapolation is shown, based on the systematics [16] (black and blue lines respectively).

clear maximum at about 9 MeV, indicating the presence of hindrance.

It is interesting to note that the cross section corresponding to the hindrance threshold is very high (σ_s=1.6mb), probably the highest measured so far. This is reported in Fig. 5 together with several other medium-light systems. The high cross section at threshold has allowed to identify rather clearly the onset of hindrance, and it may also be possible to extend the measurements further down in energy.

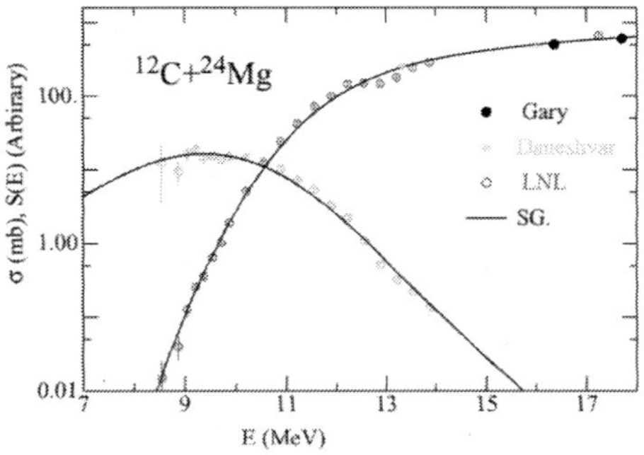

Figure 4. Excitation function and S factor for ^{12}C + ^{24}Mg, previous data are also reported [17, 18]. The curves are least-square fits of the excitation function.

4 Towards light systems and possible consequences for astrophysics

Concerning the two systems that I have just discussed in the previous Section, their energy threshold for hindrance (E_s) nicely fits in the phenomenological systematics that

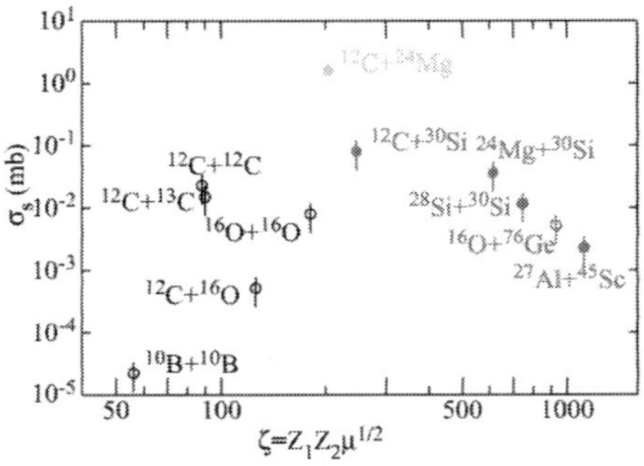

Figure 5. The threshold cross section vs. the system parameter ζ for ^{12}C + ^{24}Mg and several other cases for comparison.

was developed by Jiang et al. a few years ago [16], and that is shown in Fig. 6. This gives us confidence that the extrapolation towards lighter systems as ^{12}C + ^{12}C and ^{16}O + ^{16}O using that systematics, is reliable.

I want to stress another point concerning the difference of logarithmic derivatives between heavy and light systems, using the representation of Fig. 7. In each case the Q value for fusion is indicated, as well as the system parameter ζ. The dashed line is $L_{CS}(E)$, the red one is a simple extrapolation and the green lines are standard CC calculations. For lighter systems, $L(E)$ and $L_{CS}(E)$ are two nearly parallel or overlapping curves so the crossing point is rather undetermined and the S factor maximum becomes broader and more difficult to be identified. For example, we notice that the slope of ^{64}Ni + ^{64}Ni is very steep and crosses L_{CS} with a large angle. Here the existence and the threshold of hindrance are quite well identified. On the contrary, in the light case of ^{10}B + ^{10}B the

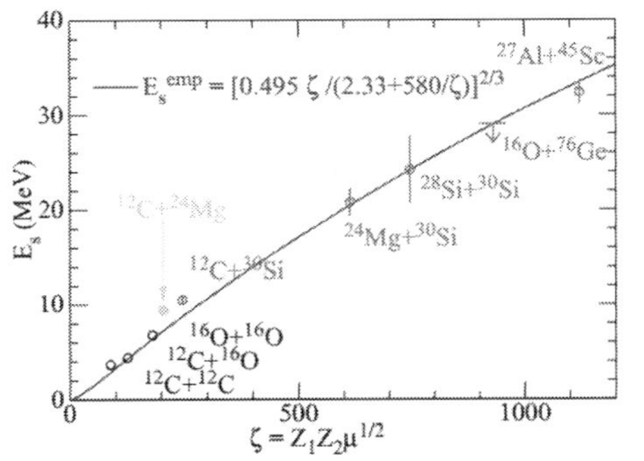

Figure 6. Systematics of E_s in several light- and medium-light mass systems [16]. The locations for ^{12}C + ^{30}Si and ^{12}C + ^{24}Mg are very close to the astrophysically relevant cases, for which several data sets exist, but they are sometimes contradictory and the errors large. This results in large uncertainties in the expected values for their hindrance threshold. The corresponding points (open symbols) have therefore been obtained from extrapolations.

two curves $L(E)$ and L_{CS} tend to overlap with each other at low energies.

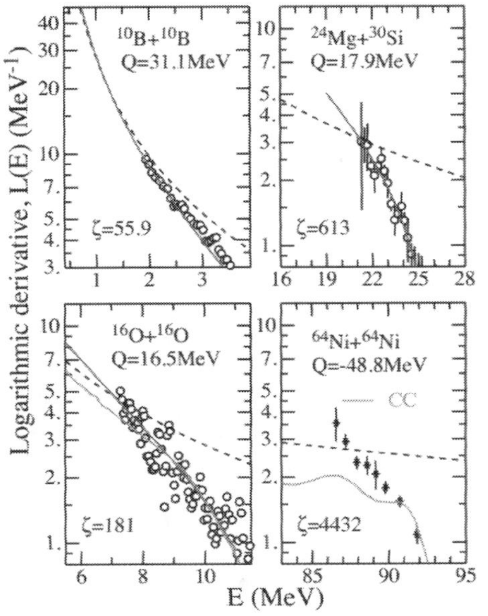

Figure 7. Logarithmic derivative for various heavy and light systems. In each case the Q-value for fusion is indicated, as well as the system parameter $\zeta = Z_1 Z_2 \mu^{1/2}$. The dashed line is L_{CS}(E), the red one is a simple extrapolation and the green lines are standard CC calculations (courtesy of C.L. Jiang).

By introducing the quantity [9]

$$RR = \frac{dL(E)/dE}{dL_{CS}(E)/dE} \tag{1}$$

this trend is well represented (see Fig. 8, upper panel). RR is the ratio of the energy derivatives of the slopes $L(E)$ and L_{CS}(E) at the crossing point. A sharp intersection point is observed for the heavier systems so that a well defined S factor maximum develops. Instead, RR approaches unity for the lighter systems for which a less well defined maximum shows up. This does not imply that the fusion hindrance disappears in such cases, however, its existence may be more easily evidenced from comparing with standard CC calculations.

Shrivastava et al. [20] made recent studies of the way fusion hindrance shows up and evolves with increasing mass and charge of light projectiles like $^{6,7}Li$ and ^{12}C on the heavy target ^{198}Pt. The case of ^{16}O + ^{208}Pb [21, 22] was also re-analysed. They measured fusion cross sections down to very small values (≈100 nb) using an off-line method by detecting coincidences between characteristic X- and γ-rays (see also [23]).

Fig. 8 (lower panel) shows that the hindrance effect is not observed for $^{6,7}Li$, but appears and becomes more evident when going to the heavier ^{12}C and ^{16}O projectiles. On the basis of this observed trend, it seems very important to study light and more symmetric systems where the existence of the hindrance phenomenon is still unclear and a matter of debate.

The very important case, from the astrophysical point of view, is the system ^{12}C + ^{12}C for which several sets of experimental data already exist. For this case, in a recent experiment performed at Argonne [24] coincidences between α and p particles, and γ-rays were detected. These data suggest that a maximum of the S factor seems to develop around 3.5-4 MeV, in agreement with the phenomenological estimate of Ref. [16]. The S factor has been also recently deduced, using the indirect Trojan Horse Method, down to very low energies [25]. The many resonances observed in this experiment produce an overall increase of the S factor below the energy range covered by direct measurements and no indication of hindrance appears.

5 Summary and conclusions

The phenomenon of hindrance in sub-barrier heavy-ion fusion was discovered about 15 years ago and it is presently established to be a general phenomenon. It is recognised in many cases by the trend of the logarithmic slope and of the S factor, that generally develops a maximum vs. energy far below the barrier. In any case the comparison of the excitation function with standard CC calculations is a more quantitative evidence for the existence of the hindrance effect. For medium-heavy systems where the Q value for fusion is negative, the existence of hindrance is necessary because the S factor vanishes at $E = -Q$. The interest in these cases is related mainly to the threshold of the effect that may strongly depend on the nuclear structure of the colliding ions.

Light systems mostly have positive Q value for fusion so that the S factor does not necessarily show a maximum at some energy. Several experiments have been recently performed to clarify this point because the existence of

 https://doi.org/10.1051/epjconf/201922301041

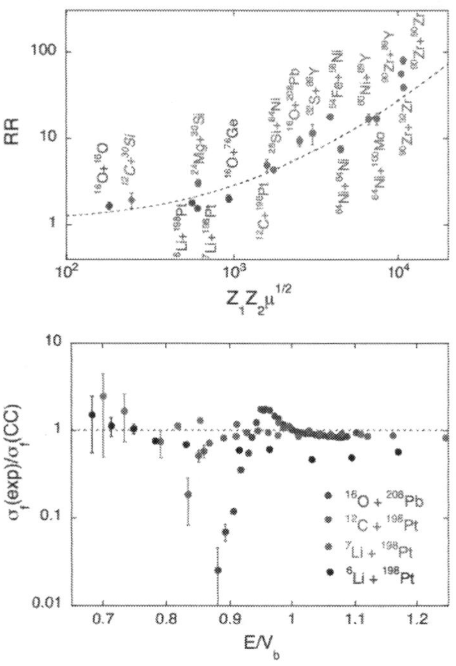

Figure 8. (upper panel) Ratio (RR) of energy derivatives of the slopes $L(E)$ and L_{CS}(E) at their crossing points, vs. the system parameter $\zeta = Z_1 Z_2 \mu^{1/2}$. The red dots are extracted from experimental data while the blue ones are extrapolated from data at higher energies. The red diamond is the value for ^{12}C + ^{30}Si [15]. The dashed line is the curve resulting from the systematics discussed in Ref. [9]. (lower panel) Measured vs. calculated fusion cross sections as a function of the energy relative to the Coulomb barrier for ^{6,7}Li + ^{198}Pt [19, 20], ^{12}C + ^{198}Pt [20] and ^{16}O + ^{208}Pb [21, 22] systems. The calculated values (σ_f(CC)) correspond to CCFULL calculations using a standard WS potential. Figure redrawn from Ref. [20].

hindrance in such light cases may have relevant consequences in astrophysics. Indeed the phase of carbon burning in massive stars might be substantially influenced, with a reduction of predicted astrophysical reaction rates.

So far hindrance has been possibly observed in various medium-light systems, independent of the sign of the Q-value. The hindrance effect is small but it has been clearly recognised in the two recent experiments on ^{12}C + ^{30}Si, ^{24}Mg that are very close to the astrophysically relevant systems, and its threshold nicely follows a systematics proposed a few years ago. The consequences for the dynamics of stellar evolution have to be clarified by further experimental and theoretical work.

Acknowledgments

I am grateful to all colleagues who have collaborated in the experiments presented in this talk for their invaluable help. Special thanks are addressed to XTU Tandem staff for their professional work. I also wish to acknowledge many other experimental and theoretical colleagues for their precious contributions to several aspects of the work presented here.

References

[1] N. Rowley, G.R. Satchler and P.H. Stelson, Phys. Lett. B **254**, 25 (1991).

[2] W. Reisdorf, F.P. Hessberger, K.D. Hildebrand, S. Hofmann, G. Münzenberg, K.-H. Schmidt, J.H.R. Schneider, W.F.W. Schneider, K,Sümmerer, G. Wirth, J.V. Kratz and K. Schlitt, Nucl. Phys. A **438**, 212 (1985).

[3] A. M. Stefanini, D. Ackermann, L. Corradi, D. R. Napoli, C. Petrache, P. Spolaore, P. Bednarczyk, H. Q. Zhang, S. Beghini, G. Montagnoli, L. Mueller, F. Scarlassara, G. F. Segato, F. Soramel and N. Rowley, Phys. Rev. Lett. **74**, 864 (1995).

[4] A. M. Stefanini, B. R. Behera, S. Beghini, L. Corradi, E. Fioretto, A. Gadea, G. Montagnoli, N. Rowley, F. Scarlassara, S. Szilner, and M. Trotta, Phys. Rev. C **76**, 014610 (2007).

[5] L.R. Gasques, E.F. Brown, A. Chieffi, C.L. Jiang, M. Limongi, C. Rolfs, M. Wiescher, and D.G. Yakovlev, Phys. Rev. C **76**, 035802 (2007).

[6] B.B. Back, H. Esbensen, C.L. Jiang, and K.E. Rehm, Rev. Mod. Phys. **86**, 317 (2014).

[7] C. L. Jiang, H. Esbensen, B. B. Back, R. V. F. Janssens, and K. E. Rehm, Phys. Rev. **C 69**, 014604 (2004).

[8] C. L. Jiang, B. B. Back, H. Esbensen, R. V. F. Janssens, and K. E. Rehm, Phys. Rev. **C 73**, 014613 (2006).

[9] C. L. Jiang, B. B. Back, R. V. F. Janssens, and K. E. Rehm, Phys. Rev. **C 75**, 057604 (2007).

[10] C.L.Jiang, K. E. Rehm, R.V. F. Janssens, H. Esbensen, I. Ahmad, B. B. Back, P. Collon, C. N. Davids, J. P. Greene, D. J. Henderson, G. Mukherjee, R. C. Pardo, M. Paul, T.O. Pennington, D. Seweryniak, S. Sinha, and Z. Zhou, Phys. Rev. Lett. 93, 012701 (2004).

[11] A.M. Stefanini, G. Montagnoli, L. Corradi, S. Courtin, E. Fioretto, A. Goasduff, F. Haas, P. Mason, R. Silvestri, Pushpendra P. Singh, F. Scarlassara, and S. Szilner, Phys. Rev. C **82**, 014614 (2010).

[12] A.M. Stefanini, G. Montagnoli, L. Corradi, S. Courtin, D. Bourgin, E. Fioretto, A. Goasduff, J. Grebosz, F. Haas, M. Mazzocco,T. Mijatović, D. Montanari, M. Pagliaroli, C. Parascandolo, F. Scarlassara, E. Strano, S. Szilner, N. Toniolo, and D. Torresi, Phys. Rev. C **92**, 064607 (2015).

[13] G. Montagnoli and A.M. Stefanini, Eur. Phys. J. A **53**, 169 (2017).

[14] T. Ichikawa, Phys. Rev. C **92**, 064604 (2015).

[15] G. Montagnoli, A. M. Stefanini, C. L. Jiang, K. Hagino, F. Galtarossa, G. Colucci, S. Bottoni, C. Broggini, A. Caciolli, P. Čolović, L. Corradi, S. Courtin, R. Depalo, E. Fioretto, G. Fruet, A. Gal, A. Goasduff, M. Heine, S. P. Hu, M. Kaur, T. Mijatović, M. Mazzocco, D. Montanari, F. Scarlassara, E. Strano, S. Szilner, and G. X. Zhang, Phys. Rev. C **97**, 024610 (2018).

[16] C. L. Jiang, K. E. Rehm, B. B. Back and R. V. F. Janssens, Phys. Rev. C **79**, 044601 (2009).

[17] S.Gary and C. Volant, Phys. Rev. C **25**, 1877 (1982).

[18] K. Daneshvar, D. G. Kovar, S. J. Krieger, and K. T. R. Davies, Phys. Rev. C **25**, 1342 (1982).

[19] A. Shrivastava, A. Navin, A. Lemasson, K. Ramachandran, V. Nanal, M. Rejmund, K. Hagino, T. Ichikawa, S. Bhattacharyya, A. Chatterjee, S. Kailas, K. Mahata, V. V. Parkar, R. G. Pillay, and P. C. Rout, Phys. Rev. Lett. **103**, 232702 (2009).

[20] A. Shrivastava, K. Mahata, S.K. Pandit, V. Nanal, T. Ichikawa, K. Hagino, A. Navin, C.S. Palshetkar, V.V. Parkar, K. Ramachandran, P.C. Rout, Abhinav Kumar, A. Chatterjee, and S. Kailas, Phys. Lett. B **755**, 332 (2016).

[21] M. Dasgupta, D. J. Hinde, A. Diaz-Torres, B. Bouriquet, Catherine I. Low, G. J. Milburn, and J. O. Newton, Phys. Rev. Lett. **99**, 192701 (2007).

[22] C.R. Morton, A.C. Berriman, M. Dasgupta, D.J. Hinde, J.O. Newton, K. Hagino and I.J. Thompson, Phys. Rev. C **60**, 044608 (1999).

[23] A. Lemasson, A. Shrivastava, A. Navin, M. Rejmund, V. Nanal, S. Bhattacharyya, A. Chatterjee, S. Kailas, K. Mahata, V.V. Parkar, R.G. Pillay, K. Ramachandran, and P.C. Rout, Nucl. Instr. Meth. Phys. Res. A **598**, 445 (2009).

[24] C. L. Jiang, D. Santiago-Gonzalez, S. Almaraz-Calderon, K. E. Rehm, B. B. Back, K. Auranen, M. L. Avila, A. D. Ayangeakaa, S. Bottoni, M. P. Carpenter, C. Dickerson, B. DiGiovine, J. P. Greene, C. R. Hoffman, R. V. F. Janssens, B. P. Kay, S. A. Kuvin, T. Lauritsen, R. C. Pardo, J. Sethi, D. Seweryniak, R. Talwar, C. Ugalde, S. Zhu, D. Bourgin, S. Courtin, F. Haas, M. Heine, G. Fruet, D. Montanari, D. G. Jenkins, L. Morris, A. Lefebvre-Schuhl, M. Alcorta, X. Fang, X. D. Tang, B. Bucher, C. M. Deibel, and S. T. Marley Phys. Rev. **C 97**, 012801(R) (2018).

[25] A. Tumino, C. Spitaleri, M. La Cognata, S. Cherubini, G. L. Guardo, M. Gulino, S. Hayakawa, I. Indelicato, L. Lamia, H. Petrascu, R. G. Pizzone, S. M. R. Puglia, G. G. Rapisarda, S. Romano, M. L. Sergi, R. Spartá and L. Trache, Nature **557**, 687 (2018).

https://doi.org/10.1051/epjconf/201922301042

New Results on Excited States in the one-particle one-hole nucleus ^{56}Co measured with MINIBALL detectors

A. Montaner-Pizá[1,2,*], *B.* Rubio[1,**], *D.* Mücher[3], *S.E.A.* Orrigo[1], *E.C.* Simpson[4], *W.* Gelletly[5], *J.* Agramunt[1], *A.* Algora[1], *C.* Bender[3], *Y.* Fujita[6], *E.* Ganioglu[7], *V.* Guadilla[1], *S.* Hellgartner[3], and *S.* Reichert[3]

[1]Instituto de Física Corpuscular IFIC (CSIC-Univ. Valencia), Valencia, Spain
[2]Dipartamento di Fisica e Astronomia, Università degli Studi di Padova - INFN, Padova, Italy
[3]Physik Department, Technische Universität München (TUM), Munich, Germany
[4]Department of Nuclear Physics, The Australian National University, Canberra ACT 2601, Australia
[5]Department of Physics, Surrey University, Surrey, United Kingdom
[6]Research Centre for Nuclear Physics RCNP, Osaka, Japan
[7]Physics Department, Science Faculty, Istanbul University, Istanbul, Turkey

Abstract. The non-yrast states of the odd-odd nucleus ^{56}Co have been investigated by studying the γ-rays induced in the predominantly fusion-evaporation reaction ^{56}Fe(p,nγ)^{56}Co at an incident energy of 10 MeV. The γ-rays were measured in-beam with four high-resolution MINIBALL-triple germanium (Ge) detectors. The experiment provided excellent data in γ-γ coincidences. The complex level scheme of ^{56}Co was constructed mainly based on the analysis of these γ-γ coincidences. The angular distributions of the γ-rays were also analysed and allowed us to assign spin-parity values to most of the excited states in this nucleus. Despite the extensive work previously done studying the ^{56}Co nucleus, the analysis presented in this work has resulted in a large improvement in the knowledge of its structure.

1 Introduction

In the present work we have studied the γ-ray de-excitation of the non-yrast states in the odd-odd nucleus ^{56}Co (Z=27, N=29) populated in the mainly fusion-evaporation reaction ^{56}Fe(p,nγ)^{56}Co at an 10-MeV incident proton beam. From the shell-model point of view, in the simplest analysis, ^{56}Co is just one-proton hole and one neutron outside the doubly magic nucleus ^{56}Ni (Z=N=28). Ideally, spectroscopy of the ^{56}Co nucleus ought to yield direct information on the particle-hole residual interaction in the *fp*-shell.

The study of ^{56}Co in this work was initially motivated by the observation of two 0^+ states in its mirror nucleus ^{56}Cu (Z=29, N=27) in a β^+ decay experiment of ^{56}Zn carried out at GANIL (Caen, France) [1]. These two states are populated through Fermi-type decay and correspond to the splitting of the isobaric analogue state (IAS). Two very similar states exist in ^{56}Co and were recently investigated in the ^{56}Fe(^{3}He, t)^{56}Co charge exchange (CE) reaction performed at the Research Centre for Nuclear Physics RCNP in Osaka (Japan) with high resolution [2]. No γ detection was possible in this work.

2 The experiment

The experiment was performed at the Maier-Leibnitz-Laboratory (MLL) of the Technische Universität München (TUM, Germany). The gamma radiation emitted in the de-excitation of the excited states in ^{56}Co was measured in-beam with four high-purity germanium (Ge) MINIBALL-triple detectors. The MINIBALL array [3] is a γ-ray spectrometer optimized to achieve a high photo-peak efficiency in combination with position sensitivity γ-ray detection.

For the present experiment at the TUM, two MINIBALL clusters were located forward and two backward with respect to the beam direction (see Fig. 1). The target chamber was positioned at the centre. We used a ^{56}Fe target with a thickness of 1.1 mg/cm^2, which was mounted in an aluminium frame. The clusters were positioned at slightly different distances from the chamber, in order to maximise the solid-angle coverage.

Figure 1. The experimental setup is shown. Two MINIBALL clusters were located forward and two backward with respect to the beam direction. The target chamber was positioned at the centre.

*e-mail: ana.montaner@ific.uv.es
**e-mail: rubio@ific.uv.es

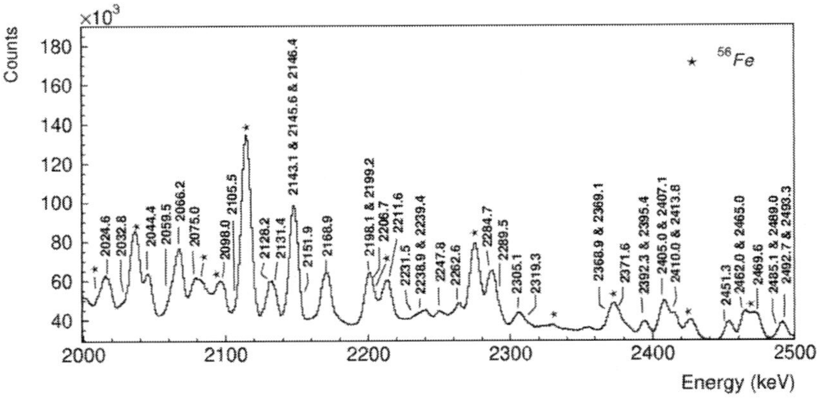

Figure 2. The γ-γ coincidence matrix-projection spectrum is shown, in the 2-2.5 MeV range. All the γ-rays observed and placed in the level scheme of ^{56}Co in this work are labelled. Note that the peaks coming from the de-excitation of excited states in ^{56}Fe are indicated with a symbol ★.

3 The level scheme of ^{56}Co

The level scheme of the ^{56}Co nucleus was constructed mainly based on the analysis of the γ-γ coincidence matrices. The γ-γ coincidence matrix-projection spectrum is shown in Fig. 2 in the 2-2.5 MeV range, which gives an idea of the complexity of the experiment.

In the present work, a large amount of new information has been obtained. In the following, we summarise the experimental results:

- A total of 223 γ-transitions, of which 169 were previously unobserved, have been observed and placed in the level scheme.
- A much higher sensitivity level for the γ-de-excitation of excited states in ^{56}Co than reported in the literature has been reached for states with spins between 0 and 6. A total of 77 excited states have been observed (up to 4.4-MeV excitation energy), 37 of which were previously known but no gamma de-excitation had been observed. In 42 cases the energy precision of the levels has been improved and 14 excited states have been observed for the first time.

  Due to the complexity of the experiment most of the γ-transition energy and intensity values have been determined in the gated-spectra of the γ-γ coincidence matrices (see [7]). The spin and parity assignments to the excited states have been made based on their γ-decay pattern and the angular distributions of the γ-ray transitions measured.

4 Theoretical interpretation

For the present discussion large-scale shell-model calculations in a truncated *fp*-shell valence space have been carried out using the NuShellX@MSU code [4] and the effective interactions KB3G [5] and GXPF1a [6]. The calculations constrained the full *fp*-shell model space by insisting that at least 2 protons and 2 neutrons are in the $1f_{7/2}$ shell.

Since the magic numbers of Z and N (=28) provide a good core for ^{56}Ni, the majority of low-lying states in ^{56}Co can be described in terms of the one-particle-one-hole (1p-1h) configurations (or multiplets), i.e., $(1f_{7/2})^{-1}_{\pi} \times (2p_{3/2}, 1f_{5/2}, 2p_{1/2})^{1}_{\nu}$. The low-lying experimental states in ^{56}Co have been identified as the main components of the 1p-1h multiplets with the help of the shell-model calculations (see [7]). Tentative identification between the experimental levels and theoretical predictions using the two previously mentioned effective interactions have been also made up to 3 MeV.

On the other hand, the study of the isospin mixing between two 0^{+} states in ^{56}Co (the IAS of the ^{56}Fe$_{g.s.}$ and a nearby state) was one of the main motivations of this work. The determination of the M1 transition intensities from both states has offered new and relevant information about the effects associated to the relative phases of the isospin-mixed states and has shown that these two states decay in a different way (see [7]).

References

[1] S.E.A. Orrigo, B. Rubio, Y. Fujita, W. Gelletly *et al.*, Phys. Rev. Lett. **112**, 222501–1 (2014)

[2] H. Fujita, Y. Fujita, T. Adachi, H. Akimune *et al.*, Phys. Rev. C **88**, 054329 (2013)

[3] P. Reiter, J. Eberth, H. Faust *et al.*, Nucl. Phys. A **701**, 209 (2002)

[4] B.A. Brown and W.D.M. Rae, Nuclear Data Sheets **120**, 115 (2014)

[5] A. Poves, J. Sánchez-Solano, E. Caurier and F. Nowacki, Nucl. Phys. A **694**, 157 (2001)

[6] M. Honma, T. Otsuka, B.A. Brown and T. Mizusaki, Phys. Rev. C **69**, 034335 (2004)

[7] A. Montaner-Pizá, Ph.D. Thesis, *Isospin mixing and in-beam study of non-yrast states in ^{56}Co* (University of Valencia, 2018)

Spontaneous fission of rutherfordium isotopes - total kinetic energies

Pavol Mosat[1,2,*], *Fritz Peter* Heßberger[3,4], *Dieter* Ackermann[3,5], *Boris* Andel[1,6], *Stanislav* Antalic[1], *Michael* Block[3,4,7], *Sigurd* Hofmann[3,8], *Zdenka* Kalaninova[1,9], *Birgit* Kindler[3], *Mustapha* Laatiaoui[3,4], *Bettina* Lommel[3], *Andrew* Mistry[3,4], *Katsuhisa* Nishio[10], *Julien* Piot[5], *Barbara* Sulignano[3,11], and *Marija* Vostinar[5]

[1]Department of Nuclear Physics and Biophysics, Comenius University in Bratislava, 84248 Bratislava, Slovakia
[2]Flerov Laboratory of Nuclear Reactions, JINR, 141980 Dubna, Russia
[3]GSI Helmholtzzentrum für Schwerionenforschung, GmbH, 64291 Darmstadt, Germany
[4]Helmholtz - Institut Mainz, 55099 Mainz, Germany
[5]GANIL, 14076 Caen CEDEX 5, France
[6]KU Leuven, Instituut voor Kern- en Stralingsfysica, B-3001 Leuven, Belgium
[7]Johannes Gutenberg-Universität Mainz, 55128 Mainz, Germany
[8]Institut für Physik, Goethe-Universität Frankfurt, 60054 Frankfurt, Germany
[9]Laboratory of Nuclear Problems, JINR, 141980 Dubna, Russia
[10]JAEA, Tokai, Ibaraki 319-1195 Japan
[11]CEA Saclay, IRFU, SPhN, 91191 Gif-sur-Yvette, France

Abstract. The isotopes ^{255,256,258}Rf were produced in the fusion-evaporation reactions ^{50}Ti + ^{207,208}Pb and ^{50}Ti + ^{209}Bi at GSI Darmstadt, using the velocity filter SHIP. Total kinetic energies of fragments from spontaneous fission for these isotopes were evaluated with a correction to pulse-height defect.

1 Introduction

Spontaneous fission (SF) is a decay mode with large impact on the stability of nuclei in the transfermium region. These isotopes are stabilized against SF mainly by microscopic shell effects [1]. Systematic studies of SF properties allow us to understand these effects and determine the production possibilities for the heaviest nuclei. Up to now, only a few results (often based on limited statistics of SF events) with measured total kinetic energy (TKE) were obtained for rutherfordium ($Z = 104$) isotopes [2–5].

2 Experiment

The experiment was performed at GSI Darmstadt in Germany. The heavy-ion ^{50}Ti beam with typical energies from 225 to 243 MeV was delivered by the UNILAC accelerator. The fusion-evaporation reactions ^{50}Ti + ^{207,208}Pb were used for production of the studied isotopes ^{255,256}Rf via xn evaporation channels from the compound nuclei ^{257}Rf and ^{258}Rf. The reaction ^{50}Ti + ^{209}Bi was used for indirect production of ^{258}Rf via EC decay of ^{258}Db.

Evaporation residues (ER) were separated from the beam and other undesired background using the velocity filter SHIP [6] and delivered to a detection setup. After passing through the time-of-flight system they were implanted $(6.5 - 6.8)\,\mu$m deep into the 16-strip position-sensitive STOP detector. The same-type detectors arranged in the "BOX" geometry were placed in front of the STOP detector to detect escaping particles. A clover detector with four Ge crystals was placed behind the STOP detector for γ- and X-ray detection.

*e-mail: pavol.mosat@fmph.uniba.sk

3 Results

In order to detect SF events from specific isotope, we used a time and position correlation search technique. In the case of ^{255}Rf, ^{256}Rf we searched for correlations between an ER implantation signal and a high-energy signal corresponding to SF. For ^{258}Rf instead of the ER implantation signal we used the low-energy signal corresponding to electrons originating from internal conversion process from de-excitation of levels populated after the EC decay $^{258}\text{Db} \xrightarrow{EC} {}^{258}\text{Rf}$. The time conditions were set to $\approx 5 \times$ half-life between signals for each isotope (8500 ms for ^{255}Rf, 35 ms for ^{256}Rf and 60 ms for ^{258}Rf) and position window of 1 mm in the detector for ^{255}Rf and ^{256}Rf. As position condition for ^{258}Rf, we required only the same strip number since for many low-energy signals from electrons the position information was not registered. We collected several hundreds of SF events for ^{255}Rf, ^{256}Rf and ^{258}Rf. The TKE distributions of fission fragments for each isotope are shown in Fig. 1. The histograms contain all registered SF events - the cases when both fragments stayed in the STOP detector ($\approx 60\%$) as well as events when one fragment escaped the STOP detector and was either detected in the BOX detector ($\approx 30\%$) or escaped the detection setup completely ($\approx 10\%$).

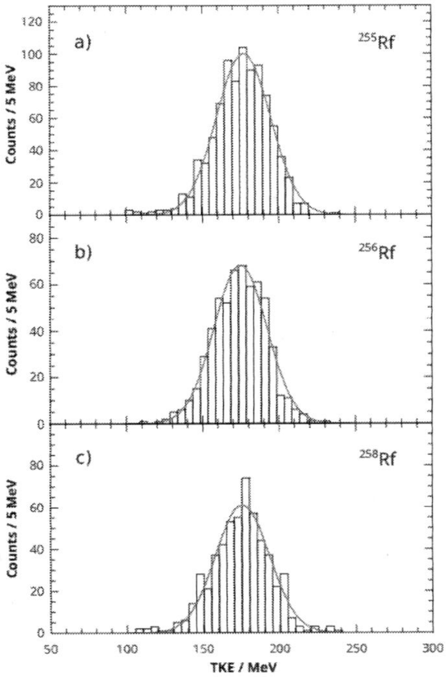

Figure 1. Total kinetic energies for ^{255}Rf, ^{256}Rf and ^{258}Rf from all observed SF events for each isotope. Due to the pulse-height defect, the positions of gaussians are shifted to lower values.

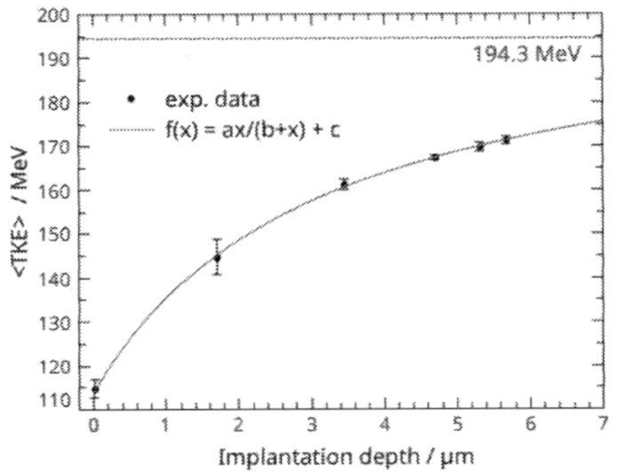

Figure 2. $\overline{TKE}$ from the SF of ^{252}No vs. implantation depth of ER in the detector. Blue dashed line: $\overline{TKE}$ = 194.3 MeV of ^{252}No from [3]. Red solid line: saturation-growth fit.

The crucial task for the evaluation of $\overline{TKE}$ using Si detectors is the correction of deficit in measured energies. There are two main effects influencing the TKE measurements, discussed in previous studies [7–9] performed at SHIP. First, the energy calibration based on α-decay energies is not valid for fission fragments due to the pulse-height defect (see e.g. [10]), resulting in a significant energy deficit. The second effect is a dependence of the detected TKE on the implantation depth of ER into the detector. In order to find the energy correction, we performed TKE measurements for ^{252}No with known $\overline{TKE}$ = 194.3 MeV [3] at six different implantation depths [8]. Our observations (shown in Fig. 2) proved the strong non-linear saturation-like dependence of detected TKE on implantation depth (see [11] for more details). The correction to the energy deficit at given implantation depth was determined as a difference of the reference $\overline{TKE}$ of 194.3 MeV and the value $\overline{TKE}_{fit}$ from saturation-growth model fit of the six points (blue dashed and red solid lines in Fig. 2) as $\Delta E = (194.3 - \overline{TKE}_{fit})$ MeV.

Table 1. In the table columns, from left to right, the isotope, $\overline{TKE}$ evaluated in this work, reference value of $\overline{TKE}$ and corresponding references are stated.

Isotope	$\overline{TKE}_{exp}$ [MeV]	$\overline{TKE}_{ref}$ [MeV]	Ref.
^{255}Rf	199.5 ± 2.7	199 ± 3	[4]
^{256}Rf	198.7 ± 2.8	198.9 ± 4.4	[2]
^{258}Rf	198.2 ± 3.0	197.6 ± 1.1	[2]

Considering the facts, that ^{252}No is close in Z, A and $\overline{TKE}$ to studied Rf isotopes, we applied this correction to $\overline{TKE}$ of ^{255}Rf, ^{256}Rf and ^{258}Rf. Corrected $\overline{TKE}$ values for each isotope are summarized in Table 1 and compared to values from previous studies. The results are in a good agreement, which supports the validity of our correction.

Acknowledgments

We thank the UNILAC staff as well as the ion source crew for delivering beams of high and stable intensity. We are also grateful to J. Steiner, W. Hartmann, and A. Hubner for production of the large area targets. We want to express our gratitude to H.G. Burkhard and J. Maurer for skillful maintenance of the mechanical and electrical components of SHIP. Four of us, BA, SA, PM, ZK, were supported by the Slovak Research and Development Agency (Contracts No. APVV-18-0268 and No. APVV-14-0524) and by the Slovak grant agency VEGA (Contract No. 1/0532/17).

References

[1] S.G. Nilsson et al., Nucl. Phys. A **131**, 1 (1969)
[2] J.F. Wild et al., J. Alloy. Comp. **213**, 86 (1994)
[3] E.K. Hulet, Phys. At. Nucl. **57**, 1099 (1994)
[4] S. Antalic et al., Eur. Phys. J. A **51**, 41 (2015)
[5] A.I. Svirikhin et al., Phys. Part. Nucl. Lett. **13**, 480 (2016)
[6] G. Münzenberg et al., Nucl. Instrum. Meth. **161**, 65 (1979)
[7] F.P. Heßberger et al., Z. Phys. A **322**, 557 (1985)
[8] K. Nishio et al., AIP Conf. Proc. **891**, 71 (2007)
[9] S. Hofmann et al., Eur. Phys. J. A **32**, 251 (2007)
[10] B. Wilkins et al., Nucl. Instrum. Meth. **92**, 381 (1971)
[11] P. Mosat et al., Acta Phys. Pol. B **49**, 605 (2018)

Coulomb Energy Density Functionals for Nuclear Systems: Recent Studies of Coulomb Exchange and Correlation Functionals

Tomoya Naito[1,2,*], *Ryosuke* Akashi[1], *Gianluca* Colò[3,4], *Haozhao* Liang[2,1], and *Xavier* Roca-Maza[3,4]

[1]Department of Physics, Graduate School of Science, The University of Tokyo, Tokyo 113-0033, Japan
[2]RIKEN Nishina Center, Wako 351-0198, Japan
[3]Dipartimento di Fisica, Università degli Studi di Milano, Via Celoria 16, 20133 Milano, Italy
[4]INFN, Sezione di Milano, Via Celoria 16, 20133 Milano, Italy

Abstract. The Coulomb exchange and correlation energy density functionals for electron systems are applied to nuclear systems. It is found that the exchange functionals in the generalized gradient approximation provide agreements with the exact-Fock energy with one adjustable parameter within a few dozen keV accuracy, whereas the correlation functionals are not directly applicable to nuclear systems due to the existence of the nuclear force.

1 Introduction

Atomic nuclei are composed of protons and neutrons that interact with one another through the nuclear and electromagnetic forces. The former, which is much stronger than the latter, dominates the properties of atomic nuclei. Nevertheless, in specific studies, it is crucial to evaluate contribution of the latter to the properties of atomic nuclei. The mass difference of the mirror nuclei and energy of the isobaric analog state are such examples. In this report, we focus on the recent studies of the Coulomb energy density functionals (EDFs) of electron systems in the context of atomic nuclei [1, 2] in the density functional theory (DFT) [3, 4].

In the DFT for electron systems, the correlation energy is considered as well as the Hartree and exchange energies, whereas in the nuclear DFT it is not considered explicitly. The correlation EDF is tested for nuclear systems, where the local density approximation (LDA) is used, and we have used the experimentally observed charge-density distribution for quantitative calculations of selected nuclei to avoid an error coming from the density [1].

To calculate Coulomb exchange energy, we carry out self-consistent Skyrme Hartree-Fock calculations by using the Perdew-Burke-Ernzerhof generalized gradient approximation (PBE-GGA) Coulomb exchange functional [5] instead of using the exact-Fock term, and the optimal value of the free parameter μ that appears in the PBE-GGA functional is also discussed [2].

2 Correlation Functional

The Coulomb correlation energies calculated by the charge density distribution [6] in the LDA, E_{Cc}, for selected nuclei from light to heavy region are shown in Table 1. For comparison, the Coulomb exchange energies calculated in the LDA, E_{Cx}, and the ratio $E_{\mathrm{Cc}}/E_{\mathrm{Cx}}$ are also shown.

It is seen that in these calculations, E_{Cc} is all around 2 % of E_{Cx}. However, Bulgac and Shaginyan [7] evaluated that in atomic nuclei, $E_{\mathrm{Cc}}/E_{\mathrm{Cx}}$ would be around −40 % to −20 %, instead of 2 %. Hypothetically, if there is only Coulomb interaction since correlation always further decreases the energy of the whole system, we have the signs of the Hartree, exchange, and correlation energies as $E_{\mathrm{Cd}} > 0$, $E_{\mathrm{Cx}} < 0$, and $E_{\mathrm{Cc}} < 0$, respectively. In reality, the correlation EDF is not separable at all. In Ref. [7] the correlation EDFs are written in terms of the response functions, and such response functions are determined by the total interaction, i.e., mainly by the attractive nuclear part, instead of the repulsive Coulomb part. The total correlation energy is still negative, mainly due to the contribution of the nuclear interaction. As a result, the contribution of Coulomb interaction becomes positive, i.e., for the Coulomb energies, E_{Cc} has the different sign as E_{Cx}. In short, the correlation energy density functionals of electron systems cannot be applied directly to atomic nuclei. It is also noted that the Coulomb correlation functional in the GGA gives around 30–80 % of E_{Cc} in the LDA.

Table 1. Coulomb correlation energies E_{Cc} for selected nuclei. Energies are shown in MeV. Data are taken from Ref. [1].

Nuclei	LDA E_{Cx}	LDA E_{Cc}	$E_{\mathrm{Cc}}^{\mathrm{LDA}}/E_{\mathrm{Cx}}^{\mathrm{LDA}}$
^{16}O	−2.638	−0.05218	1.978 %
^{40}Ca	−7.087	−0.1329	1.875 %
^{48}Ca	−7.113	−0.1332	1.873 %
^{58}Ni	−10.28	−0.1879	1.828 %
^{116}Sn	−18.41	−0.3361	1.826 %
^{124}Sn	−18.24	−0.3356	1.840 %
^{208}Pb	−30.31	−0.5524	1.823 %

*e-mail: naito@cms.phys.s.u-tokyo.ac.jp

3 Exchange Functional

The GGA Coulomb exchange functionals have been proposed as

$$E_{\mathrm{Cx}}[\rho_{\mathrm{ch}}] = -\frac{3}{4}\frac{e^2}{4\pi\varepsilon_0}\left(\frac{3}{\pi}\right)^{1/3}\int [\rho_{\mathrm{ch}}(\mathbf{r})]^{4/3} F(s(\mathbf{r}))\, d\mathbf{r}, \tag{3.1}$$

where ρ_{ch} is the charge density distribution, F is the enhancement factor depending on the density gradient, and s denotes the dimensionless density gradient

$$s = \frac{|\nabla \rho_{\mathrm{ch}}|}{2k_{\mathrm{F}}\rho_{\mathrm{ch}}}, \qquad k_{\mathrm{F}} = \left(3\pi^2\rho_{\mathrm{ch}}\right)^{1/3}. \tag{3.2}$$

In particular,

$$F(s) = 1 + \kappa - \frac{\kappa}{1 + \mu s^2/\kappa}, \tag{3.3}$$

is used in the PBE-GGA Coulomb exchange functional to satisfy some physical conditions, and $F \equiv 1$ corresponds to the LDA one, i.e., the Hartree-Fock-Slater approximation [8, 9]. The parameter $\kappa = 0.804$ is determined for any value of μ by the Hölder inequality [10]. In contrast, two different values of μ have been widely used in the studies of atoms [5] and solids [11], respectively. For the PBE-GGA functional, $\mu = 0.21951$ is determined by the random phase approximation of the homogeneous electron gas. Since this μ can be a different value for nuclear systems, the free parameter of the PBE-GGA Coulomb exchange functional, μ, is multiplied by a factor λ. For the nuclear part, the SAMi functional [12] is used in the self-consistent calculation. For comparison, the exact-Fock energies are also calculated [13].

The deviation of the Coulomb exchange energy E_{Cx} of the PBE-GGA or the exact-Fock from that of the LDA, $\Delta E_{\mathrm{Cx}}^{\mathrm{LDA}}$,

$$\Delta E_{\mathrm{Cx}}^{\mathrm{LDA}} = \frac{E_{\mathrm{Cx}} - E_{\mathrm{Cx}}^{\mathrm{LDA}}}{E_{\mathrm{Cx}}} \tag{3.4}$$

are shown as a function of mass number A in Fig. 1.

It is found that in the light-mass region, to reproduce the exact-Fock results, $\lambda = 1.50$ or more is required, while in the medium-heavy- and heavy-mass regions $\lambda = 1.25$ reproduces well the exact-Fock results. The PBE-GGA result with $\lambda = 1.00$ reproduces the exact-Fock result in the case of the super-heavy nucleus $^{310}126$ since the ratio of the surface region to the volume region in the super-heavy nuclei is smaller than that in the medium-heavy or heavy nuclei.

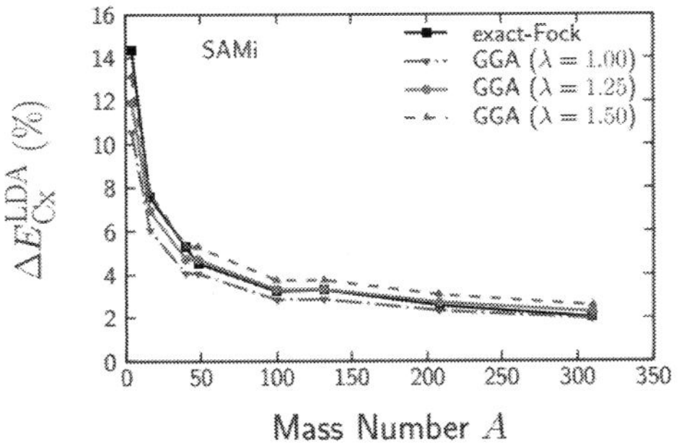

Figure 1. Deviation between the PBE-GGA and the LDA, $\Delta E_{\mathrm{Cx}}^{\mathrm{LDA}}$ defined as Eq. (3.4). For comparison the deviation between the exact-Fock and the LDA is also shown. Taken from Ref. [2].

4 Conclusion

In these works [1, 2], the Coulomb exchange and correlation EDFs in electron systems are applied to the nuclear systems. On the one hand, the Coulomb correlation energy density functionals of electron systems are not applicable for atomic nuclei, because these functionals are not separable and the nuclear interaction determines properties of atomic nuclei mainly. On the other hand, the PBE-GGA Coulomb exchange functional with $\lambda = 1.25$ reproduces the exact-Fock energy in the self-consistent Skyrme Hartree-Fock calculations for atomic nuclei. It should be emphasized that the numerical cost of the self-consistent calculations with the PBE-GGA exchange functional is $O\left(N^3\right)$, whereas that with the exact-Fock term is $O\left(N^4\right)$.

TN and HL would like to thank the RIKEN iTHEMS program and the JSPS-NSFC Bilateral Program for Joint Research Project on Nuclear mass and life for unravelling mysteries of the r-process. TN acknowledges the financial support from Computational Science Alliance, The University of Tokyo, Università degli Studi di Milano, and the JSPS Grant-in-Aid for JSPS Fellows under Grant No. 19J20543. HL acknowledges the JSPS Grant-in-Aid for Early-Career Scientists under Grant No. 18K13549. GC and XRM acknowledge funding from the European Union's Horizon 2020 research and innovation program under Grant No. 654002.

References

[1] T. Naito, R. Akashi, H. Liang, Phys. Rev. C **97**, 044319 (2018)
[2] T. Naito, X. Roca-Maza, G. Colò, H. Liang, Phys. Rev. C **99**, 024309 (2019)
[3] P. Hohenberg, W. Kohn, Phys. Rev. **136**, B864 (1964)
[4] W. Kohn, L.J. Sham, Phys. Rev. **140**, A1133 (1965)
[5] J.P. Perdew, K. Burke, M. Ernzerhof, Phys. Rev. Lett. **77**, 3865 (1996)
[6] H. De Vries, C. De Jager, C. De Vries, At. Data Nucl. Data Tables **36**, 495 (1987)
[7] A. Bulgac, V.R. Shaginyan, Nucl. Phys. A **601**, 103 (1996); Phys. Lett. B **469**, 1 (1999)
[8] P.A.M. Dirac, Proc. Camb. Phil. Soc. **26**, 376 (1930)
[9] J.C. Slater, Phys. Rev. **81**, 385 (1951)
[10] E.H. Lieb, M. Loss, *Analysis*, 2nd edn. (American Mathematical Society, 2001)
[11] J.P. Perdew, A. Ruzsinszky, G.I. Csonka, O.A. Vydrov, G.E. Scuseria, L.A. Constantin, X. Zhou, K. Burke, Phys. Rev. Lett. **100**, 136406 (2008)
[12] X. Roca-Maza, G. Colò, H. Sagawa, Phys. Rev. C **86**, 031306 (2012)
[13] X. Roca-Maza, L.G. Cao, G. Colò, H. Sagawa, Phys. Rev. C **94**, 044313 (2016)

How do we infer shell effects at high-excitation energies? A new spectroscopic probe to search for magic numbers

Cebo Ngwetsheni[1,*] and *José Nicolás* Orce[1,**]

[1]Department of Physics & Astronomy, University of the Western Cape, P/B X17 Bellville ZA-7535, South Africa

Abstract. The nuclear dipole polarizability is mainly governed by the dynamics of the giant dipole resonance and, assuming validity of the brink-Axel hypothesis, has been investigated along with the effects of the low-energy enhancement of the photon strength function for nuclides in medium- and heavy-mass nuclei. Cubic-polynomial fits to both data sets extrapolated down to a gamma-ray energy of 0.1 MeV show a significant reduction of the nuclear dipole polarizability for semi-magic nuclei, with magic numbers N =28, 50 and 82, which supports shell effects at high-excitation energies in the the quasi-continuum region. This work assigns σ_{-2} values as sensitive measures of long-range correlations of the nuclear force and provides a new spectroscopic probe to search for "old" and "new" magic numbers at high-excitation energies.

1 Motivation

Atomic masses are useful measures to study nuclear structure via nuclear binding energies. Deviations from the smooth trend presented by two-neutron separation energies can be associated with a sudden onset of deformation [1] or the rise of a magic number [2]. This information depends on high-precision measurements of atomic masses and is limited to ground and isomeric states of nuclei [3]. In this work, we propose an alternative spectroscopic probe, namely the nuclear polarizability, to investigate the rise of magic numbers at high-excitation energies. This work expands on the data analysis of our recent publications [4, 5] by doing a systematics study for the limited available information and assuming that the enhancement of the photon-strength function continues at lower gamma-ray energies of $E_\gamma \geqq 0.1$ MeV. Overall, this cutoff shows a greater sensitivity of the nuclear polarizability to the long-range correlations in the nuclear wave functions, as realized from relative sharp drops of the nuclear polarizability in semi-magic nuclei.

The nuclear polarizability is dominated by the dynamics of the isovector giant dipole resonance (GDR) [6, 7], which is microscopically described as a coherent superposition of 1 particle - 1 hole (1p - 1h) excitations [8], and, macroscopically, as the inter-penetrating motion of proton and neutron fluids out of phase. The latter results from the nuclear symmetry energy, a_{sym}, in the Bethe-Weizsäcker semi-empirical mass formula [9, 10] acting as a restoring force [6], $a_{sym}(\rho_N - \rho_Z)^2/\rho_A$, where ρ_N, ρ_Z and ρ_A are the neutron, proton and total mass densities, respectively. Using the liquid drop model, with potential energy ρ_Z, Migdal calculated the ground state (g.s.) electric dipole polarizability $\alpha_{E1} = \frac{\mathbf{P}}{\mathbf{E}}$, where **P** is the electric dipole moment and **E** the electric field strength, connecting to a_{sym} as follows,

$$\alpha_{E1} = \frac{e^2R^2A}{40a_{sym}} = 2.25 \times 10^{-3}A^{5/3}\ \mathrm{fm}^3, \quad (1)$$

where a_{sym} = 23 MeV was assumed by Migdal as well as a defined spherical surface of radius $R = 1.2A^{1/3}$ fm [6]. Hence, α_{E1} is proportional to the size and diffuseness of the nucleus. As a second-order effect in perturbation theory, α_{E1} is related to the total cross section σ_{total} and its (−2) moment, σ_{-2}, in the following manner [11],

$$\alpha_{E1} = 2e^2 \sum_n \frac{\langle i \parallel \hat{E}1 \parallel n\rangle\langle n \parallel \hat{E}1 \parallel i\rangle}{E_\gamma} = \frac{\hbar c}{2\pi^2}\sigma_{-2}, \quad (2)$$

where $\hat{E}1$ is the electric dipole operator, $|i\rangle$ and $|n\rangle$ are the ground and excited state vectors and σ_{-2} is defined as,

$$\sigma_{-2} := \int_0^{E_{\gamma max}} \frac{\sigma_{total}(E_\gamma)}{E_\gamma^2}\, dE_\gamma, \quad (3)$$

where $E_{\gamma_{max}}$ is the maximum E_γ available experimentally [12]. Combining Eqs. 1 and 2, Migdal determined,

$$\sigma_{-2} = 2.25A^{5/3}\mu\mathrm{b/MeV}. \quad (4)$$

A new empirical formula for σ_{-2} [13] has been determined from the 1988 photoneutron cross-section evaluation using monoenergetic photons [12], which agrees with Migdal's power-law formula,

$$\sigma_{-2} = 2.4\kappa\, A^{5/3}\mu\mathrm{b/MeV}, \quad (5)$$

where the polarizability parameter κ is included to account for deviations from the actual GDR effects to that predicted

*e-mail: 3344489@myuwc.ac.za
**e-mail: jnorce@uwc.ac.za

by the hydrodynamic model [6, 14]. The polarizability parameter κ can therefore be extracted for known σ_{-2} values and vice versa.

Generally, σ_{-2} values should include both electric (α_{E1}) and magnetic (χ_{M1}) dipole polarizabilities [15],

$$\sigma_{-2} = \frac{2\pi^2}{\hbar c}(\alpha_{E1} + \chi_{M1}). \tag{6}$$

Similarly, χ_{M1} is a measure of the magnetic dipole response, $\mathbf{M}$, to a magnetic field, $\mathbf{B}$, i.e. $\mathbf{M} = \chi_{M1} \cdot \mathbf{B}$ [16], and can be decomposed into paramagnetic (χ^{para}_{M1}) and diamagnetic (χ^{dia}_{M1}) susceptibilities,

$$\begin{aligned} \chi_{M1} &= \chi^{para}_{M1} + \chi^{dia}_{M1} \qquad (7) \\ &= 2\sum_n \frac{\langle i \parallel \hat{M}1 \parallel n\rangle\langle n \parallel \hat{M}1 \parallel i\rangle}{E_\gamma} - \frac{Ze^2}{6mc^2}\langle r^2\rangle. \end{aligned}$$

Permanent magnetic dipole moments, i.e. paramagnetism, is dominant for $A < 20$ nuclides and may have a substantial contribution to σ_{-2} values (e.g. ^{6}Li and ^{7}Li), whereas diamagnetism has negligible effects [15, 16]. It would be interesting to have an updated evaluation of $M1$ strengths to investigate whether paramagnetism contributes in heavy nuclei, particularly from mixed-symmetry states recently found in the $A \sim 90$ region [17–19].

Because of the $1/E_\gamma{}^2$ weighting in Eq. 3, σ_{-2} values are not significantly affected by contributions above the GDR energy (e.g. recent discrepancies found between the Livermore and Saclay photoneutron data with monochromatic photon beams [20, 21] or nucleon resonances at $E_\gamma \gtrsim 140$ MeV [22]), and are extremely sensitive measures – unlike σ_{total} – of long-range correlations in the nuclear wave functions.

2 Low-energy enhancement of the photon strength function

The photon strength function $f(E_\gamma)$ characterizes average EM decay and absorption properties of excited nuclei. Recent measurements of $f(E_\gamma)$ by the Oslo group have revealed an enhancement at low E_γ [23–26]. These measurements are performed in the quasi-continuum energy region and assume the validity of the Brink-Axel hypothesis [27, 28], which states that $f(E_\gamma)$ is independent of the particular structure of the initial and final state [29] and only depends on E$_\gamma$, i.e. GDR properties are similar for all nuclear states. To date, the EM character of the LEE remains undetermined experimentally, although polarization asymmetry measurements of γ rays in ^{56}Fe show an admixture of $M1$ and $E1$ radiation [30]. Various interpretations of the LEE have been proposed, explaining its dipole origin as $M1$ [31–36] and $E1$ [37] dipole radiation [38]. Shell-model (SM) calculations consistently support the $M1$ nature of the LEE [31–34]. The main purpose of this work is to quantify the potentially large contribution from the LEE to the nuclear polarizability and σ_{-2} values assuming dipole radiation and validity of the Brink-Axel hypothesis.

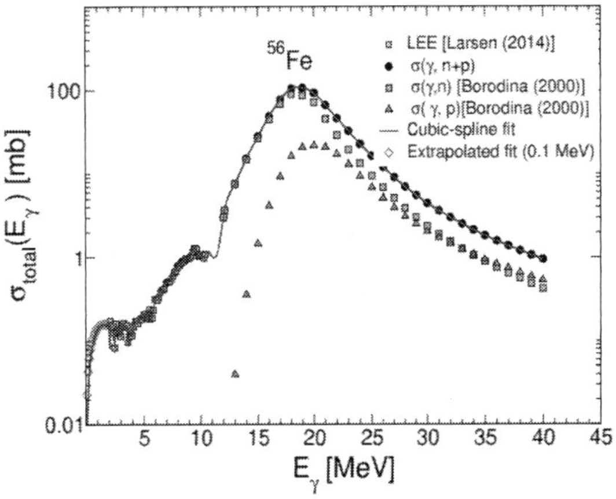

Figure 1. $\sigma_{total}(E_\gamma)$ vs E_γ on a log scale on the y-axis showing the interpolation function data (solid blue line) for ^{56}Fe, the green squares represent $\sigma_{LEE}(E_\gamma)$ and black filled circles the sum of proton and neutron emission cross sections for the GDR.

3 Systematics of total σ_{-2} values and effect in the nuclear polarizability

The LEE is generally observed in medium-mass nuclei in the $A \approx 50$ and 90 mass regions and only for ^{105}Cd, ^{138,139}La and ^{151,153}Sm [39–41] in heavy-mass nuclei. These nuclei, spanning the mass range $A = 45 - 153$, have been considered in order to obtain a systematic study of LEE + GDR effects on σ_{-2} values, which requires the combined analysis of LEE and GDR cross sections,

$$\sigma_{total}(E_\gamma) = \sigma_{GDR}(E_\gamma) + \sigma_{LEE}(E_\gamma), \tag{8}$$

where $\sigma_{GDR}(E_\gamma) = \sigma(\gamma, p) + \sigma(\gamma, n) + \sigma(\gamma, 2n) + \sigma(\gamma, np)$ is given by photo-nuclear absorption reactions at energies above nucleon threshold, and $\sigma_{LEE}(E_\gamma)$ is the cross section contribution from the LEE region below the neutron threshold. The GDR data are obtained from the experimental nuclear-reaction databases EXFOR [42] and ENDF [43], whereas the LEE data come from the Oslo compilation of $f(E_\gamma)$ [44], in units of MeV^{-3}. The LEE data can be converted to cross sections as follows [45],

$$\sigma_{LEE}(E_\gamma) = \pi^2 g_J (\hbar c)^2 f(E_\gamma) E_\gamma \ \text{[mb]}, \tag{9}$$

where g_J is the statistical factor $g_J = \frac{2J_f+1}{2J_i+1}$ with spins J_i and J_f corresponding to initial and final states, respectively. The magnitude of g_J affects σ_{-2} and polarizability values proportionally. Considering the dipole character of the LEE, an average $\langle g_J \rangle = 1$ is a reasonable approximation for dipole $\Delta J = 0$ and $\Delta J = 1$ transitions.

The combination of GDR and LEE contributions may be arguable *a priori*, because $\sigma_{GDR}(E_\gamma)$ corresponds to transitions between excited states $|n\rangle$ in the GDR region and the g.s. $|i\rangle$, whereas $\sigma_{LEE}(E_\gamma)$ results from transitions between excited states in the quasi-continuum region. Recent studies of $f(E_\gamma)$ by Guttormsen and co-workers

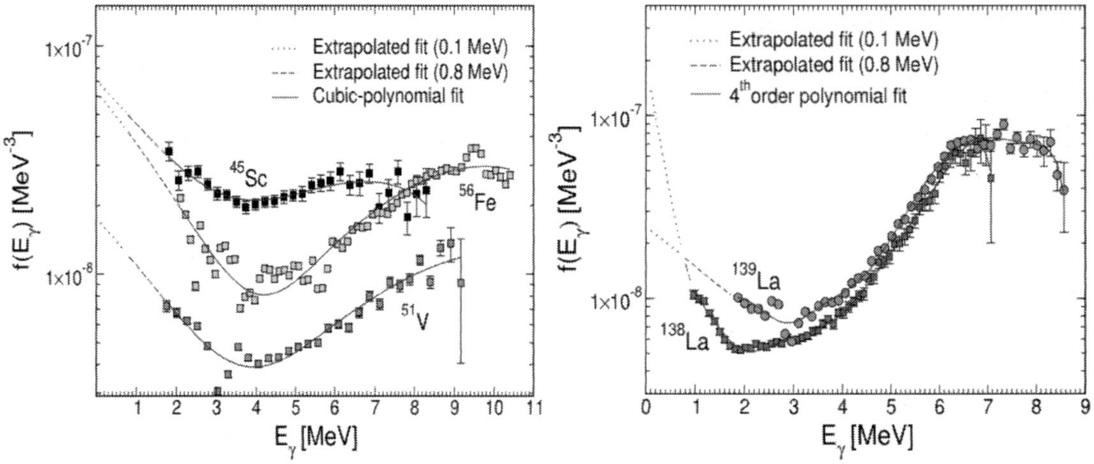

Figure 2. $f(E_\gamma)$ for ^{45}Sc, ^{51}V and ^{56}Fe (left panel) fitted with a cubic polynomial and ^{138,139}La (right panel) fitted with a 4th-order polynomial, including extrapolated data to $E_{\gamma_{min}}$ = 0.8 MeV (blue dashed line) and $E_{\gamma_{min}}$ = 0.1 MeV (red dotted line).

support the validity of the Brink-Axel hypothesis at different bin energies in the quasi-continuum region [46]. This, together with the fact that GDR studies of nuclei at relatively low temperatures T and spins J and cold nuclei in their ground state present similar features [47, 48], may allow for combining the LEE and GDR cross sections [4]. As shown in Figs. 1 and 2, it is somewhat surprising that the data from the low-energy tail of the GDR generally merge nicely with the LEE data without applying any normalization factor. Nonetheless, the validity for the combination of these data needs further experimental confirmation.

An interpolation method for calculating $\sigma_{total}(E_\gamma)$ and σ_{-2} values has been used in this work. This method is independent of any physical phenomena and operates by creating a function – cubic or 4th order polynomial – that interpolates (add data) between the fixed experimental data points. As an example, Fig. 1 shows the total cross section of ^{56}Fe with a cubic-spline interpolation function (solid blue line). The resulting functions are integrated accordingly to obtain $\sigma_{total}(E_\gamma)$, which yields the σ_{-2} values listed in Table 1. Most nuclei present an energy gap with missing data between the LEE and GDR data, which may include the $M1$ spin-flip resonance and the pygmy dipole resonances (PDR) in neutron-rich nuclei. Therefore, data from ENDF [43] – when available – have been used to fill the gap [4, 5]. Additionally, data near nucleon threshold energies generally present large uncertainties and have been excluded.

Because of minimal RMS errors, a cubic-polynomial interpolation has been selected as the interpolating function throughout this work. Similar results are often obtained using a 4th-order polynomial interpolation. Lower and higher order polynomial interpolations present unexpected structures (bumps) in the energy-gap region and above. The errors associated to σ_{-2} values are calculated from upper and lower *loci* limits of $\sigma_{total}(E_\gamma)$, including LEE and GDR contributions, which yield an uncertainty of 7%. Uncertainties for the interpolation data spanning the gap and extrapolated data are treated as three points standard deviation of the mean, in order to determine the upper and lower limits for these data sets. Most of the considered nuclei are stable, except ^{50}V, ^{138}La and ^{153}Sm with no experimental GDR cross sections. Therefore, GDR data of stable neighboring isotopes were used, i.e. ^{51}V, ^{139}La and ^{152}Sm, under the assumption that neighboring nuclei present similar $f(E_\gamma)$ [65] and considering the fact that σ_{-2} values show a strong dependence on nuclear mass A.

Setting up the low-energy cut-off, $E_{\gamma_{min}}$, for the LEE is not obvious. In our previous study, we extrapolated the LEE data down to 800 keV from experimental observations which show $E_{\gamma_{min}} \approx 1$ MeV for most nuclei, except for ^{153}Sm where measurements were carried out down to $E_{\gamma_{min}}$ = 645 keV. Recent SM studies [30, 33, 35] explore, however, the behavior of $f(E_\gamma)$ at very low E_γ, supporting the continuation of the LEE down to $E_{\gamma_{min}}$ = 0. We have investigated this situation extrapolating the data to $E_{\gamma_{min}}$ = 0.1 MeV. It is not obvious to us the physical meaning of going down to $E_{\gamma_{min}}$ = 0 without any further experimental evidence. Consequently, Fig. 2 shows extrapolated fits of $f(E_\gamma)$ data down to $E_{\gamma_{min}}$ = 0.1 MeV. Similar fits were done for all the considered nuclei to explore the evolution of $f(E_\gamma)$ towards $E_{\gamma_{max}} \approx 0$. The results are listed in Table 1.

A large enhancement of σ_{-2} values is found for ^{45}Sc and ^{56}Fe and ^{153}Sm as compared with semi-magic nuclei ^{51}V and ^{139}La. If these predictions of $f(E_\gamma)$ for $E_{\gamma_{min}} \rightarrow 0$ MeV are consistent with experimental findings, reaction rates in nucleosynthesis following rapid-neutron capture – the r-process – may strongly be affected together with the predicted abundances of nuclei [40, 69–71]. Additional fits using Eq. 4 in Ref. [69] have been performed, which

Nucleus	$E_{\gamma(min)}$ (MeV)	$E_{\gamma(max)}$(GDR) (MeV)	$E_{\gamma(max)}$(LEE) (MeV)	σ_{-2}(total) (μb/MeV)	σ_{-2}(LEE) (μb/MeV)	**C** %	κ
$^{45}_{21}$Sc*	0.1	28.1	3.2	2488(602)	873	35.1 %	1.82(4)
$^{50}_{23}$V	0.1	27.8	3.1	1597(105)	187	11.7 %	0.98(1)
$^{51}_{23}$V	0.1	27.8	3.1	1564(273)	140	9.1 %	0.92(2)
$^{56}_{26}$Fe*	0.1	40.0	3.8	2658(354)	568	21.4 %	1.35(2)
$^{76}_{32}$Ge	0.1	26.5	2.3	3291(1279)	135	4.1 %	1.00(4)
$^{92}_{40}$Zr	0.1	27.8	2.2	3166(528)	199	6.3 %	0.70(1)
$^{95}_{42}$Mo	0.1	27.8	2.5	5017(673)	340	6.8 %	1.05(1)
$^{138}_{57}$La	0.1	24.3	1.9	8580(1240)	319	7.3%	0.97(1)
$^{139}_{57}$La	0.1	24.3	2.5	8183(837)	221	2.7%	0.91(1)
$^{153}_{62}$Sm	0.1	20.0	1.6	12998(1638)	3093	23.9%	1.23(2)

Table 1. Contributions of the LEE with $E_{\gamma_{min}} = 0.1$ MeV represented by **C**. $E_{\gamma(max)}$(LEE) is the energy at which the LEE starts.

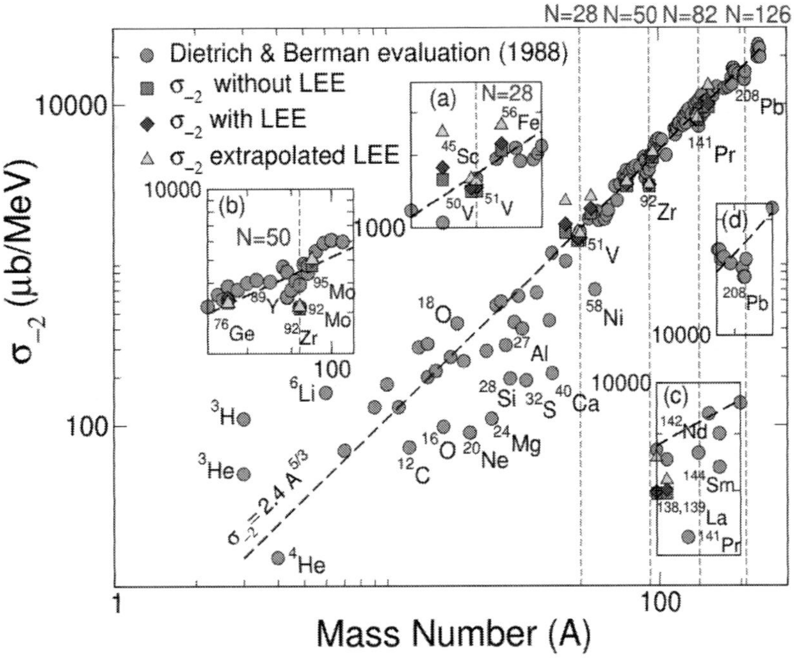

Figure 3. σ_{-2} *vs* A on a log-log scale from the photoneutron cross-section evaluation (solid circles) [12] and σ_{-2} data listed in Table 1 excluding (squares) and including (diamonds) the LEE contributions down to $E_{\gamma} = 0.1$ MeV. For comparison, the empirical formula $\sigma_{-2} = 2.4(1)\,A^{5/3}\mu$b/MeV (dashed line) [13] is plotted. Shell effects are noticeable for $N = 28$ (inset (a)), $N = 50$ (inset (b)), 82 (inset (c)) and 126 (inset (d)) isotones.

show an even larger LEE contribution, but no indications of shell effects when comparing neighboring nuclei.

4 Discussion

Shell effects arising from σ_{-2} values are evident in nuclei with or near magic numbers $N = 28$, 50, 82 and 126. As shown in Fig. 3, a more distinct trend of σ_{-2} values emerges when the LEE contributions are extrapolated to $E_{\gamma_{min}} = 0.1$ MeV, with a relatively smaller contribution to σ_{-2} values in ^{51}V ($N = 28$), ^{92}Zr ($N = 50$ and $Z = 40$) and ^{139}La ($N = 82$), as compared with neighboring nuclides. As listed in Table 1, these semi-magic nuclei present smaller polarizability parameters of $\kappa \approx 0.90$ for ^{51}V and ^{139}La and an even smaller $\kappa = 0.70$ for ^{92}Zr. These findings, assuming validity of the Brink-Axel hypothesis, support the continuation of shell effects in the quasi-continuum region where the LEE of $f(E_\gamma)$ occurs. The fact that the nuclear polarizability can change at higher excitation energies from its ground-state value is supported by different κ values predicted by consistent no-core-shell-model calculations in ^{12}C – with a larger κ value determined for the first 2^+ excitation at 4.439 MeV with respect to the ground state – using both $2N$ and $2N + 3N$ chiral forces [72].

Furthermore, Table 1 shows considerable LEE contributions to σ_{-2} values for more deformed nuclei such as ^{45}Sc, ^{56}Fe and ^{153}Sm with $\kappa > 1$ values, while other nuclei close to magic numbers present consistent values of $\kappa \approx 1$. Certainly from these results, $E_{\gamma_{min}}$ is crucial for determining how the LEE affects the nuclear polarizability. Additionally, the LEE contribution is affected by the smaller $E_{\gamma_{max}}$(LEE) for heavy nuclei and the inverse mass dependence of E_{GDR} [7, 68].

The sudden drop of σ_{-2} values and $\kappa < 1$ for nuclei with neutron magic numbers $N = 28$, 50 and 82 supports the $M1$ nature of the LEE predicted by large-scale SM calculations and the validity of the generalized Brink-Axel hypothesis, which seems to allow for structural changes. This work emphasizes the need for additional $f(E_\gamma)$ measurements at low energies below 1 MeV, and opens a new research avenue to investigate the existence and evolution of magic numbers at high-excitation energies from σ_{-2} measurements [4].

Finally, it is relevant to note that the combination of data sets with such a distinct nature in this study, namely LEE and GDR data, remains unsettling and needs to be explored further; for instance, the Brink-Axel hypothesis could additionally be supported with the presence of LEE associated with GDRs built on excited states, preferably around the quasi-continuum region. More experimental $f(E_\gamma)$ measurements and theoretical work are clearly needed to validate these interesting findings.

Acknowledgements

This work was supported by the National Research Foundation of South Africa under Grant 93500, the MANuS/MatSci program and the SA-CERN collaboration.

References

[1] U. Hager *et al.*, Phys. Rev. Lett. **96**, 069901 (2006).
[2] F. Wienholtz *et al.*, Nature **498**, 346 (2013).
[3] J. Dilling, K. Blaum, M. Brodeur, and S. Eliseev, Annu. Rev. Nucl. Part. Sci. **68**, 45 (2018).
[4] C. Ngwetsheni and J. N. Orce, Phys. Lett. B **792**, 335 (2019).
[5] C. Ngwetsheni and J. N. Orce, to be published in Hyperfine Interactions (2019).
[6] A. Migdal, J. Exptl. Theoret. Phys. U.S.S.R. **15**, 81 (1945).
[7] H. Steinwedel, J. H. D. Jensen, and P. Jensen, Phys. Rev. **79**, 1019 (1950).
[8] M. Danos, B. S. Ishkhanov, N. P. Yudin, and R. A. Eramzhyan, Phys. Uspekhi **37**(12), 1297 (1995).
[9] H. A. Bethe and R. F. Bacher, Rev. Mod. Phys. **8**, 82 (1936).
[10] C. F. von Weizsäcker, Z. Phys. **96**, 431 (1935).
[11] A. B. Migdal, A. A. Lushnikov, and D. F. Zaretsky, Nucl. Phys. A **66**, 193 (1965).
[12] S. S. Dietrich and B. L. Berman, Atom. Data Nucl. Data Tables **38**, 199 (1988).
[13] J. N. Orce, Phys. Rev. C **91**, 064602 (2015).
[14] J. S. Levinger, Phys. Rev. **107**, 554 (1957).
[15] W. Knüpfer and A. Richter, Phys. Lett. B **107**, 325 (1981).
[16] W. Knüpfer and A. Richter, Z. Phys. A **320**, 253 (1985).
[17] N. Pietralla *et al.*, Phys. Rev. Lett. **83**, 1303 (1999).
[18] C. Fransen *et al.*, Phys. Rev. C **67**, 024307 (2003).
[19] J. N. Orce *et al.*, Phys. Rev. Lett. **97**, 062504 (2006).
[20] B. S. Ishkhanov *et al.*, Phys. At. Nucl. **67**, 1664 (2004).
[21] I. Gheorghe *et al.*, Phys. Rev. C **96**, 044604 (2017).
[22] J. Ahrens, H. Gimm, A. Zieger and B. Ziegler, Il Nuovo Cimento A **32**, N. 3, 364 (1976).
[23] A. Voinov *et al.*, Phys. Rev. Lett. **93**, 142504 (2004).
[24] A. Schiller *et al.*, Nucl. Instr. Meth. Phys. Res. A **447**, 498 (2000).
[25] M. Guttormsen *et al.*, Nucl. Instr. Meth. Phys. Res. A **255**, 518 (1987).
[26] M. Wiedeking *et al.*, Phys. Rev. Lett. **108**, 162503 (2012).
[27] D. Brink, doctoral thesis, Oxford University, 1955 (unpublished).
[28] P. Axel, Phys. Rev. **126**, 671 (1962).
[29] D. Martin *et al.*, Phys. Rev. Lett. **119**, 182503 (2017).
[30] M. D. Jones *et al.*, Phys. Rev. C **97**, 024327 (2018).
[31] R. Schwengner, S. Frauendorf and A. C. Larsen, Phys. Rev. Lett. **111**, 232504 (2013).
[32] B. A. Brown and A. C. Larsen, Phys. Rev. Lett. **113**, 252502 (2014).
[33] K. Sieja, Phys. Rev. Lett. **119**, 052502 (2017).
[34] K. Sieja, EPJ Web of Conferences **146**, 05004 (2017).
[35] J.E. Midtbo *et al.*, Phys. Rev. C **98**, 064321 (2018).
[36] A. C. Larsen *et al.*, Phys. Rev. C **97**, 054329 (2018).
[37] E. Litvinova and N. Belov, Phys. Rev. C **88**, 031302(R) (2013).
[38] J. D. Jackson, *Classical Electrodynamics*, John Wiley & Sons, Inc. (1962).
[39] A. C. Larsen *et al.*, Phys. Rev. C **87**, 014319 (2013).
[40] B. V. Kheswa *et al.*, Phys. Lett. B **744**, 268 (2015).
[41] A. Simon *et al.*, Phys. Rev. C **93**, 034303 (2016).
[42] https://www-nds.iaea.org/exfor/exfor.htm
[43] https://www.nndc.bnl.gov/ensdf/
[44] http://www.mn.uio.no/fysikk/english/research/about/infrastructure/OCL/nuclear-physics-research/compilation/
[45] M. Baranger, E. Vogt, Adv. in Nucl. Phys. **7**, 229 (2012).
[46] M. Guttormsen *et al.*, Phys. Rev. Lett. **116**, 012502 (2016).
[47] A. Schiller and M. Thoennessen, Atom. Data Nucl. Data Tables **93**, 549 (2007).
[48] J. J. Gaardhøje, Annu. Rev. Nucl. Part. Sci. **42**, 483 (1992).

[49] A. Vessiere *et al.*, Nucl. Phys. A **227**, 513 (1974).
[50] A. C. Larsen *et al.*, Phys. Rev. C **76**, 044303 (2007).
[51] K. Shoda, Nucl. Phys. A **239**, 397 (1975).
[52] S. C. Fultz *et al.*, Phys. Rev **128**, 2345 (1962). ,
[53] A. C. Larsen *et al.*, Phys. Rev. C **73**, 064301 (2006).
[54] S. S. Borodina *et al.*, Moscow State Univ. Inst. of Nucl. Phys. Rep. **6** (2000).
[55] A. C. Larsen *et al.*, Phys. Rev. Lett. **111**, 242504 (2013).
[56] P. Carlos *et al.*, Nucl. Phys. A **258**, 365 (1976).
[57] A. Spyrou *et al.*, Phys. Rev. Lett. **113**, 232502 (2014).
[58] B. L. Berman *et al.*, Phys. Rev. **162**, 1098 (1967).
[59] M. Guttormsen *et al.*, Phys. Rev. C **96**, 024313 (2017).
[60] H. Utsunomiya *et al.*, Phys. Rev. Lett. **100**, 162502 (2008).
[61] H. Beil *et al.*, Nucl. Phys. A **227**, 427 (1974).
[62] H. Utsunomiya *et al.*, Phys. Rev. C **88**, 015805 (2013).
[63] H. Beil *et al.*, Nucl. Phys. A **172**, 426 (1971).
[64] P. Carlos *et al.*, Nucl. Phys. A **225**, 171 (1974).
[65] M. Guttormsen *et al.*, Phys. Rev. C **71**, 044307 (2005).
[66] B. L. Berman and S. C. Fultz, Rev. Mod. Phys. **47**, 713 (1975).
[67] J. Tian, H. Cui, K. Zheng, and N. Wang, Phys. Rev. C **90**, 024313 (2014).
[68] O. Bohigas, N. van Giai, and D. Vautherin, Phys. Lett. B **102**, 105 (1981).
[69] A. C. Larsen and S. Goriely, Phys. Rev. C **82**, 014318 (2010).
[70] M. Arnould, S. Goriely and K. Takahashi, Phys. Rep. **450**, 97 (2007).
[71] A. C. Larsen, A. Spyrou, S. N. Liddick and M. Guttormsen, Prog. Part. Nucl. Phys. **107**, 69 (2019).
[72] M. K. Raju *et al.*, Phys. Lett. B. **777**, 250 (2018).

EPJ Web of Conferences **223**, 01046 (2019) https://doi.org/10.1051/epjconf/201922301046
NSD2019

Coexistence and evolution of shapes: mean-field-based interacting boson model

Kosuke Nomura[1,*]

[1]Department of Physics, Faculty of Science, University of Zagreb, 10000 Zagreb, Croatia

Abstract. A method of deriving the Hamiltonian of the interacting boson model, that is based on the microscopic framework of the nuclear energy density functional, is presented. The constrained self-consistent mean-field calculation with a given energy density functional provides potential energy surface within the relevant collective coordinates, which is subsequently mapped onto the expectation value of the interacting-boson Hamiltonian in the boson condensate state. This procedure completely determines the strength parameters of the IBM, and the diagonalization of the mapped Hamiltonian yields excitation spectra and transition rates for a given nucleus. Two recent applications of the method are discussed, that is, the descriptions of the intruder states in Cadmium isotopes and the octupole correlations in neutron-rich odd-mass Barium isotopes.

1 Introduction

The nuclear shapes and collective excitations have been one of the most prominent and studied themes of nuclear structure physics. Experiments using radioactive-ion beams allow to study exotic nuclei that exhibit variety of shapes and related excitations, e.g., evolution and coexistence of quadrupole deformed shapes, pear-shaped (or octupole) deformation, etc., and necessitate timely systematic, as well as reliable, theoretical investigations. The interacting boson model (IBM) [1], a model where correlated $J = 0^+$ (S) and 2^+ (D) nucleon pairs are approximated to (s and d) bosons, has been remarkably successful in phenomenological description of low-lying states in nuclei. On the other hand, microscopic foundation of the IBM, i.e., an attempt to derive the bosonic Hamiltonian from nucleonic degrees of freedom, has been extensively pursued. Conventionally, this has been done within the seniority scheme of the nuclear shell model [2], but this approach has been somewhat limited to nearly spherical nuclei, due to the fact that in deformed systems the seniority classification breaks down in the presence of non-degenerate shells. In this context, a completely different method of deriving the Hamiltonian of the IBM was developed [3]: the results of the self-consistent mean-field (SCMF) calculation within the framework of the nuclear density functional theory (DFT) are used as a microscopic input to determine the Hamiltonian of the IBM. Since the DFT framework allows for a global mean-field description of many nuclear properties over the entire region of the nuclear chart, it has become possible to derive the IBM Hamiltonian for any arbitrary nuclei with a variety of shapes: strongly axially-deformed [4], γ-soft [5], and octupole [6] shapes, and shape coexistence [7]. This facilitates a detailed, computationally feasible, and systematic prediction of spectroscopy in nuclei, including those far from the stability. Here we describe the basic notions of this method and highlight two of its recent applications: the descriptions of intruder states in even-even Cd isotopes [8], and the onset of octupole deformation in neutron-rich odd-mass Ba isotopes [9].

*e-mail: knomura@phy.hr

2 Mean-field derivation of the IBM

The first step is a standard constrained SCMF calculation based on a given relativistic [10] or non-relativistic [11] EDF. In this case, the constraints are on mass quadrupole moments related to the axial deformation β and non-axial deformation γ in the geometrical collective model [12]. The SCMF calculations are performed to obtain a potential energy surface (PES) for a given set of collective coordinates (β, γ). Examples in Fig. 1 show the SCMF PESs for the axially deformed Sm nuclei, which are obtained from the constrained Hartree-Fock+BCS method. The PES for ^{148}Sm exhibits a nearly spherical minimum, typical of the vibrational nucleus. It becomes soft in β and γ in ^{150}Sm, which is then considered a transitional nucleus. Finally, on the PES for ^{154}Sm one sees well defined prolate minimum, indicating that this nucleus is strongly deformed.

Having the SCMF PES for a given deformation (β, γ), denoted by $E_{\rm SCMF}(\beta, \gamma)$, for each nucleus, we subsequently map it to the IBM PES, $E_{\rm IBM}(\beta, \gamma)$. To be more specific, the IBM Hamiltonian is constructed, i.e., the IBM strengths parameters are determined, so as to reproduce the SCMF surface at each set of (β, γ) as close as possible.

We consider the version of the IBM, that distinguishes between neutrons and protons (IBM-2) [2]. In the IBM-2, number of neutron (proton) bosons N_ν (N_π) equals that of pairs of valence neutrons (protons). The IBM

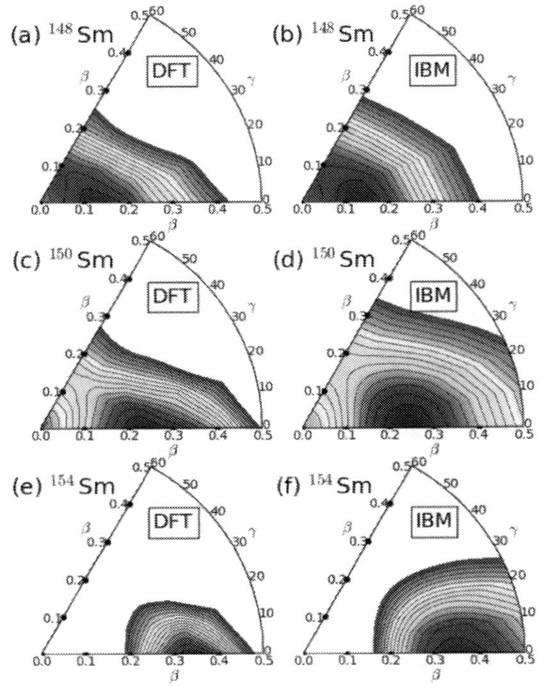

Figure 1. (Color online) The SCMF and mapped IBM PESs for the ^{148,152,154}Sm isotopes. The Skyrme-SkM* functional and the density-dependent zero-range pairing force with strength 1250 MeV fm^3 were used. Energy difference between neighbouring contours is 200 keV.

PES is obtained as the expectation value of a given IBM Hamiltonian in the boson coherent state [13]. The coherent state in the IBM-2 is in general defined as the product of the neutron λ_ν and proton λ_π boson condensates $(N_\nu!N_\pi!)^{-1/2}(\lambda_\nu)^{N_\nu}(\lambda_\pi)^{N_\pi}|0\rangle$, where $\lambda_\tau^\dagger = s_\tau^\dagger + \beta_\tau \cos\gamma_\tau d_{\tau,0}^\dagger + \beta_\tau \sin\gamma_\tau (d_{\tau,+2}^\dagger + d_{\tau,-2}^\dagger)/\sqrt{2}$ ($\tau = \nu$ or π) and $|0\rangle$ is the inert core, and is parametrized by the deformation parameters for neutrons and protons, $(\beta_\nu, \beta_\pi, \gamma_\nu, \gamma_\pi)$, and the Euler angles $\Omega = (\theta_1, \theta_2, \theta_3)$ for the relative orientation of the neutron condensate with respect to that of the proton condensate. The $(\beta_\nu, \beta_\pi, \gamma_\nu, \gamma_\pi)$ deformations could in principle vary independently but, in realistic cases, introducing all of these variables entails too much complexity. Therefore, and as is usual in the collective model [12], we assume equal deformations of the proton and neutron systems, i.e., $\beta_\nu = \beta_\pi \equiv \beta_B$ and $\gamma_\nu = \gamma_\pi \equiv \gamma_B$. In addition, the variable β_B in the boson system is proportional to β deformation in the collective model, $\beta_B = C\beta$ with C being coefficient, while γ_B and γ are supposed to be identical $\gamma_B = \gamma$. It is also assumed that the neutron and proton condensates are aligned, that is, $\Omega = 0$. Under these conditions, we equate the SCMF to the IBM PESs: $E_{\mathrm{SCMF}}(\beta,\gamma) \sim E_{\mathrm{IBM}}(\beta,\gamma)$. Note that it represents an approximate equality as it is fulfilled within a limited range of (β,γ) plane, i.e., in the vicinity of the global minimum. We restrict ourselves to this range because it is most relevant to the low-energy collective states. On the other hand, one should not try to reproduce the region far from the energy minimum as the topology of the microscopic energy surface around the region is determined by single-nucleon configurations that are, by construction, outside the model space of the IBM consisting of only collective pairs of valence nucleons. For this reason, the IBM PES is, as shown on the right column of Fig. 1, always flat in the region far from the global minimum. The topology of the SCMF PES around global minimum should reflect essential features of fermion many-body systems, such as the Pauli principle, the antisymmetrization, and the underlying nuclear forces. Through the mapping procedure, these effects are supposed to be incorporated in the IBM.

The IBM Hamiltonian that embodies the essentials of the underlying nucleonic interactions reads:

$$\hat{H}_{\mathrm{B}} = \epsilon \hat{n}_d + \kappa \hat{Q}_\nu \cdot \hat{Q}_\pi + \kappa' \hat{L} \cdot \hat{L}, \tag{1}$$

where the definitions of each term are found, e.g., in Refs. [3, 4]. The parameters ϵ, κ, χ_ν, and χ_π (which appear in the quadrupole operators $\hat{Q}_\nu$ and $\hat{Q}_\pi$), plus the coefficient C, are to be fitted to the fermionic PES. Only the parameter κ' should be determined independently of the other parameters in such a way [4], that a rotational response of nucleonic system, i.e., cranking moment of inertia that is obtained from the SCMF calculation at the equilibrium minimum, be reproduced in the boson system.

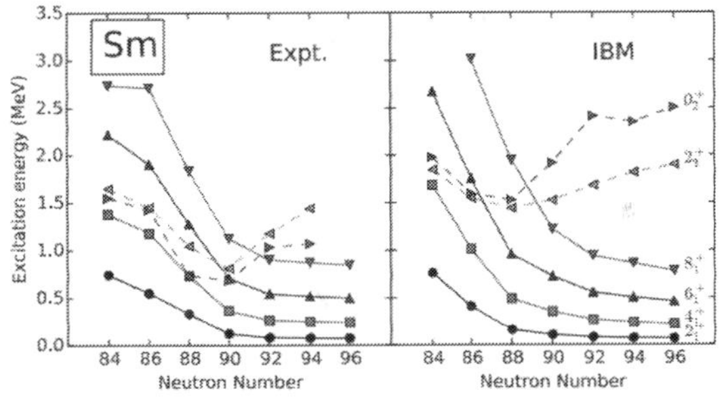

Figure 2. (Color online) Experimental and theoretical low-energy levels in Sm isotopes as functions of the neutron number.

Having determined its strength parameters by the above procedure, the resultant IBM Hamiltonian is diagonalized in the laboratory frame, which provides excitation energies and electromagnetic transition rates for a given nucleus. The comparison of the calculated low-lying spectra to the experimental ones in Fig. 2 indicates that the mapped IBM has nicely reproduced the evolution of nuclear structure against the neutron number from nearly spherical vibrational to strongly deformed states.

3 Structure of even-even Cd isotopes

The even-even Cd isotopes have been traditionally a well-established example for vibrational nuclei. Experiments

 https://doi.org/10.1051/epjconf/201922301046

carried out more recently, however, identified several additional 0^+ and 2^+ at low energy, and the question as to whether these states arise from intruder excitation from other major shell has been a long-standing problem. It is then the purpose of Ref. [8] that we addressed this question within the SCMF-IBM calculation that includes intruder configuration in the boson system.

3.1 IBM with configuration mixing

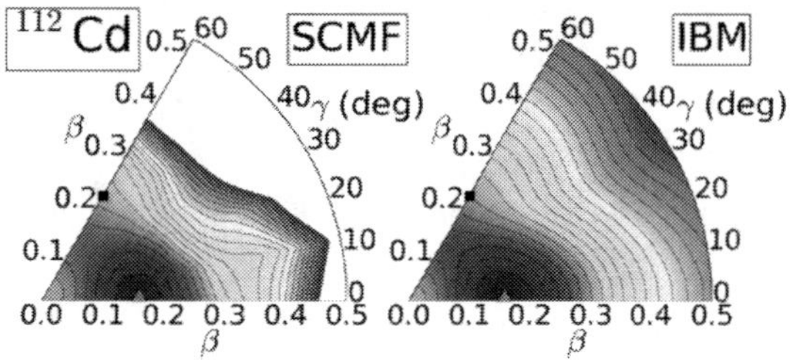

Figure 3. (Color online) Left: The (β, γ) PES for ^{112}Cd, obtained from the SCMF calculation using the Skyrme SLy6 EDF with the density-dependent zero-range pairing interaction with the strength 1000 MeV fm^3. Right: the corresponding IBM-2 PES. Energy difference between neighbouring contours is 250 keV. The global minimum is indicated by solid triangle, while the local minimum is identified by solid squares.

We draw on the left-hand side in Fig. 3 the (β, γ) SCMF PES for the ^{112}Cd, based on the Skyrme-SLy6 EDF. For all the considered Cd nuclei, a prolate global minimum is found with moderate axial deformation $\beta \approx 0.15$. We also observe on the oblate side ($\gamma \approx 60°$) a much less pronounced local minimum between $\beta = 0.2$ and 0.3.

We then construct from the SCMF result the IBM-2 Hamiltonian that includes configuration mixing. A method to incorporate intruder configuration was proposed in Ref. [14], that the different shell-model spaces of $0p-0h$, $2p-2h$, $4p-4h$, ... excitations be associated with the corresponding boson spaces comprising N, $N+2$, $N+4$, ... bosons, with N being the total boson number $N = N_\pi + N_\nu$, and that the different boson subspaces be subsequently mixed. In most of the configuration mixing IBM calculations, particle-like and hole-like bosons are not distinguished and, as the excitation of a pair (or boson) increases the boson number by two, the $0p-0h$, $2p-2h$, etc. configurations differ in boson number by two. Here we take the doubly-magic nuclei ^{100}Sn and ^{132}Sn (for ^{116}Cd) as inert cores for the bosons. Also we take into account up to the $2p-2h$ intruder excitation of protons across the $Z = 50$ shell closure. Accordingly, $N_\pi = 1$ and 3 for the normal (i.e., $0p-0h$) and intruder ($2p-2h$) configurations, respectively, while $5 \leq N_\nu \leq 8$. The configuration mixing IBM-2 Hamiltonian is then written as:

$$\hat{H} = \hat{\mathcal{P}}_1 \hat{H}_1 \hat{\mathcal{P}}_1 + \hat{\mathcal{P}}_3 (\hat{H}_3 + \Delta) \hat{\mathcal{P}}_3 + \hat{H}_{\text{mix}}, \quad (2)$$

where $\hat{H}_1$ ($\hat{H}_3$) and $\hat{\mathcal{P}}_1$ ($\hat{\mathcal{P}}_3$) are the Hamiltonian of and the projection operator onto the normal and intruder configuration spaces, respectively. Δ stands for the energy needed to promote a proton boson across the shell closure. $\hat{H}_{\text{mix}}$ in the above equation is the term that is allowed to mix the two configurations.

The coherent state for the configuration mixing IBM is defined as the direct sum of the coherent state for each unperturbed configuration [15]. The bosonic PES is obtained as the lower eigenvalue of the 2×2 coherent-state matrix. The parameters for the Hamiltonian for each configuration are determined by associating the Hamiltonian with each mean-field minimum, i.e., $0p-0h$ Hamiltonian for the prolate minimum, and $2p-2h$ one for the oblate local minimum, and the energy offset Δ and the mixing strength in $\hat{H}_{\text{mix}}$ are determined so that the energy difference between the two mean-field minima and the barrier height for these minima, respectively, are reproduced. On the right-hand side of Fig. 3 we show the mapped IBM-2 PES for ^{112}Cd. One notices that the topology of the corresponding SCMF PES in the neighborhood of the minimum is well reproduced by the IBM one.

3.2 Results for the spectroscopic properties

Figure 4 shows the comparison of the predicted and experimental excitation spectra for $^{108-116}$Cd. The predicted energies of the intruder states can be found in those of the third or higher 0^+ states. The SCMF-IBM calculation does predict the approximately right excitation energies and the dependence on neutron number with the lowest 0_3^+ energy at the mid-shell nucleus ^{114}Cd. The description of the spacing between both 0^+ states is overestimated in ^{108,110}Cd, but agrees quite well with experiment in $^{112-116}$Cd. However, phenomenological calculations assign for ^{114}Cd the experimental 0_2^+ state to the intruder configuration. Therefore, the energy difference between intruder and normal states is generally overestimated. The predicted normal states are systematically too deformed as can be seen by the energies of the first 2^+, 4^+ and especially 6^+ states. Experimentally states are observed with more vibrational energies although there are serious problems with the electric quadrupole transitions. Phenomenological calculations identified in $^{110-114}$Cd the 2_3^+ state as the intruder state, while the present calculation yield as main component the much higher lying 2_4^+ state. In contrast to the normal states the spacing between the 0^+ and 2^+ intruder states is somewhat large. The description of the non-yrast states could be improved, e.g., by including higher-order terms in the IBM Hamiltonian.

We have also studied the $B(E2)$ transition rates. The E2 operator in the consistent-Q formalism [1] has been employed, with the effective charges taken from earlier IBM-2 fitting calculations. In view of the absence of fitting the data, there has been good agreement, and changing trends in the $B(E2)$ values have been well described [8]. However, we found a few exceptions, mostly involving the 0_2^+, 0_3^+, 2_2^+ and 2_3^+ states as could be expected from the different nature of these states compared to the phenomenological IBM-2 calculations. As an example, the $B(E2; 0_2^+ \to 2_1^+)$ are ten times underpredicted in ^{112,114}Cd and ten times overpredicted in ^{116}Cd [8]. This confirms that energies of these states, which are the lowest where

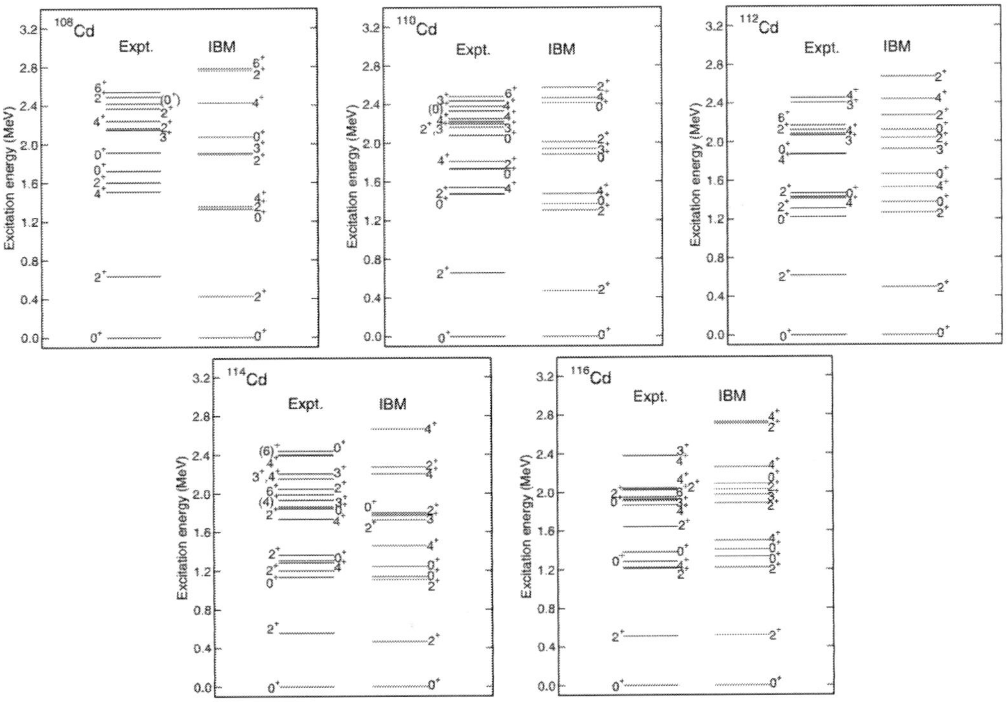

Figure 4. (Color online) Experimental and predicted excitation spectra for the $^{108-116}$Cd nuclei. Based on Ref. [8].

normal and intruder states mix, were not satisfactorily described. Another reason could be that at the SCMF level the prolate minimum, from which the normal states are mainly constructed, was predicted to be too deformed. Besides the $B(E2)$ transitions, an overall reasonable description of the $E0$ transition rates in the Cd nuclei has also been obtained (see, Ref. [8]).

4 Octupole correlations in neutron-rich odd-mass Ba isotopes

Octupole deformation attracts a renowned interest in nuclear structure, and is also a declared objective of the RIB experiments. In addition, nuclei with octupole deformation are expected to show large electric dipole moment, which would imply the CP violation and thus points to physics beyond the standard model of elementary particles. In Ref. [9] we have extended the SCMF-IBM method to investigated the octupole correlations in neutron-rich odd-mass Ba isotopes in the mass $A \approx 140-150$ region,where octupole correlation is expected to be most pronounced. In this case, two additional degrees of freedom should be introduced to the *sd*-IBM system: (i) octupole (f) boson with spin and parity $J = 3^-$; (ii) an unpaired nucleon (fermion) degree of freedom for describing odd-mass system. For the sake of simplicity, we do not distinguish neutron and proton degrees of freedom, and assume axial symmetry.

4.1 Octupole deformation in even-even nuclei

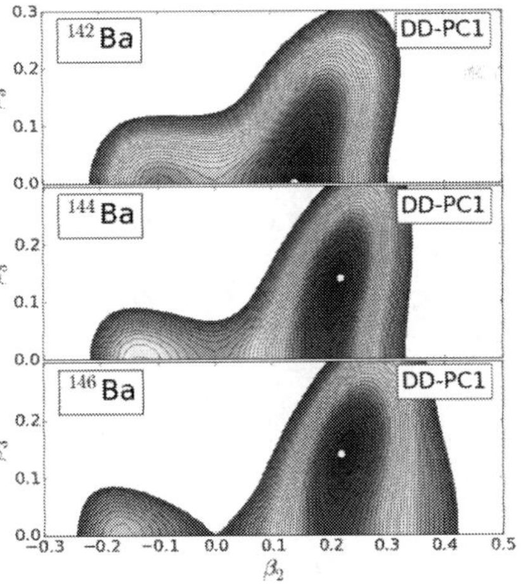

Figure 5. (Color online) The SCMF (β_2,β_3) PESs for ^{142,144,146}Ba, obtained with the DD-PC1 functional and a separable pairing force of finite range. The energy difference between neighbouring contours is 200 keV. Equilibrium minima are identified by open circles. Based on Ref. [9].

Figure 5 depicts the PESs in terms of the axially-symmetric quadrupole β_2 and octupole β_3 deforma-

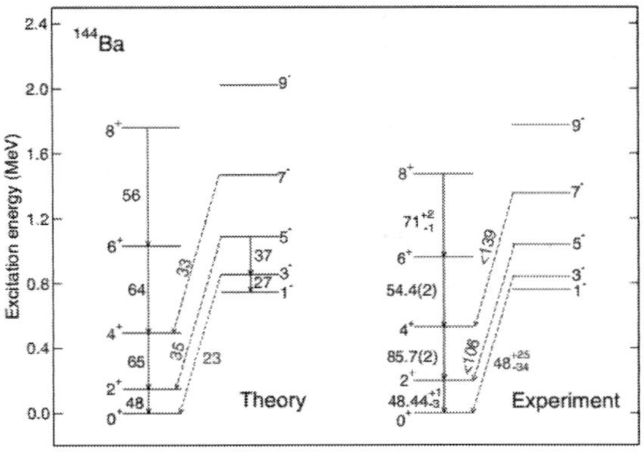

Figure 6. (Color online) Low-energy positive- and negative-parity spectra of the even-even boson core nuclei ^{142,144,146}Ba. The $B(E2)$ (numbers along arrows within each band) and $B(E3)$ (inter-band, dashed arrows) values are given in Weisskopf units. Experimental values are taken from Ref. [16].

tions, computed by the constrained relativistic Hartree-Bogoliubov method [10]. For ^{142}Ba the equilibrium minimum is found on the $\beta_3 = 0$ axis, indicating that it has a weakly-deformed quadrupole shape. In ^{144,146}Ba a minimum with non-zero β_3 deformation ($\beta \approx 0.1$) appears. The minimum is not very pronounced and is rather soft in the β_3 direction, suggesting the occurrence of octupole vibrational states in these nuclei.

The excitation spectra and transition rates for the octupole-deformed nuclei are computed by diagonalizing the IBM Hamiltonian comprising $J = 0^+$ (s), 2^+ (d), and 3^- (f) boson degrees of freedom, which is determined from the SCMF (β_2,β_3) PES. As an example, the low-energy level scheme for the ^{144}Ba isotope is displayed in Fig. 6. In general, the theoretical predictions are in good agreement with the experimental results [16], not only for the excitation energies but also for the $E2$ and $E3$ transition strengths. We note the very low-lying negative-parity band, which conforms to the results that the corresponding SCMF PES in Fig. 5 exhibits a rather pronounced $\beta_3 \neq 0$ minimum, when compared to the one for ^{142}Ba. A rather large value $B(E3; 3^- \to 0^+)$ is predicted for the ^{144}Ba nucleus, but is still considerably smaller than the experimental value [16]. Note, however, the large uncertainty of the latter.

4.2 Particle-boson coupling

In order to compute the spectroscopy of the odd-mass nuclei, we resort to the method developed in Ref. [17], that is based on the SCMF calculation and the particle-core coupling. Here the even-even core is described by the IBM, and the interaction between the odd particle and the even-even boson core is modelled in terms of the interacting boson-fermion model (IBFM) [18]. The Hamiltonian of the IBFM is composed of the IBM Hamiltonian $\hat{H}_{\rm B}$, the Hamiltonian for a single fermion $\hat{H}_{\rm F}$, and the Hamiltonian $\hat{H}_{\rm BF}$ that couples fermion and boson spaces:

$$\hat{H}_{\rm IBFM} = \hat{H}_{\rm B} + \hat{H}_{\rm F} + \hat{H}_{\rm BF}. \tag{3}$$

The single-fermion Hamiltonian $\hat{H}_{\rm F} = \sum_j \epsilon_j(-\sqrt{2j+1})(a_j^\dagger \tilde{a}_j)^{(0)}$, where ϵ_j is the single-particle energy for the orbital j. For the sake of simplicity, we present the form of the Hamiltonian $\hat{H}_{\rm BF}$ within sd-IBFM. Based on the generalized seniority scheme, the following three terms of $\hat{H}_{\rm BF}$ have been shown to be most essential to the low-lying spectra in odd-mass nuclei [18]:

$$\Gamma_0 \sum_{jj'} \gamma_{jj'} \hat{Q}_B \cdot (a_j^\dagger \tilde{a}_{j'})^{(2)} \tag{4}$$

$$-2\Lambda_0 \sum_{jj'j''} \sqrt{\frac{5}{2j''+1}} \beta_{jj''}\beta_{j'j''} : ((d^\dagger \tilde{a}_j)^{(j'')} (a_{j'}^\dagger \tilde{d})^{(j'')})^{(0)} : \tag{5}$$

$$-A_0 \sum_j \sqrt{2j+1} (a_j^\dagger \tilde{a}_j)^{(0)} \hat{n}_d. \tag{6}$$

The first, second and third terms are referred to as the dynamical quadrupole, exchange, and monopole interactions, respectively. The quantities $\gamma_{jj'} = (u_j u_{j'} - v_j v_{j'}) Q_{jj'}$ and $\beta_{jj'} = (u_j v_{j'} + v_j u_{j'}) Q_{jj'}$, are given in terms of the occupation probabilities u_j and v_j for the orbital j (satisfying $u_j^2 + v_j^2 = 1$) and the matrix element of the quadrupole operator in the single-particle basis $Q_{jj'} = \langle j || Y^{(2)} || j' \rangle$. Γ_0, Λ_0, and A_0 denote strength parameters. The single-particle energies ϵ_j and occupation numbers v_j^2 are computed fully microscopically by the SCMF calculation constrained to zero deformation [17]. The three strength parameters of $\hat{H}_{\rm BF}$, i.e., Γ_0, Λ_0, and A_0, are considered free parameters, and are fitted to reasonably reproduce experimental low-lying levels in a given nucleus. For the sdf-IBFM system, we have derived similar formulas to those in Eqs. (4)–(6) from the generalized seniority scheme. More details can be found in Ref. [9].

4.3 Excitation spectra for the odd-mass nucleus

Experimental information is available for the isotope ^{145}Ba, that is enough to compare with the theoretical prediction. Since the corresponding even-even boson core nucleus ^{144}Ba exhibits an octupole-soft potential at the SCMF level (cf. Fig. 5), we also expect that octupole correlations play an important role in the low-energy spectra of this odd-mass nucleus. In Fig. 7 the calculated excitation spectrum for ^{145}Ba is compared to the experimental bands available in [19]. The bands built on the $3/2_1^-$, $5/2_2^-$, and $5/2_1^+$ states are predicted, and it would be interesting if they can be confirmed by experiments. The lowest two $\pi = -1$ bands in ^{145}Ba are built on the $5/2_1^-$ and $7/2_1^-$ states, which are characterized by the coupling of the unpaired neutron in the $1h_{9/2}$ single-particle orbital to the sd boson space. The lowest $\pi = +1$ state $9/2_1^+$ is described by the coupling of the $1i_{13/2}$ orbital to sd-boson states.

The theoretical negative-parity band built on the $15/2_4^-$ state (calculated at 1167 keV) is dominated by the coupling of the $1i_{13/2}$ single-particle orbital to states with one f-boson. Theoretically the $15/2_4^-$ state appears to

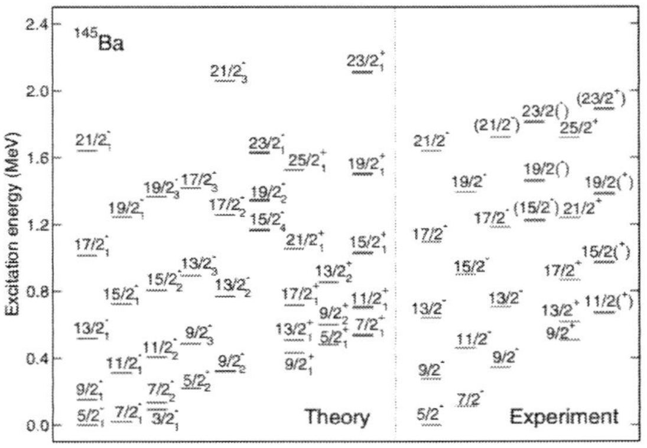

Figure 7. (Color online) Theoretical and experimental [19] energy spectra for the nucleus ^{145}Ba. Those theoretical levels that are suggested to contain one *f* boson, as well as the experimentally suggested octupole bands, are marked as thick lines.

be an octupole state and is located close to the experimental $15/2^-$ state found at 1226 keV [19]. Moreover, rather strong E3 transitions from the $J^\pi = 15/2^-_4$ band to the corresponding low-lying positive-parity bands are predicted: for instance $B(E3; 15/2^-_4 \to 9/2^+_1) = 25$ and $B(E3; 19/2^-_2 \to 13/2^+_1) = 31$ (in W.u), comparable to the $B(E3; 3^-_1 \to 0^+_1)$ value of 23 W.u. in the even-even core nucleus ^{144}Ba. To verify model predictions experimental information on the $B(E2)$ and $B(E3)$ values is needed.

5 Summary

I have presented a method of deriving the IBM Hamiltonian based on the SCMF framework. The constrained SCMF calculation with a given EDF is performed to obtain the PES in terms of the relevant collective coordinates. The SCMF PES is then mapped onto the expectation value of the IBM Hamiltonian in the boson coherent state. This procedure completely determines the strength parameters of the IBM, and the diagonalization of the mapped IBM Hamiltonian provides excitation spectra and transition probabilities. Two recent applications of the method are highlighted, that involve the configuration mixing of normal and intruder states, the octupole deformation, and the odd nucleon degree of freedom. As the EDF framework, on which the mapping procedure is built, provides a global and reliable mean-field description of intrinsic properties of nuclei, it has been made possible to determine the IBM Hamiltonian for arbitrary nuclei, starting only from the nucleonic degrees of freedom. This points to a crucial step toward the microscopic foundation of the IBM within a unified theoretical framework. In addition, both theoretical approaches, i.e., IBM and DFT, can be complementary to each other, in such a way that the latter gives microscopic input with the former, and the former facilitates spectroscopic calculations, which are, from both computational and methodological points of view, highly demanding with the latter approach. On the practical side, contrary to the conventional IBM studies, using the present method we have gained the capability of predicting the spectroscopic properties of nuclei in an accurate, systematic, and mathematically simple way. This opens up a new possibility to study heavy exotic nuclei that are now under rigorous investigations at RIB facilities around the world.

The results reported here are based on the works with J. Jolie, T. Nikšić, T. Otsuka, N. Shimizu, and D. Vretenar. This work is financed within the Tenure Track Pilot Programme of the Croatian Science Foundation and the École Polytechnique Fédérale de Lausanne and the Project TTP-2018-07-3554 Exotic Nuclear Structure and Dynamics, with funds of the Croatian-Swiss Research Programme.

References

[1] F. Iachello, and A. Arima, *The interacting boson model* (Cambridge University Press, Cambridge, 1987)
[2] T. Otsuka, A. Arima, and F. Iachello, Nucl. Phys. A **309**, 1 (1978)
[3] K. Nomura, N. Shimizu, and T. Otsuka, Phys. Rev. Lett. **101**, 142501 (2008)
[4] K. Nomura et al., Phys. Rev. C **83**, 041302 (2011)
[5] K. Nomura, N. Shimizu, D. Vretenar, T. Nikšić, and T. Otsuka, Phys. Rev. Lett. **108**, 132501 (2012)
[6] K. Nomura et al., Phys. Rev. C **89**, 024312 (2014)
[7] K. Nomura et al., J. Phys. G **43**, 024008 (2016)
[8] K. Nomura, and J. Jolie, Phys. Rev. C **98**, 024303 (2018)
[9] K. Nomura, T. Nikšić, and D. Vretenar, Phys. Rev. C **97**, 024317 (2018).
[10] D. Vretenar et al., Phys. Rep. **409**, 101 (2005)
[11] M. Bender et al., Rev. Mod. Phys. **75**, 121 (2003)
[12] A. Bohr, and B. R. Mottelson, *Nuclear Structure* (Benjamin, N.Y., 1975)
[13] J. N. Ginocchio and M. W. Kirson, Nucl. Phys. A **350**, 31 (1980).
[14] P. D. Duval and B. R. Barrett, Phys. Lett. B **100**, 223 (1981).
[15] A. Frank et al., Phys. Rev. C **69**, 034323 (2004).
[16] B. Bucher et al., Phys. Rev. Lett. **116**, 112503 (2016)
[17] K. Nomura et al., Phys. Rev. C **93**, 054305 (2016).
[18] F. Iachello, and P. Van Isacker, *The Interacting Boson-Fermion Model* (Cambridge University Press, Cambridge, England, 1991).
[19] T. Rzaca-Urban et al., Phys. Rev. C **86**, 044324 (2012).

Hidden symmetries in the HFB norm overlap functions

Makito Oi[1,*], *Takahiro* Mizusaki[1], *Noritaka* Shimizu[2], and *Yang* Sun[3,4]

[1]Institute of Natural Sciences, Senshu University, 3-8 Kanda-Jinbocho, Chiyoda-ku, Tokyo 101-8425, Japan
[2]Center for Nuclear Study, The University of Tokyo, 7-3-1 Hongo, Bunkyo-ku, Tokyo 113-0033, Japan,
[3]School of Physics and Astronomy, Shanghai Jiao Tong University, Shanghai 200240, China,
[4]Institute of Modern Physics, Chinese Academy of Sciences, Lanzhou 730000, China

Abstract. A brief consideration about hidden symmetries in the HFB norm overlap functions is presented, in particular, in association with the presence of pair-wise degeneracy of the eigenvalue spectrum of a product of two antisymmetric matrices.

1 Introduction

The mean field approximation, such as the Hartree-Fock-Bogoliubov (HFB) method, has been successful to describe various nuclear properties such as the moment of inertia of the ground-state rotational band. There are types of physics, however, demanding higher order correlations beyond the mean field approximation. Collective phenomena, such as surface vibration and pairing vibration, are such. The generator coordinate method (GCM) is one of the methods that can take such dynamical collective degrees of freedom into account. However, there had been a problem in implementing the GCM to numerical calculations, which is the sign problem in obtaining the norm overlap function for the GCM equation.

In 1983, Neergård and Wüst proposed a method [1] how to calculate the HFB norm overlap functions without the sign ambiguity, which damages the Onishi formula [2], the first exact and compact formula to enable the calculation of the HFB overlap. This method exploits the pairwise degeneracy of the norm overlap function. For some technical reasons, however, the Neergård-Wüst method was not widely employed by many authors [3]. The primary reason is that their method requires the diagonalisation of a matrix without any symmetries and special properties, such as hermiticity and antisymmetry. Many feared that this lack of speciality may result in serious numerical errors in the evaluation of the HFB overlap.

But we recently found that this "general matrix" possess a hidden symmetry, which can allow numerical calculation with high precision for the pairwise degeneracy obtained from a general matrix.

*e-mail: m.oi@isc.senshu-u.ac.jp

2 The Onishi formula and the Neergård-Wüst method

The Onishi formula reads

$$\langle\Psi_0|\Psi_1\rangle = \sqrt{\det(I - Z_0^* Z_1)}, \qquad (1)$$

where the n-by-n identity matrix is denoted as I and the HFB state Ψ_m is expressed as

$$|\Psi_\alpha\rangle = \exp\left(\frac{1}{2}\sum_{i,j}^{n}(Z_\alpha)_{ij}c_i^\dagger c_j^\dagger\right)|0\rangle. \qquad (2)$$

This form of the HFB state is also referred to as the Thouless ansatz. Here, the single-particle creation and annihilation operators for fermion are $(c, c^\dagger)$ and its model-space dimension is n, which is an even integer for even-even nuclear systems. The variational parameter Z_α is an n-by-n antisymmetric matrix, $Z_\alpha^T = -Z_\alpha$.

In general, the Onishi formula has a form of the square root of a complex function, so that it has the sign ambiguity. Neergård and Wüst believed that the HFB norm overlap function should not possess such a complicated mathematical structure because the Thouless ansatz is essentially an exponential. Based on this consideration, they believed that the matrix $Z_0^* Z_1$, which does not hold any special properties and symmetries, must give rise to pair-wise eigenvalues. In other words, the pair-wise degeneracy can be a solution to avoid the mathematical complexity, but they did not demonstrate that the other ways are excluded theoretically. Only a numerical demonstration was given to support their claim. In accordance with their approach, once the diagonalisation is done, the norm overlap function should be expressed as

$$\langle\Psi_0|\Psi_1\rangle = \prod_{i=1}^{n/2}(1 - \lambda_i), \qquad (3)$$

where λ_i is one of the paired eigenvalues. Due to absences of any symmetries and special properties, λ_i is a complex number.

There were many who doubted that such a general matrix can produce the well-structured pairs of eigenvalues, but our recent numerical investigation showed that such a structure is well maintained for a practical HFB calculations with the pair-plus-quadrupole model. We then decided to analyse why this pair-wise eigenvalues are guaranteed for such a general matrix $Z_0^* Z_1$.

3 Demonstration of the pair-wise degeneracy

Although the matrix $Z_0^* Z_1$ does not hold any symmetries and special properties, it is obvious that it is a product of two antisymmetric matrices Z_0^* and Z_1. It is natural to suspect that the pair-wise degeneracy is a consequence of this "hidden" symmetric properties of $Z_0^* Z_1$. In other words, the necessity of the pair-wise degeneracy is naturally derived from the "hidden symmetries" that I noticed in the present work, which will be demonstrated in the following.

For the sake of simplicity, we will employ simpler expressions in the consideration of the eigenvalues of a product of two antisymmetric matrices. Such antisymmetric matrices are denoted as A and B hereafter, so that the eigenvalue equation of our interest becomes

$$AB\boldsymbol{v}^i = \lambda^i \boldsymbol{v}^i, \tag{4}$$

and we want to demonstrate that the eigenvalue spectrum has pair-wise degeneracies. It can be justified that the rank of the matrix AB is n, because $\det(AB) = \det A \cdot \det B \neq 0$.

First of all, a vector $\boldsymbol{w}^i = B\boldsymbol{v}^i$ is introduced. The transpose of this new vector corresponds to the left eigenvector of AB,

$$(\boldsymbol{w}^i)^T AB = \lambda^i (\boldsymbol{w}^i)^T, \tag{5}$$

which can be easily verified by multiplying eq.(4) with B from left. Operating A to $\boldsymbol{w}^i$ from left gives rise to the equivalent expression to the original eigenvalue equation, eq.(4),

$$A\boldsymbol{w}^i = \lambda^i \boldsymbol{v}^i. \tag{6}$$

Here, we introduce collective expressions for eigenvalues λ^i and the eigenvectors $\boldsymbol{v}^i$ and $\boldsymbol{w}^i$ in the form of n-by-n matrices,

$$\Lambda = \mathrm{diag}(\lambda^1, \cdots, \lambda^n), \tag{7}$$

$$V = (\boldsymbol{v}^1 \ \boldsymbol{v}^2 \ \cdots \ \boldsymbol{v}^n), \tag{8}$$

$$W = (\boldsymbol{w}^1 \ \boldsymbol{w}^2 \ \cdots \ \boldsymbol{w}^n). \tag{9}$$

(10)

Then, the above expressions are rewritten

$$W = BV, \tag{11}$$

$$AW = V\Lambda. \tag{12}$$

From the first equation in the above, that is, eq.(11), we can derive a antisymmetric relation

$$V^T W = -(V^T W)^T, \tag{13}$$

by demanding B to be antisymmetric.

From the second equation in the above, that is, eq.(12), we can also derive a commutation relation

$$\left[\Lambda, W^T V\right] = 0, \tag{14}$$

by demanding A to be antisymmetric. This commutation relation can be equivalently expressed as

$$(\lambda_i - \lambda_j)(W^T V)_{ij} = 0, \tag{15}$$

Let us assume that there is no degeneracy in the eigenvalue spectrum. Then, we have $W^T V = 0$, which implies that $B = 0$. which is contradictory from the original condition that B is antisymmetric. Therefore, the eigenvalue spectrum of AB must be degenerate.

Let us now suppose that the eigenvalues of $i = 1$ and of $i = 2$ are pair-wise degenerate, $\lambda^1 = \lambda^2$. Then, $(W^T V)_{12} = -(W^T V)_{21} \neq 0$, and the antisymmetric matrices B is expressed as

$$B_{pq} = -(W^T V)_{12} \begin{vmatrix} (V^{-1})_{1p} & (V^{-1})_{1q} \\ (V^{-1})_{2p} & (V^{-1})_{2q} \end{vmatrix}. \tag{16}$$

The eigenvalue equation now reads

$$(ABV)_{ij} = (W^T V)_{12} \left\{-(V^{-1}A)_{2i}\delta_{1j} + (V^{-1}A)_{1i}\delta_{2j}\right\}. \tag{17}$$

This means that if $j \neq 1, 2$, we have $AB\boldsymbol{v}^j = 0$, which contradicts the fact that AB is a full-rank matrix.

Acknowledgement: This research is supported by a Senshu-University grant (Kenkyu Josei) for the study of "Ryoshi tatai-kei ni okeru tsui-sokan oyobi koji-sokan no suri-teki katsu suchi-teki toriatukai ni tsuite no kenkyu" (H.31), which is appreciated by one of the authors (MO).

References

[1] K Neergård and E. Wüst, Nucl. Phys. A **402**, (1983).
[2] N. Onishi and S. Yoshida, Nucl. Phys. **80**, (1966).
[3] for instance, L. Robledo, Phys. Rev. C**79**, (2009).

Lifetime measurements in ^{105}Sn: the puzzle of B(E2) strengths in Sn isotopes

Giorgia Pasqualato[1,2,*], *Andrea* Gottardo[3], *Alain* Goasduff[1,2], *Josè Javier* Valiente Dobón[3], and *Daniele* Mengoni[1,2]

[1]INFN, Sezione di Padova, Padova, Italy
[2]Università di Padova, Padova, Italy
[3]INFN, Laboratori Nazionali di Legnaro, Legnaro (Padova), Italy

Abstract. One of the most studied case to explore the evolution of shell closures far from stability is the doubly-magic and self-conjugated ^{100}Sn nucleus. Information on its doubly-magic nature can be extracted from the systematic study of the tin isotopic chain. In this context, the lifetimes of the neutron deficient ^{105}Sn have been investigated with the coincidence Recoil Distance Doppler Shift (RDDS) technique through the reaction ^{50}Cr(^{58}Ni,2pn)^{105}Sn. Preliminary results concerning the lifetimes of ^{105}In excited states are in good agreement with the adopted lifetimes, demonstrating the feasibility to extract lifetimes with this experimental technique.

1 Introduction

The study of the nuclear structure of atomic nuclei with a number of protons and neutrons close to magic numbers is an effective mean to test different nucleon-nucleon interaction sets used as an input for the shell model codes. A prominent nucleus in this context is the doubly-magic and self-conjugated ^{100}Sn. Decay-spectroscopy provided indirect indication that the Z=50 shell closure is present for this system [1]. To test the robustness of the proton shell closure approaching N=50 the spectroscopy of the Sn isotopic chain therefore becomes useful [2–4].

The generalized seniority scheme predicts a parabolic behaviour of the B(E2; $2^+ \to 0^+$) values for Sn isotopes. Nevertheless, the experimental values for the B(E2; $2^+ \to 0^+$) in neutron-deficient isotopes do not follow a parabola, at least up to ^{114}Sn. The study of the reduced transition probabilities via fusion-evaporation reaction is hampered by the presence of low-lying isomers in even-even Sn isotopes. Therefore, B(E2; $2^+ \to 0^+$) values have been extracted from measurements of Coulomb excitation, yet the results suffer from large uncertainties. Recently, a direct measurement of the lifetime of the 2_1^+ and the 4_1^+ excited states in ^{106,108}Sn was performed via a multi-nucleon transfer reaction for the very first time [5].

Shell-model calculations with the CD-Bonn interaction in the full N=50–N=82 valence space show that the quadrupole strength of core-coupled states in odd-even nuclei closely follow the general trend of the corresponding even-even isotopes for the 2^+ states. In this regard, the neutron-deficient odd-even ^{105}Sn isotope provides the opportunity for direct lifetime measurements of low-lying levels in odd Sn isotopes close to ^{100}Sn.

Another indication of the decreasing of collectivity going towards ^{100}Sn comes from the study of the M1 transition strength for the first excited states in odd-even tin isotopes. Shell model calculations predict a dominant M1 character for this transition, which is highly suppressed in case of a single-particle nature of this state, being an *l*-forbidden M1. This leads to a lifetime of several ns for this state, which is in a great disagreement with the ^{105}Sn measured value of only several hundreds ps. Actually, this value has been measured only once and it was never published but only presented in a short annual report [11]. A precise re-measurement of the lifetime of the $7/2^+$ first excited state in ^{105}Sn is needed. Moreover, it will possibly shed light on the controversy regarding the spin-parity assignments of the ground state and first excited state, $5/2^+$ and/or $7/2^+$, for odd Sn isotopes in this region [12], [13].

Since lifetimes of the order of ps are expected, the coincidence Recoil Distance Doppler Shift (RDDS) technique can be applied [6]. This method is based on the distinction, for a detected γ-ray transition, of a Doppler shifted component, emitted by nuclei in-flight, and a stopped component, emitted by nuclei at rest.

2 Experimental Details

The ^{105}Sn nucleus was populated via a fusion evaporation reaction using a 180 MeV ^{50}Cr beam and a ^{58}Ni target of 1 mg/cm^2 thickness. Based on a previous spectroscopy investigation of this nucleus [7], such a reaction was chosen to optimize the production cross section of the channel of interest. The 16 mg/cm^2 ^{197}Au stopper was mounted in the plunger device behind the target. Plunger was placed at the center of the reaction chamber.

The experiment was performed at Legnaro National Laboratories using the GALIELO array [8] coupled to the EUCLIDES detector [9] and the GALILEO plunger device [10]. GALILEO is a γ-ray spectrometer composed of

*e-mail: giorgia.pasqualato@pd.infn.it

25 Compton-suppressed HPGe detectors. The position of the GALILEO detectors at different angles with respect to the beam direction (152°, 129°, 119°, 61° and 51°), allows one to estimate excited-state lifetimes for each ring, as the energy shift of the γ ray emitted in-flight depends on the direction of emission. In this way, an average lifetime can be extracted, leading to a more precise measurement. The ΔE-E silicon array EUCLIDES was used to detect evaporated light charged particles and select the reaction channel of interest. The array was placed only at forward angles in the plunger configuration. A total of 12 target-stopper distances in the range between 10 μm and 8000 μm were chosen to be able to measure the very short lifetimes of the high-spin magnetic band of ^{105}Sn as well as its long-lived $7/2^+$ first excited state.

3 Analysis and Preliminary Results

The selection of the channel of interest between all the possible fusion-evaporation and the Coulomb-excitation events is possible requesting γ coincidences with light charged particles. As an example, Figure 1 shows how the Coulomb-excitation events, due to the interaction between the beam and the stopper, can be significantly reduced in the spectrum requesting a 2-protons coincidence. A further selection is possible by gating on a $2p$-coincidence γ-γ matrix on the 200-keV, $7/2^+ \rightarrow 5/2^+$, ground state transition of ^{105}Sn. The resulting spectrum is shown in Figure 2 and presents the identification of many ^{105}Sn transitions among a few peaks owing to the main contaminants: ^{105}In ($3p$), ^{104}Cd ($4p$) and ^{102}Cd ($1\alpha2p$).

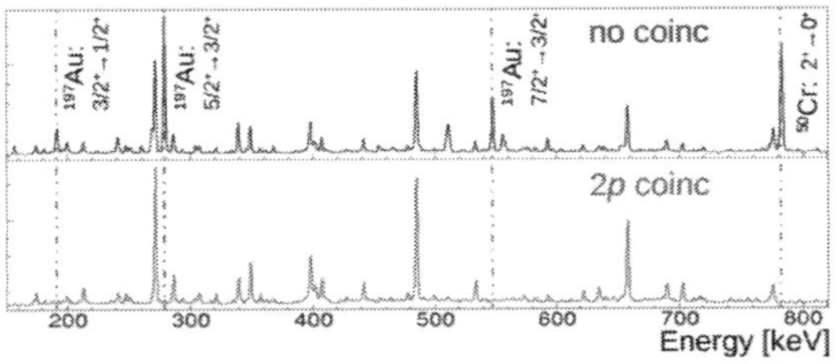

Figure 1. Comparison between the spectra in coincidence with two protons (bottom) and the total γ spectrum (top) detected at 152° with a target-stopper distance of 10 μm. The suppression of the Coulomb excitation events between the ^{50}Cr beam and the ^{197}Au stopper are shown.

The lifetime of an excited state can be measured with the RDDS technique. The nucleus populated in the target from the reaction flies with a velocity v until the stopper. The energy of the γ ray emitted in-flight is Doppler shifted depending on the velocity of the nucleus and on the angle of the γ-ray emission. Depending on the distance between the target and the stopper, a different ratio between the intensities of the in-flight peak and the stopped peak is observed.

In this initial state of the analysis some of the excited states of ^{105}In, which are known in the literature, are being studied, to validate the adopted experimental technique. The analyzed spectra are obtained by gating on the in-flight component of the direct feeding transition, in a $2p$-coincidence γ-γ matrix. The lifetime of the excited states of interest are estimated with the Differential Decay-Curve Method (DDCM) [6]. Preliminary results show a good agreement of our measurements with the adopted ^{105}In lifetimes.

Data analysis is now oriented to the measurements of the lifetimes of ^{105}Sn excited states. With the collected statistics it will be possible to precisely measure the lifetimes of the long-lived $7/2^+$ and $25/2^+$ excited states, the last corresponding to a possible core-breaking nature. A careful analysis is needed to extract the lifetimes of other excited states in ^{105}Sn, which are characterized by shorter lifetimes.

Thanks to the GALILEO layout in 5 different rings it will be also possible to extract the mixing ratio of the $7/2^+ \rightarrow gs$ transition in ^{105}Sn by the ratios of angular distributions from oriented states.

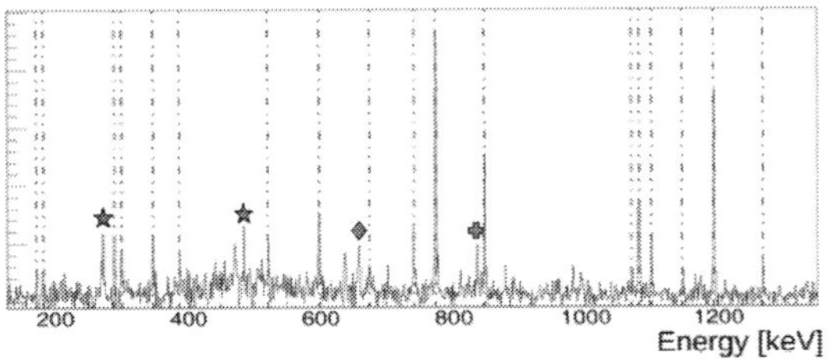

Figure 2. From γ-ray detected at 152° with a target-stopper distance of 10 μm, projection from a $2p$-coincidence γ-γ matrix on the 200-keV energy ground state transition of ^{105}Sn. The peaks corresponding to the main ^{105}Sn transitions are marked with vertical dashed lines. The peaks corresponding to the main contaminants are also pointed out: ★ ^{105}In, ◆ ^{102}Cd, ✚ ^{104}Cd.

References

[1] C. Hinke et al., Nature 486, 341345 (2012).
[2] V.M. Bader et al., Phys. Rev. C 88, 051301 (2013).
[3] G. Guastalla et al., Phys. Rev. Lett. 110, 172501 (2013).
[4] Doornenbal et al., Phys. Rev. Lett. 90, 061302 (2014).
[5] M. Siciliano et al., Phys. Rev. Lett. (2019), submitted.
[6] A. Dewald et al., Prog. Part. Nucl. Phys. 679, 786 (2012).
[7] A. Gadea et al., Phys. Rev. C 55, R1 (1997).
[8] J.J. Valiente-Dobón et al., LNL-INFN Ann.Rep. (2014).
[9] D. Testov et al., Eur. Phys. J.A, 55 (2019).
[10] C. Müller-Gatermann et al., Nucl. Inst. and Met. A920, 95 (2019).
[11] T. Ishii et al., Japan Atomic Energy Res.Inst.Tandem VDG Ann.Rep., p.11, (1993).
[12] I.G. Darby et al., Phys. Rev. Lett. 105, 162502 (2010).
[13] D. Seweryniak et al., Phys. Rev. Lett. 99, 022504 (2007).

Thick-target inverse kinematic method in order to investigate alpha-clustering in ^{212}Po

Maria Grazia Pellegriti[1,*], *Agatino* Musumarra[1,2], *Enrico* De Filippo[1], *Marzio* De Napoli[1], *Alessia* Di Pietro[3], *Pierpaolo* Figuera[3], *Maria* Fisichella[3], *Concettina* Maiolino[3], *Danilo* Rifuggiato[3], *Domenico* Santonocito[3], *Valentina* Scuderi[3], *Emanuele* Strano[4,5], *Domenico* Torresi[3]

[1]Istituto Nazionale di Fisica Nucleare – Sezione di Catania, via Santa Sofia n. 64, 95123 Catania, Italy
[2]Dipartimento di Fisica ed Astronomia, Università di Catania, via Santa Sofia n.64, 95123 Catania, Italy
[3]Istituto Nazionale di Fisica Nucleare – Laboratori Nazionali del Sud, via Santa Sofia n. 62, 95123 Catania, Italy
[4]Dipartimento di Fisica, Università di Padova, via F. Marzolo 8, I-35131 Padova, Italy
[5]Istituto Nazionale di Fisica Nucleare – Sezione di Padova, via F. Marzolo 8, I-35131 Padova, Italy

Abstract. The inverse-kinematic thick-target method has been used in order to investigate ^{212}Po alpha-structure by the elastic scattering of ^{208}Pb on ^{4}He target. A ^{208}Pb beam, accelerated by the Superconducting Cyclotron (CS) of Laboratori Nazionali del Sud - INFN, at the incident energy of 10.1 A MeV was impinging onto a specifically designed ^{4}He gas cell, two meter long. The gas cell was acting both as target and as beam degrader, stopping the beam before reaching the alpha-particle detection system placed at 0° with respect to the beam axis. In order to disentangle the elastic contribution from other reaction channels (e.g. inelastic scattering) a microchannel plate was used to measure the Time of Flight (ToF) of both the ^{208}Pb beam particles and the ejectiles along the gas cell. The ^{208}Pb stopping power in the ^{4}He gas target was also measured, as a key ingredient in order to establish the interaction point inside the gas cell, in turn determining the solid angle covered by the detector. In the following, the experimental technique will be described, and the results of a preliminary data analysis will be shown.

1 Introduction

The α-particle is considered as one of the basic building blocks of nuclei because of its large binding energy [1]. Opposite to this picture is the single particle description of nuclei, based on the hypothesis of a common mean field (shell model). A strong revival of the α-cluster model occurred in the 1960s when both experimental and theoretical studies revealed that the concept of α-clustering is essential for understanding the structure of light nuclei. Nuclear states based on α clusters have been mainly observed as excited states, close to the decay thresholds into the constituent clusters [1].
The persistence of clustering in heavy-nuclei is much less documented. The interesting question is whether the alpha cluster structure can survive in a mean field approach. It is traditionally believed that the mean field dominated in heavy nuclei. For medium nuclei, it was found in [2,3] that there exist the inversion doublet bands in the fp-shell ^{40}Ca and ^{44}Ti, thus supporting the hypothesis they could have an α+core cluster structure, like the well-known structure of the light ^{16}O and ^{20}Ne nuclei. Furthermore, a heavy closed-shell core plus alpha can also be described in terms of a cluster structure. In this framework, a possible candidate is the ^{208}Pb+α structure characterizing ^{212}Po [4], a nucleus exhibiting two protons and two neutrons outside the doubly magic ^{208}Pb core. The enhancement of the α-decay width in ^{212}Po was studied in [5] and it was found that the larger configuration mixing plays a very important role in the explanation of the width enhancement.
New interest, from both experimental [6] and theoretical point of view [7-9], raised around ^{212}Po and recently a review on alpha-clustering effects in heavy nuclei has been published [10].
In the work of Astier et al. [6], the discovery of low-excited negative parity states in ^{212}Po is claimed in the E_x region below 3 MeV, with evidence of very enhanced E_1 transitions. According to [6], these states can e considered as fingerprints of the α-^{208}Pb structure, a signature of the α-core oscillatory motion. Later Suzuki et al. [7] has claimed that the experimental evidence of [6] has to be attributed to the strong contribution of an α-^{208}Pb (3^-, 2.615MeV) cluster state.
In order to further investigate ^{212}Po α-structure, the elastic scattering ^{4}He(^{208}Pb, α)^{208}Pb has been measured at Laboratori Nazionali del Sud - INFN by using the thick-target inverse kinematics (TTIK) resonant elastic scattering method in the E_{cm} energy range below 30 MeV.

* Corresponding author: mariagrazia.pellegriti@ct.infn.it

2 The method

The TTIK resonant elastic scattering method [11-14] makes use of the sensitivity of the detected recoil yield to the presence of a resonant state in the compound nucleus. By using the beam energy loss in the target, it is possible to continuously induce the elastic scattering in a given range of E_{CM} energies. The inverse kinematics scattering (projectile heavier than the target nucleus), produces forward focused recoils (typically protons or α particles) with much smaller energy loss in the target than the beam particles. The detected recoil yield shows interference effects due to the presence of both nuclear and Coulomb contribution in the elastic cross section.

The choice of a proper target thickness is essential for implementing this method and it will provide in one step a general overview of the level scheme in a wide centre-of-mass energy range.

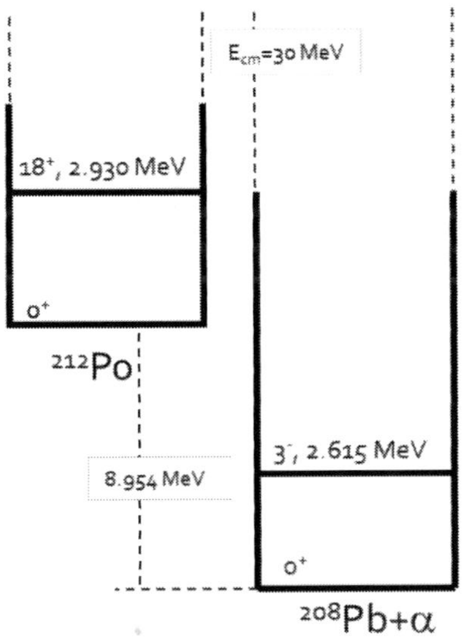

Fig. 1. ^{208}Pb and ^{212}Po level scheme.

The experimental limitation of the thick target method is represented by the contamination, coming from other possible reaction mechanisms (like inelastic scattering), of the final elastic scattering excitation function.

This means that recoil particles reaching the detectors at the same energy can be generated by different reaction mechanisms. This implies different paths inside the target for different classes of events. In order to experimentally disentangle the elastic cross section excitation function, an appropriate target thickness is provided [13] or, as a further option for gaseous target, ToF measurement is performed [14].

As already stated, we have investigated the ^{212}Po α-^{208}Pb structure around the Coulomb barrier by using the TTIK; Fig. 1 reports a simplified scheme of the involved levels configuration in ^{212}Po and ^{208}Pb nuclei and the investigated E_{cm} energy range.

As it has been demonstrated in [14], the knowledge of target stopping power for the traversing beam is crucial in performing elastic scattering experiments by using the thick target method.

For this reason, a measurement of ^{4}He stopping power for a ^{208}Pb beam in the energy range spanning from 1.588 GeV down to zero has been performed.

3 Set-up

The ^{208}Pb beam (31^{+} charge state), extracted from the SERSE ECR source, has been accelerated by the CS at the beam energy of 10.1 A·MeV. ^{208}Pb particles were impinging on a Micro Channel Plate (MCP) detector. The MCP foil has been placed at 45° with respect to the beam direction. The MCP signal provides a reference for time-of-flight measurements and, at the same time, allows beam particle counting.

The beam was then impinging on the ^{4}He gas target (see Fig. 2 and Fig. 3) separated by a beam-line by an Al window, 20 μm thick (see Fig. 4). The gas target length, from the Al window to the detector, placed at the end of the gas target (see Fig. 2), was 1834±1 mm. The ^{4}He gas pressure inside the beam pipe was measured by using an absolute pressure gauge (MKS Baratron 122A). The temperature was also measured by using a Pt100 probe.

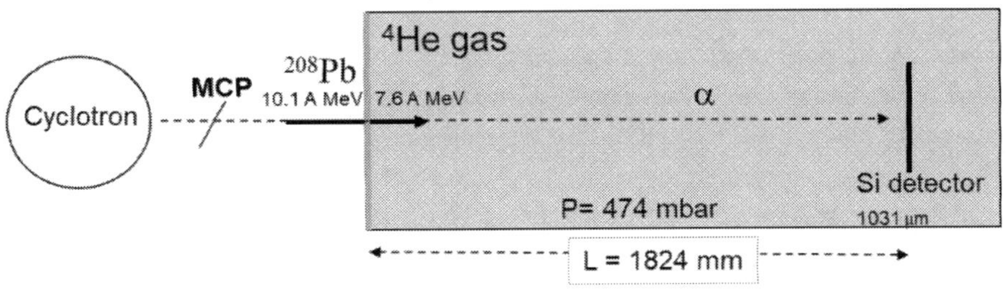

Fig. 2. Experimental set-up.

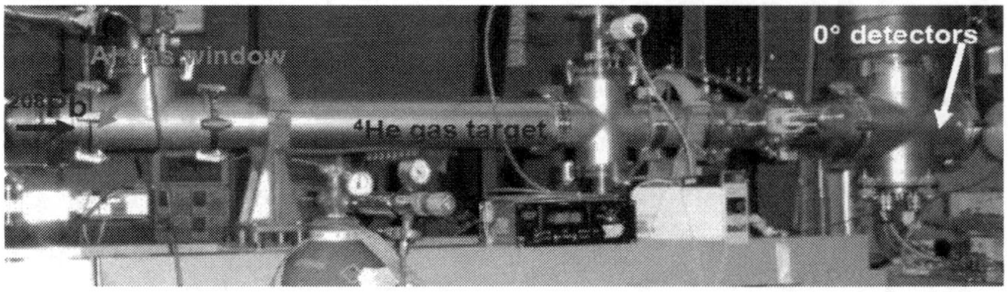

Fig. 3. Picture of ^{4}He gas chamber.

During the experiment two different detector set-ups were used. In the following we will refer to the result obtained by the set-up based on an ORTEC silicon surface-barrier detector, 1031 μm thick, placed at zero degree detecting the recoiling alpha-particles.

Fig. 4. Al gas window.

The detector was collimated (ϕ = 6 mm) and used as start detector (trigger) for ToF measurement, stop was provided by the MCP detecting the primary ^{208}Pb beam particles. Overall ToF time resolution has been measured to be ~ 1 ns.

Stopping power measurements of ^{208}Pb in ^{4}He have also been performed. As it has been underlined by Zadro et al. [14], this is essential in order to recover the beam energy for a given scattering event.

4 Preliminary data analysis

The detector energy calibration was performed by using an ^{241}Am-^{244}Cu-^{239}Pu alpha-source, the energy corresponding to the alpha particles punching through the silicon detector and the ADC pedestal. The ToF was calibrated by using different delay, namely: 8, 16 and 24 ns.

ToF versus detected energy 2D-spectrum is plotted in Fig. 5. Elastic scattered alpha particles appear as the main contribution.

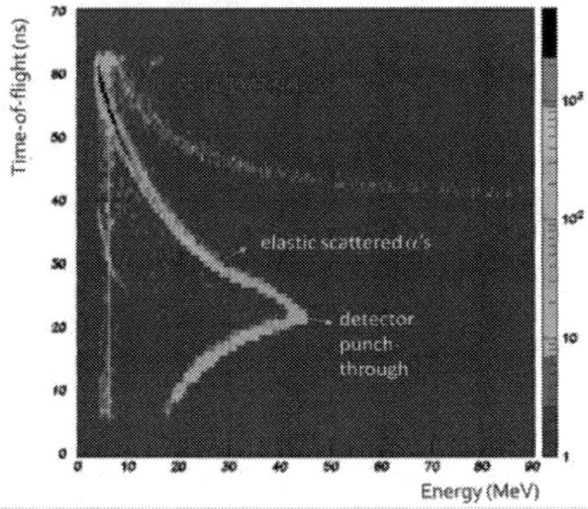

Fig. 5. Time-of-flight vs detected energy 2D-spectrum. The elastic scattering events in the lower part refer to alpha particles scattered close to the entrance Al-window.

These are clearly disentangled from a contamination related to heavier ejectiles produced in the ^{27}Al window (upper part) and protons (left side). The vertical, time-uncorrelated line, has been generated by alpha particles produced by the decay of heavy residues implanted in the Al-window. A full GEANT-4 simulation is in progress in order to take into account the setup geometry and all the possible sources of contamination.

A preliminary data analysis has been performed for the low energy alpha particles which stop in the detector. For these events the centre-of-mass energy have been recovered and, consequently, the cross-section deduced. A function correlating centre of mass energy vs lab energy (alpha's detected energy) has been calculated by using the LISE++ [15] tools. It has been used during the analysis in order to convert the detected alpha energy into centre-of-mass energy. At the same time the detector solid angle as a function of the centre-of-mass energy has been calculated in order to obtain the elastic cross-section. Finally, the experimental cross-section has been normalized to the Rutherford cross section at Θ_{cm}=180°, thus to obtain the absolute cross section shown in Fig. 6.

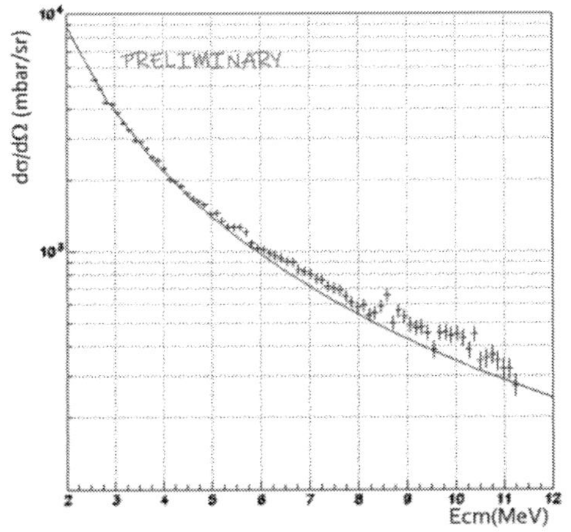

Fig. 6. Experimental centre-of-mass cross-section distribution. Absolute cross section was obtained by normalizing experimental low energy data to Rutherford cross section (solid line).

As shown in Fig. 6, the experimental trend agrees with the Rutherford cross section at low energy, but in the energy region starting from 8 MeV, the experimental cross section exceeds the theoretical one. Must be underlined that the Coulomb-barrier height of the ^{208}Pb-α system is 26 MeV (classical Coulomb barrier), thus it is striking to have evidence of structures at such low E_{cm} energies. Results from [16,17], obtained by using an effective alpha-particle structure model, where the particle is bound to a doubly magic nucleus, show that α clustering can occur near the surface of the heavy core. This means, the α-like clustering can be found in nuclear system at a low density region. Furthermore, Tohsaki and Itagaki [18,19] assume that some heavy nuclei can have specific α-clustering geometrical structure where the α cluster is considered as the building block. These geometric configurations can be favoured in very heavy (super-heavy nuclei) when α clusters form a nucleus and the Coulomb repulsion, most important at large distances, is strongly suppressed by forming the hollow structures [18].

5 Conclusions

In order to investigate ^{212}Po alpha-structure, the thick-target inverse kinematic method has been used in order to study elastic scattering of ^{208}Pb on ^{4}He target. The preliminary data analysis has shown an interesting deviation of the elastic cross-section at energies well below the Coulomb barrier. Such behaviour can be explained in the framework of a strong α-clustering in ^{212}Po. A full simulation of the experimental set-up and a further data analysis, on data obtained during the same experiment by using a different detector set-up, will be performed in order to confirm such a preliminary evidence.

References

1. K. Ikeda, T. Marumori, R. Tamagaki, H. Tanaka Suppl. Progr. Theor. Phys. **52**, 1 (1972)
2. T. Tamaya, K. Katory, M. Fujiwara, S. Kato and S. Ohkhubo Prog. Theor. Phys. **132**, 73 (1998)
3. T. Sakuda and S. Ohkubo Prog. Theor. Phys. **132**, 103 (1998)
4. R. G. Lovas, R. J. Liotta, A. Insolia, K. Varga and D. S. Dellion Phys. Rep. **294**, 265 (1998)
5. I. Tonozuka and A. Arima Nucl. Phys. A**323**, 45 (1979)
6. A. Astier, P. Petkov, M.-G. Porquet, D.S. Delion and P. Schuck Phys. Rev. Lett. **104**, 0427701 (2010)
7. Y. Suzuki and S. Ohkubo Phys. Rev. C **82**, 041303(R) (2010)
8. T.T. Ibrahim, S. M. Perez, and S. M. Wyngaardt Phys. Rev. C **82**, 034302 (2010)
9. Dongdong Ni and Zhongzhou Ren Phys. Rev. C **83**, 014310 (2011)
10. Zhongzhou Ren and Bo Zhou Front. Phys. **13**(6) 132110 (2018)
11. V. Z. Goldberg, G. V. Rogachev, W. H. Trzaska, J. J. Kolata, A. Andreyev, C. Angulo, M. J. G. Borge, S. Cherubini, G. Chubarian, G. Crowley, P. Van Duppen, M. Gorska, M. Gulino, M. Huyse, P. Jesinger, K.-M. Källman, M. Lattuada, T. Lönnroth, M. Mutterer, R. Raabe, S. Romano, M. V. Rozhkov, B. B. Skorodumov, C. Spitaleri, O. Tengblad, and A. Tumino, Phys. Rev. C **69**, 024602 (2004)
12. C. Angulo, G. Tabacaru, M. Couder, M. Gaelens, P. Leleux, A. Ninane, F. Vanderbist, T. Davinson, P. J. Woods, J. S. Schweitzer, N. L. Achouri, J. C. Angélique, E. Berthoumieux, F. de Oliveira Santos, P. Himpe, and P. Descouvemont Phys. Rev. C **67**, 014308 (2003)
13. M. G. Pellegriti, N.L. Achouri, C.Angulo, J.-C. Angélique, E.Berthoumieux, E. Casarejos, M. Couder, T. Davinson, C. Ghag, A. St. Murphy N. A. Orr, I. Raye, I. G. Stefane, P. Descouvemont Phys. Lett. **B** 659 (2008) 864
14. M. Zadro, A. Di Pietro, P. Figuera, M. Fisichella, M. Lattuada, A. Maggio, F. Pansini, M. Papa, V. Scuderi, O.Yu. Goryunov, V.V. Ostashko Nucl. Instr. Meth. B**259**, 836 (2007)
15. LISE++, http://lise.nscl.msu.edu/lise.html
16. G. Röpke, P. Schuck, Y. Funaki, H. Horiuchi, et al., Phys. Rev. C **90**, 034304 (2014)
17. C. Xu, Z. Ren, G. Röpke, P. Schuck, et al., Phys. Rev. C **93**, 011306 (2016)
18. A. Tohsaki and N. Itagaki, Phys. Rev. C **97**, 011301 (2018)
19. A. Tohsaki and N. Itagaki, Phys. Rev. C **98**, 014302 (2018)

Lights and (some) shadows in the comparison among experimental data of heavy ion collisions at Fermi energies and the dynamical model AMD

Silvia Piantelli[1], *Alessandro* Olmi[1], *Paolo R.* Maurenzig[1,2], *Akira* Ono[3], *Maurizio* Bini[1,2], *Giovanni* Casini[1], *Alessio* Mangiarotti[4], *Gabriele* Pasquali[1,2], *Giacomo* Poggi[1,2], *Andrea* Stefanini[1,2], *Sandro* Barlini[1,2], *Alberto* Camaiani[1,2], *Caterina* Ciampi[1,2], *Catalin* Frosin[1,2], *Pietro* Ottanelli[1,2], and *Simone* Valdré[1]

[1]INFN, Sezione di Firenze, Italy
[2]Dipartimento di Fisica, Università di Firenze, Firenze, Italy
[3]Department of Physics, Tohoku University, Sendai, Japan
[4]Instituto de Fisica da Universidade de Sao Paulo, Sao Paulo, Brazil

Abstract. The simulation of heavy ion collisions in the Fermi energy region is a challenge for the theoretical models; in particular it is difficult to obtain a coherent description in all the impact parameter range and to reproduce all the experimental observables. In this contribution we will show the very good job done by the dynamical model AMD [1] followed by the statistical code GEMINI [2, 3] as an afterburner. The model is able to reproduce the main characteristics of peripheral and semiperipheral collisions, although some discrepancies still persist.

1 Introduction

Heavy ion collisions in the Fermi energy domain present a very rich phenomenology, ranging from binary collisions and midvelocity emission in peripheral collisions up to multifragmentation in central reactions. As a consequence they are a challenge for the theoretical description.

Satisfactory results have been obtained by means of transport models, which can be grouped in two main classes: BUU-like models and Quantum Molecular Dynamics (QMD) models.

AMD (Antisymmetrized Molecular Dynamics [1]) is a QMD model which proved its reliability in the description of central reactions [4, 5]. It is therefore interesting to compare its predictions with experimental results concerning peripheral collisions, as done in [6] for the systems ^{93}Nb+^{93,116}Sn at 38 MeV/nucleon on data collected with the FIASCO setup [7].

FIASCO was equipped with 24 position sensitive Parallel Plate Avalanche Counters, able to measure the velocity vector of heavy ($Z \geq 10$) fragments with very low identification thresholds (0.1 MeV/nucleon), covering the 70% of the forward hemisphere; as a consequence, the setup was particularly well suited for the investigation of peripheral and semiperipheral collisions. 96 ΔE-E silicon telescopes allowed to measure also the charge of the QuasiProjectile QP below 6° and 186 three layer phoswich telescopes, identifying and measuring the time of flight of Light Charged Particles LCP and Intermediate Mass Fragments IMF (up to Z=26), with a coverage corresponding to 30% of the forward hemisphere, completed the setup.

Details on the used version of the AMD model can be found in [6]. Here we only recall that the main free parameter of the model is the in-medium nucleon-nucleon cross section $\sigma = y\rho^{\frac{2}{3}} tanh\left(\frac{\sigma_{free}}{y\rho^{-2/3}}\right)$, ruled by the y parameter, whose standard value is $y = 0.85$. In our analysis also $y = 0.42$ was tested.

Primary AMD data have been de-excited with GEMINI as an afterburner, in both the C++ version [2] and the Fortran90 one [3]. The switching time between the dynamical and the statistical phase has been tested between 200 fm/c and 10000 fm/c.

Experimental QuasiProjectile properties and emitted particle multiplicities for binary events have been compared with the simulation results as a function of the centrality. Before comparison, simulated data have been filtered with a software replica of the setup.

The collision centrality has been estimated as $C_b = \frac{1}{2} M v_{QP}^{cm} v_{QT}^{cm}$, where $v_{QP(QT)}$ is the center of mass secondary velocity of the QP (QuasiTarget QT); in the experimental case such velocity corresponds to the measured one, for the simulation it is the velocity after the application of the afterburner and of the geometrical filter (except for the 4π simulation). M is the total mass of the system. C_b is well correlated to the impact parameter, as shown in Fig. 2 of [6], at least in the range 5 fm - grazing impact parameter. Below 5 fm C_b loses its reliability as impact parameter estimator because of the increasing weight of ternary events, coming from QP/QT fissions.

2 Results

As shown in Fig. 5 of [6], the simulation is able to well reproduce the behaviour of the center of mass polar angle of the QP as a function of the centrality (Wilczynski

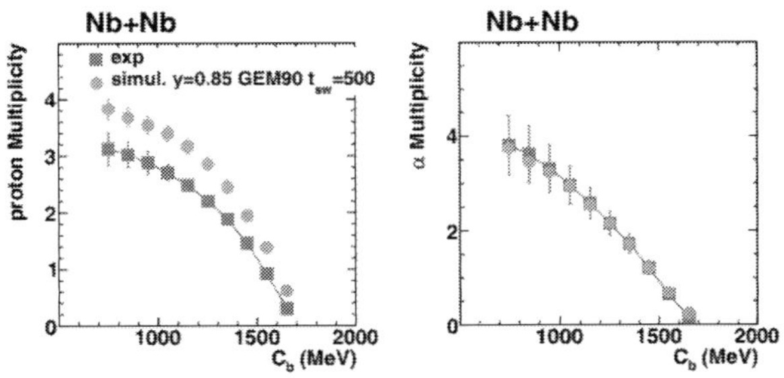

Figure 1. Experimental and simulated (best result: switching time 500 fm/c, GEMINI Fortran90 as afterburner, screening parameter $y = 0.85$) multiplicity for protons (left) and α particles (right) as a function of the centrality (estimated starting from the velocities of QP and QT, as explained in the text) in binary reactions.

plot) both in terms of average values and of standard deviation of the distribution. All the tested configurations (afterburner, switching time, y parameter) are summarized in table II of [6] with a quality factor indicating the degree of agreement between experimental data and simulation (0 means perfect reproduction of the experimental data). All the tested configurations of the parameters give a satisfactory reproduction of the experimental data.

Also the charge of the QP is well reproduced by the simulation, as shown in Fig. 6 of [6], at least in terms of average value; for the standard deviation, on the contrary, there are some systematic differences. The use of GEMINI Fortran90 clearly improves the quality of the agreement; concerning the switching time, no significant improvement is obtained for times longer than 500 fm/c, while it seems important to prolong the calculation beyond 200 fm/c, as summarized in table III of [6].

The total emitted charge, shown in Fig. 9 of [6], is well reproduced, almost independently of the in-medium cross section, of the afterburner and of the AMD switching time, with the possible exception of 200 fm/c, thus meaning that for such time dynamical effects are still important (and they cannot be mimicked by an afterburner). On the contrary, the chemistry of the emitted particles is strongly dependent on the switching time and the afterburner and -to a minor extent- on the in-medium cross section. For all the particles (Fig 7 of [6], panels b) c) d) e)) except for the protons (Fig 7 of [6] panel a)) the multiplicities are quite well reproduced for a switching time of 500 fm/c and GEMINI f90 as afterburner, while worse results are obtained when GEMINI++ is used. In case of protons simulated data always overestimate the experimental values, although the use of GEMINI f90 strongly reduces the discrepancy. In Fig. 1 the results concerning proton and α particle multiplicities for the best simulation (i.e. switching time 500fm/c, y=0.85, GEMINI f90 as afterburner) are shown. While α multiplicities are very well reproduced, in case of protons, mainly for the most dissipative reactions, the discrepancy between the simulation and experimental data is of the order of 25%. The switching time seems to be the most critical parameter for the particle multiplicities mainly for protons. In fact, as shown in Fig. 8 of [6], while for the other particles the multiplicities directly produced by AMD (i.e. the primary ones) tend to saturate for times longer than 2500 fm/c, in case of protons the primary multiplicity goes on increasing and at 10000 fm/c it reaches the experimental value. Since the excitation energy of QP and QT is still different from 0 at such extreme switching time, it is obvious that the application of whatever afterburner produces a significant overestimation of the final proton multiplicity. Moreover, the behaviour of the two versions of the afterburner is different: while for GEMINI++ the final particle multiplicies seem little depending on the switching time, this is not true for GEMINI Fortran90 concerning protons and deuterons. This different behaviour can be interpreted in the hypothesis that the emission of simple particles from AMD is somehow compatible with that from GEMINI++, while it is different from that of GEMINI f90.

3 Conclusions

The AMD model coupled to GEMINI as an afterburner proved to be able to reproduce the main properties of the QP and of the emitted particles in binary peripheral and semiperipheral events. In particular, the average center of mass polar angle of the QP and its charge as a function of the reaction centrality are well reproduced by the simulation, almost independently of the switching time, of the in-medium cross section and of the version of the afterburner. On the contrary, for light particles better results are obtained if the dynamical code is stopped at 500 fm/c and GEMINI f90 is used as an afterburner. The obtained results are very promising, since they suggest that AMD followed by GEMINI can be used as a predictive tool in the study of heavy ion collisions in the Fermi energy regime.

References

[1] A.Ono et al., Phys. Rev. Lett. **68**, 2898 (1992).
[2] R. J. Charity, Phys. Rev. C **82**, 014610 (2010).
[3] R.J.Charity et al., Nucl. Phys. A **483** 371 (1988).
[4] A.Ono et al., Phys. Rev. C **66** 014603 (2002).
[5] A.Ono, J. Phys. Conf. Series **420** 012103 (2013).
[6] S.Piantelli et al., Phys. Rev. C **99** 064616 (2019).
[7] M.Bini et al., NIMA **515** 497 (2003).

Intermediate Mass Fragments production at low energy: reaction mechanism and isospin influence

Sara Pirrone[1,*], *Giuseppe* Politi[2,1], *Brunilde* Gnoffo[2,1], *Enrico* De Filippo[1], *Paolo* Russotto[3], *Giuseppe* Cardella[1], *Elena* Geraci[2,1], *Luigi* Lo Monaco[2], *Nunzia Simona* Martorana[2,3], *Angelo* Pagano[1], *Emanuele Vincenzo* Pagano[3], *Ernest* Piasecki[4], *Francesca* Rizzo[2,3], *Antonio* Russo[3], and *Marina* Trimarchi[5,1]

[1]INFN, Sezione di Catania -Catania, Italy
[2]Dipartimento di Fisica e Astronomia "Ettore Majorana", Università degli Studi di Catania - Catania, Italy
[3]INFN, Laboratori Nazionali del Sud- Catania, Italy
[4]Heavy Ion Laboratory, University of Warsaw, Warsaw, Poland
[5]Dipartimento di Scienze Matematiche e Informatiche, Scienze Fisiche e Scienze della Terra, Università, Messina - Messina, Italy

Abstract. In this work, we present a recent study concerning the break-up of the Projectile-Like (PLF) into two fragments, following more violent deep-inelastic collisions in the reactions ^{78}Kr + ^{40}Ca and ^{86}Kr + ^{48}Ca at 10 A MeV, realized at INFN-Laboratori Nazionali del Sud by using the 4π multi detector CHIMERA. For these systems, we have already analyzed the fusion-evaporation and fission-like processes. A selection method has been developed, in order to discriminate PLF break up events from those due to other mechanisms which populate the same region of the phase- space. A preference for PLF aligned break-up, along the direction of the PLF-TLF separation axis with the light fragment emitted in the backward part, has been put in light, suggesting the presence of dynamical effects. Also a comparison between the neutron-rich ^{86}Kr + ^{48}Ca and neutron-poor ^{78}Kr + ^{40}Ca systems is presented, to put in evidence the isospin role in these processes.

1 Introduction

The energy domain E/A $\leq$ 15 AMeV, the so called low enery regime, is characterized from the competition between the Compound Nucleus formation and its consequent de-excitation through fusion-fission and fusion-evaporation, and other binary processes like quasi elastic scattering and deep inelastic collisions.
These reaction mechanisms populate a wide range of mass region that include Light Charged Particles, Fission Fragments, Evaporation Residues and Intermediate Mass Fragments (Z$\geq$ 3).

Mainly the reaction mechanisms responsible of the IMF production are the fission - like processes and the damped deep inelastic collision followed by projectile-like break -up. These processes, at this energy, populate the same region of the phase space, and in order to disentangle among them, a careful selection method has to be developed.
We have already studied the fusion-evaporation and fission-like processes in the reactions ^{78}Kr + ^{40}Ca and ^{86}Kr + ^{48}Ca at 10 AMeV [1–4] with the 4π multidetector CHIMERA [5–7].

In particular, considering the mass, charge and angular distributions of the reaction products, well identified by CHIMERA detector, we were able to measure the evaporation residues and fission-like cross sections for the two reactions. The comparison between the two systems, that differ by 16 neutrons, the maximum difference in neutrons number achievable by using these stable nuclei, put in light that the neutron enrichment seems to limit the formation of the composite system and to inhibit the fission decay channel.

A brief summary of the main results of this analysis are presented. Then we are going to present a selection method for the analysis of projectile–like break-up events. We would like to point out that in previous experiments performed at higher energy, it was observed an evident influence of the isospin on the nature of the break-up of the PLF [8–10], and now we explore if also at low energy the isospin plays a crucial role.

2 Fusion-evaporation and fission-like processes

The capability of detection and the high accuracy of the measurements obtained by using the CHIMERA array, allowed us to identify and kinematically distinguish the reaction products coming from different reaction mechanisms[1].

In particular, we were able to separate the evaporation residues and the fission like fragments coming from the decay of the compound nucleus, as it is shown in Fig.1, in which are reported the results obtained selecting "complete events" for the two systems. We define "complete events", those selected by imposing the condition of a total measured momentum, $P_{tot} \geq 0.6 P_{beam}$, where P_{beam}

*e-mail: sara.pirrone@ct.infn.it

https://doi.org/10.1051/epjconf/201922301051

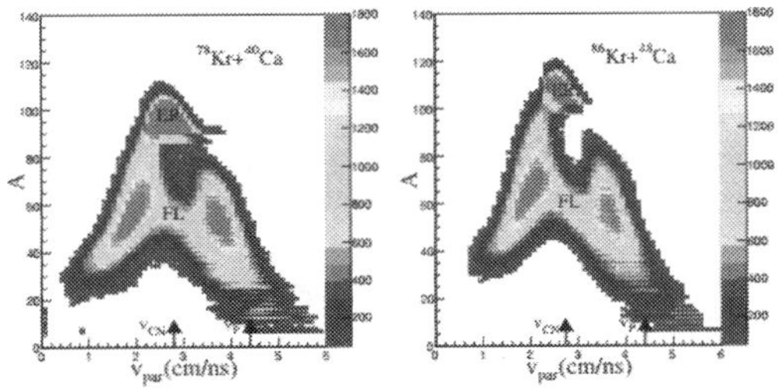

Figure 1. Mass versus parallel velocity for reaction products detected in "complete events" for the two systems; the arrows indicate compound nucleus (v_{CN}) and projectile velocity (v_P). (Color scale of the Z-axis is in relative units).

is the total beam momentum, and a total detected mass $A_{tot} \geq 0.75A_{CN}$, for each system.

In this figure the mass A versus the parallel velocity v_{par} is reported for each reaction product and for the two studied systems; we can distinguish the fragments coming from the different decay modes of the composite system formed in the reactions, as the Evaporation Residues (ER with $90 \leq A \leq 110$, for the neutron poor system and $100 \leq A \leq 120$ for the neutron rich one) centered around the compound nucleus velocity v_{CN} and the fission-like fragments (FL with $10 \leq A \leq 85$ for the neutron poor system and $10 \leq A \leq 95$ for the neutron rich one) distributed along the two branches, coming from the corresponding possible kinematic solutions.

For the two systems, we built the angular distributions of the IMF, and by their integration, and by using the study of the Rutherford and elatic scattering, we calculated the production cross sections for each fragment. The charge distribution is presented in Fig.2.

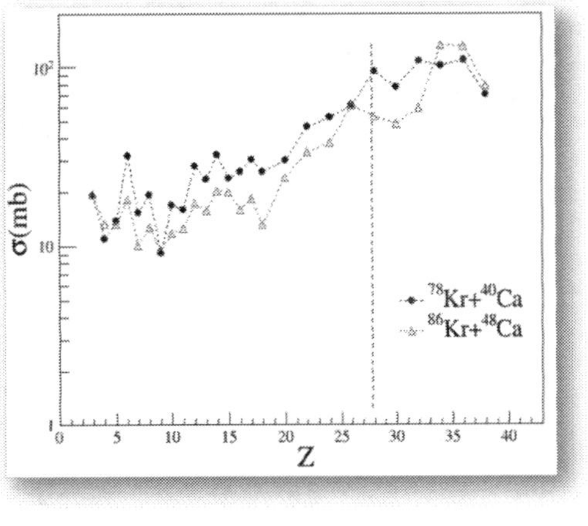

Figure 2. Charge distributions for the two systems

We note that for both systems the distribution presents a staggering effect, more evident for the neutron poor system, and that is not symmetric with respect to Z=26/28, that indicates a presence of a non equilibrated component of the production mechanism of the heavier fragments.

Table 1. Evaporation residues, fission-like, fusion and total quarter point reaction cross sections for the two reactions.

	σ_{ER} (mb)	σ_{FL} (mb)	σ_{fus} (mb)	σ^{qp}_{reac} (mb)
^{78}Kr+^{40}Ca	455±70	850±120	1305±190	2390±250
^{86}Kr+^{48}Ca	400±60	530±85	930±145	2520±260

In the Table 1 is reported the measured cross sections for different reaction mechanisms and for both systems. From these results we can deduce that the fusion-fission process is favoured for the neutron poor system. Besides, looking at the difference $\sigma^{qp}_{reac} - \sigma_{fus}$, we note that this quantity is bigger for the neutron rich system, meaning a larger contribution of non equilibrated reactions in this case.

3 Project-like break-up processes

To discriminate PLF coming from the break-up events from the other processes which populate the same region in phase space, first of all we have to select events with three IMF in the exit channel. In Fig.3 is shown, for this class of events, the correlation between the fragment mass and parallel velocity of the reaction products in the case of the ^{78}Kr + ^{40}Ca system.

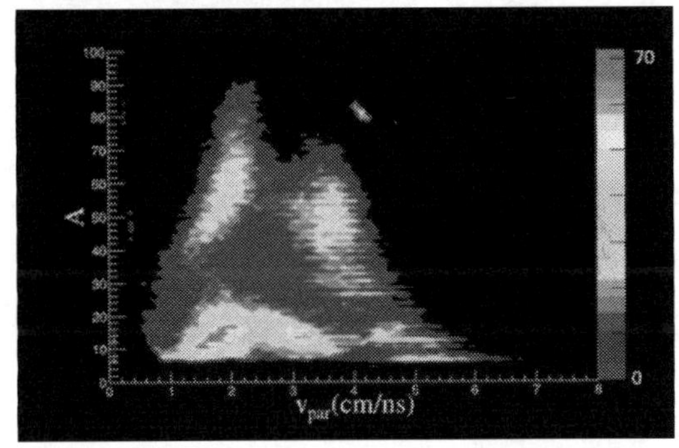

Figure 3. Correlation between the fragment mass and parallel velocity of the reaction products in events with 3 IMFs well detected and identified

Different mechanisms populate this plot: the break-up of the slow fission fragment from Compound Nucleus, the break-up of the fast fission fragment from Compound Nucleus and the break-up of the PLF.

In fact, after its formation the Compound Nucleus can decay through fission and then the slow fission fragment can break-up into two fragments. At the end of this process we have three body: the fast fragment emitted in the fission of the Compound Nucleus (F_1) and the two fragments produced in the break-up of the slow- fission fragment ($F_2^I and F_2^{II}$), for threshold reasons, in the kinematical solution where the lighter fragment is forward emitted.

Similarly, if after the fission of the Compound Nucleus the fast fragment produced breaks up into two fragments, at the end we observe: the slow fragment emitted in the fission of the Compound Nucleus (F_2) and the two fragments produced in the break-up of the fast- fission fragment (F_1^I and F_1^{II}), with the lighter fragment backward or forward emitted.

Finally, the break-up of the projectile mechanism can occur. This mechanism is originated from a a two step mechanism, compatible with the hypothesis of a binary deep inelastic interaction, followed by the further fission-like decay of the PLF. The three body produced in this reaction are: a part of the Target-Like Fragment (T^*) and the two fragment coming from the break-up of the PLF (P^I and P^{II}), with the lighter backward or forward emitted.

In Fig.4 is reported the reconstructed source of each combination of two fragments produced in events with 3 IMFs. They are the slow fragment emitted in the fission of the Compound nucleus (F_2) or the PLF (P^*) or the fast fission fragment (F_1).

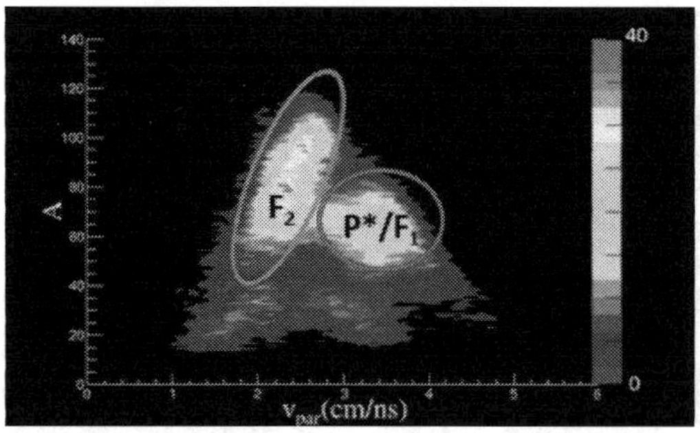

Figure 4. The mass versus parallel velocity plot of the reconstructed source of each combination of two fragments in events with 3 IMFs well detected and identified.

In order to be sure to take into account only events due the PLF break-up, we required that the reconstructed source of two fragments is $(P^*)^*/\ (F_1)$ and that the third fragment does not belong to the slow branch of fission from Compound Nucleus.

The figures are relative to the ^{78}Kr + ^{40}Ca system, but similar results were obtained for the ^{86}Kr + ^{48}Ca system. The correlation between the mass and the parallel velocity of the fragments produced int the PLF break-up, is shown in Fig. 5 on the left for ^{78}Kr + ^{40}Ca and on the right for the ^{86}Kr + ^{48}Ca system.

Very important information about the nature of the break-up process can be extracted from the angular distribution in the PLF reference frame. A main signature of a dynamical break- up is an alignment of the break- up axis with the separation direction of the two primary fragments (PLF and TLF), with the lighter breakup fragment backward emitted. The aligned break-up demonstrates the persistence of the memory of the previous deep inelastic step and in particular of the direction of the separation axis between PLF and TLF.

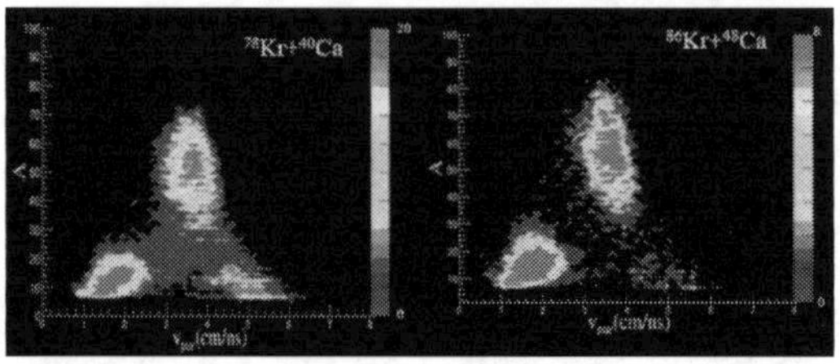

Figure 5. Mass versus parallel velocity lot for the PLF break-up on the left for $^{78}Kr +^{40}Ca$ and on the right for the ^{86}Kr + ^{48}Ca system.

A very nice observable, that allows to disentangle between statistical and dynamical break-up, is the cosine of θ proximity. The proximity angle is defined as the angle between the separation axis between PLF and TLF and the break-up axis oriented from the Light to the Heavy fragment.

The distributions in θ proximity for different asymmetry parameters, A_H/A_L, and for the two systems, are reported in Fig.6.

In these distributions, we observe the presence of two components, one isotropic, coming from relaxed process, and the other one, peaked around 1, with aligned fragments, suggesting dynamical effects. Dynamical break-up component prevails for the neutron rich system, in agreement with the results obtained in previous experiments performed at higher energy[8–10].

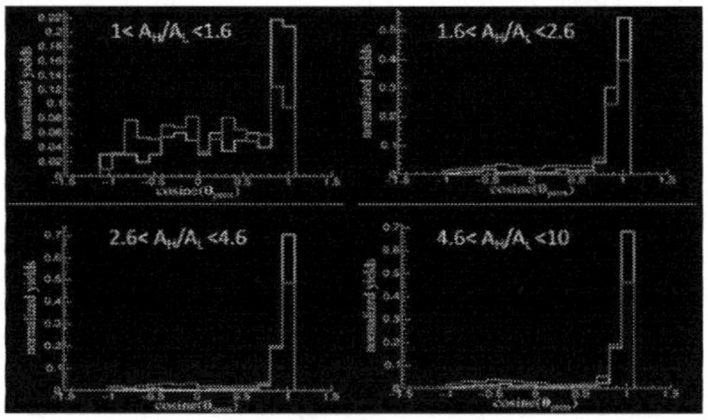

Figure 6. Distribution in cosine of θ proximity for the two studied systems for different asymmetry parameters.

4 Conclusion

Analisys relative to the IMF production in the reactions ^{78}Kr + ^{40}Ca and ^{86}Kr + ^{48}Ca were presented. In particular we reported the obtained results from the Compound nucleus reactions, with production of fission like fragments, evaporation residues and light charge particles.Then we presented a selection method, in order to discriminate PLF break-up from events due to other mechanisms, fusion-fission like processes, which populate the same region of the phase-space. A preference for PLF aligned break-up, along the direction of the PLF-TLF separation axis with the light fragment emitted in the backward part, has been evidenced, suggesting dynamical-non equilibrium effects. The results seem to put in evidence that the dynamical effects are more pronounced for the reactions ^{86}Kr + ^{48}Ca, respect to ^{78}Kr + ^{40}Ca one, for the different asymmetry parameters, in agreement with the results obtained at higher energy.

References

[1] S.Pirrone et al., Eur. Phys. J.**55**, 22 (2019)
[2] B.Gnoffo, Il Nuovo Cimento C,**39** 275 (2016)
[3] S.Pirrone et al., Journal of Physic: Conf. Series **515** 012018 (2014)
[4] G.Politi et al., JPS Conf. Proc.**6** 030082 (2015)
[5] A.Pagano et al., Nucl. Phys. A **681** 331 (2001)
[6] G.Politi et al., IEEE NSS Conference Record **N28-5** 1140 (2005)
[7] M.Alderighi et al., IEEE Transactions on Nuclear Science **53** 279 (2006)
[8] E.De Filippo, A.Pagano, Eur. Phys. J. **A50** 32 (2014)
[9] P.Russotto et al., Phys. Rev. C **81** 064605 (2010)
[10] P.Russotto et al., Phys. Rev. C **91** 014610 (2015)

https://doi.org/10.1051/epjconf/201922301052

Fusion probability of massive nuclei in reactions leading to heavy composite nuclear systems

Roman Sagaidak*

Flerov Laboratory of Nuclear Reactions, Joint Institute for Nuclear Research, Dubna 141980, Russia

Abstract. Reactions between massive nuclei show a considerable reduction in fusion-evaporation cross-sections at the Coulomb barrier according to the comparison of experimental values with those calculated by barrier passing (BP) and statistical model (SM) approximations. Reduced fusion cross-sections corresponding to fusion probability P_{CN}<1 are accompanied by a high probability of deep-inelastic and quasi-fission processes arising on the way to fusion. At the same time, the excitation functions for evaporation residues (ERs) obtained in very mass-asymmetric projectile-target combinations are well described in the framework of the BP model (assuming P_{CN}=1) and SM approximations. In the framework of SM, the survivability of produced heavy nuclei can be described with the use of adjusted macroscopic fission barriers. Fusion suppression appears in less asymmetric combinations, for which P_{CN} values can be estimated using survivability obtained for very asymmetric ones leading to the same CN. An attempt was made to systemize the P_{CN} data derived from different projectile-target combinations leading to ERs in the range from Pb to the most heavies, which are compared with P_{CN} values obtained in fission experiments.

1 Motivation and approach

Reactions with massive nuclei show a considerable reduction in fusion at the Coulomb barrier. It follows from the comparison of experimental cross-sections with those calculated using a barrier passing (BP) model. Reduced fusion cross sections are accompanied by a high probability of deep-inelastic and quasi-fission (QF) processes arising on the way to fusion. The detection of evaporation residues (ERs) resulting from a compound nucleus (CN) formation is an unambiguous sign of the complete fusion of projectile and target nuclei, whereas detected fission (fission-like) events do not specify the CN formation since CN-fission strongly interferes with the QF process. Theoretical models describing ER cross-sections σ_{ER} treat them as the product of i) capture cross-section σ_{cap} relating to the formation of a composite (di-nuclear) system, ii) fusion probability P_{CN} corresponding to the CN formation from the composite system, and iii) survivability against fission W_{sv} while the CN decays. Most of the models reproduce measured σ_{ER} quite well, but they give P_{CN} values differed from each other within several orders of the magnitude [1]. Such a difference implies a similar distinction in W_{sv}.

At the same time, available cross-section data on the fusion, fission and ER production, which are obtained in very mass-asymmetric projectile-target combinations, can be well described in the framework of the BP and statistical model (SM) approximations realized in the HIVAP code [2] (see examples in [3]). In that case, the BP cross-section is associated with σ_{cap} equaled to the fusion cross-section with a reasonable assumption that P_{CN}=1. In the calculations of σ_{cap} at sub-barrier energies, the effect of coupling the entrance channel to other reaction channels is taken into account via fluctuations of radius-parameter r_0. These fluctuations are generated around average value r_0=1.12 fm with a Gaussian distribution and barrier fluctuation parameter $\sigma(r_0)$ [4]. Variations of strength V_0 and fluctuation parameter $\sigma(r_0)/r_0$ in the exponential nuclear potential [4] allow one to reproduce the experimental cross-section data for the capture (fusion), CN-fission and ER production in calculations for very asymmetric systems.

The survivability is calculated in the framework of SM approximations with the Reisdorf's expression for calculations of macroscopic level-density parameters in fission and evaporation channels [2]. The macroscopic components of fission barriers adjusted with scaling factor k_f at rotating liquid-drop (LD) fission barriers $B_f^{LD}(L)$ [5] are used in the expression for fission barrier height $B_f(L) = k_f B_f^{LD}(L) - \Delta W_{gs}$. The empirical masses [6] are used to calculate shell correction energies ΔW_{gs} (determined as the difference between the empirical and LD masses), as well as for the calculations of excitation and separation energies.

Fitting thus calculated excitation functions to the measured ones obtained in very asymmetric projectile-target combinations by adjusting fission barriers, one can get estimates of W_{sv}. Fusion suppression corresponding to P_{CN}<1 appears in less asymmetric combinations. It can be derived using W_{sv} obtained for very asymmetric combinations leading to the same CN and σ_{cap} measured or obtained with the BP model calculations.

* Corresponding author: sagaidak@jinr.ru

2 P_{CN} from fission study

In heavy ion (HI) experiments, P_{CN} values can be derived with the detection of fission (fission-like) fragments (FFs) and subsequent comparison of a total FF yield including deep-inelastic events with the FF yield assigned to true CN-fission. The events relating to CN-fission are extracted with an appropriate decomposition of the obtained FF angular distributions [7] and with the decomposition of the measured total kinetic energy and mass distributions for FFs [8]. In Fig. 1 P_{CN} values derived from fission studies in reactions with W, Au, and Pb targets [8] are shown as a function of an excess of the interaction energy over the Bass barrier [9]. As one can see, the $^{28}Si+^{208}Pb$ and $^{30}Si+^{197}Au$ data, corresponding to nearly the same mass-asymmetry in the entrance channel, are in sharp disagreement with each other. The same is for the $^{32}S+^{197}Au$ and $^{36}S+^{197}Au$ data. So as a result, P_{CN} values obtained in the ^{30}Si, $^{36}S+^{197}Au$ study were omitted in subsequent analysis (see below).

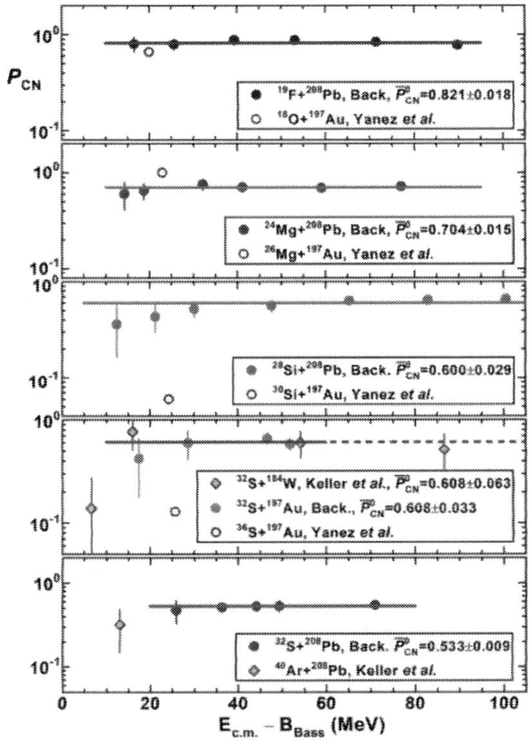

Fig. 1. P_{CN} values (symbols) derived from fission studies in reactions with W, Au, and Pb targets [8] are shown as a function of an excess of the interaction energy over the Bass barrier [10]. A constant fit to the data (lines) and appropriate mean values are also indicated.

In Fig. 2 P_{CN} values derived from fission studies in the interaction of ^{238}U with the Mg to Ca target nuclei and obtained in the $^{40,48}Ca+^{238}U$ and $^{26}Mg+^{248}Cm$ reactions [8] are shown as the same function of the energy as shown in Fig. 1. Inconsistency of the data in the vicinity of the barrier [9] only allows one to consider P_{CN} values at energies well above the barrier.

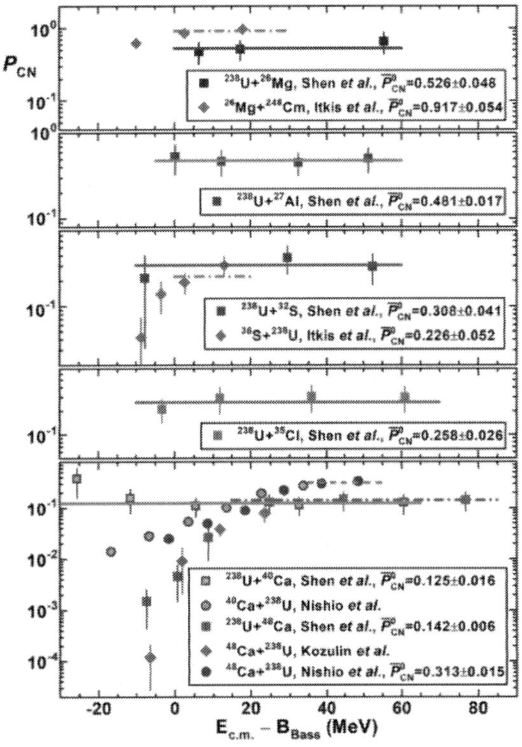

Fig. 2. The same as in Fig. 1 but in the cases of the interaction of ^{238}U with the Mg to Ca target nuclei and for the $^{40,48}Ca+^{238}U$ and $^{26}Mg+^{248}Cm$ reactions [8].

3 P_{CN} from ER cross sections

The analysis described in Section 1, was applied for the first time to the data obtained in ^{12}C and ^{48}Ca reactions leading to $^{216,218}Ra^*$ compound nuclei [10]. Then it was used to estimate of P_{CN} using ER cross-section data obtained in some selected very asymmetric and less asymmetric (up to nearly symmetric) projectile-target combinations leading to $^{202}Pb^*$, $^{220}Th^*$, $^{248}Fm^*$ and trans-fermium compound nuclei [11]. These results have been used for further P_{CN} data systemizing (see below).

In Figs. 3 and 4 ER cross-section data obtained recently in reactions induced by $^{44,48}Ca$ and ^{50}Ti on rare-earth elements [12] are compared with the ER and fission excitation functions obtained in very asymmetric reactions [13] leading to the same compound nuclei $^{202}Po^*$ and $^{210}Rn^*$, respectively. As one can see in Fig. 3, ER and fission cross sections obtained in reactions induced by ^{16}O and ^{34}S are well described using the same macroscopic component of fission barriers (the same k_f at the LD values). At the same time, in order to reproduce the excitation functions for the sum of xn evaporation channels $\sum\sigma_{xn}$ obtained in $^{44,48}Ca$ reactions, the magnitude of P_{CN}=0.27 has to be introduced. A similar situation is observed for the ^{48}Ca induced reaction leading to $^{210}Rn^*$ (see Fig. 4). Despite nearly the same excitation energies at the fusion (Bass) barriers for the reactions with ^{48}Ca and ^{50}Ti, $\sum\sigma_{xn}$ drops by an order of magnitude for the latter that corresponds to P_{CN}=0.03.

https://doi.org/10.1051/epjconf/201922301052

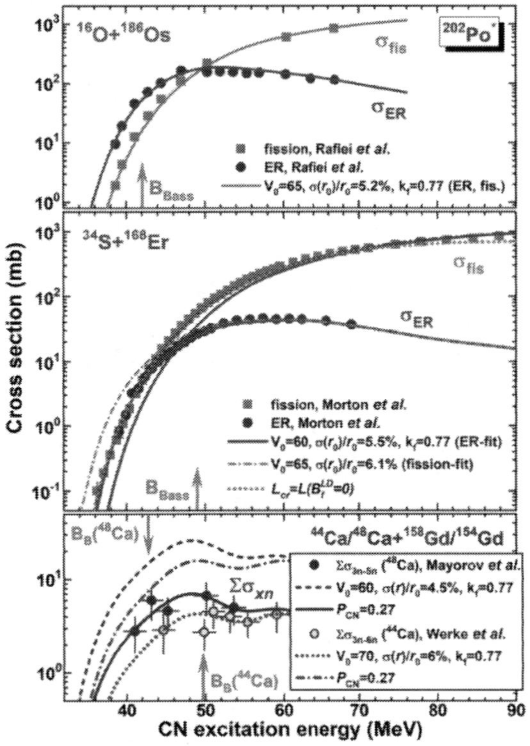

Fig. 3. ER and fission cross sections obtained in reactions induced by ^{16}O, ^{34}S [13] and ^{44,48}Ca [12] that lead to the ^{202}Po* CN (symbols) are compared with the calculations [4] using the same scaling parameter k_f = 0.77 at the LD fission barriers (lines). In the cases of ^{44,48}Ca, the magnitude of P_{CN}=0.27 has to be introduced to reproduce the ER cross-section data [13].

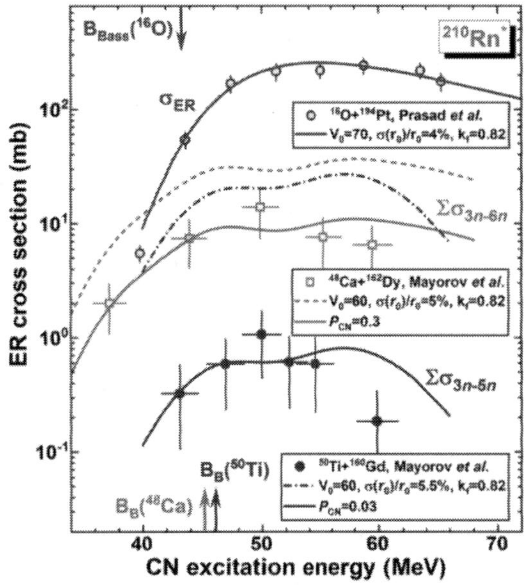

Fig. 4. The same as in Fig. 3 but for ER cross sections obtained in reactions leading to the ^{210}Rn* CN [12, 13] and calculations with k_f=0.82. In the cases of ^{48}Ca and ^{50}Ti, P_{CN}=0.3 and 0.03, respectively, have to be introduced to describe the data.

It was revealed, according to the analysis of ER excitation functions, that the decay of compound nuclei from Fm* to Rf* formed in very asymmetric reactions could be described with k_f=1.2 [11]. That is in contrast to k_f<1, with which the decay of compound nuclei with $Z \leq 98$ can be described. Applying this finding to the description of the survivability of even heavier compound nuclei ^{268}Sg* and ^{274}Hs* formed in asymmetric reactions with ^{30}Si and ^{26}Mg, respectively, one can arrive at P_{CN}<1 for both the reactions, as shown in Fig. 5.

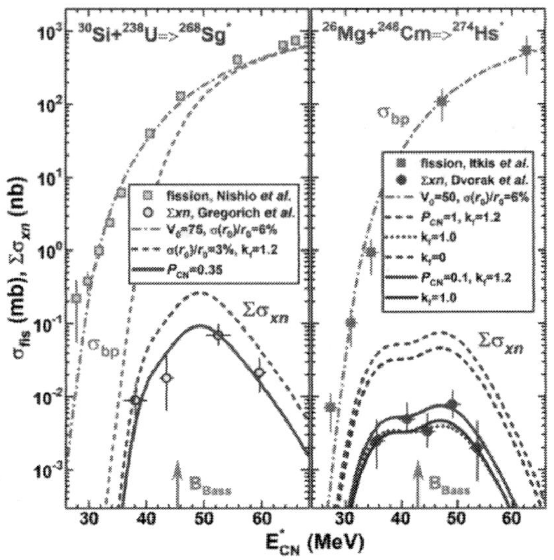

Fig. 5. The same as in Figs. 3 and 4 but for the cross sections obtained in reactions leading to the ^{268}Sg* and ^{274}Hs* compound nuclei [14, 15]. Calculations with P_{CN}=0.35 and k_f= 1.2, and with the values indicated in the right panel were used to describe the ^{30}Si+^{228}U and ^{26}Mg+^{248}Cm data, respectively.

As one can see in Fig. 5, small variations in the macroscopic component of fission barriers have a small effect on the production cross sections for the heaviest nuclei. It is the result of a small value of this component (~0.4 MeV) for Hs fission barriers used in calculations. Neglecting this component leads to P_{CN}=1, but this assumption is in contradiction to a smooth drop of the macroscopic fission barriers to zero with an increase in the CN fissility and to a trend implying a general reduction in P_{CN} with the same change (see below).

4 P_{CN} systematics and summary

Several approaches to the P_{CN} data scaling were tested with argument x corresponding to Coulomb factor $Z_pZ_t/(A_p^{\frac{1}{3}} + A_t^{\frac{1}{3}})$, equilibrated mean fissility X_{mean}^{eq} and effective fissility X_{eff} (the last two were proposed earlier, within the application of the extra-push model [16] to data analysis). P_{CN} values obtained with fission and ER data were separately fitted using $f(x) = 1/\{1 + exp[k(x - x_c)]\}$ function, with k and x_c as fitted parameters. As in the case of fission data, some ER P_{CN}

data had been in a significant deviation from a general trend. These data corresponding to the formation of ^{202}Po*, ^{210}Rn*, ^{248}Fm* and ^{274}Hs* in reactions with ^{34}S, ^{50}Ti and ^{26}Mg (see Figs. 3–5 and [11]) were omitted in all fitting procedures. The least χ^2 value was obtained with the equilibrated mean fissility as the argument of x. The result of the fitting of both the fission and ER P_{CN} data considered in this work are shown in Fig. 6.

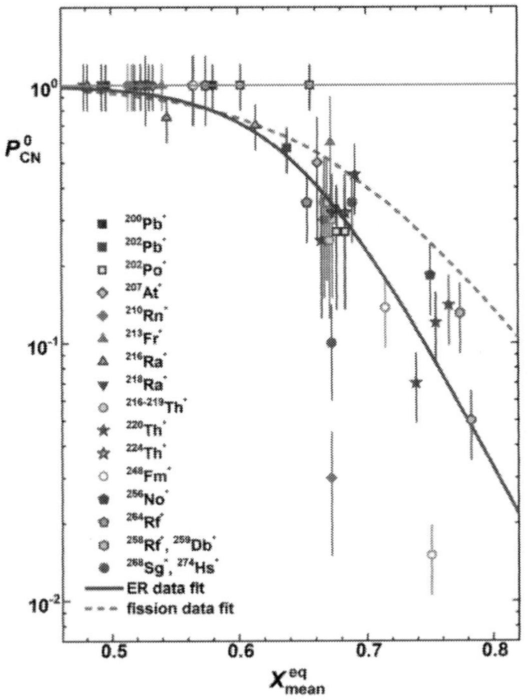

Fig. 6. The P_{CN} data derived from the analysis of ER cross sections obtained in complete fusion reactions with the different mass-asymmetry in the entrance channel and leading to the same CN (designated by the corresponding symbol) are fitted with a function of the equilibrated mean fissility (a solid line). The result of the same function fit to the P_{CN} data obtained in fission studies (see Figs. 1 and 2) is shown by a dashed line. See the text for more details.

As one can see, the fitted P_{CN} values obtained with the ER data decrease faster than those corresponding to fission data as the equilibrated mean fissility increases. The former could be applied to the estimate of the drop in the P_{CN} value at the transition from ^{48}Ca to ^{50}Ti induced reactions leading to the same ^{288}Fl* CN. This drop should not exceed a factor of two, implying the same survivability in both reactions.

Summarizing one has to mention that

- P_{CN} and survivability W_{sv} in the complete fusion reactions leading to the heaviest nuclei are correlating values in the calculations of ER cross-sections. Available fission and ER cross-section data were used to consider P_{CN} and W_{sv}. ER data could be described in the framework of the barrier passing model for capture and the statistical model (SM) for a CN-decay using P_{CN} as an adjustable parameter.
- P_{CN} values obtained in reactions corresponding to fission of heavy composite system formed in nucleus-nucleus collisions were scaled with the Coulomb factor and fissility parameters proposed in the framework of the extra-push model.
- P_{CN} values were also derived by comparing ER cross-sections obtained in very asymmetric projectile-target combinations (having P_{CN}=1) and those obtained in less asymmetric ones, for which P_{CN} must be obtained. The survivability of heavy nuclei produced in very asymmetric reactions was reproduced by adjusting the macroscopic component of fission barriers within SM approximations. These barriers were used for the P_{CN} estimates in more symmetric reactions leading to the same CN.
- P_{CN} values obtained with the ER cross-sections were also scaled in the same way as fission data. A comparison of both dependencies shows that a drop in P_{CN} values deduced with the ER data as functions of the Coulomb factor and fissility occurs faster than the one for similar values obtained with fission data.

References

1. R.S. Naik *et al.*, Phys. Rev. C **76**, 054604 (2007); W. Loveland, EPJ Web of Conf. **131**, 04003 (2016)
2. W. Reisdorf, Z. Phys. A **300**, 227 (1981); W. Reisdorf, M. Schädel, Z. Phys. A **343**, 47 (1992)
3. L. Corradi *et al.*, Phys. Rev. C **71**, 014609 (2005); R.N. Sagaidak, A.N. Andreyev, Phys. Rev. C **79**, 054613 (2009)
4. W. Reisdorf *et al.*, Nucl. Phys. A **438**, 212 (1985)
5. S. Cohen *et al.*, Ann. Phys. (NY) **82**, 557 (1974)
6. G. Audi *et al.*, Nucl. Phys. A **729**, 337 (2003)
7. B.B. Back, Phys. Rev. C **31**, 2104 (1985); R. Yanez *et al.*, Phys. Rev. C **88**, 014606 (2013)
8. W.Q. Shen *et al.*, Phys. Rev. C **36**, 115 (1987); E.M Kozulin *et al.*, Phys. Lett. B **686**, 227 (2010); I.M. Itkis *et al.*, Phys. Rev. C **83**, 064613 (2011); K. Nishio *et al.*, Phys. Rev. C **86**, 034608 (2012)
9. R. Bass, Lect. Notes Phys. **117**, 281 (1980)
10. R.N. Sagaidak *et al.*, Phys. Rev. C **68**, 014603 (2003)
11. R.N. Sagaidak, Eur. Phys. J. D **45**, 59 (2007); R.N. Sagaidak, EPJ Web of Conf. **21**, 06001 (2012)
12. D.A. Mayorov *et al.*, Phys. Rev. C **90**, 024602 (2014); D.A. Mayorov *et al.*, Phys. Rev. C **92**, 054601 (2015); T.A. Werke *et al.*, Phys. Rev. C **92**, 054617 (2015)
13. R. Rafiei *et al.*, Phys. Rev. C **77**, 024606 (2008); C.R. Morton *et al.*, Phys. Rev. C **62**, 024607 (2000); E. Prasad *et al.*, Phys. Rev. C **84**, 064606 (2011)
14. K. Nishio *et al.*, Eur. Phys. J. A **29**, 281 (2006); K.E. Gregorich *et al.*, Phys. Rev. C **74**, 044611 (2006)
15. J. Dvorak *et al.*, Phys. Rev. Lett. **100**, 132503 (2008); I.M. Itkis *et al.*, Phys. Rev. C **83**, 064613 (2011)
16. H. Gäggeler *et al.*, Z. Phys. A **316**, 291 (1984); J.P. Blocki *et al.*, Nucl. Phys. A **459**, 145(1986)

Single and Double Charge Exchange Excitations of Spin-Isospin Mode

Hiroyuki Sagawa[1,2,*]

[1]RIKEN Nishina Center, Wako, Saitama 351-0198, Japan
[2]Center for Mathematics and Physics, University of Aizu, Aizu-Wakamatsu, Fukushima 965-8580, Japan

Abstract. We study the sum rules of double Gamow-Teller (DGT) excitations through double spin-isospin operator $(\sigma t_-)^2$. In general, 2^+ states in the granddaughter nuclei have dominant transition strength in DGT excitations and 0^+ states are weak, except in $T = 1$ mother nuclei in which 0^+ strength is competitive with 2^+ strength. A possibility to extract the unit cross section for the DGT transition strength is pointed out in the $(\sigma t_-)^2$ excitation of double isobaric analog state (DIAS) in $T = 1$ nuclei.

1 Introduction

The possibility to induce double charge-exchange (DCX) excitations by means of heavy-ion beams at intermediate energies [1, 2] has recently fostered the interest on new collective excitations such as double isobaric analog states (DIAS) and double Gamow-Teller giant resonances (DGTR). In the 1980s, DCX reactions were performed by using pion beams, i.e., (π^+, π^-) and (π^-, π^+) reactions have been studied. Through these experimental investigations, the DIAS, the dipole giant resonance built on the isobaric analog state (IAS) and the double dipole resonance states were identified [3–6]. However, the DGTRs were not found in the pion double charge-exchange spectra. In the middle of the 1990s, the heavy-ion DCX experiments were performed at energies of 76 and 100 MeV/u, with the hope that the DGTR might be observed in the ^{24}Mg(^{18}O, ^{18}Ne)^{24}Ne reaction [7]. However, no clear evidence of DGTR was found in this reaction. This is mainly because the (^{18}O, ^{18}Ne) reaction is a $(2n, 2p)$ type reaction, and even the single GTR in the t_+ channel induced by the (n, p) reaction is weak in $N = Z$ nuclei such as ^{24}Mg. A research program based on a new reaction, namely (^{12}C, ^{12}Be(0_2^+)) has been planned at the RIKEN RIBF facility with high intensity heavy-ion beams at the optimal energy of $E_{lab} = 250$ MeV/nucleon to excite the spin-isospin response [8]. A big advantage of this reaction is based on the fact that it is a $(2p, 2n)$ type DCX reaction and one can use a neutron-rich target to excite DGTR strength. Although many theoretical efforts have been devoted to studies of double β-decays, DGTR strengths corresponding to the double β-decays are still too small to be identified in these experiments. Recently, shell-model calculations were performed to study the DGTR of ^{48}Ti [9], and also Ti-isotopes [10]. At the same time, DGTR strength distributions in several nuclei have been studied by using the sum rule approach [11–15]. In ref.[14], the DGT and DIAS states in $T = 1$ nuclei were studied in the same framework in order to establish a possible unit cross section of DGTR in comparison with the DIAS. The sum rule approach is further extended in ref. [15] to evaluate the energy difference between DGT and DIAS by using separable-type interactions. Minimally-biased theoretical predictions based on sum rules will provide a robust and global view of the DGTR, and can provide a good guideline for the future experimental studies.

*e-mail: sagawa@ribf.riken.jp

2 Sum Rules for Double Spin-Isospin Phonon States

The sum rule for the single GT transitions is well known and proportional to the neutron excess,

$$S_- - S_+ = \sum_f |\langle f|\hat{O}_-(\mathrm{GT})|i\rangle|^2 - \sum_f |\langle f|\hat{O}_+(\mathrm{GT})|i\rangle|^2 = 3(N - Z) \quad (1)$$

where the GT transition operators read

$$\hat{O}_\pm(\mathrm{GT}) = \sum_\alpha \sigma(\alpha)\mathrm{t}_\pm(\alpha). \quad (2)$$

The GT sum rule is model independent and gives a good guidance to perform the single charge exchange reactions such as (p,n) and (^{3}He,t) reactions, to observe GTR strength in many nuclei (see for example a review article ref. [16]). Calculated results of single charge exchange GT strength for t_- and t_+ channels are shown in Fig. 2. A self-consistent Hartree-Fock (HF) + random phase approximation (RPA) are applied with Skyrme EDF SAMi [17] and SAMi-T [18] without and with tensor interactions, respectively. Two peak positions of t_- channels are reasonably well described by RPA calculations. The peak for t_+ channel is also reproduced well by the RPA calculations. For both channels, however, the absolute magnitudes are over-estimated, i.e., by 35% for t_- channel and

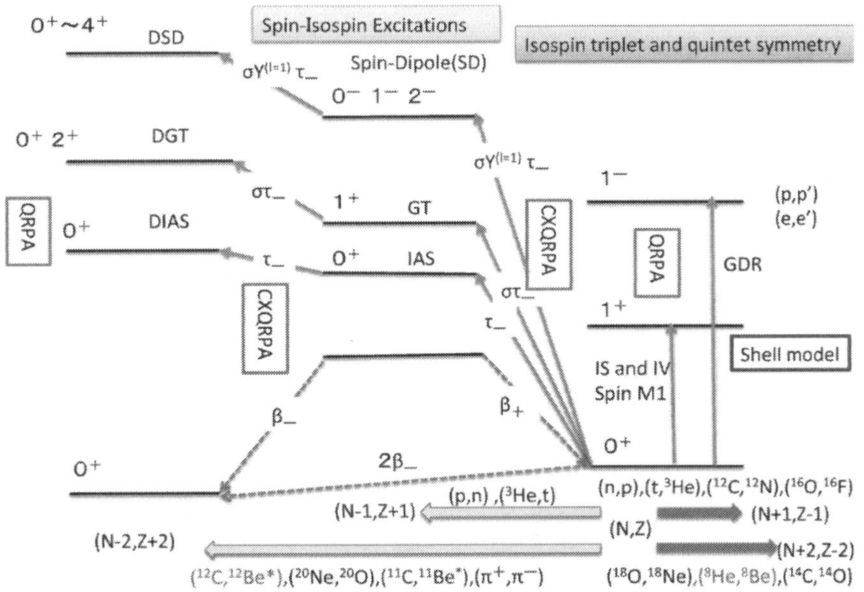

Figure 1. Single- and double-charge exchange excitations and their possible experimental probes. Spin-isospin mode is excited by various charge exchange reactions. (p,n) and $(^3\text{He},t)$ are well adopted for single-charge exchange reactions in experimental studies. Heavy Ion charge exchange reactions are also available to perform double charge exchange reactions. From theoretical sides, RPA, Quasi-particle RPA (QRPA) and charge exchange QRPA (CXQRPA) are often used to study these excitations and modern large scale shell model calculations are also available for the theoretical study.

more for t_+ channel. The quenching factor $qf = 0.64$ is multiplied on the GT strength of t_- channel. These calculations certainly give a confidence for the application of HF+RPA model for double charge exchange excitations.

The DGT transition operator $\hat{O}_\pm(\text{GT})^2$ can be projected to good multipole states to be

$$[\hat{O}_\pm(\text{GT}) \times \hat{O}_\pm(\text{GT})]^{(J)}, J = 0, 2 \tag{3}$$

The sum rule strength is given by the reduced matrix element,

$$D_\pm^{(J)} = \frac{1}{2J_i + 1} \sum_f |\langle f \| [\hat{O}_\pm(\text{GT}) \times \hat{O}_\pm(\text{GT})]^{(J)} \| i \rangle|^2, \tag{4}$$

where J_i is the angular momentum of the initial state and the double bars mean the reduced matrix element for the angular momentum. The sum rule values are evaluated by using the double commutator as

$$\begin{aligned} D_-^{(J=0)} - D_+^{(J=0)} &= \langle i | \left[[\hat{O}_+ \times \hat{O}_+]^{(J=0)}, [\hat{O}_- \times \hat{O}_-]^{(J=0)} \right] | i \rangle \\ &= 2(N-Z)(N-Z+1) \\ &\quad + \frac{4}{3}\left[(N-Z)S_+ - \langle i | [i\hat{\Sigma} \cdot (\hat{O}_- \times \hat{O}_+) + \hat{\Sigma} \cdot \hat{\Sigma}] | i \rangle \right] \end{aligned} \tag{5}$$

where

$$\hat{\Sigma} = \sum_\alpha \sigma(\alpha) \tag{6}$$

and $(\hat{O}_- \times \hat{O}_+)$ is the vector product of two operators. In the last line of Eq. (5), we use an identity

$$(\hat{O}_+ \times \hat{O}_-) \cdot \hat{\Sigma} = -\hat{\Sigma} \cdot (\hat{O}_- \times \hat{O}_+) + 2i\hat{\Sigma} \cdot \hat{\Sigma} \tag{7}$$

which makes the final formula of the sum rule (5) different from those in the references [11, 13], but it is equivalent to the one in ref. [12]. It is important for quantitative study to rewrite the formula by using the identity (7) since the terms including $\hat{O}_+$ on the right can be omitted in $N > Z$ nuclei in a good accuracy. The sum rule for $J^\pi = 2^+$ final states can be obtained in a similar way to that for $J = 0$ states;

$$\begin{aligned} &D_-^{(J=2)} - D_+^{(J=2)} \\ &= \langle i | \sum_\mu (-1)^\mu \left[[\hat{O}_+ \times \hat{O}_+]_\mu^{(J=2)}, [\hat{O}_- \times \hat{O}_-]_{-\mu}^{(J=2)} \right] | i \rangle \\ &= 10(N-Z)(N-Z-2) \\ &+ \frac{10}{3}\left[2(N-Z)S_+ + \langle i | [i\hat{\Sigma} \cdot (\hat{O}_- \times \hat{O}_+) + \hat{\Sigma} \cdot \hat{\Sigma}] | i \rangle \right] \end{aligned} \tag{8}$$

It should be noticed that the sum rule for $J = 2$ final states are not given in refs. [11, 12], but given in ref. [13] in a different form.

Next, we consider the DGT transition to the DIAS state. The DIAS state is defined as

$$|\text{DIAS}\rangle = \frac{T_- T_-}{N_f} |i\rangle \tag{9}$$

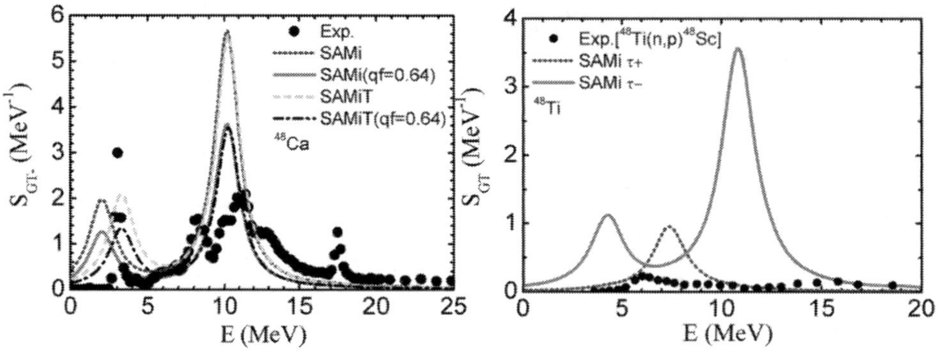

Figure 2. Calculated results of single charge exchange GT strength for t_- and t_+ channels. Skyrme EDF SAMi and SAMi+T are used for self-consistent HF+RPA calculations. For t_- channel, the quenching factor $qf = 0.64$ are introduced for GT transition strength. Experimental data are taken from ref. [19]. Courtesy of L.G Cao.

Table 1. Various sum rule values for DGT transitions to $J^\pi = 0^+, 2^+$ and DIAS states given in Eqs. (5), (8) and (11), respectively. The initial state is listed in the first column with the isospin T. The double Fermi transitions to DIAS (10) are listed in the fourth column. For Eqs. (5) and (8), the spin sum rule $\hat{\Sigma}\cdot\hat{\Sigma}$ terms in the last line is obtained by shell model calculations except ^{90}Zr. The single particle value of Σ_{sum} is used for ^{90}Zr. The single particle values from Eq. (13) are given in the bracket in the sixth column. In the second and third columns, the value in the bracket is the upper limit for $J = 0$, while that is the lower limit for $J = 2$. The values Σ_{sum} for DIAS$^{(J=0)}$(DGT) in Eq. (11) are calculated by using the same shell model calculations as for the DGT sum rules. The value DIAS$^{(J=0)}$(DGT) with the single particle value of Σ_{sum} in Eq. (13) is also listed in the fifth column with brackets. See the text for more details.

Initial state (T)	($J = 0$)	($J = 2$)	DIAS	DIAS(DGT)	Σ_{sum}
^{6}He (1)	12.0(12)	0.0 (0)	4	3.70(3.70)	0.211×10^{-3} (0.667)
^{8}He (2)	39.7(40)	80.7 (80)	24	7.71(2.47)	0.052 (1.333)
^{14}C (1)	8.98(12)	7.55 (0)	4	4.56 (0.15)	0.566 (1.333)
^{18}O (1)	10.4(12)	3.96 (0)	4	7.72 (2.61)	0.297 (0.80)
^{20}O (2)	35.5(40)	91.3 (80)	24	4.13 (1.74)	0.845 (1.60)
^{42}Ca (1)	8.50(12)	8.75 (0)	4	3.80 (2.18)	0.66 (0.86)
^{44}Ca (2)	32.6(40)	98.5(80)	24	2.31 (1.47)	1.38 (1.71)
^{46}Ca (3)	72.3(84)	269.3(240)	60	1.89 (1.32)	2.20 (2.57)
^{48}Ca (4)	135.5(144)	501.2(480)	112	3.70 (1.25)	1.59 (3.43)
^{90}Zr (5)	196.3(220)	859.2(800)	180	(1.11)	(4.44)

where $T_- = \sum_\alpha t_-(\alpha)$ is the isospin lowering operator and the normalization factor is $N_f = \sqrt{2(N-Z)(N-Z-1)}$. The double Fermi transition to DIAS state is evaluated to be

$$|\langle \text{DIAS}|T_-T_-|i\rangle|^2 = 2(N-Z)(N-Z-1) \qquad (10)$$

in which the mother state is assumed to have a good isospin quantum number $T = T_z = (N-Z)/2$. The DGT transition to DIAS is also expressed as

$$\begin{aligned} D_-^{(J=0)}(\text{DIAS}) &= \left|\langle \text{DIAS}|\left[\hat{O}_- \times \hat{O}_-\right]^{(J=0)}|i\rangle\right|^2 \\ &= \left|\frac{1}{N_f}\langle \hat{0}|T_+T_+\left[\hat{O}_- \times \hat{O}_-\right]^{(J=0)}|i\rangle\right|^2 \\ &= \frac{1}{3N_f^2}\left[8\Sigma_{sum} - 2(S_- + S_+)\right]^2 \\ &= \frac{1}{6(N-Z)(N-Z-1)}\left[8\Sigma_{sum} - 2(S_- + S_+)\right]^2 \end{aligned} \qquad (11)$$

where Σ_{sum} is the sum rule of the isovector spin transition,

$$\Sigma_{sum} = \langle i|\sum_{\alpha,\mu}\sigma_\mu(\alpha)t_z(\alpha)\sum_{\beta,\mu'}\sigma_{\mu'}(\beta)t_z(\beta)|i\rangle. \qquad (12)$$

For a single-j configuration with the occupation probability v^2, the value Σ_{sum} is evaluated to be

$$\begin{aligned} \Sigma_{sum} &= \sum_{p,h,\mu}|\langle (ph)1\mu|\sigma_\mu t_z|\hat{0}\rangle|^2 \\ &= \frac{v_h^2}{4}\frac{(2j_h+1)(2j_h-1)}{j_h} \quad \text{for } j_h = j_p + 1 \end{aligned} \qquad (13)$$

where we use a filling approximation for the last occupied orbit.

Calculated sum rule values for DGT transitions to $J^\pi = 0^+$ and 2^+ in Eqs. (5) and (8), respectively, are listed in Table 1 for 10 nuclei, together with those for DIAS states by Fermi and GT operators (10) and (11). In $N > Z$ nuclei with the good isospin quantum number, the vector product

$(\hat{O}_- \times \hat{O}_+)$ and the sum S_+ in Eqs. (5) and (8) vanish because of the isospin raising operator in $\hat{O}_+$. The spin sum rule $\hat{\Sigma}\cdot\hat{\Sigma}$ term is obtained by shell model calculations. The single particle value is used for ^{90}Zr assuming the neutron excitation from $1g_{9/2}$ to $1g_{7/2}$. The value in the bracket is obtained without the spin sum rule term, i.e., the upper limit for $J = 0$, while the lower limit for $J = 2$. The values Σ_{sum} for DIAS$^{(J=0)}$(DGT) in Eq. (11) are calculated by using the same shell model wave functions as for $\Sigma \cdot \Sigma$ and shown in the sixth column in Table 1. The single particle values of Σ_{sum} in Eq. (13) are given in the bracket in the sixth column. The value DIAS$_-^{(J=0)}$(DGT) with the single particle value of Σ_{sum} is also listed in the fifth column with brackets. In all the nuclei in the present study, the value of Σ_{sum} is one-fourth of $\hat{\Sigma}\cdot\hat{\Sigma}$ because of semi-closed configuration (either proton or neutron configuration is a closed shell).

The values Σ_{sum}, equivalently $\hat{\Sigma}\cdot\hat{\Sigma}$, are almost negligible for p-shell nuclei ^{6}He and ^{8}He, because of good SU(4) spin-isospin symmetry in which the spin transitions are allowed only for spin and isospin flip transition between $(S,T) = (0,1) \to (1,0)$. This selection turns out to be the selection of transitions $(J^\pi, T) = (0^+, 1) \to (1^+, 0)$ in ^{6}He and $(J^\pi, T) = (0^+, 2) \to (1^+, 1)$ in ^{8}He, corresponding to the charge exchange GT transitions. The shell model values Σ_{sum} are smaller than the single-particle values in other nuclei by a factor 1.2 ~ 2.5. The effect of $\hat{\Sigma}\cdot\hat{\Sigma}$ term for $J = 0$ and $J = 2$ sum rules are negligible for ^{6}He and ^{8}He, while it amounts at most 20% in other nuclei with $T > 1$ and it becomes much larger in $T = 1$ nuclei such as ^{14}C, ^{18}O and ^{42}Ca.

In $N >> Z$ nuclei, the sum rule values of DGT transitions are approximately proportional to (2J+1) factor, i.e., the value for $J = 2$ is 5 times larger than that for $J = 0$ in the same nucleus. However, this proportionality is strongly modified in $N \sim Z$ nuclei. In the extreme, the sum rule for $J = 2$ transitions is smaller than that for $J = 0$ in the nuclei ^{6}He, ^{14}C, and ^{18}O with $N = Z + 2$, and the two values are almost equal in ^{42}Ca.

It is noticed that the DGT transition to DIAS state is much smaller than the sum rule values of DGT transitions for $J = 0$ and $J = 2$ states in nuclei with $N > Z + 2$. However, in $T = 1$ nuclei, the DGT transition to DIAS state is comparable with the DGT sum rules (5) and (8). This characteristics turns out the possibility to extract the unit cross section for DGT transition in $T = 1$ nuclei, especially ^{14}C and ^{18}O, since the DIAS state might have a narrow width and may be distinguished clearly from other DGT states in double charge exchange reactions [3]. This argument is entirely benefitted from the fact that the spin-isospin response is much favorable that the isospin response in the medium energy charge exchange reactions with E_{lab} ~200MeV/u.

3 Summary

In summary, we studied the sum rules of DGT excitations by the operator $(\sigma t_-)^2$. In nuclei $N >> Z$, the $J^\pi = 2^+$ excitations dominate the DGT strength, due to the multipole factor $(2J + 1)$, more than 0^+ excitations. However, in nuclei with $N \sim Z$, the 0^+ excitations become substantially strong, even larger than 2^+ excitations in light nuclei with $T = 1$, ^{14}C an ^{18}O. The excitation to DIAS is also studied through $(\sigma t_-)^2$ and t_-^2 operators to investigate the possibility to extract the unit cross sections for DGT strength. We pointed out that the strength of DIAS excitations by $(\sigma t_-)^2$ operator is competitive to DGT strength in the light $T = 1$ nuclei. This characteristic feature may give a good opportunity to extract the unit cross section for DGT strength.

Acknowledgments

This work was supported in part by JSPS KAKENHI Grant Numbers JP19K03858.

References

[1] M. Takaki, T. Uesaka et al., Proposal for experiment at RCNP, "Search for double Gamow Teller giant resonances in ^{48}Ti via the heavy-ion double charge exchange ^{48}Ca(^{12}C, ^{12}Be(0_2^+)) reaction" (2015).

[2] F. Cappuzzello, C. Agodi, M. Bond? D. Carbone, M. Cavallaro, A. Foti, Journal of Physics: Conference Series 630, 012018 (2015).

[3] M. Kaletka, K. K. Seth, A. Saha, D. Barlow, and D. Kielczewska, Physics Letters B 199, 336 (1987).

[4] S. Mordechai and C.F. Moore, Nature 352, 393 (1991)

[5] H. Ward et al. , Phys. Rev. Lett. 70, 3209 (1993)

[6] Ph. Chomaz and N. Frascaria, Phys. Rep. 252, 275 (1995).

[7] J. Blomgren et al., Phys. Lett. B362, 34 (1995).

[8] T. Uesaka, M. Takaki et al., Proposal for Nuclear Physics Experiment at RI Beam Factory " Search for Double Gamow-Teller Giant Resonances in $\beta\beta$−decay nuclei via the heavy-ion double charge exchange^{48}Ca(^{12}C, ^{12}Be(0_2^+)) reaction" (2016).

[9] N. Shimizu, J. Menendez and K. Yako, Phys. Rev. Lett. 120, 142502 (2018).

[10] N. Auerbach and B. M. Loc, Phys. Rev. C98, 064301 (2018).

[11] P. Vogel, M. Ericson and J. D. Vergados, Phys. Lett. B212, 259(1988).

[12] D. C. Zheng, L. Zamick and N. Auerbach, Phys. Rev. C40, 936 (1989).

[13] K. Muto, Phys. Lett. B277, 13 (1992).

[14] H. Sagawa and T. Uesaka, Phys. Rev. C 94, 064325 (2016).

[15] X. Roca-Maza, H. Sagawa and G. Colò, arXiv:1907.06368(2019).

[16] M. Ichimura, H. Sakai, and T. Wakasa, Prog. in Part. Nucl. Phys. 56, 446 (2006).

[17] X. Roca-Maza, G. Colò, and H. Sagawa, Phys. Rev. C 86, 031306(R)(2012).

[18] S. H. Shen, G. Colò and X. Roca-Maza, Phys. Rev. C 99, 034322(2019).

[19] K. Yako et al., Phys. Rev. Lett. 103, 012503 (2009).

EPJ Web of Conferences **223**, 01054 (2019) https://doi.org/10.1051/epjconf/201922301054
NSD2019

Probing the nuclear structure in the vicinity of ^{78}Ni

E. Sahin[1,*], *K.* Hadyńska-Klęk[1,2,3,4], *F.L.* Bello Garrote[1], and *A.* Görgen[1] for the EURICA and SEASTAR collaborations

[1]University of Oslo, Department of Physics, Oslo, Norway.
[2]Laboratori Nazionali di Legnaro dell'INFN, Legnaro, Italy
[3]University of Surrey, Department of Physics, Surrey, UK
[4]Heavy Ion Laboratory, University of Warsaw, Poland

Abstract. Theoretical and experimental studies of neutron-rich nuclei have shown that the general concept of shell structure is not as robust and universal as earlier thought, but can exhibit significant changes as a function of neutron excess. New magic numbers appear and some other conventional ones disappear mainly because of a different ordering of the single-particle orbitals. In the present contribution, recent experimental studies of neutron-rich Cu isotopes, performed at RIKEN using β decay and one-proton knockout reactions, will be discussed. Neutron-rich nuclei near ^{78}Ni were populated through in-flight fission of ^{238}U on thick ^{9}Be targets in both experiments. In the β-decay study, ^{75,77}Ni nuclei were implanted into the WAS3ABi silicon array, while γ rays from excited states in ^{75,77}Cu emitted after β decay of the implanted ions were detected with the EURICA Ge detector array that was surrounding the active stopper. In a second experiment within the SEASTAR campaign at RIKEN, the same ^{75,77}Cu nuclei were produced in (p,2p) knockout reactions from ^{76,78}Zn beam particles at around 250 MeV/nucleon impinging onto the MINOS liquid hydrogen target. In the latter experiment the DALI2 NaI array was used to detect de-excitation γ rays measured in coincidence with Cu nuclei identified in the Zero Degree Spectrometer. Both studies are complimentary and greatly contribute to our understanding on the nuclear structure in the ^{78}Ni region.

1 Introduction

Over the last two decades, investigations at radioactive ion beam facilities have revealed that atomic nuclei undergo significant changes in their shell structure when moving towards neutron-rich isotopes. This so-called *shell evolution* for nuclei with large neutron-to-proton asymmetry is due to specific properties of the nucleon-nucleon interaction, which can lead to a different ordering of single-particle orbitals and, hence, to the appearance or disappearance of shell gaps. The phenomenon is by now well established in lighter mass regions with $Z < 28$ and $N < 40$.

As an example, Figure 1a shows the energies of the first excited 2^+ states, $E(2_1^+)$, as a function of neutron number N for isotopes of Si, S, Ar, and Ca. One can clearly see the disappearance of the neutron $N = 28$ shell gap in ^{42}Si [1], ^{44}S [2], and ^{46}Ar [3], as evidenced by their low $E(2_1^+)$ energies. On the other hand, the excitation energies for ^{52,54}Ca are very high, indicating the appearance of new shell gaps at $N = 32$ and 34. The shell gap at $N = 34$ appears to be rather robust, as illustrated by the high excitation energy of the 2_1^+ state in ^{52}Ar [4]. The reduction or even collapse of shell closures when going far away from stability is connected with an onset of deformation. The phenomenon where deformation is favored due to particle-hole excitations across major shells has been termed "island of inversion". In general, the coexistence of intruder and normal configurations leads to shape coexistence, which consequently can be expected in regions of disappearing shell closures. As an illustrative example ^{44}S can be mentioned, where a prolate 0_1^+ ground state coexists with a spherical excited 0_2^+ state [5]. In this case the reduction of the $N = 28$ shell gap enables quadrupole excitations from the $1f_{7/2}$ to the $2p_{3/2}$ orbitals.

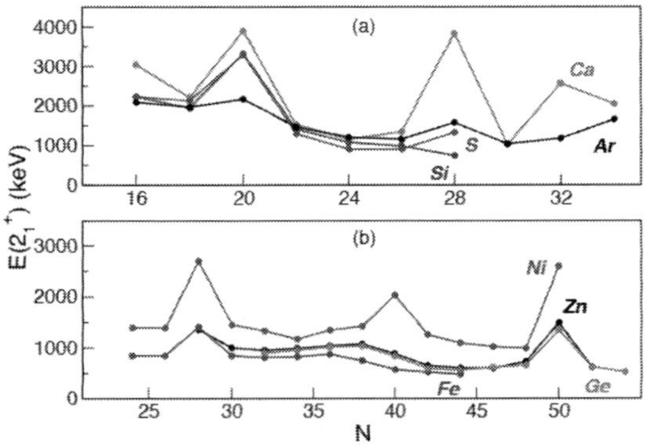

Figure 1. First excited 2^+ states as a function of neutron number (a) for Si, S, Ar, Ca and (b) for Fe, Ni, Zn, and Ge isotopes. Data is taken from References [4, 6, 7].

*e-mail: eda.sahin@fys.uio.no

In the light of these results, questions arise concerning the robustness of the $N = 50$ shell gap for neutron-rich nuclei, the magic character of ^{78}Ni, and possible intruder states and an onset of deformation in the region around ^{78}Ni. Experimental information in this region of the nuclear chart is very sparse due to the extreme proton-to-neutron ratio, which makes production of these nuclei difficult and experimental studies in general very challenging. The energy of the 2^+_1 state in ^{78}Ni was only very recently measured to be 2.6 MeV [6]. This high energy supports the notion of doubly magic character for ^{78}Ni and the persistence of the $Z = 28$ and $N = 50$ shell gaps (see Figure 1b). However, detailed studies on the evolution of single-particle orbitals in this mass region are still very much incomplete. The aim of the experiments that are presented here was to contribute to a more quantitative understanding of the shell structure near ^{78}Ni. More specifically, the evolution of the proton single-particle energies will be discussed as a function of neutron number, together with the consequences for the occupation of intruder configurations and the onset of deformation and collectivity.

2 Evolution of the Z=28 shell gap

Changes in the single-particle structure are strongly related to certain properties of the (residual) interactions between nucleons. In particular the tensor component of the monopole interaction was found to be responsible for the changes in the single-particle energies (SPEs) [8]. The tensor force is attractive between two orbits with relative spin orientations $j_> = l + 1/2$ and $j'_< = l' - 1/2$ (or $j_< = l - 1/2$ and $j'_> = l' + 1/2$) and repulsive between those with spin orientations $j_>$ and $j'_>$ (or $j_<$ and $j'_<$).

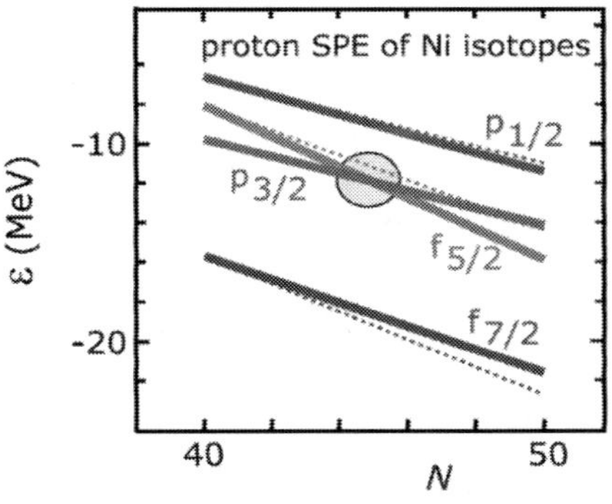

Figure 2. Calculated proton SPEs for the Ni chain. Figure is taken from Ref. [9]. The same color code is also used in Figures 3, 4, and 6.

The evolution of the $Z = 28$ shell gap was previously investigated by systematically studying the properties of the Ni isotopes from ^{68}Ni ($N = 40$) to ^{78}Ni ($N = 50$). Figure 2 shows the resulting proton SPEs as a function of neutron number from shell model calculations for the Ni isotopes [9]. Note that the SPEs change more dramatically when including the tensor component of the monopole interaction (solid lines), while the relative changes are less pronounced when the tensor component is not included (dashed lines). According to these calculations, the changes in the proton SPEs lead to an inversion of the $\pi 2p_{3/2}$ and $\pi 1f_{5/2}$ orbitals with increasing number of neutrons in the $\nu 1g_{9/2}$ orbital. The crossing of the $\pi 2p_{3/2}$ and $\pi 1f_{5/2}$ orbitals is consistent with the measured $5/2^-$ spin-parity for the ground state in ^{75}Cu [10]. Furthermore, the calculations predict a reduction for the $Z = 28$ shell gap as a result of the attraction between the $\pi 1f_{5/2}$ and the $\nu 1g_{9/2}$ and the repulsion between the $\pi 1f_{7/2}$ and $\nu 1g_{9/2}$ orbitals.

Proton SPEs as a function of neutron number can be determined by studying single-proton excitations in the Cu ($Z = 29$) isotopes, where the last uncoupled proton resides above the $Z = 28$ gap in the $f_{5/2}p$ shell ($1f_{5/2}$, $2p_{3/2}$, $2p_{1/2}$). Indeed, the neutron-rich copper isotopes from ^{69}Cu (N=40) to ^{79}Cu (N=50) have played an important role to study the single-particle structure in the ^{78}Ni region. Previous experimental and theoretical studies, however, indicated that the low-lying states in Cu nuclei can have different origins [11, 12] and can be classified into the following types:

- the $3/2^-$ and $5/2^-$ "*single-particle*" states, dominated by the single-proton configurations, $\pi 2p_{3/2}$ and $\pi 1f_{5/2}$, respectively,
- the "*particle-core coupled*" states from $9/2^-$ to $1/2^-$ and from $7/2^-$ to $1/2^-$ will appear as a result of the $\pi 2p_{3/2}$ and $\pi 1f_{5/2}$ single-particle excitations coupled to the first excited 2^+ state of the even-even Ni core, denoted as $2p_{3/2}\otimes 2^+_1$ and $1f_{5/2}\otimes 2^+_1$, respectively
- the $7/2^-$ "*intruder*" state arises from the excitation of a proton from the $1f_{7/2}$ orbital across the $Z = 28$ shell gap, leaving a hole in the $1f_{7/2}$ orbit, *i.e.* $\pi 1f^{-1}_{7/2}$.

2.1 Beta-decay of ^{75,77}Ni

Cu nuclei have been studied in detail up to ^{73}Cu via different reaction mechanisms such as β-decay, Coulomb excitation, and single- and multi-nucleon transfer reaction studies [13–21]. For the heavier isotopes ^{75}Cu and ^{77}Cu, only the spin and parity of the ground state was known [10], and, in case of ^{75}Cu, two low-lying isomeric states [22, 23]. To identify low-lying states and establish the level schemes for ^{75}Cu and ^{77}Cu, β-decay experiments were carried out at RI Beam Factory (RIBF) of the RIKEN Nishina Center [24]. The secondary beam particles near ^{78}Ni were produced via fission reaction of the ^{238}U primary beam on a thick ^{9}Be target. After particle identification in atomic number (Z) and mass-to-charge ratio (A/Q) using the $TOF - B\rho - \Delta E$ technique [25] in the BigRIPS fragment separator [26, 27], the ions were delivered to the experimental setup through the ZeroDegree Spectrometer (ZDS) [27] and implanted in a stack of eight highly pixelised double-sided silicon strip detectors (WAS3ABI [28]). Each DSSSD had 60 horizontal and 40 vertical strips of 1 mm pitch, respectively. Surrounding the implantation detector was the EURICA germanium detector array [28, 29] for the detection of γ rays emitted after β decay of the implanted ions.

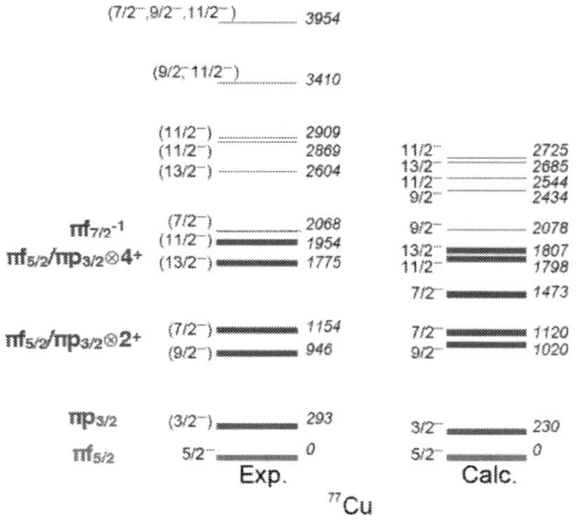

Figure 3. Excited states of ^{77}Cu compared to the MCSM calculations. The orbital configurations are indicated for the lowest states.

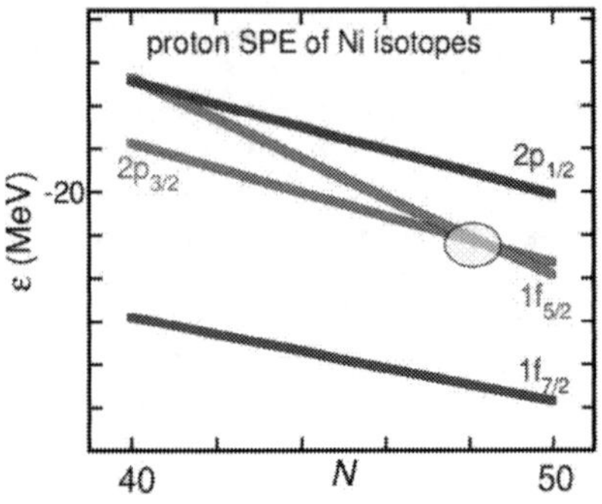

Figure 4. Calculated proton SPEs for the Ni chain. Figure is adapted from Ref. [32].

The level structure of ^{75,77}Cu was investigated through γ-ray spectroscopy following the β decay of ^{75,77}Ni. The implanted Ni ions were correlated on an event-by-event basis in time and position with β decays detected in WAS3ABI. It was required that the β-decay electrons were detected in the same DSSSD within a correlation area that covered up to two pixels away from the implantation position. The time correlation between β-decay events and detected γ rays was used to identify γ-ray transitions as originating from ^{75,77}Cu or their respective daughter decays. Due to the high efficiency of EURICA it was possible to analyze γ-γ coincidences and use the coincidence relationships to build level schemes up to around 4 MeV excitation energy for both nuclei. A total of 28 new excited states were identified in ^{75}Cu [33] and 12 in ^{77}Cu [32]. Tentative spin-parity values were assigned for all states in ^{77}Cu based on the β-decay feeding (log(ft) values) and γ-decay properties, whereas this was only possible for the lowest states in ^{75}Cu. Figure 3 shows the level scheme for ^{77}Cu obtained from the present data together with results from state-of-the-art Monte Carlo Shell Model calculations [30, 31]. The calculations used the A3DA interaction [30] with rather large valence space, *i.e.* the full fp shell ($1f_{7/2}$, $1f_{5/2}$, $2p_{3/2}$, $2p_{1/2}$) plus the $1g_{9/2}$ and $2d_{5/2}$ orbitals outside a ^{40}Ca core.

There is rather good agreement between the experimental results and the results from the shell model calculations, in particular for the states up to 2 MeV excitation energy. All low-lying states are explained as having either single-particle, core-coupled, or intruder character as explained above. With new experimental results available, the shell model calculations have been refined compared to the previous results shown in Figure 2. The SPEs from the new shell model calculations are shown in Figure 4. The results indicate that the inversion of the $\pi 2p_{3/2}$ and $\pi 1f_{5/2}$ orbitals does not occur at $N = 46$ (^{75}Cu), as was reported in 2010 in the work of Otsuka et al. [9], but occurs instead at $N = 48$ (^{77}Cu). The fact that the $5/2^-$ state is lower than the $3/2^-$ state already in ^{75}Cu is explained by correlation effects due to multipole interaction, which are stronger for the $5/2^-$ state than for the $3/2^-$ state [32]. Furthermore, with the crossing of these two orbits found to occur later than previously thought, the reduction of the $Z = 28$ shell gap is also smaller than in previously calculations. As one of the main results from the present β-decay study it can be concluded that the size of the $Z = 28$ shell gap is reduced from 6.5 MeV at $N = 40$ to 5 MeV at $N = 50$ [32]. More detailed discussions, together with extended results on ^{75}Cu and their comparison to the large-scale shell-model calculations, will be presented in a forthcoming article [33].

3 Proton excitations across $Z = 28$

The β-decay experiment discussed above gave insight into the single-particle properties in the ^{78}Ni region based on comparisons with shell model calculations. It also gave a first indication for a $7/2^-$ intruder state based on a proton excitation across the $Z = 28$ shell gap (see Figure 4). Complementary experiments using direct reaction mechanism such as one-proton knockout selectively populate excited states arising from single-proton excitations. Such experiments can give a more direct identification of single-particle states based on $2p_{3/2}$ and $1f_{5/2}$ protons, and, in particular, of the $(\pi 1f_{7/2})^{-1}$ particle-hole excitation across the $Z = 28$ shell gap. As the interpretation of the $7/2^-$ states found in the β-decay experiment is less straightforward as compared to the other low-lying states, complementary knock-out experiments can provide essential information on the evolution of the shell structure close to ^{78}Ni. The location of the $7/2^-$ $2p-1h$ intruder state along the chain of the Cu isotopes is important to provide sensitive information on the expected shape coexistence and onset of deformation in the ^{78}Ni region [31].

3.1 One-proton removal reaction

Experiments to study neutron-rich Cu isotopes in knockout reactions were performed within the SEASTAR collaboration and campaign at RIKEN. The ^{75,77}Cu nuclei were produced through one-proton knockout of ^{76,78}Zn beam particles. In terms of reaction mechanism, the one-proton knockout strength is expected to be largest for the $1f_{7/2}$ orbital since there are almost 8 protons in the ground-state configuration of the ^{76,78}Zn isotopes, while the remaining single-particle strength due to 2 protons above $Z = 28$ will be mostly shared between the $1f_{5/2}$, $2p_{3/2}$, and $1p_{1/2}$ orbitals. As a consequence, the cross section to populate the $7/2^-$ particle-hole state is expected to be large, and the corresponding state can be clearly identified.

The experiment used again an intense primary beam of ^{238}U and a primary ^{9}Be target. The BigRIPS separator was optimized to select neutron-rich Zn isotopes. Identification of the incoming projectiles and outgoing reaction products was performed on event-by-event basis similar to the technique mentioned in Sect. 2.1. For the purpose of one-proton knockout, a reaction target of 10 cm thick liquid hydrogen, surrounded by the active target time projection chamber MINOS [34], was exploited during the experiment. The DALI2 array [35], coupled to the MINOS device, was used for the detection of the γ rays emitted from the reaction products. The setup allowed performing event-by-event Doppler correction of the γ rays detected in DALI2 using the reaction vertex information reconstructed in MINOS. Details of the procedure can be found in Ref. [36].

Figure 5 shows the Doppler-corrected γ-ray spectra obtained for ^{75}Cu and ^{77}Cu resulting from the ^{76}Zn$(p,2p)^{75}$Cu and ^{78}Zn$(p,2p)^{77}$Cu knockout reactions ,respectively. Eight transitions of 490(10), 530(10), 880(10), 950(10), 990(10) 1410(20), 1480(20), and 1680(30) keV energies are visible in the spectrum for ^{75}Cu (Figure 5a) while six transitions of 290 (10), 860(20), 950(20), 1150(20), 1740(30), and 2060(20) keV are seen in in the spectrum for ^{77}Cu (Figure 5b). The knowledge of the level schemes of ^{75}Cu and ^{77}Cu from the β-decay experiments was of great help for the identification of transitions in the spectra following the knock-out reactions. In fact, all γ-ray transitions indicated in Figure 5 were also seen following the β decay of ^{75}Ni and ^{77}Ni, respectively [32, 33].

The transition at 290 keV lies in the low-energy region where a large atomic background makes its identification difficult. The spectrum shown in the inset of Figure 5b was obtained using only DALI2 detectors at forward angles. Due to a relatively long lifetime of the $(3/2^-)$ state at 290 keV excitation energy, its decay transition appears shifted to 260 keV under these conditions. Using the line-shape of this transition depopulating the level at 290 keV, a lower limit of about 0.5 ns for the half life of the $(3/2^-)$ state could be extracted in the present experiment. Figure 6 shows examples of the coincidence spectra gated on the 530- and 490-keV transitions in ^{75}Cu. The coincidence relation between 530 keV and 880 keV and 950 keV and between 490 keV and 990 keV γ-ray transitions confirms the level scheme obtained in Ref. [33]. The similar analysis has been performed for ^{77}Cu and confirms the level scheme reported in Ref. [32]

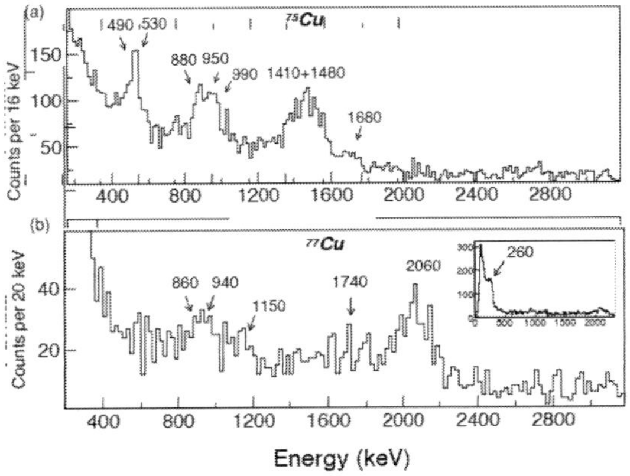

Figure 5. Doppler corrected γ-ray spectra (a) for ^{75}Cu and (b) for ^{77}Cu. Inset: The same spectrum as (b) obtained only using DALI2 detectors at forward angles.

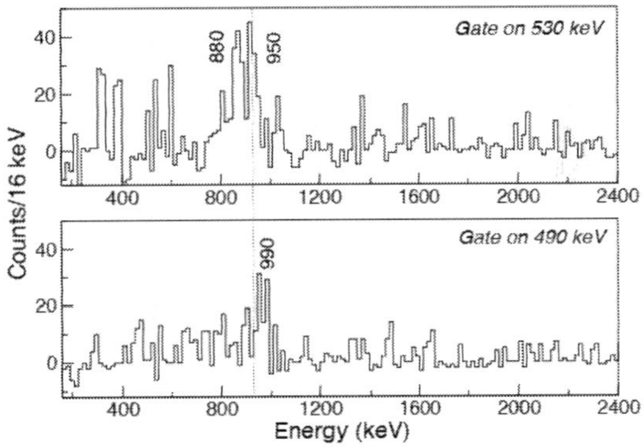

Figure 6. Background subtracted γ-γ coincidences for ^{75}Cu. The spectra gated on 530-keV and on 490-keV transitions are shown in upper and bottom panel, respectively.

From Figure 5 it is evident that the transitions at around 1.5 MeV in ^{75}Cu and 2 MeV in ^{77}Cu are more strongly populated compared to the other transitions seen in the spectra. Accordingly, the state at 1483 keV in ^{75}Cu and at 2068 keV in ^{77}Cu will have the largest population through the (p,2p) reaction and, hence, can be assigned as the states based on the $(\pi 1 f_{7/2})^{-1}$ particle-hole excitation. The detailed analysis including extraction of the exclusive cross sections and spectroscopic factors will be reported in another forthcoming article [37].

The preliminarily results from the one-proton knockout and beta-decay experiments at RIKEN can be translated into systematics of the excited states along the Cu chain. Figure 7 shows the energy of the low-lying states of the Cu isotopes as a function of the neutron number. From the systematics we can make the following observations:

- An increase in the energy of the $3/2^-$ single-particle state (green full dots) and a decrease in the $5/2^-$ single-particle state (red full squares) state is clearly visible as a result of the monopole tensor interaction, which is repulsive between the $\pi 2p_{3/2}$ and $\nu 1g_{9/2}$ and strongly attractive between the $\pi 1f_{5/2}$ and $\nu 1g_{9/2}$ orbitals, respectively.

- The particle-core coupled states (orange full diamonds) dominate the higher-energy regions. For the lighter isotopes from $N = 36$ to 44 one observes $7/2^-$ and $11/2^-$ states based largely on the $2p_{3/2}\otimes 2_1^+$ coupling, while for the heavier isotopes from $N = 46$ to 48 one observes $9/2^-$ and $13/2^-$ states based largely on the $1f_{5/2}\otimes 2_1^+$ coupling. All these core-coupled states follow very closely the trend of the 2_1^+ and 4_1^+ states in the corresponding $^{(A-1)}$Ni core.
- The $7/2^-$ "*intruder*" state (blue full triangle) exhibits a parabolic behavior indicating a possible existence of shape coexistence along the copper chain from $N = 40$ to $N = 50$. Similar parabolic behaviors have been observed in different mass regions such as the Hg (Z=80), Pb (Z=82), and Sn (Z=50) isotopic chains [38, 39].

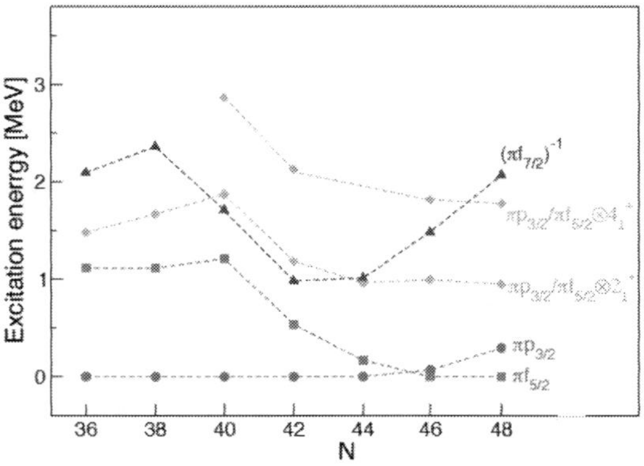

Figure 7. Systematics of the low-lying states in the Cu nuclei. Experimental values are from References [7, 12, 14, 32, 33]

4 Conclusion

In conclusion, the status of the current spectroscopic studies for ^{75}Cu and ^{77}Cu performed at RIKEN has been discussed. Level schemes up to around 4 MeV were constructed for both nuclei from the β-decay studies of ^{75}Ni and ^{77}Ni within the EURICA campaign. The single-particle character of the low-lying states was identified more selectively via one-proton knockout reactions of ^{76}Zn and ^{78}Zn projectiles on a proton target within the SEASTAR campaign. The results were compared to large-scale shell model calculations, yielding information about the single-particle structure in the ^{78}Ni region in general and the size of the $Z = 28$ shell gap in particular. The results allow a significant extension of the systematics along the chain of Cu isotopes. The identification of intruder states suggest that deformation and shape coexistence are important for the ^{78}Ni region. Further results and more detailed discussions will be presented in forthcoming publications.

References

[1] B. Bastin *et al.*, Phys. Rev. Lett. **99** (2007) 022503.
[2] T. Glasmacher *et al.*, Phys. Lett. B**395** (1997) 163.
[3] H.Scheit *et al.*, Phys.Rev.Lett. **77**, 3967 (1996)
[4] H.N. Liu *et al.*, Phys.Rev.Lett. **122**, 072502 (2019).
[5] C. Force *et al.*, Phys.Rev.Lett. **105**, 102501 (2010).
[6] R. Taniuchi *et al.*, Nature (London) **569**, 53 (2019).
[7] NNDC-National Nuclear Data Center, https://www.nndc.bnl.gov.
[8] T. Otsuka, T. Suzuki, R. Fujimoto, H. Grawe, and Y. Akaishi, Phys. Rev. Lett. **95**, 232502 (2005).
[9] T. Otsuka *et al.*, Phys. Rev. Lett. **104**, 012501 (2010).
[10] K.T. Flanagan et al., Phys. Rev. Lett. 103, 142501 (2009).
[11] A.M. Oros-Peusquens and P.F. Mantica., Nucl. Phys. A **669** (2000).
[12] T. Ishii, M. Asai,M. Makishima, I. Hossain, M. Ogawa, J. Hasegawa, M. Matsuda, and S. Ichikawa, Phys. Rev. Lett. **84**, 39 (2000).
[13] S. Franchoo *et al.*, Phys. Rev. Lett. **81**, 3100 (1998).
[14] S. Franchoo *et al.*, Phys. Rev. C **64** 054308 (2001).
[15] I. Stefanescu *et al.*, Phys. Rev. Lett. **100**, 112502 (2008).
[16] I. Stefanescu *et al.*, Phys. Rev. Lett. **100**, 112502 (2008).
[17] B. Zeidman and J. A. Nolen, Phys. Rev. **C18**, 2122 (1978).
[18] P. Morfouace *et al.*, Phys.Lett. B **751**, 306 (2015).
[19] P. Morfouace *et al.*, Phys.Rev. C **93**, 064308 (2016).
[20] M. Doncel *et al.*, Acta Phys. Pol. **B44**, 505 (2013).
[21] E. Sahin *et al.*, Phys. Rev. C **91** 034302 (2015).
[22] J.M. Daugas et al., Phys. Rev. C **81**, 034304 (2010).
[23] C. Petrone at al., Phys. Rev. C **94**, 024319 (2016).
[24] T. Onishi *et al.*, J. Phys. Soc. Jpn. **79**, 073201 (2010).
[25] N. Fukuda, T. Kubo, T. Ohnishi, N. Inabe, H. Takeda, D. Kameda, and H. Suzuki, Nucl. Instrum. Methods Phys. Res., Sect. B **317** 323 (2013).
[26] T. Kubo, Nucl. Instrum. Methods Phys. Res., Sect. B **204** 97 (2003).
[27] T. Kubo *et al.*, Prog. Theor. Exp. Phys. **2012** 3C003 (2012).
[28] S. Nishimura, Prog. Theor. Exp. Phys. **2012** 03C006 (2012).
[29] P.-A. Söderström, S. Nishimura, P. Doornenbal, G. Lorusso, T. Sumikama, H. Watanabe, Z. Xu, H. Baba, F. Browne, S. Go et al., Nucl. Instrum. Methods Phys. Res., Sect. B **317** 649 (2013).
[30] N. Shimizu, T. Abe, Y. Tsunoda, Y. Utsuno, T. Yoshida, T. Mizusaki, M. Honma, and T. Otsuka, Prog. Theor. Exp. Phys. **2012** 01A205 (2012).
[31] Y. Tsunoda, T. Otsuka, N. Shimizu, M. Honma, and Y. Utsuno, Phys. Rev. C **89**, 031301(R) (2014).
[32] E. Sahin *et al.*, Phys. Rev. Lett. **118**, 242502 (2017).
[33] F. Bello Garrote *et al.*, submitted to Phys. Rev. C.
[34] A. Obertelli *et al.*, Eur. Phys. J. A **50**, 8 (2014).

[35] S. Takeuchi, T. Motobayashi, Y. Togano, M. Matsushita, N. Aoi, K. Demichi, H. Hasegawa, and H. Murakami, Nucl. Instrum. Methods Phys. Res., Sect. A **763**, 596 (2014).
[36] C. Santamaria *et al.*, Phys. Rev. Lett. **115**, 192501 (2015).
[37] E. Sahin *et al.*, submitted to Phys. Rev. Lett.
[38] J. Elseviers *et al.*, Phys. Rev. C **84** 034307 (2011).
[39] K. Heyde and J. L. Wood, Rev. Mod. Phys. **83**, 1467 (2011).

Role of charge equilibration in multinucleon transfer in damped collisions of heavy ions

Vyacheslav Saiko[1,2,*] and *Alexander* Karpov[1,2,**]

[1]Flerov Laboratory of Nuclear Reactions, JINR, Dubna, Russia
[2]Dubna State University, Dubna, Russia

Abstract. In this work the charge equilibration process has been analyzed within the Langevin-type dynamical approach. Its duration and energy dependence are discussed. We have analyzed the isotopic distributions of final products obtained in the isospin-asymmetric ^{58}Ni,^{40}Ca + ^{208}Pb reactions. Comparison of ^{58}Ni,^{64}Ni + ^{208}Pb systems have been done in order to analyze the final yields of neutron-rich heavy nuclides.

1 Introduction

Multinucleon transfer (MNT) is one of the main processes observed in collisions of heavy ions at energies near the Coulomb barrier. It leads to the production of exotic nuclei formed in channels with a transfer of dozens of nucleons between projectile and target. Nowadays this type of reaction is considered as a promising method of production of neutron-rich isotopes of heavy elements hardly achievable by other reactions, e.g. fusion or fragmentation. These nuclides and their properties are crucial for detailed understanding of the pathway of the astrophysical r-process.

Langevin-type dynamical models are powerful tools for analyzing the processes of low-energy nuclear collisions (e.g. fusion, fission, deep-inelastic scattering) [1–3]. They provide an appropriate macroscopic description of various measurable characteristics such as angular, mass, and energy distributions of reaction products by taking into consideration the concept of friction and the statistical fluctuations in collisions of heavy nuclei.

One of such models has been recently developed and has successfully described a bulk of experimental data obtained in the following MNT reactions: ^{136}Xe + ^{198}Pt,^{208}Pb,^{209}Bi; ^{160}Gd + ^{186}W; ^{208}Pb + ^{208}Pb; ^{238}U + ^{238}U,^{248}Cm using the same set of the model parameters [4, 5]. All these combinations are characterized by close values of neutron-to-proton ratio N/Z of projectiles and targets and can be called isospin-symmetric ones. In the present work we aimed at analysis of isospin-asymmetric systems consisted of nuclei with quite a different N/Z: ^{40}Ca,^{58}Ni + ^{208}Pb (see Table 1). In this case, the process of reaching an equilibrium of the N/Z value during collision usually called the charge equilibration (or isospin transport) plays a significant role in reaction dynamics and formation of the fragments.

*e-mail: saiko@jinr.ru
**e-mail: karpov@jinr.ru

2 Model

A collision of two heavy nuclei is considered in the model in a continuous way as a three-step process. At the first stage, nuclei approach each other, then they form a mononuclear system, which evolves and finally decays into two fragments. The evolution of the system of colliding nuclei is governed by the multidimensional potential energy calculated within the macro-microscopic approach [6], based on the two-center shell model (TCSM) [7]. We use eight degrees of freedom in the model, four of them originate from the TCSM parametrization and define nuclear shapes: elongation, two ellipsoidal deformations, and mass asymmetry. The charge asymmetry $\eta_Z = (Z_2 - Z_1)/(Z_1 + Z_2)$ is the variable defining the N/Z ratio of fragments (at fixed masses) and, therefore, plays the most important role in describing the charge equilibration process, which is the topic of the article. The model also includes the rotation angle of the nuclear system as a whole θ and the angle of orientation of both fragments $\varphi_{1,2}$ relative to the beam direction.

Table 1. N/Z values of projectile and target for the studied reactions.

Combination	$(N/Z)_{projectile}$	$(N/Z)_{target}$
^{136}Xe + ^{198}Pt	1.519	1.538
^{136}Xe + ^{208}Pb	1.519	1.537
^{136}Xe + ^{209}Bi	1.519	1.518
^{160}Gd + ^{186}W	1.5	1.514
^{238}U + ^{238}Cm	1.587	1.583
^{64}Ni + ^{208}Pb	1.286	1.537
^{58}Ni + ^{208}Pb	1.071	1.537
^{40}Ca + ^{208}Pb	1	1.537

Evolution of all the degrees of freedom is treated by the set of the Langevin equations:

$$\dot{q}_i = \sum_j \mu_{ij} p_j, \qquad (1)$$
$$\dot{p}_i = F_i^{\text{driving}} + F_i^{\text{friction}} + F_i^{\text{random}}.$$

Here $\mu_{ij} = \|m_{ij}\|^{-1}$ are the inverse mass coefficients calculated within the Werner-Wheeler hydrodynamical approach [8], q_i and p_i are degrees of freedom and their momenta, respectively. The driving force is given mainly by a derivative of the potential energy. The friction coefficients γ_{ij} determining the friction force are calculated using the wall-and-window model of one-body dissipation [9]. According to the Einstein's relation, the amplitude of the random force is proportional to $\sqrt{\gamma_{ij}T}$, where T is the nuclear temperature.

The system of Langevin equations (1) is solved numerically with the following initial conditions: two nuclei start to approach each other from a distance $\approx$ 50 fm with a given impact parameter and collision energy in the center-of-mass frame. Calculation of a trajectory is terminated when the distance between formed fragments reaches initial value. Thus, the obtained trajectory keeps all characteristics of the collision and primary products, such as nucleon numbers, scattering angles, kinetic and excitation energies, etc.

All these characteristics are the input data for the statistical model, which is used to simulate a decay cascade of excited primary products in order to obtain the final ones [10]. The emission of neutrons, protons, α-particles, and γ-quanta as well as sequential fission process are considered in the calculations. In the latter case, masses and atomic numbers of the sequential fission fragments are simulated using the GEF code [11]. Thus, the calculation results can be directly compared with the experimentally measured data.

Employing the described approach, we simulate a large number of trajectories for each relevant impact parameter and get the differential cross section of various processes in the standard way:

$$\frac{d^4\sigma}{dZdAdEd\Omega}(Z,A,E,\theta) = \int_0^{b_{\max}} \frac{\Delta N(b,Z,A,E,\theta)b\,db}{N_{\text{tot}}(b)\Delta Z\Delta A\Delta E\sin\theta\Delta\theta}, \qquad (2)$$

where $\Delta N(b,Z,A,E,\theta)$ is the number of events in a given bin and $N_{\text{tot}}(b)$ is the total number of events simulated for each impact parameter. Integration of (2) allows one to obtain various distributions of reaction products. The detailed description of the dynamical and statistical approaches mentioned above can be found in [4].

3 Results and discussion

In spite of the fact that the model describes rather well the MNT for many systems (see [4, 5]), it fails to reproduce the experimental data for isospin-asymmetric combinations accurately enough. Particular example is the isotopic distributions of final projectile-like fragments (PLFs) obtained in the ^{40}Ca,^{58}Ni + ^{208}Pb reactions ($E_{\text{lab.}}$(Ca) = 249 MeV and $E_{\text{lab.}}$(Ni) = 328.4 MeV, respectively) shown in Fig. 1 by thin histograms. The angular ranges $42^\circ < \theta_{\text{lab.}} < 100^\circ$ and $70^\circ < \theta_{\text{lab.}} < 110^\circ$ covered in the corresponding experiments [12, 13] had been taken into account in the calculations. First, the heavy (neutron-rich) part of the isotopic distributions is overestimated for all values of Z in Fig. 1. It is essential to rectify this failure in order to provide reasonable predictions of yields of neutron-rich nuclides, which are a matter of priority. Second, the calculated shapes of the Ca and Ni distributions contain a plateau, while the experimental data indicate almost an exponential trend. Mainly it is due to the pickup channels of 1-3 neutrons that are strongly underestimated by the model.

The nucleon transfer is described as a collective process in the model, while the mechanism of few nucleon transfer might be a single-particle one. Therefore, calculated cross sections can significantly differ from the experimental data.

We have found that calculated isotopic distributions of target-like fragments (TLFs) obtained in mass-asymmetric reactions (not shown) have been shifted relative to the experimental data, but in the opposite direction than for PLFs. It led us to reconsideration of the division of the excitation energy between two fragments. Thus, instead of fully mass-proportional sharing of the energy, we assume the equal partition at the early stage, since nearly the same number of nucleons in a projectile and a target are initially involved in the dissipation process. Experimental investigation of mass-asymmetric collisions [14, 15], as well as microscopic calculations [16], have confirmed this assumption. We treat the transition from the equal division of the excitation energy to the mass-proportional one as

$$E_i^* = \frac{E^*}{2}\exp(-t/\tau) + \frac{A_i}{A_{tot}}E^*\left[1-\exp(-t/\tau)\right], \qquad (3)$$

with the characteristic transition time τ=2 zs. Here E^* is the total excitation energy of a nuclear system, t is the interaction time, A_{tot} and A_i are the total and the fragment masses, respectively.

This model modification leads to the lighter fragments get more the excitation energy and, therefore, evaporate more neutrons. The corresponding isotopic distributions of final PLFs shift to lighter masses. The trend is obviously opposite for the TLFs: their lower excitation energy leads to isotopic distributions shifted toward heavier masses. Thus, this correction favors the survival of neutron-rich isotopes of heavier elements in the mass-asymmetric reactions.

The problem with describing the few-neutron pickup channels has been solved efficiently by the twice increased random force for the charge-asymmetry degree of freedom at the initial stage of a collision. Calculations showed that this regime of doubled fluctuations of the charge asymmetry is appropriate until the excitation energy reaches the value of 10 MeV.

The results obtained within the modified model are shown in Fig. 1 by thick histograms. Overall agreement with the experimental data has been improved. The adopted corrections have improved the shape of the Ni

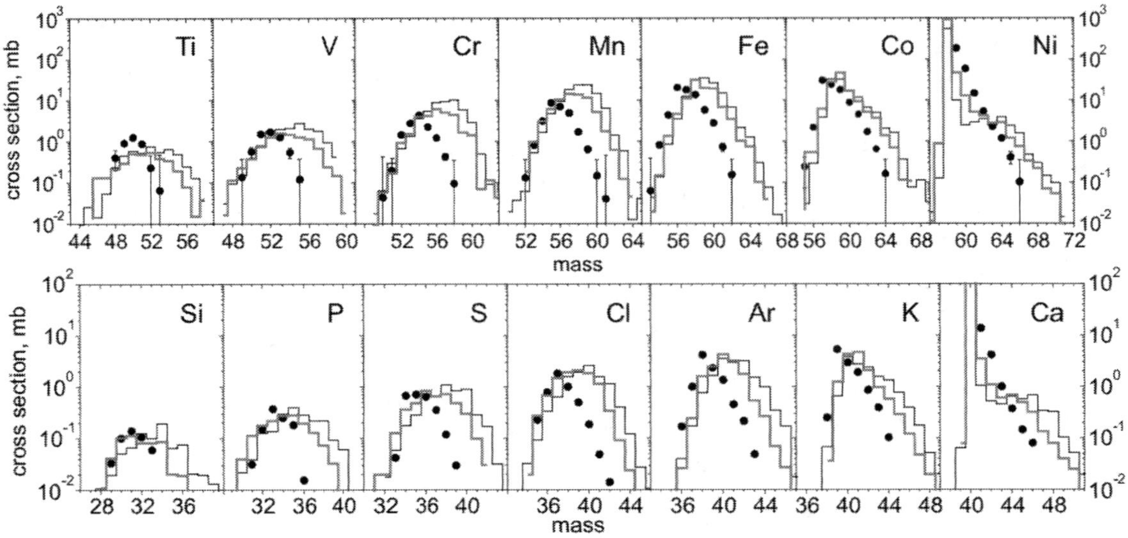

Figure 1. Final isotopic distributions of PLFs obtained in the ^{58}Ni + ^{208}Pb reaction at $E_{\mathrm{lab.}}$ = 328.4 MeV (upper panel) and in the ^{40}Ca + ^{208}Pb reaction at $E_{\mathrm{lab.}}$ = 249 MeV (bottom panel). Theoretical results obtained before and after the model improvement are shown by the thin and thick histograms, respectively. Symbols indicate the experimental data taken from [12, 13].

and Ca distributions. It has become smoother and consistent with the experimental data, however, a small deviation from the data is still presented in the calculations. The shift of the centroids of the calculated isotopic distributions toward the corresponding experimental values is negligible for the few-proton-transfer channels. However, it increases with the number of transferred protons reaching the experimental data.

Further, we have considered the charge equilibration process in the studied systems. First, it occurs via the proton and neutron transfer in opposite directions as following the derivative of the potential energy with respect to the mass and charge asymmetries. In particular, the dependence of the N/Z equilibration on the TKEL is shown in Fig. 2. The theoretical curves for the ^{58}Ni + ^{208}Pb system are consistent with the experimental data taken from [17]. According to the calculations, the charge equilibration process in the considered systems takes 0.5 zs, which coincides with the corresponding relaxation time obtained within the microscopic TDHF model [16].

Finally, the ^{58}Ni,^{64}Ni + ^{208}Pb reactions ($E_{\mathrm{lab.}}$ = 345, 350 MeV) have been investigated. The final isotopic distributions for Z-even elements are shown in Fig. 3 in comparison with the data from Ref. [18, 19].

The charge equilibration process taking place in the isospin-asymmetric ^{58}Ni + ^{208}Pb reaction favours larger yields of lightest PLFs compared to the isospin-symmetric ^{64}Ni + ^{208}Pb reaction. Comparing the two reactions, one can notice that the final distributions of PLFs become closer to each other with larger nucleon transfer, while for TLFs the situation is vice versa. It makes the production of neutron-rich isotopes of heavy elements more preferable in the isospin-symmetric ^{64}Ni + ^{208}Pb system, which has a larger neutron excess.

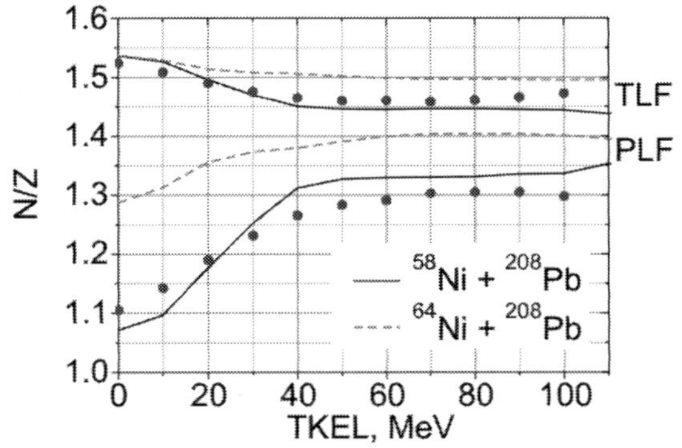

Figure 2. Charge equilibration process in the ^{58}Ni,^{64}Ni + ^{208}Pb reactions as a function of total kinetic energy losses (TKEL). Symbols are the experimental data obtained in [17] for the ^{58}Ni-induced reaction.

In conclusion, the charge equilibration process has been analyzed within the Langevin-type dynamical approach. This process is governed by the potential energy landscape in Z and N coordinates, namely, by the transfer of neutrons and protons in the opposite directions. Its time scale 0.5 zs has been extracted from the calculations for the ^{58}Ni + ^{208}Pb reaction.

The model [4] has been improved by taking into account the equal sharing of the excitation energy and increased fluctuations of the charge asymmetry degree of freedom at the initial stage of a collision. It provides a better agreement of calculated and experimental isotopic distributions of products, especially for the transfer channels of a significant number of protons (> 5p). It allows

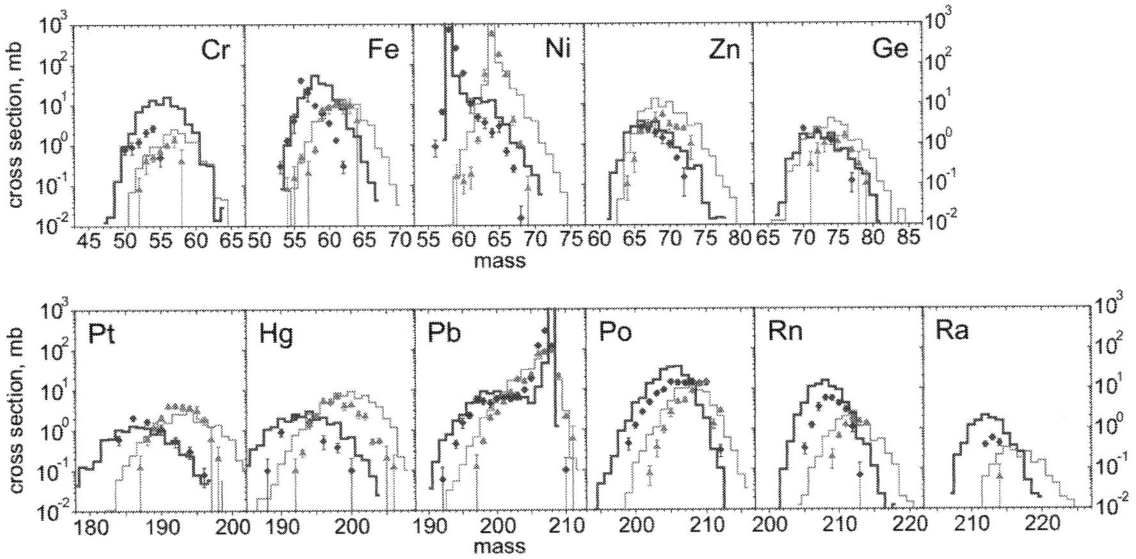

Figure 3. Final isotopic distributions of PLFs and TLFs obtained in two reactions: $^{58}Ni + ^{208}Pb$, $E_{lab.} = 345$ MeV (diamonds and thick histograms) and $^{64}Ni + ^{208}Pb$, $E_{lab.} = 350$ MeV (triangles and thin histograms). The histograms are the results of improved calculations and the symbols are the experimental data from [18, 19].

us to compare the $^{58}Ni, ^{64}Ni + ^{208}Pb$ reactions and choose a more appropriate one for production of neutron-rich nuclides. In spite of the increased yields of mass-asymmetric products in the $^{58}Ni + ^{208}Pb$ reaction, the production cross sections of neutron-rich isotopes are larger in the $^{64}Ni + ^{208}Pb$ reaction.

The work has been supported by the RSF Grant No. 19-42-02014.

References

[1] P. Fröbrich, I.I. Gontchar, Phys. Rep. **292**, 131 (1998)

[2] A.V. Karpov, P.N. Nadtochy, D.V. Vanin, and G.D. Adeev, Phys. Rev. C **63**, 054610 (2001)

[3] V. Zagrebaev, W. Greiner, J. Phys. G **31**, 825 (2005)

[4] A.V. Karpov, V.V. Saiko, Phys. Rev. C **96**, 024618 (2017)

[5] V.V. Saiko, A.V. Karpov, Phys. Rev. C **99**, 014613 (2019)

[6] V. Zagrebaev, A. Karpov, Y. Aritomo, M. Naumenko, and W. Greiner, Phys. Part. Nucl. **38**, 469 (2007)

[7] J. Maruhn, W. Greiner, Z. Phys. A **251**, 431 (1972)

[8] K.T.R. Davies, A.J. Sierk, and J.R. Nix, Phys. Rev. C **13**, 2385 (1976)

[9] A. J. Sierk, J. R. Nix, Phys. Rev. C **21**, 982 (1980)

[10] A.V. Karpov, A.S. Denikin, A.P. Alekseev, et al., Phys. At. Nucl. **79**, 749 (2016); http://nrv.jinr.ru

[11] K.-H. Schmidt, B. Jurado, C. Amouroux, and C. Schmitt, Nucl. Data Sheets **131**, 107 (2016); http://www.khserzhausen.de/GEF.html.

[12] S. Szilner, L. Corradi, G. Pollarolo, et al., Phys. Rev. C **71**, 044610 (2005)

[13] L. Corradi, A.M. Vinodkumar, A.M. Stefanini, et al., Phys. Rev. C **66**, 024606 (2002)

[14] T.C. Awes, R.I.Ferguson, R. Novotny, et al., Phys. Rev. Lett. **52**, 251 (1984)

[15] R. Vandenbosch, A. Lazzarini, D. Leach, et al., Phys. Rev. Lett. **52**, 1964 (1984)

[16] A.S. Umar, C. Simenel, and W. Ye, Phys. Rev. C **96**, 024625 (2017)

[17] K. Sapotta, R. Bass, V. Hartmann, et al., Phys. Rev. C **31**, 1297 (1985)

[18] W. Królas, R. Broda, B. Fornal, et al., Nucl. Phys. A **832**, 170 (2010)

[19] W. Królas, R. Broda, B. Fornal, et al., Nucl. Phys. A **724**, 289 (2003)

Quartet structure of self-conjugate nuclei

Michelangelo Sambataro[1,*]

[1]Istituto Nazionale di Fisica Nucleare - Sezione di Catania, Via S. Sofia 64, I-95123 Catania, Italy

Abstract. We provide a description of even-even $N = Z$ nuclei in a formalism of quartets. Quartets are four-body correlated structures characterized by isospin T and angular momentum J. We show that the ground state correlations induced by a realistic shell model interaction can be well accounted for in terms of a restricted set of $T = 0$ low-J quartets, the $J = 0$ one playing by far a leading role among them. A conceptually similar description of odd-odd self-conjugate nuclei is given in terms of two distinct families of building blocks, one formed by the same $T = 0$ quartets employed for the even-even systems and the other by collective pairs with either $T = 0$ or $T = 1$. Some applications of this formalism are discussed for nuclei in the *sd* shell.

1 Introduction

Self-conjugate nuclei are characterized by an equal number of protons and neutrons distributed over the same single particle orbits. In these nuclei, owing to the charge independence of the nuclear interaction, the isovector proton-neutron (*pn*) pairing force represents an essential component of the effective nuclear interaction. This force acts equally on proton-proton, neutron-neutron as well as *pn* pairs. In addition, *pn* pairing is also expected to occur in an isoscalar form. The competition between these two types of *pn* pairing in $N = Z$ nuclei has been matter of great debate among nuclear physicists [1].

In recent years we have carried out an analysis of the *pn* pairing in even-even $N = Z$ nuclei both in the isovector and in the isoscalar channels [2–5]. This analysis has evidenced, on the one hand, that a description of the ground state correlations induced by this interaction in terms of a condensate of collective pairs (of various form [5]) is not satisfactory and, on the other hand, that these correlations can be accounted for to a high degree of precision by approximating the ground state as a product of identical $T = 0$ quartets. $T = 0$ quartets are four-body correlated structures formed by two protons and two neutrons and, in the case of a spherical mean field, they are also characterized by a total angular momentum $J = 0$. We have also explored a more sophisticated approximation which consists in letting the quartets to be all distinct and we have verified that it leads to basically exact results in the case of *pn* isovector pairing in deformed systems [3]. In all cases the quartets have been constructed variationally for each nucleus.

In the presence of a general Hamiltonian, other quartets are reasonably expected to come into play in addition to the $T = 0, J = 0$ ones emerging from the analysis of the *pn* pairing. In Ref. [6] we have extended the approach of Refs. [2–5] to realistic even-even $N = Z$ nuclei and focused our attention not only on the ground state but also on excited states. We have explored the role of quartets of all possible isospins ($T = 0, 1, 2$) and also of various angular momenta J. It has been evidenced that only the $T = 0$ quartets play an effective role in the structure of the low-lying states of an even-even $N = Z$ nucleus and, among these, the $J = 0$ quartet is by far the most relevant one.

Quartetting is espected to play a key role also in the structure of odd-odd $N = Z$ nuclei. These nuclei exhibit a peculiar coexistence of isospin $T = 0$ and $T = 1$ states at very low excitation energies. In Ref. [7] we have shown that the basic features of the spectra of these nuclei can be reproduced by resorting to two distinct families of building blocks, one formed by collective $T = 0$ quartets and the other by collective pairs with either $T = 0$ or $T = 1$ isospin. In this contribution we will summarize the basic results that have been achieved for both even-even and odd-odd nuclei.

2 Quartetting in even-even nuclei

We work in a spherically symmetric mean field and label the single-particle states by $i \equiv \{n_i, l_i, j_i\}$, where the standard notation for the orbital quantum numbers is used. The quartet creation operator is defined as

$$Q^+_{\alpha,JM,TT_z} = \sum_{i_1 j_1 J_1 T_1} \sum_{i_2 j_2 J_2 T_2} C^{(\alpha)}_{i_1 j_1 J_1 T_1, i_2 j_2 J_2 T_2} \times \Big[[a^+_{i_1} a^+_{j_1}]^{J_1 T_1} [a^+_{i_2} a^+_{j_2}]^{J_2 T_2} \Big]^{JT}_{MT_z},$$

where a^+_i creates a fermion in the single particle state i and $J(T)$ and $M(T_z)$ denote, respectively, the total angular momentum(isospin) and the relative projections. No restrictions on the intermediate couplings $J_1 T_1$ and $J_2 T_2$ are introduced in the calculations. The criterion adopted for the selection of the quartets has been that of choosing, as representative of the quartets with a given isospin

*e-mail: michelangelo.sambataro@ct.infn.it

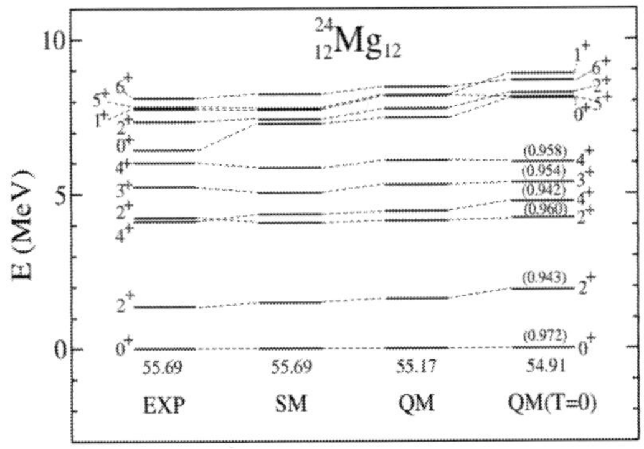

Figure 1. Spectrum of ^{24}Mg obtained in the quartet model (QM) compared to experimental data (EXP) and shell model (SM) results. QM(T=0) denotes the results obtained only with T=0 quartets; the numbers on top of these levels are the overlaps between the QM and SM eigenfunctions. The number below each spectrum gives the ground state correlation energy, namely the difference between the total ground state energy and the energy in the absence of interaction.

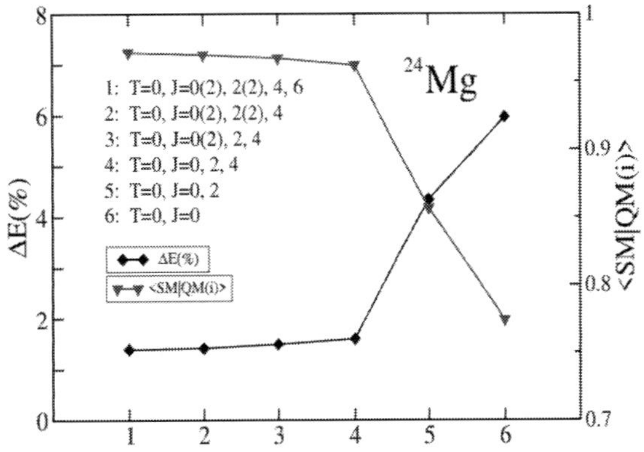

Figure 2. Relative errors (with respect to the SM value) in the ground state correlation energy of ^{24}Mg obtained within the QM with the sets of quartets indicated in the figure. For each set we also show the overlaps between SM and QM wave functions.

T, those describing the lowest levels with that isospin in nuclei with four active particles outside the inert core of reference [6]. For applications within the *sd* shell, the inert core is represented by ^{16}O and the nuclei which have therefore been considered for the definition of the quartets are $^{20}_{10}$Ne$_{10}$, $^{20}_{9}$F$_{11}$ and $^{20}_{8}$O$_{12}$. The lowest states of these nuclei are characterized by T =0, 1 and 2, respectively. Each of these states therefore identifies a quartet with given T,J (and a projection $T_z = (N-Z)/2$). We have performed configuration interaction calculations in spaces built in terms of these selected sets of quartets. We have also explored reductions of these sets to identify the most relevant quartets in the structure of the nuclei under investigation. The USDB interaction [8] has been employed throughout the calculations.

We start our analysis by examining the self-conjugate nucleus ^{24}Mg. In Fig. 1, we compare the experimental spectrum of this nucleus with that resulting from a shell model (SM) calculation and with the spectrum obtained in the quartet model (QM) when all the selected T=0, 1, 2 quartets are taken into account. The quartet approach is seen to reproduce well both the SM ground state correlation energy and the SM excited states up to an energy of about 9 MeV. The SM is in turn able to fit well the experimental levels.

Having verified that the selected sets of T=0, 1, 2 quartets are sufficient to describe the low-energy spectrum of ^{24}Mg, it becomes of interest to investigate the role of the different quartets. We begin by focusing on isospin. In Fig. 1, on the right hand side, we show the theoretical spectrum obtained in the quartet formalism when only T=0 quartets are retained. On top of the lowest levels we also show the overlaps with the corresponding SM eigenstates. One can see that this spectrum does not exhibit relevant differences with respect to the full QM calculation and that the above overlaps are pretty large, what confirms the good quality of the QM(T = 0) wave functions. This result provides a clear evidence of the marginal role played by the T=1 and T=2 quartets in the structure of these states.

As a next step, we concentrate on the ground state of ^{24}Mg by employing only T = 0 quartets. In Fig. 2, we show how the error in the correlation energy of this state, relative to the SM result, varies by reducing, one quartet at a time and starting from the highest one in energy, the set of T=0 quartets. The correlation energy remains basically unchanged up to the point where only the lowest J=0,2,4 quartets are left. From this point on further reductions in the set of quartets lead to significant variations in the energy. The overlaps between the SM and QM ground states which are shown in the same figure exhibit a behavior which is consistent with that of the energies. Thus, these calculations indicate that the T=0 quartets with J=0,2,4 play a major role in the ground state of ^{24}Mg. Among these quartets, the T=0, J=0 quartet is by far the one which contributes most to the correlation energy since, as it can be seen in Fig. 2, an approximation in terms of only this quartet accounts for about 94% of the total energy. A very similar conclusion has emerged also from an analysis of the ground state of the three-quartet system ^{28}Si [6]. It is also worthy noticing that the role of the T=0, J=0 quartet can be further increased in a sig-

 https://doi.org/10.1051/epjconf/201922301056

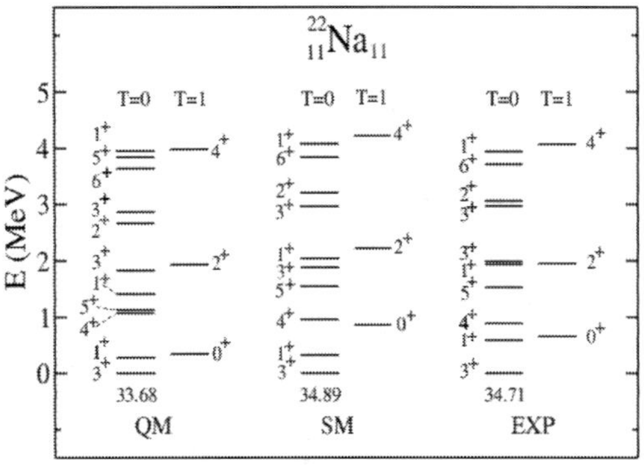

Figure 3. The spectrum of ^{22}Na obtained with the quartet formalism (QM) compared with shell model (SM) and experimental (EXP) spectra. The number below each spectrum is the ground state correlation energy (see Fig. 1).

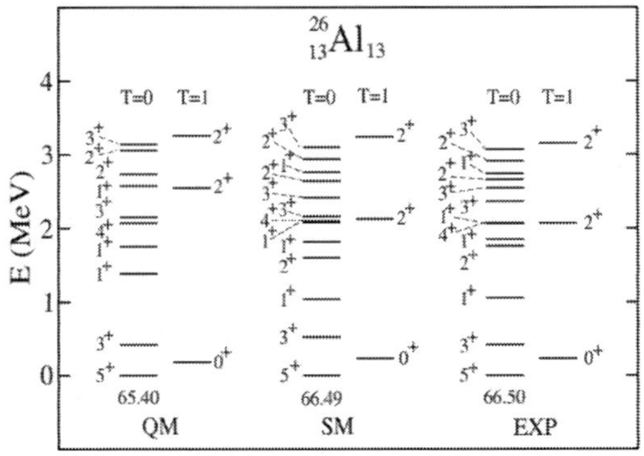

Figure 4. The same as in Fig. 3 for ^{26}Al.

nificant way if this is constructed variationally for each nucleus [9].

3 Quartetting in odd-odd nuclei

The starting point of our analysis of odd-odd $N = Z$ nuclei is the assumption that a state with isospin T can be represented as a superposition of products of one or more $T = 0$ quartets and one extra pair with isospin T. Thus, within this scheme, the total isospin of the state coincides with that of the pair. Based on the conclusions of the previous analysis on even-even nuclei, we involve in the calculations only three $T = 0$ quartets, namely the $J = 0, 2, 4$ quartets describing the lowest three states of ^{20}Ne. In analogy, as collective pairs with isospin T, we assume those describing the lowest three states with that isospin in ^{18}F (one proton and one neutron outside the ^{16}O core). These states are characterized by angular momenta $J = 1, 3, 5$ for $T = 0$ and $J = 0, 2, 4$ for $T = 1$. Once fixed, these quartets and pairs are no longer modified. By defining the creation operator for a collective pair as

$$P^{+}_{JM,TT_z} = \sum_{ij} p_{ij}[a^{+}_i a^{+}_j]^{JT}_{MT_z}, \tag{1}$$

where $M(T_z)$ stands for the projection of $J(T)$, we perform configuration interaction calculations in the space (we adopt the m-scheme)

$$\{Q^{+}_{J_1M_1} P^{+}_{JM,T0}|0\rangle\} \tag{2}$$

for ^{22}Na and in the space

$$\{Q^{+}_{J_1M_1} Q^{+}_{J_2M_2} P^{+}_{JM,T0}|0\rangle\} \tag{3}$$

for ^{26}Al, with $|0\rangle$ representing the reference vacuum.

Fig. 3 illustrates the results obtained in the case of ^{22}Na. In this figure we compare the positive-parity experimental spectrum of ^{22}Na up to an energy of about 4 MeV (EXP) with the spectrum obtained with the present quartet formalism (QM) and with the shell model result (SM). The agreement between QM and SM spectra looks good. This agreement is further emphasized by the values of the overlaps between corresponding QM and SM states. These are: 0.94 ($J = 3$), 0.93 ($J = 1$), 0.97 ($J = 0$), 0.92 ($J = 4$), 0.96 ($J = 5$). In Fig. 4 we show the analogous comparison for ^{26}Al. The agreement is satisfactory also in this case although one may notice that three $T = 0$ states (with $J = 1, 2, 3$) around 2 MeV are missing in the QM spectrum in comparison with the SM one. This suggests that the inclusion of some extra element either in the set of quartets or in that of $T = 0$ pairs would be welcome. We have evaluated also in this case the overlaps between the lowest QM and SM eigenstates and found the following values: 0.94 ($J = 5$), 0.94 ($J = 0$), 0.95 ($J = 3$), 0.86 ($J = 1$). These values support the good quality of our approximation scheme.

4 Conclusions

We have reported on a description of both even-even and odd-odd $N = Z$ nuclei in the *sd* shell in a formalism of $T = 0$ quartets. For all the nuclei that have been analyzed the quartet formalism has provided a description of the low-energy spectra comparable in accuracy with that of shell model calculations. This fact confirms the importance of quartet degrees of freedom in $N = Z$ nuclei and validates the present quartet formalism as the appropriate tool for treating them.

References

[1] S. Frauendorf and A.O. Macchiavelli, *Progr. Part. Nucl. Phys.* **78**, 24 (2014)

[2] N. Sandulescu, D. Negrea, J. Dukelsky, and C.W. Johnson, *Phys. Rev.* C **85**, 061303(R) (2012)

[3] M. Sambataro and N. Sandulescu, *Phys. Rev.* C **88**, 061303(R) (2013)

[4] M. Sambataro, N. Sandulescu, and C.W. Johnson, *Phys. Lett.* B **740**, 137 (2015)

[5] M. Sambataro and N. Sandulescu, *Phys. Rev.* C **93**, 054320 (2016)

[6] M. Sambataro and N. Sandulescu, *Phys. Rev. Lett.* **115**, 112501 (2015)

[7] M. Sambataro and N. Sandulescu, *Phys. Lett.* B **763**, 151 (2016)

[8] B.A. Brown and W.A. Richter, *Phys. Rev.* C **74**, 034315 (2006)

[9] M. Sambataro and N. Sandulescu, *J. Phys.* G: *Nucl. Phys.* **40**, 055107 (2013)

EPJ Web of Conferences **223**, 01057 (2019) https://doi.org/10.1051/epjconf/201922301057
NSD2019

Variation of NMEs of $0\nu\beta\beta$ for ^{48}Ca with different components of NN interaction

Shahariar Sarkar[1,*], *Pawan* Kumar[1], *Kanhaiya* Jha[1], and *P.K.* Raina[1]

[1]Indian Institute of Technology Ropar, Rupnagar, Punjab-140001, INDIA

Abstract. We examine the sensitivity of nuclear matrix elements (NMEs) for light-neutrino exchange mechanism of neutrinoless double beta decay ($0\nu\beta\beta$) for ^{48}Ca to the various components of two-nucleon interaction, GXPF1A, in *fp* model space. It is found that the contribution in NMEs coming from the central component is close to contribution from total interaction. The spin-orbit and tensor components are found canceling the contribution of each other.

1 Introduction

Neutrinoless double beta decay ($0\nu\beta\beta$) is a rare weak process, known mainly for determining the nature of neutrino and neutrino mass [1]. The decay rate of this process depends on nuclear matrix elements (NMEs), which are calculated using theoretical nuclear many-body models. In the literature, the nuclear shell model (NSM) has been widely used to calculate NMEs [1].

It is well known that the total two-nucleon interaction is the sum of central (C), spin-orbit (SO), and tensor (T) interaction. The earlier calculations of NMEs in NSM have been done using total two-nucleon interaction [2–5]. In the present work, we examine the variation of NMEs for light neutrino exchange mechanism of $0\nu\beta\beta$ of ^{48}Ca with different components of two-nucleon interaction-GXPF1A [6] of *fp*-model space using closure approximation. The employed interaction is decomposed into its central, spin-orbit, and tensor components using spin-tensor decomposition (STD) method [7–10].

This article is organized as follows. In section 2, the theoretical formalism for NMEs for light neutrino exchange mechanism of $0\nu\beta\beta$ is given. The STD is discussed in section 3. Results and discussion are presented in section 4. The summary of this work is given in section 5.

2 Nuclear Matrix Elements of $0\nu\beta\beta$

The NMEs for light neutrino exchange mechanism of $0\nu\beta\beta$ can be presented as a sum of Gamow-Teller (M_{GT}), Fermi (M_F), and tensor (M_T) matrix elements [4];

$$M^{0\nu} = M_{GT} - \left(\frac{g_V}{g_A}\right)^2 M_F + M_T \tag{1}$$

where, $g_V = 1$ and $g_A = 1.27$. $M_\alpha = \langle f|\tau_{-1}\tau_{-2}O^\alpha_{12}|i\rangle$ with $\alpha = GT, F, T$ can be written in terms of two-body transition density (TBTD) and two-body matrix elements ($\langle k_1', k_2', J|\tau_{-1}\tau_{-2}O^\alpha_{12}|k_1, k_2, J\rangle$) [4];

$$M_\alpha = \sum_{J,k_1'\le k_2',k_1\le k_2} TBTD(f,i,k,J) \langle k_1', k_2', J|\tau_{-1}\tau_{-2}O^\alpha_{12}|k_1, k_2, J\rangle \tag{2}$$

where, τ_- is isospin lowering operator, O^α_{12} is transition operator of $0\nu\beta\beta$ defined with spin (σ) and neutrino potential operator ($H_\alpha(r, E_\kappa)$) [2]; $O_{GT} = (\sigma_1.\sigma_2)H_{GT}(r, E_\kappa)$, $O_F = H_F(r, E_\kappa)$ and $O_T = (3(\sigma_1.\hat{\mathbf{r}})(\sigma_2.\hat{\mathbf{r}}) - (\sigma_1.\sigma_2))H_T(r, E_\kappa)$. In Eq.(2), k stands for the set of quantum numbers ($n; l; j$), $|i\rangle$ and $|f\rangle$ refer to the ground state (0^+) of ^{48}Ca and ^{48}Ti, respectively. Neutrino potential is defined as [3];

$$H_\alpha(r, E_\kappa) = \frac{2R}{\pi}\int_0^\infty \frac{j_p(qr)h_\alpha(q^2)qdq}{q + E_\kappa - (E_i + E_f)/2} \tag{3}$$

Where, E_κ , E_i and E_f are the energies of ^{48}Sc, ^{48}Ca and ^{48}Ti respectively. q is the momentum of the virtual Majorana neutrino and r is the distance between the nucleons. $j_p(qr)$ is the spherical Bessel function with p =0 and 2. In closure approximation one replaces $E_\kappa - (E_i + E_f)/2 \longrightarrow \langle E\rangle$, where, $\langle E\rangle$ is the closure energy which takes care the effects of large number of excitation energy of states of intermediate nuclei (^{48}Sc). Closure approximation removes the complications of calculating large number of states of ^{48}Sc. As neutrino momentum (q) in the decay is high (~100-200 MeV), NMEs are not much sensitive with the excitation energy of ^{48}Sc. So by replacing all excitation energy with a constant closure energy $\langle E\rangle$ gives NME with around 90% accuracy [3].

3 Spin Tensor Decomposition (STD)

Nucleons are intrinsic spin 1/2 fermions; therefore, the interaction between two-nucleon can be written as the linear sum of the scalar product of configuration space operator Q and spin space operator S of rank k [7, 8];

*e-mail: shahariar.sarkar@iitrpr.ac.in

$$V = \sum_{k=0}^{2} V(k) = \sum_{k=0}^{2} Q^k.S^k \tag{4}$$

where, rank $k = 0$, 1 and 2 represent central, spin-orbit and tensor force, respectively. Using the LS-coupled two-nucleon wave functions, the matrix element for each V^k can be calculated from matrix element for V [9];

$$\langle (ab), LS; JM|V(k)|(cd), L'S'; JM\rangle = (2k+1)(-1)^J \times \begin{Bmatrix} L & S & J \\ S' & L' & k \end{Bmatrix} \sum_{J'} (-1)^{J'}(2J'+1) \begin{Bmatrix} L & S & J' \\ S' & L' & k \end{Bmatrix} \times \langle (ab), LS; J'M|V|(cd), L'S'; J'M\rangle \tag{5}$$

here, a includes quantum numbers n_a and l_a.

4 Results and Discussion

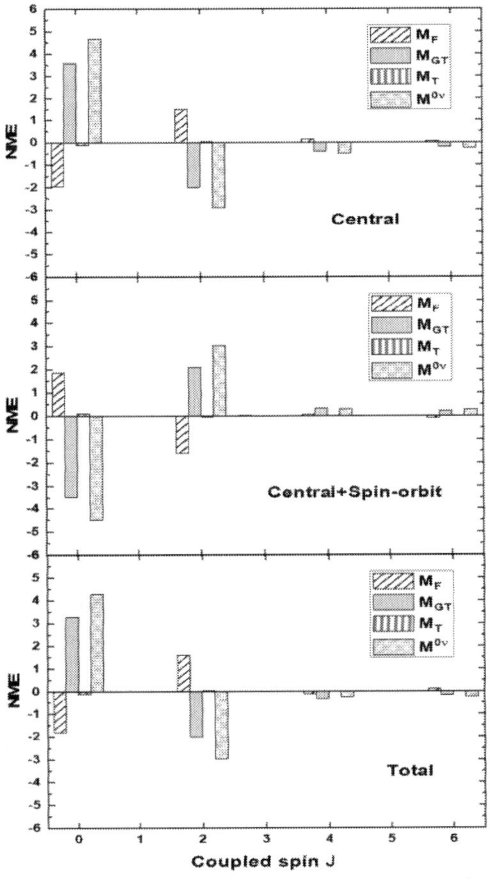

Figure 1. Contribution of various spin-parity (J^π) of decaying neutrons or final protons to the NMEs.

We have calculated TBTD in terms of two-nucleon transfer amplitudes (TNA) with 50 states of ^{46}Ca using method described in Ref. [4]. The TBMEs have been calculated with closure energy $\langle E\rangle = 0.5$ MeV [3]. The effect of finite nucleon size (FNS), higher order currents (HOC) [2] have also been considered. The calculated NMEs are given in Table 1.

Table 1. NMEs for ^{48}Ca calculated with different components of two-nucleon interaction.

NME	Types	C	C+SO	C+SO+T
M_F	FNS+HOC	-0.273	0.234	-0.217
M_{GT}	FNS+HOC	0.933	-0.828	0.790
M_T	FNS+HOC	-0.081	0.070	-0.076
$M^{0\nu}$	FNS+HOC	1.021	-0.904	0.848

It is found that the NMEs calculated with C interaction are near to NMEs calculated with total interaction. On the addition of SO part to C part, the sign of NMEs gets change but in absolute value they remain almost same. Similar effects are also seen when we add T part to C+SO part of two-nucleon interaction. Thus, we infer that SO and T parts negate the effect of each other.

Results of NMEs as a function of coupled spin-parity of decaying nucleons are shown in Fig. 1. It is found that the dominant contribution in NMEs comes from $J^\pi = 0^+$ and 2^+. But, their contribution are present with the opposite effect resulting in a small value of NMEs. Negligible contributions comes from other J^π.

5 Summary

We have examined the sensitivity of NME for light-neutrino exchange mechanism of $0\nu\beta\beta$ for ^{48}Ca with various components we get using STD for GXPF1A interaction. It is found that the NMEs calculated with C part and total interaction are close to each other. SO and T parts negate the contribution of each other in the NMEs. Dominating contribution to NMEs comes from 0^+ and 2^+ spin-parity staes of decaying nucleons.

Acknowledgements

Shahariar Sarkar thanks Indian Institute of Technology Ropar for providing the travel support to attend the NSD-2019 conference.

References

[1] Vergados et al., Reports on Progress in Physics **75**, 106301 (2012)
[2] Horoi et al., Phys. Rev. C **81**, 024321 (2010)
[3] Sen'Kov et al., Phys. Rev. C **88**, 064312 (2013)
[4] Brown et al., Phys. Rev. Lett. **113**, 262501 (2014)
[5] Iwata et al., Phys. Rev. Lett. **116**, 112502 (2016)
[6] Honma et al., EPJ A **86**, 034314 (2005)
[7] M. W. Kirson, Phys. Lett. **47B**, 110-114 (1973)
[8] Kumar et al., Phys. Rev. C **100**, 024328 (2019)
[9] Elliott et al., Nucl. Phys. **121**, 241 (1968)
[10] Kenji et al.,Nuclear Physics A **333**, 67–76 (1980)

Study of continuum excitation by light weakly bound projectiles on proton target

O. Sgouros[1,2,*], *V.* Soukeras[1,2], *A.* Pakou[1], *F.* Cappuzzello[2,3], *L.* Acosta[4,5], *C.* Agodi[2], *N.* Alamanos[6], *S.* Calabrese[2], *D.* Carbone[2], *M.* Cavallaro[2], *A.* Foti[3,5], *N.* Keeley[7], *I.* Martel[8], *K.* Rusek[9], and *D.* Torresi[2]

[1]Department of Physics and HINP, The University of Ioannina, Ioannina, Greece
[2]INFN Laboratori Nazionali del Sud, Catania, Italy
[3]Dipartimento di Fisica e Astronomia, Universita di Catania, Catania, Italy
[4]Instituto de Fisica, Universidad Nacional Autonoma de Mexico, Mexico D.F, Mexico
[5]INFN – Sezione di Catania, Catania, Italy
[6]CEA-Saclay, DAPNIA-SphN, Gif-sur-Yvette, France
[7]National Centre for Nuclear Research, Otwock Warsaw, Poland
[8]Departamento di Fisica Aplicada, Universidad de Huelva, Huelva, Spain
[9]Heavy Ion Laboratory, University of Warsaw, Warsaw, Poland

Abstract. Elastic scattering and breakup angular distribution measurements for the systems ^{6,7}Li + p were performed at the MAGNEX facility of the **I**stituto **N**azionale di **F**isica **N**ucleare-**L**aboratori **N**azionali del **S**ud (INFN-LNS) in Catania, in the energy range of (2.3-5.4)AMeV. The breakup channel was identified and quantified adopting the algorithm MULTIP. Within this algorithm which is a Monte Carlo simulation code, the history of the breakup fragments can be tagged from the rest frame of the decay nucleus itself to the laboratory frame. Angular distribution data of both elastic scattering and breakup were analyzed under the same theoretical model and the influence of continuum on the elastic channel was investigated.

1 Introduction

Weakly bound nuclei are characterized by relatively small binding energies and thus, during a nuclear collision with a target nucleus, they can easily break into their cluster constituents. Therefore, breakup may be used as a tool to assess coupling effects on the elastic channel, when weakly bound nuclei are involved. The most widespread theoretical model which was developed to describe the effect of couplings of the breakup channel on elastic scattering is the **C**ontinuum **D**iscetized **C**oupled **C**hannels (CDCC) model. From the experimental point of view, is not always an easy task to obtain a clear signature of the breakup mechanism by studying energy spectra in singles, since due to their continuous shape they may interfere with events originating from different sources of background or with events triggered by a different reaction mechanism. Therefore, we have performed elastic scattering and exclusive breakup measurements with the weakly bound nuclei ^{6,7}Li on a proton target. Both data are considered into a CDCC framework. The goal is twofold: first to test the realism behind the discretization of the continuum phase space by comparing experimental and simulated energy spectra. In this respect the CDCC reaction model is coupled to a recently developed simulation algorithm, named MULTIP [1]. The second task is to probe coupling channels effects of the breakup on elastic channel in a CDCC approach.

*e-mail: onoufrios.sgouros@lns.infn.it

2 Experimental Setup

The two experiments were performed at the MAGNEX facility [2, 3] of the INFN-LNS. The ^{6}Li and ^{7}Li beams were delivered by the TANDEM accelerator and impinged on a thin CH_2 target. For the elastic scattering measurement [4–6], the ^{6}Li (^{7}Li) ejectiles were detected by the MAGNEX spectrometer while, for the breakup measurement [6–8] the α-particles were measured in MAGNEX in coincidence with ^{2}H (^{3}H) measured in a silicon detector, appropriately masked by a tantalum foil to avoid elastic scattering deterioration. The angular range covered by MAGNEX was $0^o \leq \theta_{lab} \leq 10^o$, while for the silicon detector was $4^o \leq \theta_{lab} \leq 6^o$.

3 Identification of the breakup channel

The Monte Carlo simulation algorithm, MULTIP [1] was employed in order to identify the kinematical loci of the breakup channel and to determine the detection efficiency of our experimental setup. Bellow are presented the main steps, adopted in the algorithm for the simulation of the energy spectra corresponding to the breakup of ^{6}Li. The same philosophy is also adopted in case of the ^{7}Li breakup.

In the first step of the simulation, the momentum vector of ^{6}Li excited nucleus in the laboratory reference frame is determined, by treating the continuum excitation as a 2-body like reaction ^{6}Li+p $\Rightarrow$ ^{6}Li*+p, using for each continuum bin the excitation energies and the reaction probabilities predicted by the CDCC calculation.

Then, the breakup of the ^{6}Li* in its rest frame is considered. The ^{6}Li nucleus acquires randomly an excitation energy inside the energy bin as specified in the CDCC framework and breaks into two fragments, an ^{4}He and a ^{2}H. It is assumed arbitrarily that the ^{4}He is emitted randomly with a specific energy and momentum, while the energy and momentum of the deuteron is determined by applying the conservation laws in the rest frame of ^{6}Li*.

In the last step, having determined the momentum vector of the excited ^{6}Li nucleus (P,θ), the momentum vectors of the breakup fragments are transformed from the ^{6}Li* rest frame to the laboratory one, by imposing a Galilean transformation followed by the appropriate rotation [1, 9]. A comparison between the experimental and simulated coincidence spectra is presented in Figure 1. The simulated data denoted with the magenta and orange points correspond to the simulation of the resonant and non-resonant breakup respectively. It is seen that the simulations are in very good agreement with the experimental data, validating the philosophy of the discretization of the continuum phase space behind the CDCC approach.

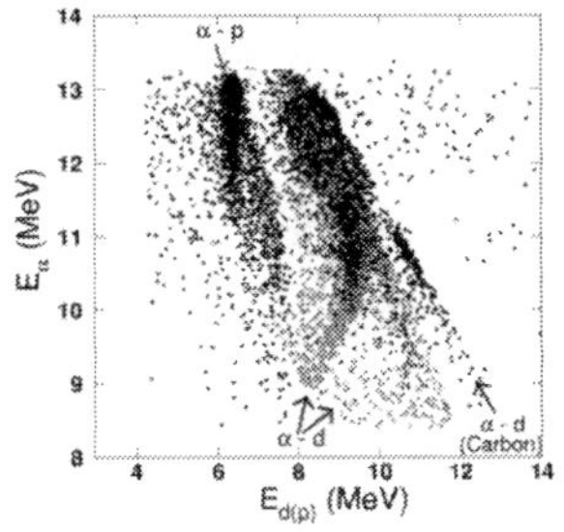

Figure 1. E_α-$E_{d(p)}$ correlation plot for the α-particles detected in MAGNEX in coincidence with deuterons (protons) detected in the detector mounted at θ_{lab} ~5° at the beam energy of 25 MeV (See text for details). Data are also reported in Ref.[6, 7].

4 Results and Discussion

Having determined the efficiency of our detection system via code MULTIP, the angular distributions for the breakup of ^{6,7}Li on a proton target were deduced. The resulted breakup data together with the elastic scattering ones were analyzed within the CDCC framework following the prescription described in Refs.[10, 11]. As an example, elastic scattering and breakup data for ^{6}Li + p system at the energy of 25 MeV are presented in Figure 2. Results for the rest of the energies may be found in [7]. As it is seen, coupling to resonant breakup is the dominant one, while coupling to the non-resonant (direct) breakup has a minor effect on the reproduction of the elastic scattering data. The same conclusion remains valid even at the lower energy of 16 MeV where although the resonant breakup is alsmost zero, its coupling influence on elastic channel is the strongest one. The analysis of the ^{7}Li+p reaction at 38 MeV [8] yielded the same results, where strong coupling to the $7/2^-$ resonance of ^{7}Li is inferred althought the resonance is barely excited. This systematic behavior for both ^{6,7}Li + p systems, suggests that the strength of the coupling to the continuum excitation is not always correlated with the magnitude of the breakup cross section.

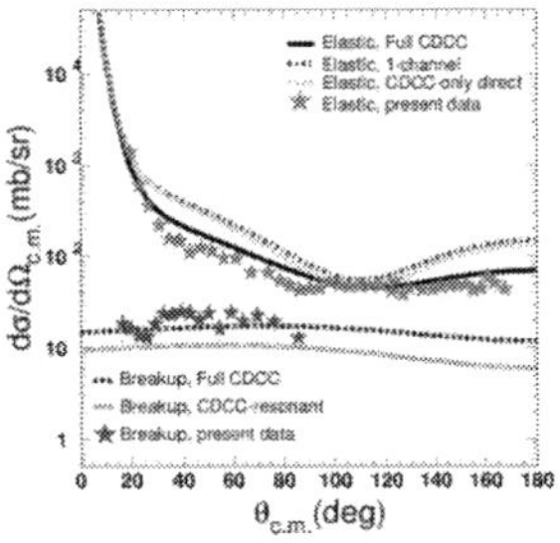

Figure 2. Experimental and theoretical angular distributions for the elastic scattering and breakup of ^{6}Li+p system at the energy of 25 MeV. Data and calculations from Ref.[6, 7].

The research leading to these results was partially funded by the European Union Seventh Framework Programme FP7/2007-2013 (Grant Agreement No. 262010-ENSAR) and also by the European Research Council (ERC) under the European Union's Horizon 2020 research and innovation programme (Grant Agreement No. 714625).

References

[1] O. Sgouros et al., Eur. Phys. J. A **53**, 165 (2017)
[2] F. Cappuzzello et al., Eur. Phys. J. A **52**, 167 (2016)
[3] M. Cavallaro et al., POS **302**, (BORMIO2017) 015 (2017)
[4] V. Soukeras et al., Phys. Rev. C **91**, 057601 (2015)
[5] A. Pakou et al., Phys. Rev. C **94**, 014604 (2016)
[6] V. Soukeras, PhD thesis, Univ. of Ioannina (2017)
[7] V. Soukeras et al., Phys. Rev. C **95**, 054614 (2017)
[8] A. Pakou et al., Phys. Rev. C **95**, 044615 (2017)
[9] K. K. Olimov et al., Int. J. Mod. Phys. E **25**, 1650021 (2016)
[10] K. Rusek et al., Phys. Rev. C **56**, 1895 (1997)
[11] K. Rusek et al., Phys. Rev. C **64**, 044602 (2001)

Evolution of nuclear spin-orbit splittings with Skyrme functional SAMi-T

Shihang Shen[1,2], *Gianluca* Colò[1,2,*], and *Xavier* Roca-Maza[1,2]

[1]Dipartimento di Fisica, Università degli Studi di Milano, Via Celoria 16, I-20133 Milano, Italy
[2]INFN, Sezione di Milano, Via Celoria 16, I-20133 Milano, Italy

Abstract. A new Skyrme functional has been developed with tensor term guided by *ab initio* relativistic Brueckner-Hartree-Fock (RBHF) studies on neutron-proton drops. Instead of extracting information on the tensor force from experimental single-particle energy splittings, the RBHF calculations do not contain beyond mean-field effects such as particle-vibration coupling and therefore the information on the tensor force can be obtained without ambiguities. The new functional gives a good description of nuclear ground-state properties as well as various giant resonances. The description for the evolution of single-particle energy splittings is also improved by the new functional.

1 Introduction

Nuclear density functional theory is a powerful tool to describe various nuclear phenomena along a large part of the nuclear chart [1, 2]. However, there are still many open questions concerning current nuclear energy density functionals such as the tensor force term [3]. When constructing the functional by fitting its parameters against bulk properties such as binding energies and charge radii, tensor force plays relatively small role and is difficult to be determined. Even though it shows important effects in describing the evolution of single-particle levels along isotopic (or isotonic) chains [4–10], it is very much mixed with the beyond-mean-field effect, such as particle-vibration coupling [11].

Taking advantage of the advances of nuclear *ab initio* calculations, a clear sign of tensor force has been shown in the evolution of spin-orbit (s.o.) splittings in the neutron drops [12, 13]. The adopted *ab initio* framework, the relativistic Brueckner-Hartree-Fock theory [14–17], has no beyond-mean-field effects such as particle-vibration coupling and therefore the information of tensor force has no ambiguities. It provides a important guide to constrain the tensor force in the nuclear energy density functional. Along this direction, in order to extract also the information of neutron-proton tensor force we have extended the study to neutron-proton drops, which is also an ideal system confined in an external field without consideration of center-of-mass correction nor Coulomb interaction [18]. Then we developed a new Skyrme functional, SAMi-T, with tensor term by fitting to the evolution of s.o. splittings of the neutron-proton drops calculated by RBHF [18].

In Sec. 2 we briefly give the formalism of the tensor term in Skyrme functional. Results of evolution of s.o. splittings along the $Z = 50$ isotopic chain and $N = 82$ isotonic chain are given in Sec. 3. Finally, we give the summary in Sec. 4.

*e-mail: Gianluca.Colo@mi.infn.it

2 Formalisms

The Skyrme effective interaction with two-body tensor term V_T is written in the standard form as [19, 20].

$$\begin{aligned} V(\mathbf{r}_1,\mathbf{r}_2) = {} & t_0(1+x_0P_\sigma)\delta(\mathbf{r}) + \frac{1}{2}t_1(1+x_1P_\sigma)\left[\mathbf{P}'^2\delta(\mathbf{r}) + \delta(\mathbf{r})\mathbf{P}^2\right] \\ & + t_2(1+x_2P_\sigma)\mathbf{P}'\cdot\delta(\mathbf{r})\mathbf{P} + \frac{1}{6}t_3(1+x_3P_\sigma)\rho^\gamma(\mathbf{R})\delta(\mathbf{r}) \\ & + iW_0(\boldsymbol{\sigma}_1+\boldsymbol{\sigma}_2)\cdot[\mathbf{P}'\times\delta(\mathbf{r})\mathbf{P}] + V_T(\mathbf{r}_1,\mathbf{r}_2), \end{aligned} \tag{1}$$

$$\begin{aligned} V_T(\mathbf{r}_1,\mathbf{r}_2) = {} & \frac{T}{2}\left\{\left[(\boldsymbol{\sigma}_1\cdot\mathbf{P}')(\boldsymbol{\sigma}_2\cdot\mathbf{P}') - \frac{1}{3}(\boldsymbol{\sigma}_1\cdot\boldsymbol{\sigma}_2)\mathbf{P}'^2\right]\delta(\mathbf{r}) \right.\\ & \left. +\delta(\mathbf{r})\left[(\boldsymbol{\sigma}_1\cdot\mathbf{P})(\boldsymbol{\sigma}_2\cdot\mathbf{P}) - \frac{1}{3}(\boldsymbol{\sigma}_1\cdot\boldsymbol{\sigma}_2)\mathbf{P}^2\right]\right\} \\ & + U\left\{(\boldsymbol{\sigma}_1\cdot\mathbf{P}')\delta(\mathbf{r})(\boldsymbol{\sigma}_2\cdot\mathbf{P}) - \frac{1}{3}(\boldsymbol{\sigma}_1\cdot\boldsymbol{\sigma}_2)\left[\mathbf{P}'\cdot\delta(\mathbf{r})\mathbf{P}\right]\right\}, \end{aligned} \tag{2}$$

where $\mathbf{r} = \mathbf{r}_1 - \mathbf{r}_2$, $\mathbf{R} = \frac{1}{2}(\mathbf{r}_1 + \mathbf{r}_2)$, $\mathbf{P} = \frac{1}{2i}(\nabla_1 - \nabla_2)$, and $\mathbf{P}'$ is the hermitian conjugate of $\mathbf{P}$ acting on the left. The spin-exchange operator reads $P_\sigma = \frac{1}{2}(1 + \boldsymbol{\sigma}_1\cdot\boldsymbol{\sigma}_2)$, and ρ is the total nucleon density.

The Hartree-Fock equations for each s.p. level can be obtained by the variational method with respect to the HF total energy as

$$\left[-\frac{\hbar^2}{2M}\nabla^2 + U_q(\mathbf{r})\right]\psi_k(\mathbf{r}) = e_k\psi_k(\mathbf{r}), \tag{3}$$

where e_k is the single-particle energy, ψ_k is the corresponding wave function, and $q = 0(1)$ labels neutrons (protons). The single-particle potential $U_q(\mathbf{r})$ is a sum of central,

 https://doi.org/10.1051/epjconf/201922301059

Coulomb and spin-orbit terms,

$$U_q(\mathbf{r}) = U_q^{(c)}(\mathbf{r}) + \delta_{q,1} U_C(\mathbf{r}) + \mathbf{U}_q^{(s.o.)}(\mathbf{r}) \cdot (-i)(\nabla \times \sigma). \quad (4)$$

The spin-orbit term reads [3, 20]

$$\mathbf{U}_q^{(s.o.)}(\mathbf{r}) = \frac{1}{2}\left[W_0 \nabla\rho + W_0' \nabla\rho_q\right] + \left[\alpha \mathbf{J}_q + \beta \mathbf{J}_{1-q}\right], \quad (5)$$

where $\mathbf{J}(\mathbf{r})$ the spin-orbit density. Starting from Eq. (1) one would derive $W_0' = W_0$. A more general form has been adopted in which W_0' can be defined and fitted independently as in the case of SAMi functional [21] among the others.

The parameters α and β in Eq. (5) include contributions from the exchange part of the central term and from the tensor term,

$$\alpha = \alpha_c + \alpha_T, \quad \beta = \beta_c + \beta_T, \quad (6)$$

where

$$\alpha_c = \frac{1}{8}(t_1 - t_2) - \frac{1}{8}(t_1 x_1 + t_2 x_2), \quad \alpha_T = \frac{5}{12}U, \quad (7a)$$

$$\beta_c = -\frac{1}{8}(t_1 x_1 + t_2 x_2), \quad \beta_T = \frac{5}{24}(T + U). \quad (7b)$$

For open-shell nuclei, HF + BCS is performed to taken into account the pairing correlations. The pairing force used in this work is a zero-range, density-dependent one of the form

$$V = V_0 \left[1 - \left(\frac{\rho\left(\frac{\mathbf{r}_1+\mathbf{r}_2}{2}\right)}{\rho_0}\right)^{\sigma}\right] \delta(\mathbf{r}_1 - \mathbf{r}_2), \quad (8)$$

with parameters the same as the ones in Ref. [7], $V_0 = 680$ MeV fm^3, $\sigma = 1$, and $\rho_0 = 0.16$ fm^{-1}.

3 Results and Discussion

The Skyrme functional SAMi-T has been developed with tensor terms guided by the RBHF calculations of the neutron-proton drops [18]. In the RBHF calculation, the Bonn A interaction has been used, which was fitted to the two-nucleon scattering data and to the deuteron properties [22]. The in-medium interaction (G-matrix) is obtained by solving the Bethe-Goldstone equation and then used in the mean-field calculation. RBHF belongs to the *ab initio* methods as it is a well-controlled many-body approximation in which a realistic interaction in input. For more details of the RBHF framework used, see Ref. [15].

The ground state properties of nuclei such as binding energies, charge radii, and spin-orbit splittings can be well described. Furthermore, SAMi-T also keeps the good merit of SAMi [21] that it provides a good description also for various collective excitations such as the giant monopole, dipole resonances, and the Gamow-Teller resonances. Especially for the spin-dipole resonances the description by SAMi-T has been improved with the inclusion of tensor term comparing with SAMi.

In Figure 1 we show the single-particle energy splittings of proton $h_{11/2}$ and $g_{7/2}$ along the $Z = 50$ isotopic chain calculated by SAMi-T and SAMi functionals, in comparison with experimental data [4]. In order to focus on the effect of tensor force, which influences the relative change of spin-orbit splittings along the isotopic (or isotonic) chains, the theoretical results have been shifted so that the splittings coincide with the experimental data at ^{132}Sn. The shift for SAMi-T is $\delta_{\text{SAMi-T}} = 6.8 - 2.8 = 4$ MeV; and for SAMi is $\delta_{\text{SAMi}} = 6.5 - 2.8 = 3.7$ MeV.

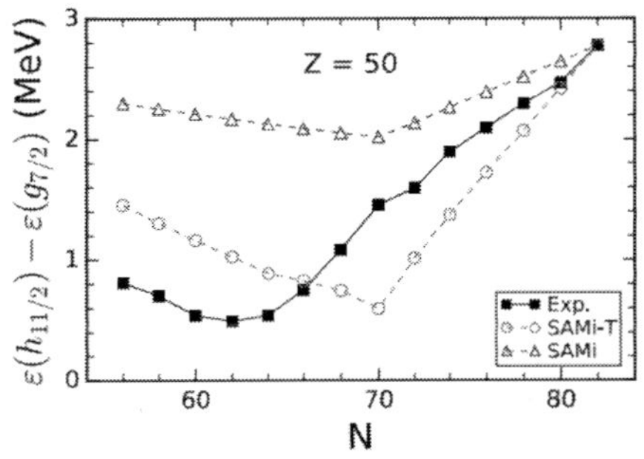

Figure 1. Single-particle energy splittings of proton $h_{11/2}$ and $g_{7/2}$ along the $Z = 50$ isotopic chain calculated by SAMi-T and SAMi functionals, in comparison with experimental data [4]. Theoretical results have been shifted so that the splittings coincide with the data at ^{132}Sn, see the text for details.

It can be seen that by including the tensor terms, the relative change of the energy splittings given by SAMi-T is larger from $N = 56$ to $N = 70$, and then to $N = 82$, comparing with SAMi. This also agrees better with the experimental data than SAMi. The energy splitting of $h_{11/2}$ and $g_{7/2}$ is determined much by the spin-orbit splittings of the spin doublets $\{h_{11/2}, h_{9/2}\}$ and $\{g_{9/2}, g_{7/2}\}$. When the s.o. splittings of these two doublets decrease, the splitting of $h_{11/2}$ and $g_{7/2}$ becomes larger and vice versa.

In the case of Fig. 1, the proton numbers are all the same as $Z = 50$ and the neutron numbers change from $N = 56$ to $N = 82$. Therefore the proton s.o. splittings are much affected by the neutron-proton tensor term, which is determined by the β parameter in Eq. (6) and the values for the two functionals are [18]: $\beta_{\text{SAMi-T}} = 102$ MeV fm^5, $\beta_{\text{SAMi}} = 32$ MeV fm^5. In other words, SAMi-T has a much stronger neutron-proton (n-p) tensor effect than SAMi. For ^{106}Sn with $N = 56$, the single-particle levels $1g_{9/2}$ and $2d_{5/2}$ are fully occupied while their spin partners $1g_{7/2}$ and $2d_{3/2}$ are empty, thus it is a spin-unsaturated system. Similarly, ^{132}Sn with $N = 82$ is also spin-unsaturated as $1h_{11/2}$ is occupied and $1h_{9/2}$ is not. ^{120}Sn, with $N = 70$, is spin-saturated as $1g_{7/2}$ and $2d_{3/2}$ are now occupied while $1h_{11/2}$ is empty. As a consequence, SAMi-T with stronger n-p tensor effect gives relatively smaller proton s.o. splittings for the neutron spin-unsaturated ^{106}Sn and ^{132}Sn, and relatively larger splittings for the spin-saturated ^{120}Sn. This leads to a relatively larger splitting between $h_{11/2}$ and $g_{7/2}$ at ^{106}Sn and ^{132}Sn and relatively smaller splitting as ^{120}Sn.

A similar analysis can be applied to the mirror case. In Fig. 2, we show the energy splittings of neutron $i_{13/2}$ and

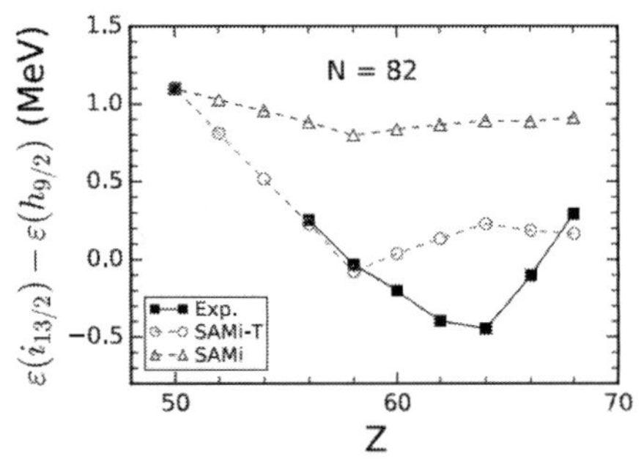

Figure 2. Energy splittings of neutron $i_{13/2}$ and $h_{9/2}$ along the N = 82 isotonic chain calculated by SAMi-T and SAMi functionals, in comparison with experimental data [4]. Theoretical results have been shifted so that the splittings coincide with the data at ^{132}Sn, see the text for details.

$h_{9/2}$ along the N = 82 isotonic chain calculated by SAMi-T and SAMi functionals, in comparison with experimental data [4]. Here the shift for SAMi-T is $\delta_{\text{SAMi-T}} = 4.7 - 1.1 = 3.6$ MeV; and for SAMi is $\delta_{\text{SAMi}} = 5.3 - 1.1 = 4.2$ MeV. With the tensor term guided by RBHF calculation, SAMi-T improves much the description of the evolution of s.o. splittings.

In both cases, there is still room to improve. For example, the minimum of the energy splittings given by the experimental data in Fig. 1 appears at N = 62 and in Fig. 2 appears at Z = 64. On the other hand, the corresponding minima given by SAMi-T (or SAMi) appear at N = 70 and Z = 58, respectively. In the single-particle pictures given by HF calculation, from N = 52 to N = 56 in the Z = 50 isotopes, the $2d_{5/2}$ orbit is being filled. From N = 58 to N = 64, the $1g_{7/2}$ orbit is being filled, which is a $j = l-1/2$ orbit and because of the tensor effect discussed above, this decreases the $\{h_{11/2}, g_{7/2}\}$ energy splitting. Furthermore, the occupation of $3s_{1/2}$ and $2d_{3/2}$ orbits will continue to decrease this energy splitting. In the end, the minimum occurs at N = 70. For N = 82 isotones in Fig. 2, the situation is slightly different as now the $1g_{7/2}$ orbit is lower than $2d_{5/2}$. Therefore, from Z = 52 on the $1g_{7/2}$ orbit will be filled before $2d_{5/2}$. Due to the tensor force, these two orbits give contrary effects on the energy splitting of $\{i_{13/2}, h_{9/2}\}$ and a minimum occurs at Z = 58. Other features of the energy density functional besides the tensor force can also affect the energy splittings. To have better agreement with the experiment, further investigations are needed.

4 Summary

A new Skyrme functional SAMi-T was developed with tensor term guided by *ab initio* relativistic Brueckner-Hartree-Fock calculations. By fitting to the spin-orbit splittings of neutron-proton drops calculated by RBHF, the tensor terms of SAMi-T are well constrained. Ground state properties and various giant resonances of finite nuclei can be well described, and tensor term improves the description in some cases [18].

The evolution of single-particle energy splittings in the Z = 50 isotopic chain and N = 82 isotonic chain have been studied with SAMi-T functional. The agreement with experimental data has been improved due to the tensor terms.

References

[1] M. Bender, P.H. Heenen, P.G. Reinhard, Rev. Mod. Phys. **75**, 121 (2003)
[2] J. Meng, *Relativistic Density Functional for Nuclear Structure*, Vol. 10 of *International Review of Nuclear Physics* (WORLD SCIENTIFIC, 2016)
[3] H. Sagawa, G. Colò, Prog. Part. Nucl. Phys. **76**, 76 (2014)
[4] J.P. Schiffer, S.J. Freeman, J.A. Caggiano, C. Deibel, A. Heinz, C.L. Jiang, R. Lewis, A. Parikh, P.D. Parker, K.E. Rehm et al., Phys. Rev. Lett. **92**, 162501 (2004)
[5] T. Otsuka, T. Suzuki, R. Fujimoto, H. Grawe, Y. Akaishi, Phys. Rev. Lett. **95**, 232502 (2005)
[6] B.A. Brown, T. Duguet, T. Otsuka, D. Abe, T. Suzuki, Phys. Rev. C **74**, 061303 (2006)
[7] G. Colò, H. Sagawa, S. Fracasso, P. Bortignon, Phys. Lett. B **646**, 227 (2007)
[8] D.M. Brink, F. Stancu, Phys. Rev. C **75**, 064311 (2007)
[9] T. Lesinski, M. Bender, K. Bennaceur, T. Duguet, J. Meyer, Phys. Rev. C **76**, 014312 (2007)
[10] W. Long, H. Sagawa, J. Meng, N. Van Giai, EPL (Europhysics Lett. **82**, 12001 (2008)
[11] A.V. Afanasjev, S.E. Agbemava, D. Ray, P. Ring, Phys. Rev. C **91**, 014324 (2015)
[12] S. Shen, H. Liang, J. Meng, P. Ring, S. Zhang, Phys. Lett. B **778**, 344 (2018)
[13] S. Shen, H. Liang, J. Meng, P. Ring, S. Zhang, Phys. Rev. C **97**, 054312 (2018)
[14] S.H. Shen, J.N. Hu, H.Z. Liang, J. Meng, P. Ring, S.Q. Zhang, Chinese Phys. Lett. **33**, 102103 (2016)
[15] S. Shen, H. Liang, J. Meng, P. Ring, S. Zhang, Phys. Rev. C **96**, 014316 (2017)
[16] S. Shen, H. Liang, J. Meng, P. Ring, S. Zhang, Phys. Lett. B **781**, 227 (2018)
[17] S. Shen, H. Liang, W.H. Long, J. Meng, P. Ring, Prog. Part. Nucl. Phys. 103713 (2019)
[18] S. Shen, G. Colò, X. Roca-Maza, Phys. Rev. C **99**, 034322 (2019)
[19] D. Vautherin, D.M. Brink, Phys. Rev. C **5**, 626 (1972)
[20] F. Stancu, D.M. Brink, H. Flocard, Phys. Lett. B **68**, 108 (1977)
[21] X. Roca-Maza, G. Colò, H. Sagawa, Phys. Rev. C **86**, 031306 (2012)
[22] R. Machleidt, in *Advances in Nuclear Physics*, edited by J.W. Negele and E. Vogt (springer, New York, 1989), Vol. 19, 189-376

EPJ Web of Conferences **223**, 01060 (2019) https://doi.org/10.1051/epjconf/201922301060
NSD2019

Nuclear structure in the neutron-deficient Sn nuclei

TKEL effects on lifetime measurements

Marco Siciliano[1,2,*], *José-Javier* Valiente-Dobón[2], and *Alain* Goasduff[3,4]

[1]Irfu/CEA, Université de Paris-Saclay, Gif-sur-Yvette, France.
[2]INFN, Laboratori Nazionali di Legnaro, Legnaro, Italy.
[3]Dipartimento di Fisica e Astronomia, Università di Padova, Padova, Italy.
[4]INFN, Sezione di Padova, Padova, Italy.

Abstract. The presence of seniority-like isomers along the $Z = 50$ isotopic chain have been an experimental limitation to the investigation of the electromagnetic properties of the low-lying states in the light Sn nuclei. Combining a multi-nucleon transfer reaction with the Recoil-Distance Doppler-Shift technique, the lifetimes of the 2_1^+ and 4_1^+ excited states have been directly measured in the neutron-deficient ^{106,108}Sn isotopes for the very first time. The emitted γ rays were detected by the AGATA array, while the reaction products were uniquely identified by the VAMOS++ magnetic spectrometer. The control of the direct feeding of the states by gating on the Total Kinetic Energy Loss, together with the unique capabilities of the two spectrometers, was crucial for the measurement in ^{108}Sn.

1 Introduction

The investigation of nuclear structure close to $Z = 50$ shell closure have been a hot and discussed topic for both theoretical and experimental groups. The Sn region is the longest isotopic chains between two experimentally accessible neutrons shell closures, namely $N = 50$ and $N = 82$. Hence, it provides a unique opportunity for systematic studies of the evolution of basic nuclear properties when going from very neutron-deficient to very neutron-rich species.

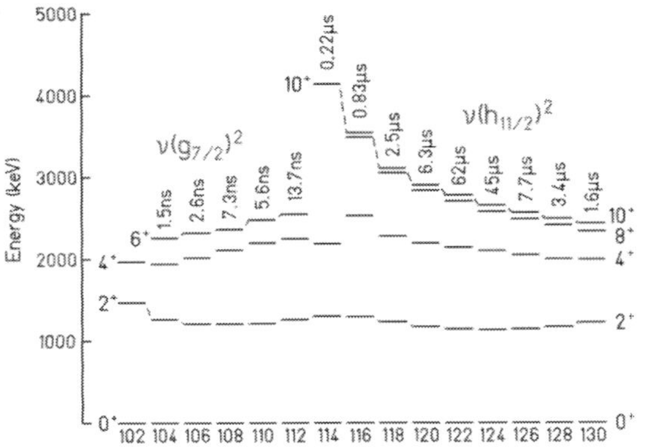

Figure 1. Excitation energy of the low-lying states along the Sn isotopic chain. The rather constant energy of the 2_1^+ and 4_1^+ excited states and the low-lying isomers, whose features can be attributed to seniority isomers, suggest the presence of pairing dominance. Adapted from Ref. [1].

In this context, the Sn nuclei have been considered a paradigm of pairing dominance. As shown in Fig. 1, the excitation energies of the first 2^+ and 4^+ states are rather constant along the Sn isotopic chain, while the low-lying isomeric states have features in common with seniority isomers due to the break of a $g_{7/2}$ and a $h_{11/2}$ neutron pair for the neutron-deficient and neutron-rich isotopes, respectively. Moreover, for isotopes with $A > 116$ the $B(E2; 2_1^+ \rightarrow 0_1^+)$ values present a parabolic behavior, that is expected for the seniority scheme. On the other hand, as presented in Fig. 2, for the lighter Sn nuclei, experimental results on transition probabilities are scarce. In fact, such a neutron-deficient region cannot be investigated with the typical combination of fusion-evaporation reactions with lifetime measurement because of the presence of low-lying isomeric states, which hindered direct measurements of lifetimes below them. Only with the advent of radioactive ion-beam facilities, Coulomb-excitation measurements have allowed to extract the reduced transition probabilities between the first excited 2^+ state and the ground state [1–8]. Within experimental uncertainties, the results suggest a rather-constant behavior for $106 \leq A \leq 110$, instead of the expected parabolic trend. Unfortunately, the lack of information on the $B(E2; 4_1^+ \rightarrow 2_1^+)$ strengths in light Sn nuclei [9], combined with large experimental uncertainties on the $B(E2; 2_1^+ \rightarrow 0_1^+)$ values, prevents firm conclusions on the shell evolution in the vicinity of the heaviest proton-bound $N = Z$ doubly-magic nucleus ^{100}Sn.

To remedy such an experimental limitation, the first lifetime measurement in neutron-deficient Sn isotopes was carried out using the Recoil Distance Doppler-Shift (RDDS) technique, providing a complementary informa-

*e-mail: marco.siciliano@cea.fr

tion to the previous Coulomb-excitation studies. Thanks to the unusual application of a Multi-Nucleon Transfer (MNT) reaction, that is commonly used to investigate neutron-rich nuclei [20–22], together with unprecedented capabilities of the powerful AGATA and VAMOS++ spectrometers, the lifetimes of the 2_1^+ and 4_1^+ states in ^{106,108}Sn have been directly measured for the very first time [23]. The employment of a MNT reaction not only allowed to directly populate the excited states below the isomers, but permitted also to control such a feeding by an appropriate gate on the Total Kinetic Energy Loss (TKEL) [24, 25].

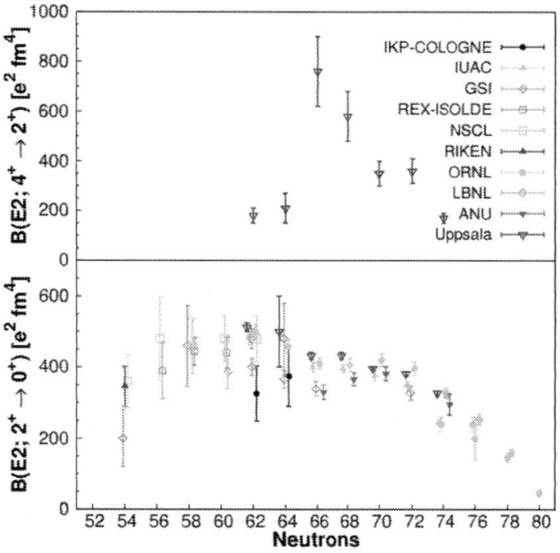

Figure 2. Experimental reduced transition probabilities $B(E2)$ for the (top) $4_1^+ \rightarrow 2_1^+$ and (bottom) $2_1^+ \rightarrow 0_1^+$ transitions along the Sn isotopic chain. Data taken from Refs. [1–19].

In this contribution the features of this unconventional procedure are investigated by showing the effects of the TKEL gate on the lifetime measurements.

2 Experiment

The light Sn isotopes were obtained in the collision of a ^{106}Cd beam, provided by the separated-sector cyclotron of the GANIL facility at an energy of 770 MeV, and a 0.7 mg/cm^2 ^{92}Mo. The lifetime measurement was performed with the Recoil Distance Doppler-Shift (RDDS) technique [26, 27] by mounting the target and a 1.6 mg/cm^2 thick ^{nat}Mg degrader on the differential Cologne plunger. The complete (A,Z) identification of the projectile-like reaction products was obtained on an event-by-event basis using the VAMOS++ spectrometer [28, 29]. In coincidence with the magnetic spectrometer, the γ rays were detected by the AGATA array [30, 31], consisting of 8 triple-cluster detectors placed at backward angles. More details about the ion identification and the analysis procedure can be found in Refs. [23, 32, 33].

Thanks to the fine position sensitivity of both the spectrometers, Doppler correction was applied on an event-by-event basis. Such a sensitivity was essential for performing the measurement because for each γ-ray transition two peaks were observed, related to its emission before and after the Mg foil. The γ rays emitted after the degrader are properly Doppler corrected (I_u), while those emitted before are shifted to lower energies (I_s) because of the different velocity of the reaction fragment. The relative intensities of the peaks area as a function of the target-degrader distance are related to the lifetime of the state of interest [26, 27].

3 Results

Thanks to the simultaneous measurement of the angle (θ_{bl}) and of the energy (E_{bl}) of the beam-like reaction fragments entering in VAMOS++, the TKEL of the reaction can be extracted under the assumption of a binary reaction without particle evaporation. For the reconstruction it was assumed, that the reaction occurs in the centre of the target: the energy of the beam ion at the centre of the target (E_r) was calculated by taking into account the energy loss in the ^{92}Mo material and the same procedure was adopted to calculate the energy E_{bl} at the centre of the target. The TKEL of the reaction was obtained by using the non-relativistic formula [34]

$$TKEL = \frac{m_{tl} - m_b}{m_{tl}} E_r - \frac{m_{tl} + m_{bl}}{m_{tl}} E_{bl} + \frac{2}{m_{tl}} \sqrt{m_b m_{bl} E_r E_{bl}} \cos\theta_{bl} \quad (1)$$

where m_t, m_{tl} and m_{bl} are the target, the target-like and beam-like masses respectively.

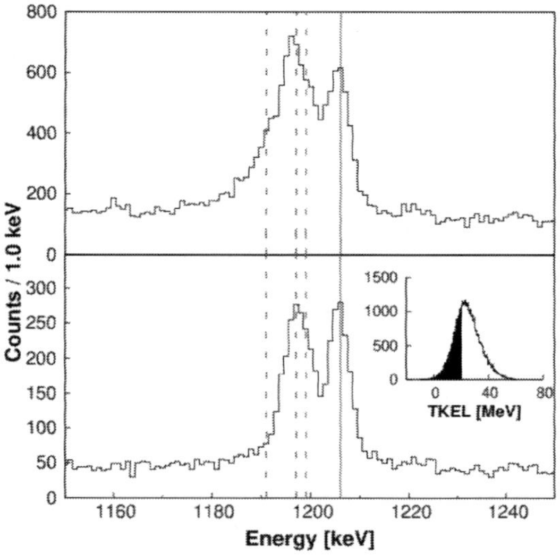

Figure 3. Doppler-corrected γ-ray energy spectra of ^{108}Sn with (bottom) and without (top) the TKEL$\leq$ 21 MeV condition, obtained by summing up the statistics of all the distances. The shifted (dashed lines) and unshifted (solid lines) components of $8_1^+ \rightarrow 6_1^+$ (green) and $2_1^+ \rightarrow 0_1^+$ (red) are marked. In the inset the TKEL distribution and the required condition are shown.

For the lifetime measurement of the low-lying states in ^{108}Sn, the adoption of the TKEL-gate procedure was crucial. Indeed, as shown in the spectra of Fig. 3, for this nucleus the energies of $8^+_1 \rightarrow 6^+_1$ and $2^+_1 \rightarrow 0^+_1$ transitions are very similar and their different components cannot be distinguished, so traditional methods cannot be used to measure the lifetime of the 2^+_1 state. However, since the population of higher (lower) excited states in the final nucleus correspond to higher (lower) values of TKEL, it is possible to reduce the population of the excited states above the 6^+ isomer by imposing the TKEL$\leq$ 21 MeV condition: the two peaks related to the $8^+_1 \rightarrow 6^+_1$ transition became negligible. Such a procedure allowed us to take into account just the 6^+_1, 4^+_1 and 2^+_1 states in the lifetime measurement via the Decay-Curve Method [23].

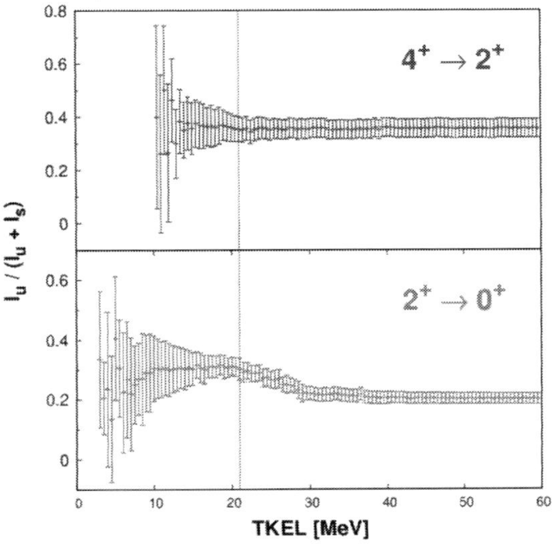

Figure 4. Decay curve ratio as a function of the TKEL-gate upper limit for the $4^+_1 \rightarrow 2^+_1$ and $2^+_1 \rightarrow 0^+_1$ transitions of ^{108}Sn. The values are obtained by summing up all the spectra obtained for the different distances. The defined threshold of the TKEL is shown with a purple solid line.

The 21 MeV upper limit was obtained via a systematic investigation on the effects of the TKEL-gate on the decay curve, displayed in Fig. 4: the threshold was chosen in order to minimize the area of the $8^+_1 \rightarrow 6^+_1$ transition peaks until they become negligible and the measured lifetime of the 2^+_1 excited state, here shown as the $I_u/(I_u + I_s)$ ratio, remained constant even for more restricted conditions. In fact, for the $2^+_1 \rightarrow 0^+_1$ transition the ratio slightly increases from 0.20 to 0.31 and then remain constant for lower TKEL values. For the $4^+_1 \rightarrow 2^+_1$ transition, instead, the ratio is practically independent on the TKEL gate and this is due to the presence of the 6^+ state: reducing the population of the higher-spin states does not affect the lifetime of the 4^+_1 excited states, since their feeding is anyway blocked by the presence of the isomer. Moreover, in Fig. 4 one should also notice that for TKEL$\approx$ 12 MeV the population of the 4^+_1 state starts to be not sufficient for measuring the lifetime, causing the evident fluctuations in the ratio; for the 2^+_1 state, instead, the fluctuations start around TKEL$\approx$ 7 MeV and the intensities ratio slightly decrease with very restrictive conditions. This is a clear example of the sensitivity of the technique for the used reaction: the gate on the reconstructed TKEL is capable to reduce so much the population of the 4^+_1 state, whose excitation energy is 905 keV larger than the one of the 2^+_1 state, that its feeding becomes negligible, making the effective lifetime of the 2^+_1 state shorter. Unfortunately, for such a restrictive condition, the statistics in the $2^+_1 \rightarrow 0^+_1$ transition components was not sufficient to perform the lifetime measurement.

4 Conclusion

In this work deep-inelastic collisions, in the specific multi-nucleon transfer reactions, have shown to be a powerful tool to investigate not only the neutron-rich nuclei but also the neutron-deficient species. Indeed, the employment of a multi-nucleon transfer reaction with the plunger device allowed to investigate the electromagnetic properties of the low-lying states close to ^{100}Sn, overcoming the experimental limitations caused by the presence of low-lying isomers. Moreover, thanks to the combination of such technique with the capabilities of both the AGATA and VAMOS++ spectrometers, the Total Kinetic Energy Loss can be reconstructed and then used to control the feeding from higher-lying states, simplifying the considered decay cascade while measuring the lifetime with the Decay Curve Method.

References

[1] R. Kumar, P. Doornenbal, A. Jhingan, R.K. Bhowmik, S. Muralithar, S. Appannababu, R. Garg, J. Gerl, M. Górska, J. Kaur et al. Physical Review C **81**, 024306 (2010).

[2] A. Banu, J. Gerl, C. Fahlander, M. Górska, H. Grawe, T. R. Saito, H.-J. Wollersheim, E. Caurier, T. Engeland, A. Gniady et al. Physical Review C **72**, 061305(R) (2005).

[3] C. Vaman, C. Andreoiu, D. Bazin, A. Becerril, B.A. Brown, C. M. Campbell, A. Chester, J.M. Cook, D.C. Dinca, A. Gade et al. Physical Review Letters **99**, 162501 (2007).

[4] J. Cederkäll, A. Ekström, C. Fahlander, A.M. Hurst, M. Hjorth-Jensen, F. Ames, A. Banu, P.A. Butler, T. Davinson, U. Datta Pramanik et al. Physical Review Letters **98**, 172501 (2007).

[5] A. Ekström, J. Cederkäll, C. Fahlander, M. Hjorth-Jensen, F. Ames, P.A. Butler, T. Davinson, J. Eberth, F. Fincke, A. Görgen et al. Physical Review Letters **101**, 012502 (2008).

[6] V.M. Bader, A. Gade, D. Weisshaar, B.A. Brown, T. Baugher, D. Bazin, J.S. Berryman, A. Ekström, M. Hjorth-Jensen, S.R. Stroberg et al. Physical Review C **88**, 051301(R) (2013).

[7] P. Doornenbal, S. Takeuchi, N. Aoi, M. Matsushita, A. Obertelli, D. Steppenbeck, H. Wang, L. Audirac,

H. Baba, P. Bednarczyk et al. Physical Review C **90**, 061302(R) (2014).

[8] R. Kumar, M. Saxena, P. Doornenbal, A. Jhingan et al. Physical Review C **96**, 054318 (2017).

[9] N. Jonsson, A. Bäcklin, J. Kantele, R. Julin, M. Luontama, A. Passoja. Nuclear Physics A **371**, 333 (1981).

[10] D.C. Radford et al. Nuclear Physics A **746**, 83 (2004).

[11] J.N. Orce, S.N. Choudry, B. Crider, E. Elhami, S. Mukhopadhyay, M. Scheck, M.T. McEllistrem, S.W. Yates. Physical Review C **76**, 021302(R) (2007).

[12] P. Doornenbal, P. Reiter, H. Grawe, H. J. Wollersheim, P. Bednarczyk, L. Caceres, J. Cederkäll, A. Ekström, J. Gerl, M. Górska et al. Physical Review C **78**, 031303(R) (2008).

[13] J.M. Allmond, D.C. Radford, C. Baktash, J.C. Batchelder, A. Galindo-Uribarri, C.J. Gross, P.A. Hausladen, K. Lagergren, Y. Larochelle, E. Padilla-Rodal et al. Physical Review C **84**, 061303(R) (2011).

[14] A. Jungclaus, J. Walker, J. Leske, K.H. Speidel, A.E. Stuchbery, M. East et al. Physics Letters B **695**, 110 (2011).

[15] G. Kumbartzki et al. Physical Review C **86**, 034319 (2012).

[16] G. Guastalla, D.D. Di Julio, M. Górska, J. Cederkäll, P. Boutachkov, P. Golubev, S. Pietri, H. Grawe, F. Nowacki, K. Sieja et al. Physical Review Letters **110**, 172501 (2013).

[17] J.M. Allmond, A.E. Stuchbery, A. Galindo-Uribarri, E. Padilla-Rodal, D.C. Radford, J.C. Batchelder, C.R. Bingham, M.E. Howard, J.F. Liang, B. Manning et al. Physical Review C **92**, 041303(R) (2015).

[18] G.J. Kumbartzki, N. Benczer-Koller, K.-H. Speidel, D.A. Torres, J.M. Allmond, P. Fallon, I. Abramovic, L.A. Bernstein, J.E. Bevins, H.L. Crawford et al. Physical Review C **93**, 044316 (2016).

[19] M. Spieker, P. Petkov, E. Litvinova, C. Müller-Gatermann, S. G. Pickstone, S. Prill, P. Scholz, A. Zilges. Physical Review C **97**, 054319 (2018).

[20] S. Szilner, C. A. Ur, L. Corradi, N. Mărginean, G. Pollarolo, A. M. Stefanini, S. Beghini, B.R. Behera, E. Fioretto, A. Gadea et al. Physical Review C **76**, 024604 (2007).

[21] L. Corradi, G. Pollarolo, and S. Szilner. Journal of Physics G **36**, 113101 (2009).

[22] J.J. Valiente-Dobón. Basic Concepts in Nuclear Physics: Theory, Experiments and Applications **182**, 87 (2016).

[23] M. Siciliano, J.J. Valiente-Dobón, A. Goasduff, F. Nowacki, A.P. Zuker. arXiv:1905.10313v2 (2019).

[24] J.J. Valiente-Dobón, D. Mengoni, A. Gadea, E. Farnea, S.M. Lenzi, S. Lunardi, A. Dewald, T. Pissulla, S. Szilner, R. Broda et al. Physical Review Letters **102**, 242502 (2009).

[25] D. Mengoni, J.J. Valiente-Dobón, E. Farnea, A. Gadea, A. Dewald, A. Latina. European Physics Journal A **42**, 387 (2009).

[26] A. Dewald, S. Harissopulos, and P. Von Brentano. Zeitschrift für Physik A **334**, 163 (1989).

[27] A. Dewald, O. Möller, P. Petkov. Progress in Particle and Nuclear Physics **67**, 786 (2012).

[28] S. Pullanhiotan, A. Chatterjee, B. Jacquot, A. Navin, M. Rejmund. Nuclear Instruments and Methods in Physics Research B **266**, 4148 (2008).

[29] M. Rejmund, B. Lecornu, A. Navin, C. Schmitt et al. Nuclear Instruments and Methods in Physics Research A **646**, 184 (2011).

[30] S. Akkoyun et al., Nuclear Instruments and Methods in Physics Research A **668**, 26 (2012).

[31] E. Clément, C. Michelagnoli, G. de France, H. J. Li, A. Lemasson, C. B. Dejean, et al. Nuclear Instruments and Methods in Physics Research A **855**, 1 (2017).

[32] M. Siciliano, J. J. Valiente-Dobón, A. Goasduff, D. Bazzacco et al. Acta Physica Polonica B **48**, 331 (2017).

[33] M. Siciliano. Nuovo Cimento C **40**, 84 (2017).

[34] A.B. Brown, C.W. Snyder, W.A. Fowler and C.C. Lauritsen. Physical Review **82**, 159 (1951).

The ^{3}H(α, γ)^{7}Li and ^{3}He(α, γ)^{7}Be Radiative Capture Reactions in a Wide Energy Range

Alexander Solovyev*

Dukhov Research Institute of Automatics (VNIIA), 22 Sushchevskaya street, Moscow 127055, Russia

Abstract. The mirror ^{3}H(α, γ)^{7}Li and ^{3}He(α, γ)^{7}Be reactions are considered from the microscopic viewpoint within a developed approach based on the multiscale algebraic version of the resonating group model. The total astrophysical *S* factors and the branching ratios for these reactions are calculated at low and intermediate energies covering the lowest resonances of the final fused nuclei. The results from the calculations are in an agreement with data from experiments.

1 Introduction

The ^{3}H(α, γ)^{7}Li and ^{3}He(α, γ)^{7}Be mirror radiative capture reactions have a great importance for astrophysical applications. In particular, both reactions are of interest to study Big Bang nucleosynthesis. The latter reaction is also important to study stellar nucleosynthesis, especially in the Sun. Despite a long story of experimental and theoretical investigations of these reactions (see references cited in works [1, 2]), the required precision has not been achieved yet. As a result, the problems still remain.

The aim of the present work is to demonstrate the energy dependences of the total astrophysical *S* factors and the branching ratios *R* obtained for the considered reactions in a wide energy range from a microscopic approach based on the multiscale algebraic version of the resonating group model (AVRGM). This approach was proposed in Ref. [1] and developed in Ref. [2].

2 Results

Formalism and details of the multiscale AVRGM approach can be found in work [1]. In order to describe the nuclear interaction, the modified Hasegawa–Nagata potential is used [3]. The intensities of the central Majorana force g_c and the spin-orbit interaction g_{ls} involved in this potential as adjustable parameters are equal to 0.977 and 3.469 respectively. These values were set to describe nuclear phase shifts of elastic scattering in the entrance channels of the reactions [2].

The total astrophysical *S* factor and the branching ratio *R* are defined by

$$S(E_{\mathrm{c.m.}}) = E_{\mathrm{c.m.}} \exp\left(\sqrt{E_G / E_{\mathrm{c.m.}}}\right)\sigma(E_{\mathrm{c.m.}}), \qquad (1)$$

$$R(E_{\mathrm{c.m.}}) = \sigma_1(E_{\mathrm{c.m.}}) / \sigma_0(E_{\mathrm{c.m.}}), \qquad (2)$$

where E_G is the Gamow energy of the colliding nuclei, $E_{\mathrm{c.m.}}$ is their relative motion energy in the center-of-mass system, σ_0 and σ_1 are the total cross sections related to the captures to the ground and first excited states of the final nucleus respectively, and σ is the total cross section of the reaction [2].

2.1 The ^{3}He(α, γ)^{7}Be total astrophysical *S* factor and branching ratio

The total astrophysical *S* factor calculated for the ^{3}He(α, γ)^{7}Be radiative capture within the multiscale AVRGM approach is shown in Fig. 1. References to experimental data denoted by the symbols in this figure are given in Ref. [2].

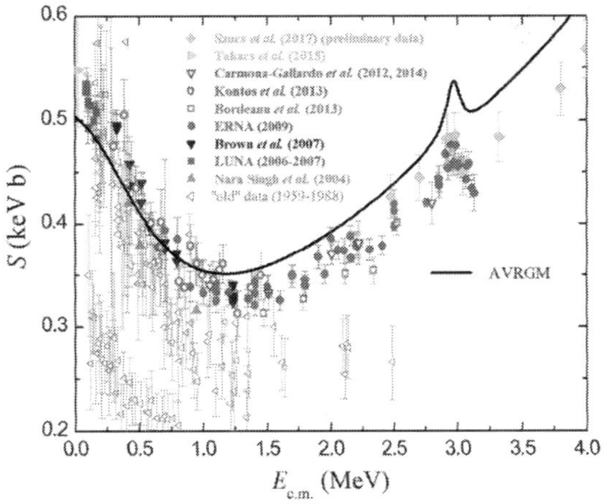

Fig. 1. Energy dependence of the ^{3}He(α, γ)^{7}Be total astrophysical *S* factor.

The obtained curve agrees reasonably well with the modern data up to vicinity of the 7/2$^-$ resonance. The calculation reproduces the position of this resonance

* alexander.solovyev@mail.ru

(2.98 MeV) but the calculated values of the astrophysical *S* factor in vicinity of the corresponding peak are slightly overestimated compared to the experimental ones.

The ^{3}He(α, γ)^{7}Be branching ratio from the multiscale AVRGM approach is drawn in Fig. 2. The presented curve agrees adequately with the data. These data were extracted from direct measurements only.

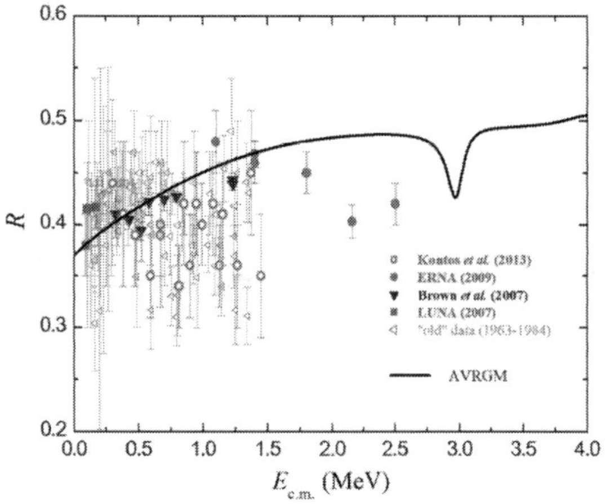

Fig. 2. Energy dependence of the ^{3}He(α, γ)^{7}Be branching ratio.

2.2 The ^{3}H(α, γ)^{7}Li total astrophysical *S* factor and branching ratio

For the ^{3}H(α, γ)^{7}Li radiative capture, the total astrophysical *S* factor and the branching ratio obtained in the framework of the multiscale AVRGM approach are depicted in Fig. 3 and Fig. 4 respectively. Experimental data are also marked in these figures (see Ref. [2] for references). The calculated curves are in a very good agreement with the data [4] in the whole energy range. The obtained 7/2$^-$ resonance position (2.11 MeV) is also very close to its experimental value (2.18 MeV). Unfortunately, experimental data on the ^{3}H(α, γ)^{7}Li total astrophysical *S* factor in vicinity of the peak corresponding to the 7/2$^-$ resonance are quite absent.

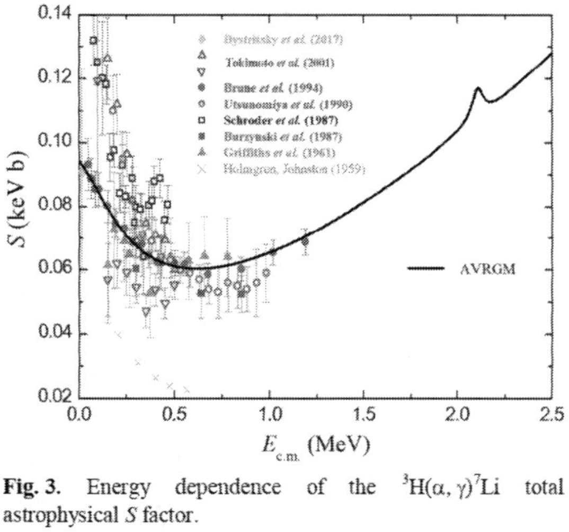

Fig. 3. Energy dependence of the ^{3}H(α, γ)^{7}Li total astrophysical *S* factor.

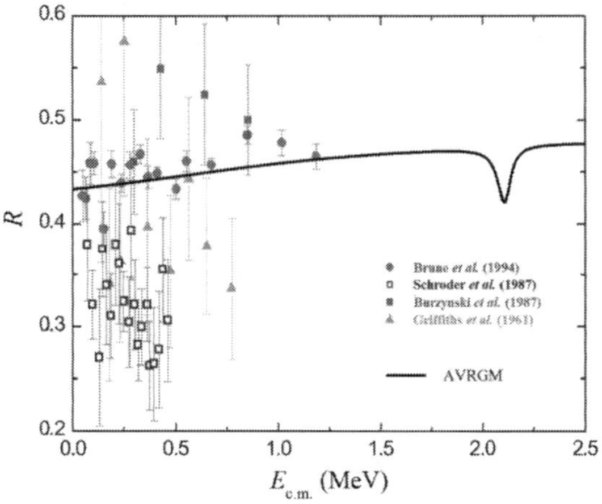

Fig. 4. Energy dependence of the ^{3}H(α, γ)^{7}Li branching ratio.

3 Conclusions

The multiscale AVRGM approach enables one to describe a wide enough set of the data on the mirror ^{3}H(α, γ)^{7}Li and ^{3}He(α, γ)^{7}Be reactions simultaneously in the unified way. In particular, the energy dependences of the calculated total astrophysical *S* factors and the branching ratios for both reactions agree with the data well. Moreover, the calculated nuclear phase shifts for the elastic scattering in the entrance channels of the reactions together with the calculated electromagnetic properties of the final fused nuclei are in a good correspondence with data extracted from experiments (see Ref. [2] for details).

On the whole, the presented curves cover a wide enough energy region, including the lowest 7/2$^-$ resonances of the ^{7}Li and ^{7}Be nuclei. In the calculations, all allowed E1, E2, and M1 transitions from the partial *s*, *p*, *d*, and *f* waves of the entrance channels of the reactions to the ground and first excited states of the final nuclei were taken into account. Undoubtedly, the E1 captures play the dominating role at all considered energies. However, the peaks in the energy dependences of the total astrophysical *S* factors are caused by the E2 transition from the 7/2$^-$ partial *f* wave to the final ground state. The M1 transitions turn out to be negligible.

Finally, the approach capabilities demonstrated in this work for ^{3}H(α, γ)^{7}Li and ^{3}He(α, γ)^{7}Be could be considered as the essential reasons for further development and applications of this approach.

References

1. A.S. Solovyev, S.Yu. Igashov, Phys. Rev. C **96**, 064605 (2017)
2. A.S. Solovyev, S.Yu. Igashov, Phys. Rev. C **99**, 054618 (2019)
3. H. Kanada, T. Kaneko, S. Nagata, M. Nomoto, Prog. Theor. Phys. **61**, 1327 (1979)
4. C.R. Brune, R.W. Kavanagh, C. Rolfs, Phys. Rev. C **50**, 2205 (1994)

Fusion Hindrance and Pauli Blocking in ^{58}Ni + ^{64}Ni

Alberto M. Stefanini[1,*], *Giovanna* Montagnoli[2], *Mirco* Del Fabbro[2], *Giulia* Colucci[2], *Petra* Čolović[3], *Lorenzo* Corradi[1], *Enrico* Fioretto[1], *Franco* Galtarossa[1], *Alain* Goasduff[2], *Jerzy* Grebosz[4], *Marcel* Heine[5], *Grzegorz* Jaworski[6], *Marco* Mazzocco[2], *Tea* Mijatović[3], *Suzana* Szilner[3], *Martin* Bajzek[3], *Daniele* Brugnara[2], *Marco* Siciliano[1], and *Irene* Zanon[7]

[1]INFN, Laboratori Nazionali di Legnaro, Padova, Italy
[2]University of Padova and INFN, Sez. di Padova, Italy
[3]Ruđer Bošković Institute, Zagreb, Croatia
[4]Institute of Nuclear Physics, Polish Academy of Sciences, Cracow, Poland
[5]IPHC, CNRS-IN2P3, Université de Strasbourg, Strasbourg Cedex 2, France
[6]Heavy Ion Laboratory, Univ. of Warsaw, Warsaw, Poland
[7]Università of Ferrara, Ferrara, Italy and INFN, Laboratori Nazionali di Legnaro, Padova, Italy

Abstract. ^{58}Ni +^{64}Ni is the first case where the influence of positive Q-value transfer channels on sub-barrier fusion was evidenced, in a very well known experiment by Beckerman et al., by comparing with the two systems ^{58}Ni +^{58}Ni and ^{64}Ni +^{64}Ni. Subsequent measurements on ^{64}Ni +^{64}Ni showed that fusion hindrance is clearly present in this case. On the other hand, no indication of hindrance can be observed for ^{58}Ni +^{64}Ni down to the measured level of 0.1 mb. In the present experiment the excitation function has been extended by two orders of magnitude downward. The cross sections for ^{58}Ni + ^{64}Ni continue decreasing very smoothly below the barrier, down to $\simeq$1 μb. The logarithmic slope of the excitation function increases slowly, showing a tendency to saturate at the lowest energies. No maximum of the astrophysical S-factor is observed. Coupled-channels (CC) calculations using a Woods-Saxon potential and including inelastic excitations only, underestimate the sub-barrier cross sections by a large amount. Good agreement is found by adding two-neutron transfer couplings to a schematical level. This behaviour is quite different from what already observed for ^{64}Ni+^{64}Ni (no positive Q-value transfer channels available), where a clear low-energy maximum of the S-factor appears, and whose excitation function is overestimated by a standard Woods-Saxon CC calculation. No hindrance effect is observed in ^{58}Ni+^{64}Ni in the measured energy range. This trend at deep sub-barrier energies reinforces the recent suggestion that the availability of several states following transfer with Q >0, effectively counterbalances the Pauli repulsion that, in general, is predicted to reduce tunneling probability inside the Coulomb barrier.

1 Introduction

The sequence of stable nickel isotopes from ^{58}Ni to ^{64}Ni offers several opportunities of studying fusion dynamics near and below the Coulomb barrier. The early experiments on fusion of Ni + Ni systems [1] are well-known and indicated for the first time the possible influence of transfer reactions on near- and sub-barrier cross sections. The excitation functions of the three systems ^{58}Ni + ^{58}Ni, ^{58}Ni + ^{64}Ni and ^{64}Ni + ^{64}Ni, besides the trivial differences due to the varying Coulomb barriers, show a remarkable feature, that is, the contrasting slope of the asymmetric system ^{58}Ni + ^{64}Ni, when compared to the other two symmetric cases. Indeed, the cross sections of ^{58}Ni + ^{64}Ni decrease much slower with decreasing energy. Shortly after, this was associated [2] with the availability, only in this system, of neutron transfer channels with positive Q-values. Later experiments for ^{58}Ni + ^{64}Ni [3] confirmed the flat shape of the excitation function, but the measured cross sections were anyway limited to $\sigma \geq$0.1 mb.

In more recent years it was found for many systems [4] that, at deep sub-barrier energies, the cross section decreases very rapidly [5], so that the excitation function is much steeper than the prediction of standard coupled-channels (CC) calculations. This phenomenon was named fusion hindrance. One of the first systems where this effect was clearly identified is ^{64}Ni + ^{64}Ni [6].

The original data of Beckerman et al. [1] were extended down to the level of $\approx$ 25 nb and the threshold of hindrance is around 0.1 mb. The effect was recognised also in the case of ^{58}Ni + ^{58}Ni at the level of 0.05 mb from the data of Ref. [1].

Low-energy hindrance is a matter of continuing experimental and theoretical interest. In the sudden approach proposed by Misicu and Esbensen [7, 8], a double folding potential is adopted (M3Y+repulsion), producing a shallow pocket as a consequence of the incompressibility of nuclear matter. This CC model has been quite successful in reproducing the hindrance behavior in a number of cases [9].

Alternatively, Ichikawa et al. [10], proposed an adiabatic neck formation between the colliding nuclei in

*e-mail: stefanini@lnl.infn.it

the overlap region, leading to hindrance. More recently, Simenel et al. [11] introduced a new microscopic model and demonstrated, on the basis of density-constrained frozen Hartree-Fock calculations, that the main effect of Pauli repulsion is to reduce tunnelling probability inside the Coulomb barrier. It has been pointed out as well that when positive Q-value transfer channels are available to the system, this effect of Pauli blocking may be reduced or disappear altogether [12], because several final states can be populated, and valence nucleons can flow between the two nuclei, thus initiating fusion. This corresponds to what observed for the system ^{40}Ca+^{96}Zr that was investigated a few years ago [13, 14]. The flat shape of its sub-barrier fusion excitation function is very peculiar and was suggested to originate from the couplings to several $Q > 0$ neutron pick up channels. ^{40}Ca+^{96}Zr was studied down to very small cross sections (2μb) and the phenomenon of fusion hindrance does not show up.

If that interpretation is correct, we expect a similar behaviour for ^{58}Ni +^{64}Ni which is a very attractive case in this sense because the Q-values for the neutron transfer channels are +3.9 MeV, for 2n and 4n pick up, and where a clear evidence of transfer couplings was already indicated below the barrier. The appearance of fusion hindrance in this case, would put serious doubts on the suggestion that Pauli blocking is not effective (or weakened) in systems with $Q > 0$ transfer channels.

The experiment has been performed very recently, and the low-energy part of the excitation function for ^{58}Ni + ^{64}Ni has been extended down by about two orders of magnitude. The results of these measurements will be reported in this contribution.

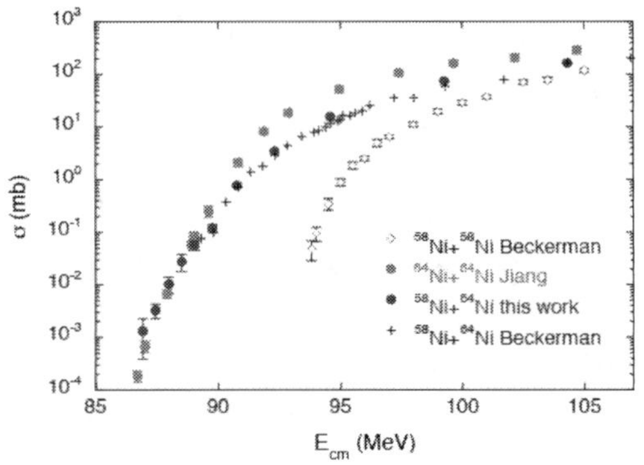

Figure 1. Fusion excitation functions for several Ni + Ni systems from the present and previous measurements (see text for setails).

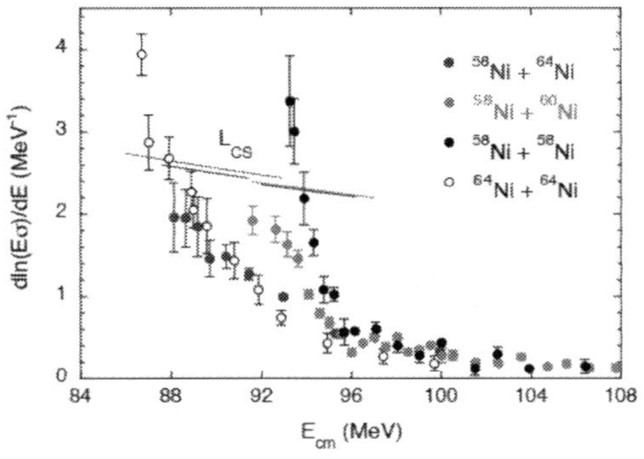

Figure 2. Logarithmic derivatives for several Ni + Ni systems.

2 Experimental set-up and Results

Fusion-evaporation cross sections have been measured for the system ^{58}Ni + ^{64}Ni at several energies near and below the Coulomb barrier, using the ^{58}Ni beam provided by the XTU Tandem accelerator of the Laboratori Nazionali di Legnaro (LNL) of INFN in the energy range 167-201 MeV. The beam intensity was 3-4 pnA.

Fusion-fission is negligible for ^{58}Ni + ^{64}Ni in the measured energy range, hence fusion cross sections were obtained by detecting at forward angles the evaporation residues (ER) following compound nucleus formation. The ER were separated from the beam by using the electrostatic deflector (see [15] and Refs. therein) usually employed for fusion measurements at LNL

The set of fusion cross sections we have obtained for ^{58}Ni + ^{64}Ni are shown in Fig. 1 with blue dots, together with the previous results for the same system and for ^{58}Ni + ^{58}Ni [1] (black crosses and open dots, respectively), and with the more recent measurements of Jiang et al. [6] on ^{64}Ni + ^{64}Ni (red dots). It is evident that the cross sections for ^{58}Ni + ^{64}Ni continue decreasing very smoothly below the barrier, while the two symmetric systems have much steeper excitation functions. This trend is clearly observed down to the lowest-measured cross section of $\simeq$1.3 μb.

Fig. 2 shows the logarithmic slope of the excitation function, derived from the measured cross sections, as the incremental ratio of two near-by points, for the various Ni+Ni systems. In this case, the existing data on ^{58}Ni + ^{60}Ni [16] have been added to the systematics. It appears that the trend of this system is similar to that of ^{58}Ni + ^{64}Ni. For both these cases, in the measured energy range, the slope increases slowly below the barrier with decreasing energy without notable irregularities. It does not reach the value expected for a constant astrophysical S-factor (L_{CS} in Fig. 2), rather it seems to saturate around 2 MeV^{-1}. Indeed, the influence of 2n-transfer couplings at the lowest energies was qualitatively suggested for ^{58}Ni + ^{60}Ni. For the other two symmetric systems Fig. 2 shows (as already known) that the slope clearly overcomes the L_{CS} value, thus presenting hindrance.

The CC calculations we are going to present in the next Section, and the comparison with ^{64}Ni + ^{64}Ni, will tell us more about this point.

3 Coupled-Channels Analysis

The CCFULL code [17] has been used to perform CC calculations for ^{58}Ni + ^{64}Ni. The ion-ion potential was a Woods Saxon parametrisation with well depth V_0= 151.85 MeV, diffuseness a= 0.67 fm and radius parameter r_o= 1.10 fm. V_0 is much deeper than what one obtains from

 https://doi.org/10.1051/epjconf/201922301062

the Akyüz-Winther potential [18]. It is used to remove unwanted oscillations of the low-energy excitation function that appear if the potential is too shallow, and consequently the incoming-wave boundary condition is not correctly applied. The parameters have been chosen to obtain a good data fit in the barrier region σ= 10-100 mb, when all channels, including the two neutron transfer (see later on) are taken into account in the calculations.

The nuclear structure information of the low-lying collective modes of ^{58}Ni and ^{64}Ni is reported in Table 1. The two nickel isotopes have quadrupole states at similar excitation energies and strengths. In the calculations we have considered up to two quadrupole phonons and only one octupole phonon (that has much higher excitation energy) in both nuclei.

Fig. 3 shows the results of the calculations when compared to the present experimental data (blue symbols). We notice that the CC results (blue line) strongly underestimate the data below the barrier. This indicates the possible effect of transfer couplings. Indeed in the recent work on ^{40}Ca + ^{96}Zr [19] it was pointed out that the effective Q-values for two-neutron as well as two-proton transfer are positive and both transfer channels can therefore influence the fusion. The situation is the same for ^{58}Ni + ^{64}Ni where the corresponding Q-values are +3.89 MeV and +2.6 MeV.

Therefore, further calculations have been performed, including a two nucleon transfer channel (+ 2N) besides the collective surface modes discussed above, using the approximate treatment of CCFULL where a pair transfer coupling between the ground states may be included. This uses the macroscopic coupling form factor given in Ref. [22]. The coupling strength F_t= 0.6 MeV has been used, adjusted for a best data fit. This rather large strength may be explained by the fact that it includes both proton and neutron transfer channels. The result is reported in Fig. 3 as a green solid line.

In the S-factor representation (Fig. 4) we can reach analogous conclusions. Indeed only including in the CC calculations the transfer coupling we have been able to reproduce the experimental S-factor trend where no maximum has been observed vs. energy.

The present results, and the comparison to CC calculations, indicate that hindrance does not show up in ^{58}Ni + ^{64}Ni in the measured energy range. At lower energies, we know that hindrance must appear, because the Q-value for fusion is negative Q_{fus}= -52.7 MeV [23].

4 Comparison with ^{64}Ni+^{64}Ni and ^{40}Ca+^{96}Zr

A comparison with the near-by system ^{64}Ni + ^{64}Ni is significant because in this case the hindrance phenomenon is present [6]. CCFULL calculations have been performed using the structure information of Table 1, and the Woods-Saxon ion-ion potential with parameters V_0= 75.98 MeV, diffuseness a= 0.676 fm and radius parameter r_o= 1.202 fm, as quoted in the original article [6]. In analogy with ^{58}Ni + ^{64}Ni, couplings to two quadrupole phonons and one octupole phonon have been considered. The result of the CC calculation is reported in Fig. 3 (see also Ref. [4]). We notice immediately that, at variance with ^{58}Ni + ^{64}Ni, the measured cross sections drop below the calculation (blue line) at low energies. Only using a shallow M3Y + repulsion potential [7, 8] (red line) one gets a good data fit, as already known.

Table 1. Excitation energies E_x, spin and parities λ^π, reduced transition probabilities and deformation parameters β_λ [20, 21] for the lowest quadrupole and octupole modes of ^{58}Ni and ^{64}Ni (see text). Nuclear and Coulomb deformation parameters have been taken to be the same in the present CC analysis.

Nucleus	E_x(MeV)	λ^π	$B(E\lambda)_{W.u.}$	β_λ
^{58}Ni	1.454	2^+	10.0	0.183 (3)
	4.475	3^-	12.6	0.20 (1)
^{64}Ni	1.346	2^+	10.0	0.179 (9)
	3.560	3^-	15.0	0.203 (20)

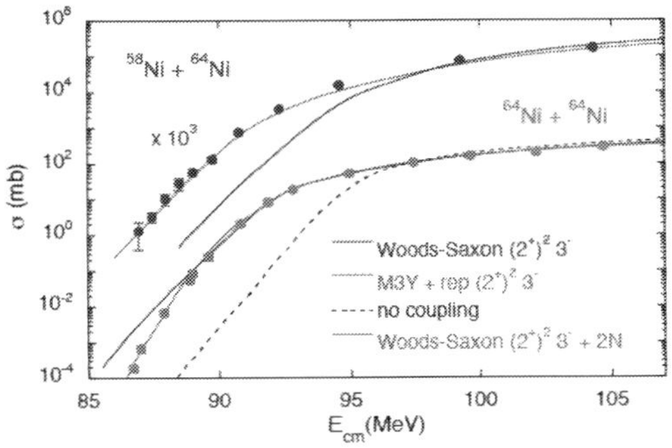

Figure 3. Fusion excitation function of ^{58}Ni + ^{64}Ni (blue dots) and ^{64}Ni + ^{64}Ni (red dots), compared with CC calculation (see text)

The comparison of the excitation functions for the two systems with the corresponding CC calculations confirms that they behave quite differently in the low energy region. This is even more clear in the astrophysical S-factor representation, as reported in Fig. 4. The experimental S-factors of the two systems are found on opposite sides with respect to the corresponding calculations using standard WS potentials. In particular, the maximum of S observed for ^{64}Ni + ^{64}Ni but not for ^{58}Ni + ^{64}Ni, is not reproduced by the CC calculations.

The case of ^{58}Ni + ^{64}Ni, on the other hand, is very similar to ^{40}Ca + ^{96}Zr [13, 14] because of the flat shape of the two sub-barrier fusion excitation functions, probably originating in both cases from the couplings to several Q >0 neutron pick-up channels.

In ^{40}Ca + ^{96}Zr fusion hindrance does not show up and this rather unusual behaviour is also due to the Q >0 transfer couplings. Indeed, the barrier distribution of this system displays a very long tail towards low energies [14], and we know the hindrance phenomenon shows up below the energy where the barrier distribution vanishes. This led to the suggestion [12] that the reaction mechanism involves the Pauli exclusion principle that in general pro-

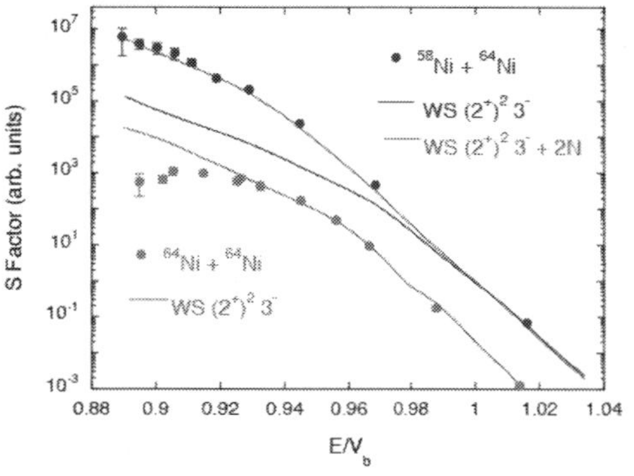

Figure 4. S-factor for ^{58,64}Ni + ^{64}Ni compared to CC calculations using a WS potential.

duces fusion hindrance [11], apart from the cases, as mentioned in the introduction, where several final states can be populated by nucleon transfer with $Q > 0$.

The absence of hindrance in the present case of ^{58}Ni + ^{64}Ni reinforces that suggestion. The strong dissimilarity with respect to the near-by case of ^{64}Ni + ^{64}Ni contributes to clarify the sub-barrier fusion dynamics, and prompts us to contemplate the possibility that the repulsive part of the potential in the Misicu-Esbensen model [7, 8] is actually a consequence of the Pauli exclusion principle.

5 Summary and conclusions

The results of fusion excitation function measurements for ^{58}Ni + ^{64}Ni have been presented. The experiment was performed using the ^{58}Ni beam of the XTU Tandem accelerator of the LNL. The excitation function obtained previously [1, 3] has been extended downwards by two orders of magnitude to about 1μb. We observe that the logarithmic slope of the excitation function has a slow increase, and tends to saturate at the lowest energies. The astrophysical S-factor does not show any maximum vs. energy.

CC calculations using a Woods-Saxon potential have been performed. The results underestimate the sub-barrier cross sections by a large amount, when only inelastic excitations are included. Good agreement is however found by schematically adding the coupling to the two-neutron transfer. The behaviour of ^{58}Ni + ^{64}Ni is quite different from what already observed for ^{64}Ni+^{64}Ni, where no positive Q-value transfer channels are available. In this case a clear low-energy maximum of the S-factor shows up, and the logarithmic slope exceeds the L_{CS} value by a large amount. The measured excitation function is overestimated by a standard WS CC calculation. No hindrance effect is observed in ^{58}Ni+^{64}Ni in the measured energy range. This makes the sub-barrier trend of this system quite similar to ^{40}Ca + ^{96}Zr, and corroborates the recent suggestion that the availability of several states following transfer with $Q > 0$, effectively counterbalances the Pauli repulsion that is predicted to reduce tunneling probability inside the Coulomb barrier.

Acknowledgments

Valuable discussions with H. Esbensen are gratefully acknowledged. The XTU Tandem staff provided us with high quality beams and M. Loriggiola prepared targets of excellent quality. The research leading to these results has received funding from the European Union Seventh Framework Program FP7/2007- 2013 under Grant Agreement No. 262010 - ENSAR. P.C., S.S and N.V. were partially supported by the Croatian Science Foundation under the project IP-2018-01-1257.

References

[1] M. Beckerman, et al. **45**, (1980) 1472.

[2] R.A. Broglia, C.H. Dasso, S. Landowne, and Å. Winther, Phys. Rev. C **27**, (1983) 2433.

[3] D. Ackermann et al., Nucl. Phys. A **583**, 129 (1995).

[4] G.Montagnoli, A.M.Stefanini EPJ A 53, 169 (2017).

[5] C. L. Jiang et al., Phys. Rev.Lett. **89**, 052701 (2002).

[6] C.L.Jiang, et al., Phys. Rev. Lett. **93**, (2004) 012701.

[7] S.Misicu and H. Esbensen, Phys. Rev. Lett. **96**,112701 (2006); Phys. Rev. C **75**, 034606 (2007).

[8] H.Esbensen, S.Misicu, Phys. Rev. C**76**, 054609 (2007).

[9] B. B. Back, H. Esbensen, C. L. Jiang, and K. E. Rehm, Rev. Mod. Phys. **86**, 317 (2014).

[10] T. Ichikawa, K. Hagino, and A. Iwamoto, Phys. Rev. C **75**, 057603 (2007).

[11] C. Simenel et al., Phys. Rev. C **95**, 031601(R) (2017).

[12] H. Esbensen and A. M. Stefanini, Phys. Rev. C **89**, 044616 (2014).

[13] H.Timmers et al., Phys. Lett. B **399**, 35 (1997); Nucl. Phys. A **633**, 421 (1998).

[14] A.M. Stefanini et al., Phys. Lett. B **728**, 639 (2014).

[15] G.Montagnoli et al., Phys. Rev. C **97**, 024610 (2018).

[16] A.M.Stefanini et al., Phys. Rev. Lett. **74**, (1995) 864.

[17] K.Hagino, N.Rowley and A.T.Kruppa, Comput. Phys. Commun. **123**, (1999) 143.

[18] Ö. Akyüz and Å. Winther, in Nuclear Structure and Heavy-Ion Physics, Proceedings of the International School of Physics "Enrico Fermi", Course LXXVII, Varenna, edited by R. A. Broglia and R. A. Ricci (North Holland, Amsterdam, 1981).

[19] H. Esbensen, A.M. Stefanini and G. Montagnoli, Phys. Rev. C **93**, 034609 (2016).

[20] S. Raman, C.W.Nestor, Jr., and P. Tikkanen, At. Data and Nucl. Data Tables **78**, (2001) 1.

[21] T. Kibédi and R. H. Spear, At. Data and Nucl. Data Tables **80**, (2002) 35.

[22] C.H. Dasso, G. Pollarolo, Phys. Lett. B **155**, 223 (1985).

[23] C. L. Jiang, H. Esbensen, B. B. Back, R. V. F. Janssens, and K. E. Rehm, Phys. Rev. C **69**, 014604 (2004).

Forbidden transitions in nuclear weak processes relevant to neutrino detection, nucleosynthesis and evolution of stars

Toshio Suzuki[1,2,*], *Satoshi* Chiba[3], *Takashi* Yoshida[4], *Koh* Takahashi[5], *Hideyuki* Umeda[4], and *Ken'ichi* Nomoto[6]

[1]Department of Physics, College of Humanities and Sciences, Nihon University, Setagaya-ku, Tokyo 156-8550, Japan
[2]National Astronomical Observatory of Japan, Mitaka, Tokyo 181-8588, Japan
[3]Reseach Laboratory for Nuclear Reactors, Tokyo Institute of Technology, Meguro, Tokyo 152-8550, Japan
[4]Department of Astronomy, Graduate School of Science, The University of Tokyo, Hongo, Bunkyo-ku, Tokyo 113-0033, Japan
[5]Max-Planck-Institut fur Gravitationsphysik (Albert-Einstein-Institute), Am Muhlenberg 1, D-14476, Potsdam-Golm, Germany
[6]Kavli Institute for the Physics and Mathematics of the Universe (WPI), The University of Tokyo, Kashiwa, Chiba 277-8583, Japan

Abstract. The distribution of the spin-dipole strengths in ^{16}O and neutrino-induced reactions on ^{16}O are investigated by shell-model calculations with new shell-model Hamiltonians. Charged-current and neutral-current reaction cross sections are evaluated in various particle and γ emission channels as well as the total ones at neutrino energies up to $E_\nu \approx$ 100 MeV. Effects of multiparticle emission channels, especially the αp emission channels, on nucleosynthesis of ^{11}B and ^{11}C in core-collapse supernova explosions are investigated. The MSW neutrino oscillation effects on charged-current reaction cross sections are investigated for future supernova burst. Electron capture rates for a forbidden transition ^{20}Ne ($0^+_{g.s.}$) → ^{20}F ($2^+_{g.s.}$) in stellar environments are evaluated by the multipole expansion method with the use of shell model Hamiltonians, and compared with those obtained by a prescription that treats the transition as an allowed Gamow-Teller (GT) transition. Different electron energy dependence of the transition strengths between the two methods is found to lead to sizable differences in the weak rates of the two methods.

1 Introduction

Roles of Gamow-Teller (GT) transitions in nuclear weak rates at stellar environments have been investigated in various astrophysical processes. Electron-capture and β-decay rates in sd-shell nuclei have been updated, and applied to nuclear URCA processes in O-Ne-Mg core of stars with M=8-10$M_\odot$ [1-3]. Those in fp-shell nuclei have been also updated with GXPF1J [4] and used to study synthesis of iron-group elements in Type Ia supernovae (SN) [5]. Neutrino-nucleus reaction cross sections for ^{12}C [6,7], ^{13}C [8], ^{40}Ar [9], ^{56}Fe and ^{56}Ni [10] have been updated, and applied to study ν-process nucleosynthesis [6,7,11] and ν properties [7,11]. In β-decays of N=126 isotones, an important role of first-forbidden transitions on enhancing the rates compared with the FRDM model [12] was pointed out [13,14], and the short half-lives were used to study r-process nucleosynthesis in core-collapse SN and binary neutron star mergers [14]. Here, we focus on forbidden transitions in ^{16}O and ^{20}Ne. We discuss spin-dipole strengths in ^{16}O and ν-induced reactions on ^{16}O in Sect. 2. In Sect. 3, e-capture rates for a second-forbidden transition in ^{20}Ne are evaluated with the multipole expansion method.

2 ν -induced reactions on ^{16}O

2.1. Spin-dipole strength in ^{16}O

ν-induced reactions on ^{12}C was studied with a shell-model Hamiltonian, SFO [15], which can reproduce the GT strength in ^{12}C and magnetic moments of p-shell nuclei systematically. The configuration space of the SFO is p-sd shell, and the quenching factor for q= g_A^{eff}/g_A is found to be close to 1, q=0.95, in contrast to the case within p-shell configurations such as the Cohen-Kurath Hamiltonian [16]. The monopole term in spin-isospin flip channel is enhanced in the SFO.

In case of ^{16}O, the GT strength is small and the spin-dipole strength is the dominant contribution to spin-dependent transitions. Therefore, the p-sd cross-shell matrix elements in SFO are improved by taking into account the tensor and two-body spin-orbit components properly: the tensor and two-body spin-orbit components are replaced by those of π+ρ meson-exchanges and σ+ρ+ω meson-exchanges, respectively. A new Hamiltonian thus obtained, SFO-tls [17], can reproduce low-lying energy levels of spin-dipole states in ^{16}O. Calculated spin-dipole strength,

$$B(SD\lambda)_{\mp} = \frac{1}{2J_i+1}\sum_f |< f \,||\, S_{\mp}^{\lambda} \,||\, i >|^2 \qquad (1)$$

$$S_{\mp\mu}^{\lambda} = r[Y^1 \times \vec{\sigma}]_{\mu}^{\lambda} t_{\mp}$$

in ^{16}O is shown in Fig. 1. Sum value B(SDλ)₋ is nearly proportional to 2λ + 1, while the averaged energy position is the lowest (highest) for 2^- (1^-) as explained below.

The energy-weighted sum of the strength can be formulated as the double commutator of the spin-dipole operator with the Hamiltonian as

* Corresponding author: suzuki@phys.chs.nihon-u.ac.jp

$$EWS^{\lambda}_{\mp} = \sum_{f,\mu} |\langle f\,|\,S^{\lambda}_{\mp\mu}\,|\,i\rangle|^2 \,(E_f - E_i) \quad (2)$$

$$EWS^{\lambda} = EWS^{\lambda}_{-} + EWS^{\lambda}_{+} = \frac{1}{2}\{\langle 0\,|[S_{-}^{\lambda\dagger},[H,S_{-}^{\lambda}]]|\,0\rangle + \langle 0\,|[S_{+}^{\lambda\dagger},[H,S_{+}^{\lambda}]]|\,0\rangle\}.$$

The EWS^{λ} is determined by the kinetic energy, one-body spin-orbit potential and two-body spin-dependent interactions [18]. For ^{16}O, EWS^{λ} evaluated with the kinetic energy and one-body spin-orbit potential $V_{LS} = \xi\Sigma_i \boldsymbol{\ell}_i \cdot \boldsymbol{\sigma}_i$ is given as

$$EWS^{\lambda} = \frac{3}{4\pi}(2\lambda+1)\left(\frac{\hbar^2}{2m}A + \frac{f_{\lambda}}{3}\xi\langle 0\,|\sum_i r_i^2\,|\,0\rangle\right), \quad (3)$$

where f_{λ} = 2, 1, and -1 for λ = 0, 1, and 2, respectively, and ^{16}O is taken to be a LS-closed core. For ^{16}O, the spin-orbit term splits them in the following order : $EWS^{2}_{-}/5 < EWS^{1}_{-}/3 < EWS^{0}_{-}$. The tensor interaction is attractive (repulsive) for 0^- and 2^- (1^-) [19]. The averaged energy defined by $\langle E_{\lambda}\rangle = EWS^{\lambda}/B(SD\lambda)$ results in an order ; $\langle E_2\rangle < \langle E_0\rangle < \langle E_1\rangle$.

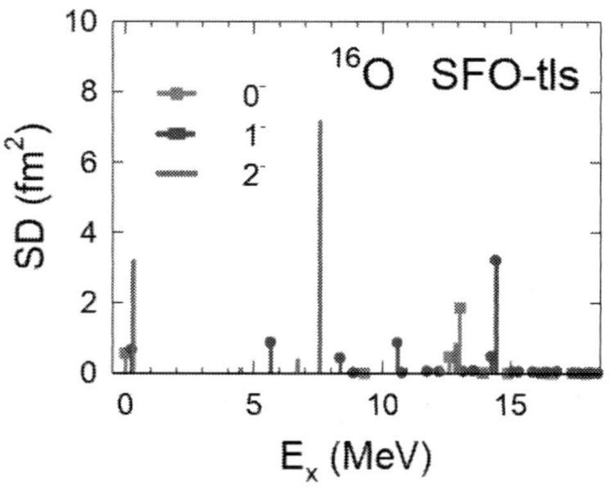

Fig. 1. Calculated spin-dipole strength in ^{16}O obtained with the SFO-tls. (Taken from Ref. [18])

2.2. ν -induced cross sections on ^{16}O

Charged- and neutral-current ν-nucleus reaction cross sections on ^{16}O are evaluated by shell-model calculations with the SFO-tls [18]. The quenching for g_A is taken to be q=0.95. The total μ-capture rate on ^{16}O obtained with q=0.95 is $\lambda = 11.20\times10^4$ s^{-1} (10.21×10^4 s^{-1}) for SFO-tls (SFO), which is close to the experimental value, $\lambda = 11.26\times10^4$ s^{-1} [20]. Total charge-exchange cross sections for ^{16}O (ν_e, e^-) ^{16}F at E_ν < 100 MeV obtained for SFO-tls as well as for SFO and previous continuum-random-phase approximation (CRPA) calculation [21] are shown in Fig. 2. The multipolarities up to J=4 are taken into account. Dominant contributions come from the transitions with $J^{\pi} = 2^-$ and 1^-. The cross sections for SFO-tls are enhanced compared with those of SFO, and found to be close to those of the CRPA except at E_ν < 30 MeV. Neutral-current cross sections for SFO-tls are also close to those of the CRPA.

Partial cross sections for various particle and γ emission channels, including multi-particle emissions in addition to single-particle ones, are evaluated by the Hauser-Feshbach model. Cross sections for p, pp, ^3He, α and αp emission channels for the excitations of 2^- states are shown in Fig. 3 for SFO-tls. The proton emission channel gives the dominant contribution, while αp and αemission channels become important at higher excita-tion energies at E_x ~30 MeV.

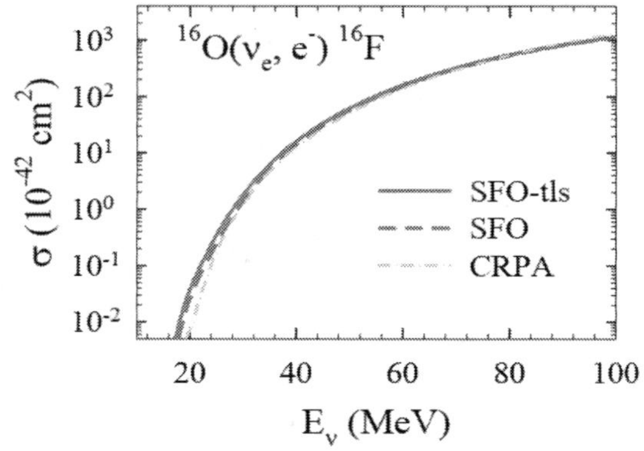

Fig. 2. Calculated total cross sections for ^{16}O (ν_e, e^-) ^{16}F obtained by shell-model calculations with the SFO-tls and SFO as well as CRPA calculation [21]. (Taken from Ref. [18])

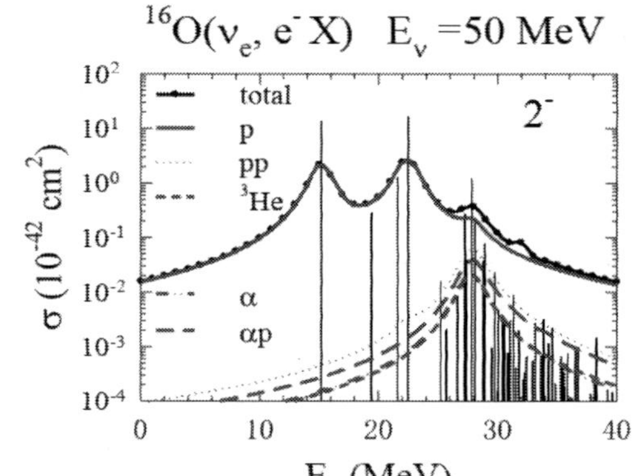

Fig. 3. Calculated partial cross sections for ^{16}O (ν_e, e^-X), with X = p, pp, ^3He, α, αp, via excitations of 2^- states in ^{16}F obtained by shell-model calculation with the SFO-tls. (Taken from Ref. [18])

Table 1. Production yields of ^{11}B and ^{11}C in supernova explosion for progenitor mass of $M=15M_{\odot}$. Here, "single-particle channels" denotes γ, p, n and α emission channels.

Production yield ($10^{-7}M_{\odot}$)	HW92 Single-particle channels	SFO-tls Single-particle channels	SFO-tls Single+Multi-particle channels
^{11}B	2.94	2.92	3.13.
^{11}C	2.80	2.71	3.20
$^{11}B+^{11}C$	5.74	5.62	6.33

A large branching ratio for the αp channel leads to the production of ^{11}B and ^{11}C by ^{16}O (ν, ν'αp) ^{11}B and ^{16}O (ν_e, e^-αp) ^{11}C reactions, respectively, in addition to the ordinary reaction channels ^{12}C (ν, ν'p) ^{11}B and ^{12}C (ν_e, e^-p) ^{11}C. The production yields of ^{11}B + ^{11}C in supernovae with 15 (20) $M_\odot$ are found to be enhanced by about 13 (12)% compared with those without the multi-particle emission channels [18] (see Table 1).

2.3 ν oscillations and detection of supernova ν

The MSW matter resonance oscillations [22] occur in C- He layer of supernovae for normal (inverted) mass hierarchy in charged-current reactions induced by $\nu_e(\bar{\nu}_e)$. Event spectra of neutrino-^{16}O charged-current reactions at Super-Kamiokande are evaluated for future super-nova neutrino bursts. The cross sections of the ^{16}O (ν_e, e^-) X and ^{16}O ($\bar{\nu}_e$, e^+) X reactions for each nuclear state with a different excitation energy are evaluated, and dependence of the cross sections on the mass hierarchies are examined [23]. Enhancement of expected event numbers for supernovae at 10 kpsc away from the earth is predicted for the $\nu_e(\bar{\nu}_e)$ -induced reaction in case of normal (inverted) hierachy [23].

3 Electron-capture rates for a second-forbidden transition in ^{20}Ne

3.1. Evolution of O-Ne-Mg cores in stars

Evolutions and final fate of a star with 8-10$M_\odot$ are sensitive to the nuclear weak rates as well as its mass. The cooling of the O-Ne-Mg core of the star produced after carbon burning occurs at certain densities by nuclear URCA processes for pairs of nuclei with A = 25 and 23 [1-3]. In later stage of the evolution, the core gets heated by γ emissions in double e-capture processes on ^{24}Mg and ^{20}Ne. It ws pointed out that a second-forbidden transition in ^{20}Ne, ^{20}Ne ($0_{g.s.}^+$) → ^{20}F (2_1^+), becomes important at high densities $\log_{10}(\rho Y_e)$ = 9.3-9.6, where Y_e is the lepton to baryon ratio, and at temperatures $\log_{10}(T)$ < 9.0 in e-capture reactions [24]. However, the transition strength has not been accurately determined [25]. The e-capture rates for the second forbidden transition in ^{20}Ne can be important to estimate the heating of the core and to determine the final fate of stars with 8-10$M_\odot$.

3.2 Electron-capture rates for ^{20}Ne ($0_{g.s.}^+$) → ^{20}F (2_1^+) in stellar environments

Recenty, the β-decay rate for the transition ^{20}F (2_1^+)→ ^{20}Ne ($0_{g.s.}^+$) was measured, and log *ft* value was obtained to be 10.47±0.11 [26], which is close to the lower limt value given in NNDC [25]. The e-capture rates for the forbidden transition in ^{20}Ne were calculated assuming the transition to be an allowed GT transition [24]. Here, we evaluate the e-capture rates accurately with the multipole expansion method [27]. In this method, there are contributions from the Coulomb, longitudinal, transverse electric and axial magnetic terms with the multipolarity $J^\pi = 2^+$. The transition strength has a dependence on the electron energy as in the cases of first-forbidden transitions in contrast to the allowed GT transition.

Calculated e-capture rates for the forbidden transition obtained with the USDB Hamitonian are shown in Fig. 4 for the temperature $\log_{10}T$ =8.6. Results obtained by the prescription assuming an allowed GT transition with the strength determined from the experimental log *ft* value [26] are also shown for comparison.

Sizable difference is found between the two methods [28]. The rates obtained by the GT prescription are found to be enhanced (reduced) compared with those with the USDB at $\log_{10}(\rho Y_e)$ < (>) 9.9. This comes from the difference in the electron energy dependence of the transition strengths between the two methods. The strength for the USDB is reduced (enhanced) at E_e < (>) 9.9 MeV compared with the GT prescription. At $\log_{10}(\rho Y_e)$ < 9.8, where the electron chemical potential is below 10 MeV, the shell-model strength is smaller than the GT one, and the rates for USDB also remain smaller than the GT one. At $\log_{10}(\rho Y_e) \geq 9.9$, on the other hand, the electron energy larger than 10 MeV can contribute to the rates due to the enhancement of the chemical potential, and the rates for USDB begin to exceed the GT rate.

Effects of the Coulomb effects, that is, the screening effects on both electrons and ions [29-32], are investigated. The Coulomb effects reduce the e-capture rates and shift them toward higher density region due to an increase of the threshold energy by $\Delta Q_C = \mu_C(Z-1) - \mu_C(Z)$, where $\mu_C(Z)$ is the Coulomb chemical potential of the nucleus with charge number Z due to the interaction of the ion with other ions in the electron background (see more details in Ref. [28]).

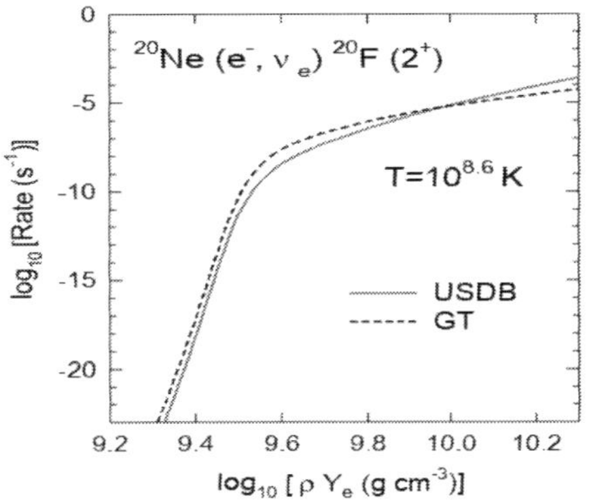

Fig. 4. Calculated e-capture rates for ^{20}Ne (e^-, ν_e) ^{20}F (2_1^+) obtained by shell-model calculations with the USDB at $\log_{10}T$ =8.6. Results obtained by the prescription assuming an allowed GT transition are also shown.

Heating of the O-Ne-Mg core due to γ emissions succeeding the double e-capture reactions, ^{20}Ne (e^-, ν_e) ^{20}F (e^-, ν_e) ^{20}O, is important in the final stage of the evolution of the core. Study of the evolution of the high density electron-degenerate core with the use of the present shell-model rates is in progress [33].

4 Summary

Forbidden transitions in ^{16}O and ^{20}Ne are studied by shell-model calculations with the use of new shell-model Hamiltonians. Spin-dipole strength in ^{16}O is investigated based on non-energy-weighted and energy-weighted sum rules. Total and partial ν-induced reactions on ^{16}O in various channels including multi-particle emissions are obtained with the SFO-tls, which can reproduce low-lying spin-dipole states in ^{16}O. Total cross sections are found to be close to those of a standard CRPA calculation, while a large branching ratio is noticed for the αp emission channel. This results in an enhancement of production yields of ^{11}B and ^{11}C in supernova explosion compared with the case without the multi-particle emission channels. Possible measurement of future supernovae at Super-Kamiokande by charged-current reactions on ^{16}O is discussed, and dependence of the cross sections on the ν mass hierarchies in MSW ν oscillations is investigated.

Electron-capture rates for a second-forbidden transition ^{20}Ne ($0_{g.s.}^+$) → ^{20}F (2_1^+) are evaluated by shell-model calculations with the USDB, and compared with those of the GT prescription, where the transition is treated as an allowed GT transition. Sizable difference is noticed in the e-capture rates in stellar environments, which is caused by the difference in the electron energy dependence of the transition strengths between the two methods.

The authos would like to thank K. Nakazato and M. Sakuda for useful discussion on the ν detection at Super-Kamiokande. They also thank S.-C. Leung and S. Zha for useful discussions and studies on the evolution of the O-Ne-Mg core. This work has been supported in part by JSPS KAKENH Grant Nos. JP15K05090 and JP19K03855. K.T. was supported by the JSPS Overseas Research Fellowships

References

1. T. Suzuki, H. Toki, K. Nomoto, ApJ **817**, 163 (2016)
2. H. Toki, T. Suzuki, K. Nomoto, S. Jones, R. Hirschi, Phys. Rev. C **88**, 015806 (2013)
3. S. Jones, R. Hirschi, K. Nomoto et al., ApJ **772**, 150 (2013)
4. M. Honma, T. Otsuka, T. Mizusaki, M. Hjorth-Jensen, B.A. Brown, JPhCS **20**, 7 (2005)
5. K. Mori, M. Famiano, T. Kajino, T. Suzuki et al., ApJ **833**, 179 (2016)
6. T. Suzuki, S. Chiba, T. Yoshida, T. Kajino, T. Otsuka, Phys. Rev. C **74**, 034307 (2006)
7. T. Yoshida, T. Suzuki, S. Chiba, T. Kajino et al., ApJ **686**, 448 (2008)
8. T. Suzuki, A.B. Balantekin, T. Kajino, Phys. Rev. C **86**, 015502 (2012)
9. T. Suzuki, M. Honma, Phys. Rev. C **87**, 014607 (2013)
10. T. Suzuki, M. Honma, K. Higashiyama, T. Yoshida, T. Kajino, T. Otsuka, H. Umeda, K. Nomoto, Phys.Rev. C **79**, 061603 (2009)
11. T. Suzuki, T. Kajino, J. Phys. **G** 40, 083101 (2013)
12. P. Möller, J.R. Nix, K.-L. Kratz, ADNDT **66**, 131 (1997) ; P. Möller, B. Pfeiffer, K.-L. Kratz, Phys. Rev. C **67**, 055802 (2003)
13. T. Suzuki, T. Yoshida, T. Kajino, T. Otsuka, Phys. Rev. C **85**, 015802 (2012)
14. T. Suzuki, S. Shibagaki, T. Yoshida, T. Kajino, T. Otsuka, ApJ **859**, 133 (2018)
15. T. Suzuki, R. Fujimoto, T. Otsuka, Phys. Rev. C **67** 044302 (2003)
16. S. Cohen, D. Kurath, Nucl. Phys. **73**, 1 (1965)
17. T. Suzuki, T. Otsuka, Phys. Rev. C **78**, 061302 (2008)
18. T. Suzuki, S. Chiba, T. Yoshida, K. Takahashi, H. Umeda, Phys. Rev. C **98**, 034613 (2018)
19. T. Suzuki, H. Sagawa, Nucl. Phys. A **637**, 547 (1998)
20. T. Suzuki, D. F. Measday, and J. P. Roalsvig, Phys. Rev. C **35**, 2212 (1987)
21. K. Langanke, P. Vogel, and E. Kolbe, Phys. Rev. Lett. **76**, 2629 (1996); E. Kolbe, K. Langanke, and P. Vogel, Phys. Rev. D **66**, 013007 (2002)
22. L. Wolfenstein, Phys. Rev. D **17**, 2369 (1978) ; **D** 20, 2634 (1979) ; S.P. Mikheyev, A.Y.Smirnov, Sov. J. Nucl. Phys. **42** 913 (1985)
23. K. Nakazato, T. Suzuki, M. Sakuda, PTEP **2018**, 123E02 (2018)
24. G. Martinez-Pinedo, Y.H. Lam, K. Langanke, R. G. Zegers, C. Sullivan, Phys. Rev. C **89**, 045806 (2014)
25. National Nuclear Data Center on-line retrieval system, http://www.nndc.bnl.gov
26. O.S. Kirsebom, et al., arXiv:1805.08149 (2018)
27. J.S. O'Connell, T.W. Donnelly, J.D. Walecka, Phys. Rev. C **6**, 719 (1972) ; J.D. Walecka, in *Muon Physics*, edited by V.M. Hughes, C.S. Wu (Academic Press, New York, 1975), pp. 113-218; N. Parr, G. Colo, E. Khan, D. Vretenar, Phys. Rev. C **80,** 055801 (2009)
28. T. Suzuki, S. Zha. S.-C. Leung, K. Nomoto, ApJ **881**, 64 (2019)
29. N. Itoh, N. Tomizawa, M. Tamamura, S. Wanajo, ApJ **579**, 380 (2002)
30. S. Ichimaru, Rev.Mod. Phys. **65**, 255 (1993)
31. W.L. Slattery, G.D. Doolen, H.E. DeWitt, Phys. Rev. A **26**, 2255 (1982)
32. A. Juodagalvis, K. Langanke, W.R. Hix et al, Nucl. Phys. A **848**, 454 (2010)
33. S. Zha, S.-C. Leung, T. Suzuki, K. Nomoto, ApJ submitted, arXiv:1907.04184 (2019)

Recent studies of heavy ion transfer reactions using large solid angle spectrometers

Suzana Szilner[1,*], *Lorenzo* Corradi[2], *Giovanni* Pollarolo[3], *Giulia* Colucci[4], *Petra* Čolović[1], *Enrico* Fioretto[2], *Franco* Galtarossa[2], *Alain* Goasduff[2,4], *Tommaso* Marchi[2], *Tea* Mijatović[1], *Giovanna* Montagnoli[4], *Daniele* Montanari[2], *Fernando* Scarlassara[4], *Alberto M.* Stefanini[2], and *Javier Jose* Valiente-Dobón[2]

[1]Ruđer Bošković Institute, Zagreb, Croatia
[2]INFN - Laboratori Nazionali di Legnaro, Legnaro, Italy
[3]Dipartimento di Fisica, Università di Torino, and Istituto Nazionale di Fisica Nucleare, Torino, Italy
[4]Dipartimento di Fisica, Università di Padova, and Istituto Nazionale di Fisica Nucleare, Padova, Italy

Abstract. We present selected results recently obtained in the study of heavy ion transfer reactions at energies close to the Coulomb barrier by employing the large solid angle magnetic spectrometer PRISMA. We discuss the production of neutron-rich heavy nuclei via multinucleon transfer processes and the related effects of secondary processes, in particular nucleon evaporation, studied in a high resolution kinematic coincidence experiment. We also present the recent results in the studies of neutron-neutron correlations for closed shell and superfluid systems.

1 Introduction

Nucleon transfer reactions at energies close to the Coulomb barrier always played an important role for nuclear structure and reaction dynamics. In the studies of the reaction mechanism, it was important to define the evolution of the reaction mechanism from the quasi-elastic regime to the more complex deep-inelastic and fusion [1]. For energies close to the Coulomb barrier, transfers constitute the largest part of the total reaction cross section and are the main source for the energy dissipation from the relative motion to the intrinsic excitation. With heavy-ions multiple transfer of nucleons becomes available in the reaction giving the possibility to study the relative role of single particle and pair transfer modes [2, 3].

The exchange of nucleons in multinucleon transfer reactions has been recently used as a suitable mechanism to populate moderately neutron-rich nuclei. Nuclear reaction models predict large primary multinucleon transfer cross sections [4, 5] that are comparable, or even larger, than those of other reactions like fragmentation at intermediate energies or low-energy fusion with radioactive beams. Being the transfer processes governed by form factors and optimum Q-value considerations, with neutron deficient projectiles on heavy targets only proton-stripping and neutron pick-up channels, for the projectiles, are available, while with neutron-rich projectiles also proton pick-up and neutron stripping channels, for the projectiles, open up. This corresponds, for the heavy partner, to the population in the south-east direction, leading to the neutron-rich heavy region, which is presently receiving peculiar attention. Due to the characteristic behavior of the binding energy, the process is essentially governed by the lighter partner of the reaction.

In order to employ multinucleon transfer reactions for studying both the production of neutron rich nuclei and the effects of the residual interaction, one needs to reach good mass and charge resolution, as well as high efficiency to measure low cross sections for massive transfer channels. All these demanding requirements are fulfilled by large solid angle spectrometers based on trajectory reconstruction [6–8]. The coupling of these spectrometers with large γ arrays [9–11] allowed, in addition, to identify γ rays coming from the decay of weak transfer channels associated with the population of nuclei moderately far from stability.

2 Generals on detection techniques

Different techniques have been employed to identify nuclei produced in transfer reactions. Most of these techniques make use of magnetic spectrographs or spectrometers for a complete identification of nuclear charge, mass and energy of final reaction products (see Ref. [1] and references therein). Magnetic spectrographs were mostly used in light ion transfer reactions, where, by combining magnetic elements of different complexity to focus momenta, it was possible to distinguish the excited states. With heavier ions more demanding conditions are required in order to keep a good resolution and to have at the same time a sufficient detection efficiency, given the large energy dynamic range of transfer products. For these ions, time-of-flight spectrometers have been designed (see for

*e-mail: szilner@irb.hr

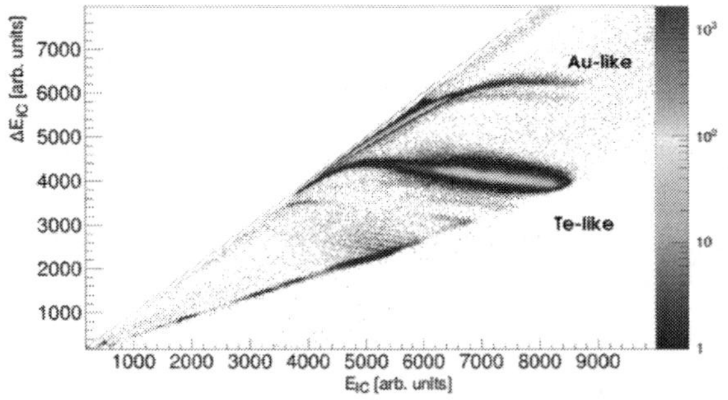

Figure 1. Matrix of ΔE versus E measured in the PRISMA ionization chamber (IC) for the ^{197}Au+^{130}Te reaction at $E_{lab}(^{197}\mathrm{Au})$ = 1.07 GeV and θ_{lab} = 37°. The most intense band corresponds to Te-like ions, while Au-like ions are located in the region where ΔE_{IC} and E_{IC} merge. One can also see the ions corresponding to fission-like events which are not stopped in the IC.

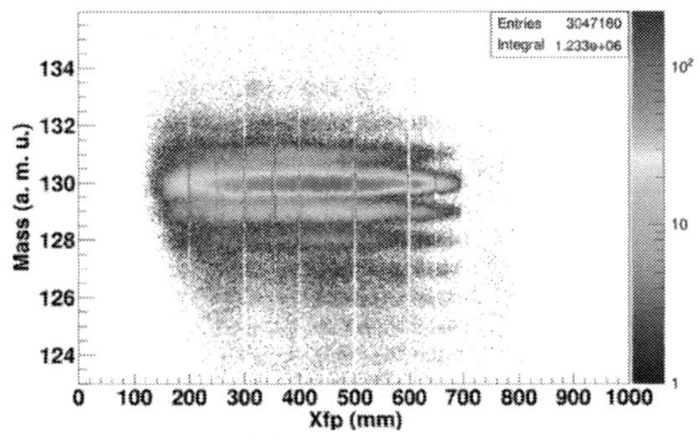

Figure 2. Mass vs position in the focal plane (Xfp) obtained in the ^{197}Au+^{130}Te [20] at E_{lab} = 1.07 GeV after the trajectory reconstruction and software-based correction of nonlinear distortions of ion trajectories in PRISMA for the Te-like ions.

example Refs. [12–14] and references therein) providing a good A and Z resolution for medium mass ions. These spectrographs or spectrometers have solid angles in the range 3-10 msr. Beyond these values, it becomes unfeasible to use complex magnetic elements to correct for the ion optical aberrations. The solution for the large solid angle spectrometers (~ 100 msr) was to simplify the magnetic element configuration and to apply the concept of trajectory reconstruction [6, 15], by using a detector system which, besides nuclear charge, energy and timing, provides the necessary position information along the ion path. With these devices [6, 7], mass and nuclear charge identification has been successfully demonstrated for ions up to A~100-130, especially when using inverse kinematics conditions. We here briefly list the main characteristics of the PRISMA spectrometer and its detector system. A position-sensitive micro-channel plate detector [16] is placed at the entrance of the spectrometer, providing a start signal for time-of-flight measurements and bi-dimensional position signals. Ions pass through the optical elements of the spectrometer (a quadrupole and a dipole) and enter a focal plane detector [17] which is made of a parallel plate detector of multiwire type, providing timing and bi-dimensional position signals with resolutions similar to the entrance detector. Located at the end of the focal plane is an array of transverse field multiparametric ionization chambers (IC), providing nuclear charge via energy loss (ΔE) and total energy (E) measurements. The large dynamical acceptance of the IC is illustrated in Fig. 1 which shows the ΔE versus E matrix for the ^{197}Au+^{130}Te reaction. In this figure, one observes the intense band corresponding to the Te-like ions (the spectrometer conditions were set for the detection of these events), as well as the Au-like and fission-like events.

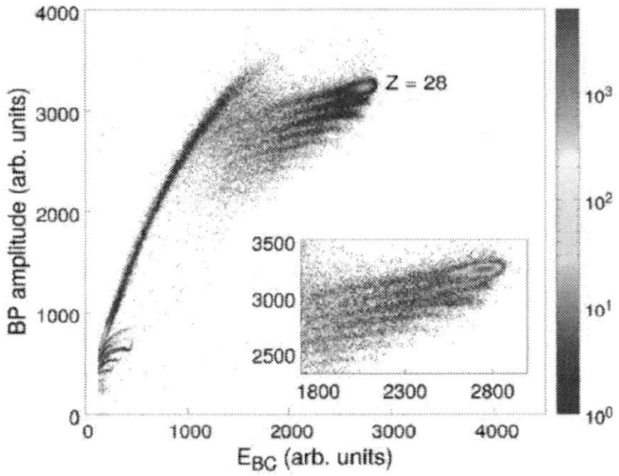

Figure 3. Two-dimensional scatter plots BP amplitude vs energy obtained for the reaction ^{58}Ni+^{208}Pb at $E_{lab}(^{58}\mathrm{Ni})$ = 400 MeV and θ_{lab} = 65° measured in the Bragg ionization chamber of the coincident detector (NOSE) [21].

The described detector system gives all the necessary information for the complete ion identification, which is performed via an event-by-event reconstruction of the trajectory inside the magnetic elements [6, 18, 19]. The result of this procedure is depicted as an example in Fig. 2, demonstrating the quality of the mass separation [20].

For the detection of the heavy partner of the reaction a dedicated set-up [21] was specifically built and coupled to PRISMA. This allowed, via a high-resolution mass-mass correlation, to study the final mass distribution of the heavy partner and the effect of secondary processes. The coincident detector (NOSE), providing information on the heavy partner of the reaction, consists of a multiwire parallel-plate avalanche counter followed by an axial-field ionization chamber (Bragg chamber) and with a geometrical solid angle about one third of PRISMA. To illustrate the resolving power of this detector, we plot in Fig. 3 the Bragg peak (BP) amplitude vs energy obtained for the ^{58}Ni+^{208}Pb reaction. This coincident detection of binary reaction fragments is very important in the context of understanding the production mechanism of heavy neutron-rich nuclei in regions of astrophysical interest, where secondary processes play a significant role.

The studies of transfer reactions at low energies greatly benefited from the use of large angle spectrometers, in particular, at energies below the barrier, where the transfer cross sections drop significantly. By employing inverse kinematics and by detecting ions at very forward angles, one has, at the same time, enough kinetic energy of the outgoing recoils (for energy and therefore mass resolution) and forward-focused angular distribution (high efficiency). This technique was utilized with the PRISMA spectrometer in the studies of the nucleon-nucleon correlations (see section 4) [22, 23].

3 The synthesis of heavy neutron-rich nuclei

The production of heavy neutron-rich nuclei in the region close to the N = 126 shell closure recently received a boost of interest being the properties of these nuclei fundamental for the understanding of the synthesis of heavy elements. The investigation of the nuclei in this region of the nuclear landscape is hampered by the difficulties in their production, as well as their direct identification with present detection techniques.

One of the promising mechanisms for the production of the heavy neutron-rich nuclei, as was already discussed in the Introduction, is nucleon transfer reactions between heavy ions at energies close to the Coulomb barrier [4]. The path leading to primary neutron-rich heavy partners, in multinucleon transfer reactions, has been investigated in recent experiments with neutron-rich stable projectiles, in ^{64}Ni+^{238}U [24], ^{40}Ar+^{208}Pb [25], ^{136}Xe+^{198}Pt [26] and ^{136}Xe+^{238}U [27]. In these reactions the light partner has been identified directly via high-resolution magnetic spectrometers while, in some of the mentioned cases, information on the heavy partner was obtained indirectly by detecting the coincident γ rays produced by the fragments. In these experiments information on secondary processes, i.e. neutron evaporation, could also be extracted as illustrated for example in Refs. [6, 28, 29]. In all these experiments it has been emphasized that secondary processes may significantly influence the final yield distributions, even though rather meager evidence of the survival probability of the heavy partner has been provided.

In order to understand and quantify the production process also for the heavy partner of the reaction, we performed an experiment with a simultaneous detection of light and heavy transfer products in the ^{197}Au+^{130}Te system [20]. We chose the neutron-rich ^{130}Te to populate neutron transfer channels leading primarily to neutron-rich Au isotopes. We exploited the performance of the PRISMA spectrometer [6] to identify isotopes in the tellurium region, while the coincident Au-like partners were detected with a dedicated set-up, NOSE [21], coupled to PRISMA. Favorable experimental conditions were achieved by employing inverse kinematics in such a way that both the light recoil and the heavy partner could have enough energy for their detection. With this set-up we were able to select binary events, Au-like events in NOSE detector were uniquely associated to the Te-like events detected in PRISMA.

For the mass identification of the reaction products, the reconstruction of their trajectory in the magnetic elements of PRISMA has been performed. The outcome of this procedure is the final mass distribution for the Te isotopes depicted in Fig. 4 (right panel) where the resolution turned out to be $\Delta A/A \sim 1/240$, allowing to separate the different isotopes. The obtained mass distribution in the coincident NOSE detector is shown in Fig. 4 (left panel), where the peaks are well centered at mass 197 and 130, corresponding to Au-like and Te-like ions, respectively. The mass resolution turned out to be $\Delta A/A \sim 1/44$, in agreement with the angle and timing resolutions of the used detectors. As already discussed, the NOSE detector was optimized for Au-like in coincidence with Te-like ions in the spectrometer. One notices that the distribution is not symmetric around mass 197, but it displays a tail towards larger values. Part of the asymmetry may be due to a transfer mechanism as well as to experimental effects (e.g. non linear response of detectors or electronics) mimicking such tails. It is only when one sets a coincidence with a high resolution spectrometer that one can construct a mass-mass correlation matrix and infer about the behaviour of the heavy partner. This is nicely demonstrated in Fig. 5, where the reconstructed mass of the Au-like ions is correlated with that of the high resolution mass distribution of the Te-like ions detected in PRISMA.

In Fig. 5 we see that the identification in mass of the light fragment with high resolution allows to separate the mass distribution of the heavy partner in well-defined bands. The black dots indicate the centroids of the projections of each band, showing how they slightly bend toward lower masses in comparison to those expected for the corresponding primary neutron transfer channels (red circles). This indicates that the primary fragments acquire significant excitation energy so that evaporation becomes relevant in defining the final yields. This analysis allowed to extract information on the average number of evaporated neutrons for each channel associated with the Te isotopes. The effect of the evaporation depends strongly on the bombarding energy and on the projectile and target combination.

A successful theoretical description of the measured transfer cross sections was achieved with the model GRAZING [30–32], which includes surface phonons for the inelastic part and a collection of single particle levels for the treatment of nucleon exchange (see for example Refs. [25, 26] and references therein). Recently, the data were also compared with other models, for example with the Time-Dependent Hartree-Fock theory [33] which has been generalized in order to be able to calculate cross sections for transfer channels [34–37]. One of the critical point in these comparisons is the accurate treatment of the secondary process. Thus, from the experimental point of view it is necessary to provide the evidence of the survival probability of the heavy partner in high-resolution measurements where the light and heavy transfer products are simultaneously detected. From the theoretical point of

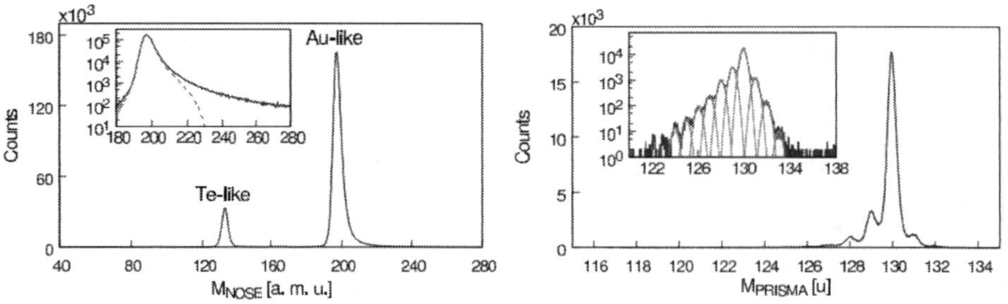

Figure 4. (left) Mass spectrum obtained employing the NOSE detector. The inset shows the mass distribution of Au-like events in logarithmic scale, with red-dashed curve with the coincidence of Te-like ions detected in PRISMA. (right) Mass distribution for the Te isotopes obtained after ion trajectory reconstruction in PRISMA. The inset shows the distribution in logarithmic scale with the multi-Gaussian fit used to evaluate the yields of the neutron transfer channels.

view, it is important to develop a fully microscopic model that incorporates the structure of the two reactants and the reaction dynamics on the same footing.

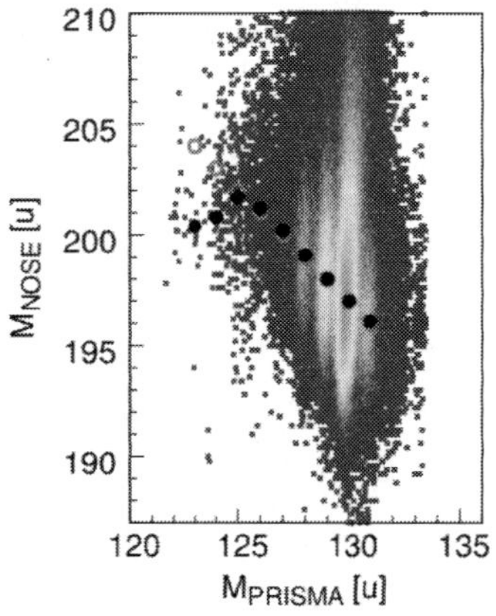

Figure 5. Mass-mass correlation matrix of Te isotopes detected in PRISMA and the heavy partner detected in coincident NOSE detector. The red (empty) circles indicate the centroids of the correlated masses of the primary neutron transfer channels, the black (full) dots indicate the experimental centroids as derived from the fits of their projections.

4 Probing nucleon-nucleon correlations in transfer reactions

Two-particle transfer reactions are one of the best probes of pairing correlations in nuclear systems. In the studies of such reactions the comparison between the cross section for two-particle transfer channels with the predictions for an independent nucleon transfer process should provide a direct measurement of nucleon-nucleon correlations. In order to study these correlations we recently measured the transfer probabilities for one- and two-nucleon transfer channels in heavy ion collisions over a wide energy range with cross sections spanning several orders of magnitude. In table 1 we list the systems measured so far with PRISMA, employing inverse or direct kinematics technique. For some of the indicates systems, PRISMA was coupled to a γ array to measure the transfer strength to specific states.

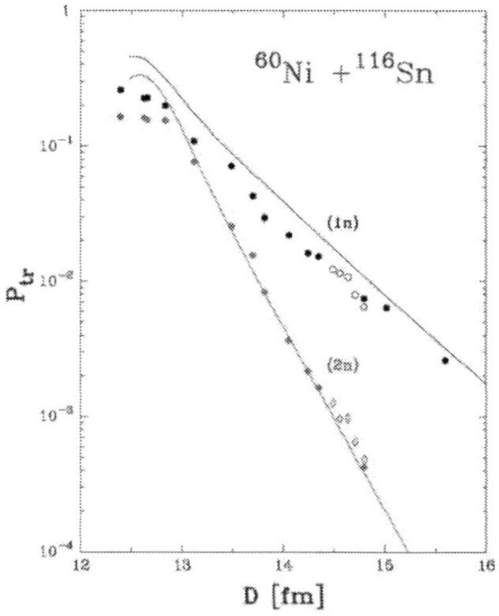

Figure 6. Experimental (points) and microscopically calculated (lines) transfer probabilities (P_{tr}) as a function of the distance of closest approach (D) for one- (^{61}Ni) and two-neutron (^{62}Ni) transfers in the ^{60}Ni+^{116}Sn system. Open symbols correspond to the probabilities obtained from angular distribution in direct kinematics [38] while solid symbols refer to the excitation function measurement performed in inverse kinematics [23].

In the measured reactions, ^{96}Zr+^{40}Ca [22] and ^{116}Sn+^{60}Ni [23], (lighter) target recoils have been detected with PRISMA, and an excitation functions at several bombarding energies have been obtained from the Coulomb barrier to 20-25% below, reaching ~ 15.5 fm of distance of closest approach. It is convenient to extract the transfer probability P_{tr}, defined as the ratio of the transfer cross section to the corresponding Rutherford cross section, and to plot these probabilities as a function of the distance of closest approach D for a Coulomb trajectory. The extrac-

Table 1. Systems studied with PRISMA, in stand-alone or coupled to a γ array, employing inverse or direct kinematics technique.

system	item	set-up	Ref.
$^{96}Zr+^{40}Ca$	excitation function, inverse kinematic	PRISMA	[22]
$^{116}Sn+^{60}Ni$	excitation function, inverse kinematic	PRISMA	[23]
	angular distribution, direct kinematic	PRISMA + AGATA	[38]
$^{206}Pb+^{118}Sn$	excitation function, inverse kinematic	PRISMA	[39]
$^{92}Mo+^{54}Fe$	excitation function, inverse kinematic	PRISMA	[40]
	angular distribution, direct kinematic	PRISMA + $LaBr_3$-array	[40]

tion of the transfer probability P_{tr} as a function of D is rather alluring because the angular distributions obtained at different bombarding energies will be all superimposed, if the Q-value window does not change significantly. Further, the transfer probability at large distances D may be obtained by measuring an excitation function at fixed center of mass angle or by measuring angular distributions at fixed energy (see Fig. 6).

The transfer probabilities for the mentioned two systems have been analyzed by employing a microscopic theory, which, for the first time in a heavy-ion collision, provided a consistent description of one- and two-neutron transfer channels, in shape and magnitude. In this microscopic theory to compute the inclusive one-neutron cross section we calculated the transfer probability for a given single particle transition and we obtained the total transfer probability by summing over all possible transitions that can be constructed from the single particle states in projectile and target. For the two-particle transfer we just mention that we diagonalized the total Hamiltonian with a model space containing only two-particle configuration coupled to 0^+ (i.e. transfer of a $J=0^+$ pair). Calculations reproduce well the experimental slope as well as the absolute values of the transfer probabilities for the one neutron channel. In the well Q-value matched system $^{60}Ni+^{116}Sn$, with the ground-to-ground state Q-value for one- and two-neutron transfers very close to the optimum, the microscopically calculated transfer probabilities well reproduce the experimental ones for (2n) channel (see Fig. 6 and Ref. [23] for details). We wish to emphasize that to calculate the (2n) channel one has to solve the well-known system of semiclassical coupled equations up to second-order Born approximation, whose amplitude consists of the simultaneous transfer of the pair of nucleons, the nonorthogonality term, and the term which represents the successive process via an intermediate channel. The ground states have been described in the BCS approximation with a standard state-independent pairing force. The nice agreement between data and calculations indicates that the two-neutron transfer channel in this system is populating essentially only the ground state, in agreement with the experimental total kinetic-energy loss distributions and consistently with particle-γ measurements [38] which provided an upper limit of 24 % for the possible excited state population. The nucleon-nucleon correlations induced by the pairing interaction, incorporated in the microscopic theory, were essential for the description of two-neutron transfers.

Recently, we also measured the probability for neutron transfer channels in the $^{206}Pb+^{118}Sn$ reaction [39]. The main purpose of this measurement was to investigate how the effect of nucleon-nucleon correlations, in the evolution of the reaction, is modified in collisions between very heavy ions. The measurement was performed in inverse kinematics by using the newly developed ^{206}Pb beam at superconducting PIAVE-ALPI accelerators of Laboratori Nazionali di Legnaro. This is the heaviest (asymmetric) semi-magic system (see table 1) with closed proton and open neutron shells, very well Q-value matched for neutron transfers. In the collisions of very heavy ions, the population of final states with high excitation energies and large angular momenta may significantly change the transfer strength of the ground-to-ground state transitions.

Besides neutron-neutron correlations, it is also important to study proton-proton and proton-neutron correlations, for which existing data are extremely scarce. To this purpose we measured the $^{92}Mo+^{54}Fe$ reaction where a nearly symmetric population of both neutron and proton pick-up and stripping channels are present. We refer to Ref. [40] for details. By using the large solid angle spectrometer PRISMA we were able to measure, beside the neutron transfer channels, also the proton transfers up to the large distances of closest approach, where the nuclear absorption is negligible.

5 Summary and outlook

The advent of the large solid angle magnetic spectrometers ensured significant advances in the field of multinucleon transfer reactions at energies close to the Coulomb barrier. The mechanism of the production of neutron rich heavy nuclei was studied in a high resolution kinematic coincidence in the $^{197}Au+^{130}Te$ system, where the extracted mass-mass correlations allowed to study the effect of secondary processes. These kind of research, besides the better understanding the transfer mechanism in itself, are of particular relevance in the optimization of multinucleon transfer mechanism for the population of neutron-rich nuclei.

The pairing correlations were discussed via transfer probabilities for one- and two-nucleon transfer channels. These probabilities have been compared with a microscopic theory that incorporates nucleon-nucleon correlation in the $^{40}Ca+^{96}Zr$ and $^{60}Ni+^{116}Sn$ systems. It is very important to continue these studies, and to compare the experimental observables with microscopic calculations where nucleon-nucleon correlations are treated together with the relevant structure and dynamics information. The issue connected with the pair correlations is presently at the focus of a renewal of interest [41–50], in particular in ongoing research with radioactive beams, where the pairing interaction may be significantly modified in nuclei with extended neutron distributions.

Acknowledgments

The material presented in this paper is the result of the work of many people who participated to different measurements and experimental preparation, whom we would like to thank. The authors acknowledge financial support from the European Regional Development Fund for the Center of Excellence for Advanced Materials and Sensing Devices (Grant No. KK.01.1.1.01.0001). This work was also partly supported by the Croatian Science Foundation under Project No. 7194 and in part under project no. IP-2018-01-1257. The authors acknowledge financial support from HORIZON2020, ENSAR2 project (No. 654002).

References

[1] L. Corradi, G. Pollarolo, and S. Szilner, J. of Phys. G: Nucl. Part. Phys. **36**, 113101 (2009).
[2] D. M. Brink and R. A. Broglia, *Nuclear Superfluidity: Pairing in Finite Systems*, Cambridge University Press (2005).
[3] R. A. Broglia and V. Zelevinsky, *Fifty Years of Nuclear BCS - Pairing in Finite Systems* , World Scientific (2013).
[4] C. H. Dasso, G. Pollarolo, and A. Winther Phys. Rev. Lett. **73**, 1907 (1994).
[5] V. Zagrebaev and W. Greiner, Phys. Rev. Lett. **101**, 122701 (2008).
[6] S. Szilner et al., Phys. Rev. C **76**, 024604 (2007).
[7] S. Pullanhiotan et al., Nucl. Instrum. Methods Phys. Res. A **593**, 343 (2008).
[8] F. Cappuzzello et al., Eur. Phys. J. A **52**, 167 (2016).
[9] A. Gadea et al., Eur. Phys. J. A **20**, 193 (2004).
[10] J. Simpson et al., Acta Physica Hungarica N.S. **11**, 159 (2000).
[11] A. Gadea et al., Nucl. Instrum. Methods Phys. Res. A **654**, 88 (2011); E. Farnea et al., Nucl. Instrum. Methods Phys. Res. A **621**, 331 (2010); S. Akkoyun et al., Nucl. Instrum. Methods Phys. Res. A **668**, 26 (2012).
[12] G. Montagnoli et al., Nucl. Instrum. Methods Phys. Res. A **454**, 306 (2000).
[13] L. Corradi et al., Phys. Rev. C **66**, 024606 (2002).
[14] S. Szilner et al., Phys. Rev. C **71**, 044610 (2005).
[15] L. Corradi et al., Nucl. Instrum. Methods Phys. Res. B **317**, 743 (2013).
[16] G. Montagnoli et al., Nucl. Instrum. Methods Phys. Res. A **547**, 455 (2005).
[17] S. Beghini et al., Nucl. Instrum. Methods Phys. Res. A **551**, 364 (2005).
[18] D. Montanari et al., Eur. Phys. J. A **47**, 4 (2011).
[19] T. Mijatović et al., Eur. Phys. J. A **52**, 113 (2016).
[20] F. Galtarossa et al., Phys. Rev. C **97**, 054606 (2018).
[21] E. Fioretto et al., Nucl. Instrum. and Methods Phys. Res. A **899**, 73 (2018).
[22] L. Corradi et al, Phys. Rev. C **84**, 034603 (2011).
[23] D. Montanari et al., Phys. Rev. Lett. **113**, 052501 (2014).
[24] L. Corradi et al., Phys. Rev. C **59**, 261 (1999).
[25] T. Mijatović et al., Phys. Rev. C **94**, 064616 (2016).
[26] Y. X. Watanabe et al., Phys. Rev. Lett. **115**, 172503 (2015).
[27] A. Vogt et al., Phys. Rev. C **92**, 024619 (2015).
[28] P. R. John et al., Phys. Rev. C **90**, 021301(R) (2014).
[29] P. R. John et al., Phys. Rev. C **95**, 064321 (2017).
[30] A. Winther, Nucl. Phys. A **572**, 191 (1994).
[31] A. Winther, Nucl. Phys. A **594**, 203 (1995).
[32] [http:/www.to.infn.it/~nanni/grazing], program GRAZING.
[33] C. Simenel and A. S. Umar, Progress in Part. and Nucl. Phys. **103**, 19 (2018).
[34] C. Simenel, Phys. Rev. Lett. **105**, 192701 (2010).
[35] K. Sekizawa and K. Yabana, Phys. Rev. C **88**, 014614 (2013).
[36] K. Sekizawa and K. Yabana, Phys. Rev. C **93**, 054616 (2016).
[37] A. V. Karpov and V. V. Saiko, Phys. Rev. C **96**, 024618 (2017).
[38] D. Montanari et al., Phys. Rev. C **93**, 054623 (2016).
[39] L. Corradi et al., JPS Conference Proceedings, in press.
[40] T. Mijatović et al., EPJ Web of Conferences in press.
[41] M. Evers et al., Phys. Rev. C **84**, 054614 (2011).
[42] A. Lemasson et al., Phys. Lett. B **697**, 454 (2011).
[43] G. Potel et al., Phys. Rev. Lett. **107**, 092501 (2011).
[44] R. A. Broglia et al., Phys. Scr. **91**, 063012 (2016).
[45] G. Scamps and D. Lacroix, Phys. Rev. C **87**, 014605 (2013).
[46] G. Scamps and K. Hagino, Phys. Rev. C **92**, 064602 (2015).
[47] G. Scamps et al., Phys. Rev. C **94**, 064606 (2016).
[48] P. Magierski, K. Sekizawa, and G. Wlazlowski, Phys. Rev. Lett. **119**, 042501 (2017).
[49] J. Dobaczewski et al., Phys. Rev. Lett. **72**, 981 (1994).
[50] F. Barranco, G. Potel, E. Vigezzi, and R.A. Broglia, Eur. Phys. J. A **55**, 104 (2019).

Nuclear Physics in Stellar Lifestyles with the Trojan Horse Method

Aurora Tumino[1,2,*], *Claudio* Spitaleri[2,3], *Marco* La Cognata[2], *Silvio* Cherubini[2,3], *Giovanni Luca* Guardo[2], *Marisa* Gulino[1,2], *Iolanda* Indelicato[2], *Livio* Lamia[2,3], *Rosario Gianluca* Pizzone[2], *Giuseppe Gabriele* Rapisarda[2], *Stefano* Romano[2,3], *Maria Letizia* Sergi[2], and *Roberta* Spartà[2]

[1]Facoltà di Ingegneria e Architettura, Università degli Studi di Enna "Kore", Enna, Italy
[2]INFN-Laboratori Nazionali del Sud, Catania, Italy
[3]Dipartimento di Fisica e Astronomia, Università degli Studi di Catania, Catania, Italy

Abstract. The Trojan Horse Method is an indirect technique to measure nuclear reactions of astrophysical relevance at the energies of interest, free of Coulomb suppression and electron screening effects. Its basic features in the framework of the theory of direct reactions will be discussed and the physics case of the $^{12}C+^{12}C$ fusion will be addressed.

1 Introduction

Coulomb repulsion between like charges is a critical issue in nuclear astrophysics, being responsible for the exponential decrease of the cross section $\sigma(E)$ of the relevant nuclear reactions at astrophysical energies. Thus, its behaviour at these low energies is quite often extrapolated from the higher energies by means of the astrophysical S(E)-factor

$$S(E) = E\sigma(E)\exp(2\pi\eta), \quad (1)$$

with η the Coulomb parameter of the colliding nuclei, and $\exp(2\pi\eta)$ the inverse of the Gamow factor that removes the Coulomb dependence of $\sigma(E)$. However, extrapolation can be source of additional uncertainties for $\sigma(E)$ due, for instance, to the presence of unexpected resonances. In the laboratory measurement of nucleosynthesis processes, another critical issue is represented by the electron screening effect that leads to an increased cross section for screened nuclei, $\sigma_s(E)$, compared to the cross section for bare nuclei $\sigma_b(E)$ [1, 2]. Therefore, the so called screening factor, defined as

$$f_{\rm lab}(E) = \sigma_s(E)/\sigma_b(E) \approx \exp(\pi\eta U_e/E)\,, \quad (2)$$

where U_e is the so-called "electron screening potential" [1, 2], has to be taken into account to determine the bare nucleus cross section. Indeed, this is the key parameter to determine a reaction rate and the only way to get it measured is via indirect methods [3–5] and references therein). They make use of direct reaction mechanisms, such as transfer processes (stripping and pick-up) and quasi-free reactions (knock-out reactions). In particular, the Trojan Horse Method (THM) ([3, 6] and references therein) has been successfully applied many times in the last two decades to reactions connected with fundamental astrophysical problems. Here we present some of the basic ideas of the THM.

*e-mail: tumino@lns.infn.it

2 Basic ideas of the Trojan Horse Method

The THM makes use of an appropriate three-body reaction $A + a \rightarrow c + C + s$ performed in quasi free (QF) kinematics at energies well above the Coulomb barrier to extract the cross section of a charged particle two-body process $A + x \rightarrow c + C$ in the Gamow energy window. This is done applying the theory for direct reactions, under the assumption that the nucleus a is described in terms of the $x \oplus s$ cluster structure. In most of the applications performed so far, where a = deuteron, its obvious $p \oplus n$ clusterization was assumed [7–10].

The QF mechanism can be sketched using a pole diagram (see Fig.1) (Shapiro et al. 1968) with two vertexes referring to a break-up (upper vertex) and to the $A + x \rightarrow c + C$ process (lower vertex), with s as spectator. The high energy in the $A + a$ entrance channel ensures that the two body interaction takes place inside the nuclear field, without experiencing either Coulomb suppression or electron screening effects. The $A + a$ relative motion is compensated for by the $x - s$ binding energy, determining the so called "quasi-free two-body energy" given by

$$E_{QF} = \frac{m_x}{m_x + m_A} E_A - B_{x-s}. \quad (3)$$

where E_A represents the beam energy, m_x and m_A are the masses of x and A particles respectively, and B_{x-s} is the binding energy for the $x - s$ system. This can be retained as a prescription to determine the best value of the beam energy to populate the relevant E_{QF}. A cutoff in the momentum distribution reflecting the Fermi motion of s inside the Trojan-horse a, defines the range of energies around E_{QF} accessible in the astrophysical relevant reaction. In the Plane Wave Approximation, the three body-cross cross section can be factorized as:

$$\frac{d^3\sigma}{dE_c d\Omega_c d\Omega_C} \propto [KF\, |\varphi_a(\mathbf{p}_{sx})|^2] \left(\frac{d\sigma}{d\Omega_{c.m.}}\right)^{\rm HOES} \quad (4)$$

https://doi.org/10.1051/epjconf/201922301065

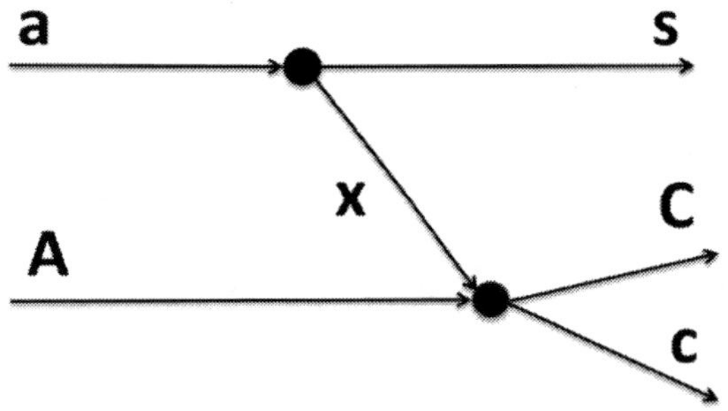

Figure 1. Pole diagram of the $a + A \to C + c + s$ quasi free reaction.

where KF is a kinematical factor containing the final state phase-space factor. It is a function of the masses, momenta and angles of the outgoing particles [5]; $\varphi_a(\mathbf{p}_{sx})$ is proportional to the Fourier transform of the radial wave function $\chi(\mathbf{r})$ for the $x - s$ inter-cluster relative motion; $(d\sigma/d\Omega_{c.m.})^{\text{HOES}}$ is the half-off-energy-shell (HOES) differential cross section for the binary reaction at the center of mass energy $E_{c.m.}$ given in post-collision prescription by

$$E_{cm} = E_{cC} - Q_{2b} \,. \tag{5}$$

Here, Q_{2b} is the Q-value of the binary reaction and E_{cC} is the relative energy of the outgoing particles c and C. Using the PWA does not change the energy dependence of the two-body cross section but only its absolute magnitude.

In a typical THM experiment the decay products (c and C) of the virtual two-body reaction of interest are detected and identified by means of telescopes (silicon detector or ionization chamber as ΔE step and position sensitive detector as E step) placed at the so called quasi free angles. After the selection of the reaction channel, a critical point is to disentangle the quasi free mechanism from other reaction mechanisms feeding the same particles in the final state, e.g. sequential decay and direct break-up. An observable which turns out to be very sensitive to the reaction mechanism is the shape of the experimental momentum distribution of the spectator. In order to reconstruct the experimental p_s distribution, the energy sharing method [11] is applied for each pair of coincidence QF angles, selecting c-C relative energy windows of 50 to 100 keV. The extracted experimental momentum distribution is then compared with the theoretical one and further data analysis is limited to the data lying in the region where the agreement between the two distributions exists (usually within few tens of MeV/c). The theoretical upper limit for the relative momentum p_{xs} between x and s (in the laboratory system $\vec{p}_{xs} = \vec{p}_x = -\vec{p}_s$) is given by $\kappa_{xs} = \sqrt{2\mu_{xs}B_{xs}}$ that represents the on-energy-shell (OES) $a = (x\,s)$ bound state wave number [12].
In the initial idea by [13], the cutoff in p_s was much higher, of hundreds of MeV/c, because it was needed to compensate for the $A + a$ relative motion. However, this condition is not suitable from the experimental point of view for a number of reasons already pointed out [3, 6]. Therefore, it is possible to derive the HOES $((d\sigma/d\Omega_{cm})^{\text{HOES}}$ from the three-body coincidence yield by simply inverting eq.5. In a final step, the HOES cross section has to be related to the relevant on-energy-shell (OES) cross section by applying the corresponding corrections. In a heuristic approach this consists essentially in replacing the Coulomb suppression in the HOES cross section, by means of the penetrability factor:

$$P_l(k_{Ax}R) = \frac{1}{G_l^2(k_{Ax}R) + F_l^2(k_{Ax}R)} \tag{6}$$

with F_l and G_l regular and irregular Coulomb wave functions. It was demonstrated that there is no Coulomb barrier in the two-body amplitude extracted from the TH reaction [9, 14, 15] and this is due to the virtuality of particle x. This seems to be the only consequence of off-energy-shell effects as suggested by the agreement between HOES and OES cross-sections for the ^{6}Li(n,α)^{3}H reaction [16]. This procedure does not allow us to extract the absolute value of the two-body cross section. However this is not a real problem since the absolute magnitude can be derived from a scaling to the direct data available at higher energies.
For resonant two-body reactions, $(d\sigma/d\Omega_{c.m.})^{\text{HOES}}$ has to be worked out to determine the corresponding on-energy-shell (OES) S(E) factor. The corresponding theoretical formalism has been recently developed in a very rigorous way [17] leading to the so called modified R-matrix approach that accounts for HOES effects due to the virtual nature of particle x. By fitting the experimental THM cross section, the reduced width amplitude γ for entrance and exit channels, energy levels and energy shifts can be deduced and used to determine the astrophysical S(E) factor, since these parameters are the same in both direct and THM data. In this way, an exact parameterization of the astrophysical S(E) factor can be obtained overcoming the extrapolation. If the resonance parameters of a single level in the relevant energy region are known, they can be fixed in the fitting procedure to obtain directly the astrophysical S(E) factor in absolute units.
Several test studies have been performed in the past years to validate the THM [35], such as the invariance of the two-body reaction amplitude with changing the Trojan Horse nucleus hiding the participant cluster x [19–21], or the use of momentum distributions from Distorted Wave Born Approximation instead of the simple PWA shape, providing same results within experimental errors [42, 44]. The THM has been applied to many reactions of astrophysical interest connected to fundamental problems in different scenarios, from BBN nucleosynthesis [7, 23–40] to AGB and more explosive sites [41–50]. In the last years, reactions involving heavier systems, such as ^{12}C and ^{16}O have been investigated [51, 52].

3 The ^{12}C+^{12}C fusion at astrophysical energies

The ^{12}C+^{12}C fusion plays a key role in a broad range of scenarios in carbon rich environments. In particular, it determines star evolution and nucleosynthesis of intermediate mass and massive stars ($\geq$8 M$_\odot$) [53]. It influences also the lower stellar mass limit for carbon ignition. This limit separates the progenitors of white dwarfs, novae and type Ia supernovae, from those of core-collapse supernovae, neutron stars, and stellar mass black holes; it constrains superbursts model with neutron and strange stars, in particular if resonances are found to contribute in the Gamow peak [54]; it influences the weak component of the s process, which produces the elements between Fe and Sr. Carbon burning during the hydrostatic phase takes place from 0.8 to 1.2 GK, corresponding to center-of-mass energies from 1 to 3 MeV. In that sub-Coulomb energy region, the cross section falls rapidly below the nanobarn range. That is why the measurement of the cross section below a center of mass energy E_{cm} of 2 MeV was never performed. The compound nucleus ^{24}Mg is formed at an excitation energy above the particle decay threshold. Alpha, proton and neutron are the dominant evaporation channels, leading respectively to ^{20}Ne, ^{23}Na and ^{23}Mg, which can also be produced in excited bound states. Below 2.5 MeV there is not enough energy to feed ^{23}Mg even in its ground state and α and p channel are the only relevant ones at low energies.

The ^{12}C+^{12}C fusion cross section at the relevant energies was thus determined from the indirect measurement of the ^{12}C(^{12}C,α)^{20}Ne and ^{12}C(^{12}C,p)^{23}Na reactions via the THM applied to the ^{12}C(^{14}N,α^{20}Ne)^{2}H and ^{12}C(^{14}N,p^{23}Na)^{2}H three-body processes in the quasi-free (QF) kinematics regime, where ^{2}H from ^{14}N is spectator to the ^{12}C+^{12}C two-body processes [52]. The experiment was performed at the INFN - Laboratori Nazionali del Sud in Catania, Italy. A ^{14}N beam accelerated at 30 MeV by the SMP TANDEM was delivered onto a 100 μg/cm^2 C target with a beam spot on target smaller than 1.5 mm. The experimental setup consisted of two telescopes (38 μm silicon detector as ΔE- and 1000 μm position sensitive detector (PSD) as E-detector) placed on both sides with respect to the beam direction in symmetric configuration (two on each side), covering angles from 7° to 30°. The ejectile of the two-body reactions (either α or p) was detected in coincidence with the spectator d particle. The angular regions covered by the detectors were optimized for the QF kinematics of the break-up process of interest, and the investigated range of deuteron momentum values was feasible to check the existence of the QF mechanism. Several steps are involved in the data analysis (see [52]) and after their completion the two-body cross section of astrophysical relevance was extracted for four channels: ^{20}Ne+α_0, ^{20}Ne+α_1, ^{23}Na+p_0 and ^{23}Na+p_1. The yield for the ^{20}Ne+α_1 channel is shown in Fig. 1 (black solid dots) projected onto the ^{12}C+^{12}C relative energy variable, E_{cm}. A modified one-level many-channel R-matrix analysis was carried out including the ^{24}Mg states reported in [52]. According to the results of [55] at $E_{c.m.} \leq 3$ MeV, and monitoring the decrease of the penetration factors for the relevant states, the fraction of the total fusion yield from α and p channels other than $\alpha_{0,1}$ and $p_{0,1}$ was neglected in the modified R-matrix analysis with estimated errors at $E_{c.m.}$ below 2 MeV lower than 1% and 2% for the α and p channels, respectively. The result for the ^{20}Ne+α_1 channel is shown in Fig. 1 as middle solid purple line with light purple band arising from the uncertainties on the resonance parameters, including correlations.

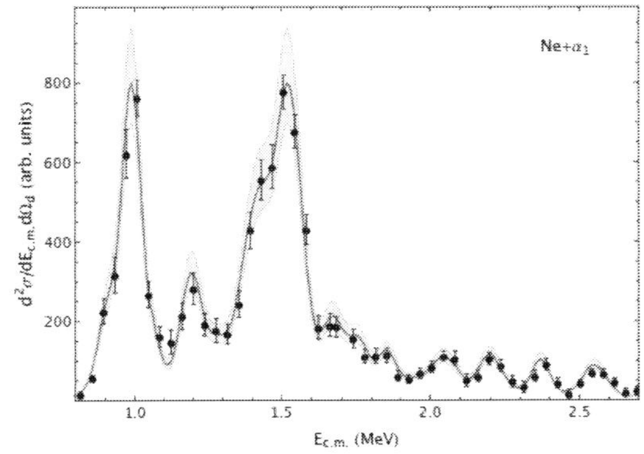

Figure 2. HOES two-body cross section for the ^{20}Ne+α_1 channel (black filled dots). Solid purple line plus shading from R-matrix calculations including uncertainties of the resonance parameters.

The resonance structure observed in the excitation functions is consistent with ^{24}Mg resonance energies reported in the literature with some tendency for the even J states to be clustered around 1.5 MeV and this might be a sign of intermediate structure of ^{24}Mg associated with a ^{12}C+^{12}C molecular configuration. The THM reduced widths thus entered a standard R-matrix code and the S(E) factors for the four reaction channels were determined.

The results are shown in Fig. 2 for the ^{20}Ne+α_1 in terms of modified S(E) factor, S(E)*, [56]. The black middle line and the grey band represent the best fit curve and the range defined by the total uncertainties, respectively. The grey band is the result of R-matrix calculations with lower and upper values of the resonance parameters provided by their errors.

The resonant structures are superimposed onto a flat non-resonant background taken from [57] . Normalization to direct data was done in the $E_{c.m.}$ window 2.5-2.63 MeV of the ^{20}Ne+α_1 channel where a sharp resonance corresponding to the level of ^{24}Mg at 16.5 MeV shows up and available data [57–60] in this region are the most accurate among those available in the full overlapping region with THM data. The resulting normalization error is 5%. All the existing direct data below $E_{c.m.}$= 3 MeV are shown as blue filled circles [57], purple filled squares [58], blue empty diamonds [59], red filled stars [60] and green filled triangles [61]. Except for the data from [57], their low energy limit is fixed by background due to hydrogen contamination in the targets. Disregarding these cases, agreement between THM and direct data is apparent within the experimental errors except for the direct low-energy limit around 2.14 MeV, where THM data do not confirm the claim of a strong resonance, rather a nearby one at 2.095 MeV about one order of magnitude less intense in the ^{20}Ne+α_1 channel (see Fig. 2) and with similar intensity in the ^{23}Na+p_1 one. The present result is in agreement with spectroscopy studies of [62, 63] with a deep at 2.14 MeV and no particularly strong α state around 2.1 MeV. Further agreement is found with the unpublished exper-

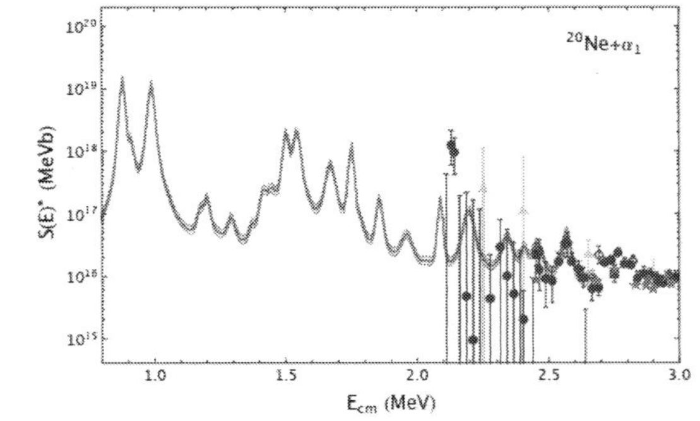

Figure 3. THM astrophysical S(E)* factor for the ^{20}Ne+α_1 channel (black solid line). The grey band represents the region spanned by R-matrix calculations with lower and upper values of the resonance parameters. Available direct data in the investigated $E_{c.m.}$ range are reported as purple filled squares [58], blue empty diamonds [59], red filled stars [60], blue filled circles [57] and green filled triangles [61].

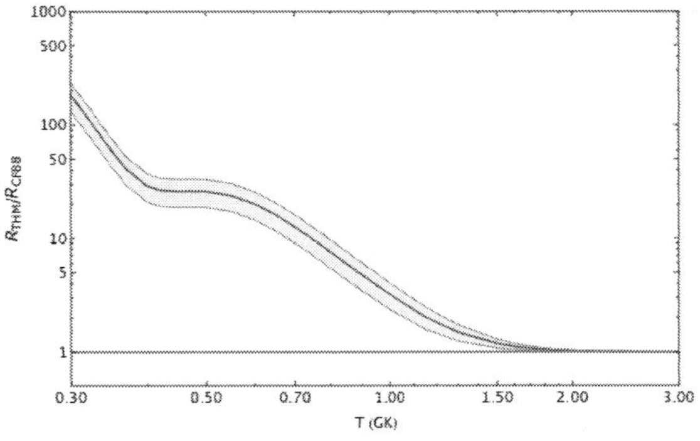

Figure 4. Ratio between the total THM ^{12}C+^{12}C reaction rate (black line) and the reference one (red line) from [67]. The grey shading defines the region spanned owing to the $\pm1\sigma$ uncertainties.

imental data down to $E_{c.m.}$=2.15 MeV from [64] for the ^{12}C(^{12}C,$p_{0,1}$)^{23}N reactions. Our result is also consistent within experimental errors with the total S(E)* from recent experiments [65, 66].

The corresponding reaction rate is shown in Fig. 3 divided by the reference rate from [67]. It experiences a variation below 2 GK with an increase from a factor of 1.18 at 1.2 GK to a factor of more that 25 at 0.5 GK. The latter increase, due mainly to the resonant structure around $E_{c.m.}$=1.5 MeV, endorses the fiducial value conjectured in [54] to reduce down to a factor of 2 the theoretical superburst ignition depths in accreting neutron stars for a realistic range of crust thermal conductivities and core Urca neutrino emissivities. This change is compatible with the observationally inferred superburst ignition depths. In other words, carbon burning can trigger superbursts. As for the hydrostatic carbon burning regime (0.6 to 1.2 GK), the present rate change will lower temperatures and densities at which ^{12}C ignites in massive post-main-sequence stars. Profiting of the stellar modeling reported in [68], for core C-burning of a star of 25M$_\odot$, the ignition temperature and density would undergo a decrease of down to 10% and 30% respectively. Recently, the impact of the new carbon fusion cross sections on Type Ia Supernovae was investigated in [69]. Their progenitors are not well understood. One popular scenario is the double-degenerate (DD) scenario, which attributes SNe Ia to WD-WD binary mergers. The resonance contribution results in a decrease of the carbon burning ignition temperature. Thus, accretion induced collapse occurs more easily and increases the birthrate of Galactic neutron stars with the contribution of the DD scenario to the SNe Ia rate becoming even smaller.

References

[1] HJ Assenbaum *et al.*, *Z. Phys.* A **327** 461 (1987)
[2] F Strieder *et al.*, *Naturwissenschaften* **88** 461 (2001)
[3] C Spitaleri *et al.*, *Phys. At. Nucl.*, **74**, 1763 (2011)
[4] A Tumino *et al.*, *Few Body Syst.*, **54**, Issue 7-10, 869 (2013)
[5] R. Tribble et al., *Rep. Prog. Phys.* 77, Issue: 10 106901 (2014)
[6] A Tumino *et al.*, *Few Body Syst.*, **54**, Issue 5-6 745 (2013)
[7] A Tumino et al., *Phys. Rev.* C **67** 065803 (2003)
[8] M La Cognata *et al.*, *Eur. Phys. J.* A **27** 249 (2006)
[9] A Tumino *et al.*, *Phys. Rev. Lett.* **98** 252502 (2007)
[10] M La Cognata *et al.*, *Phys. Rev.* C **76** 065804 (2007)
[11] J Kasagi *et al.*, *Nucl. Phys.* A **239** 233 (1975)
[12] I.S. Shapiro, *Soviet Phys. Usp.*, **10**(4), 515 (1968)
[13] Baur, G. 1986, *Nucl. Phys.* A **458**, 188 (1986)
[14] AM Mukhamedzhanov et al., *Eur. Phys. J.* A **27** 205 (2006)
[15] A Tumino *et al.*, *Phys. Rev.* C **78** 064001 (2008)
[16] A Tumino *et al.*, *Eur. Phys. J.* A **25** 649 (2005)
[17] AM Mukhamedzhanov, *Phys. Rev.* C **84** 044616 (2011)
[18] C Spitaleri *et al.*, *Phys. Rev.* C **91** 024612 (2015)
[19] A Tumino *et al.*, *Eur. Phys. J. A* direct, 1 (2006) DOI: 10.1140/epja/i2006-08-038-1
[20] RG Pizzone *et al.*, *Phys. Rev.* C **83**, 045801 (2011)
[21] RG Pizzone *et al.*, *Phys. Rev.* C **87**, 025805 (2013)
[22] M La Cognata *et al.*, *Astrophysical J. Lett.* **739**, L54 (2011)
[23] RG Pizzone *et al.*, *Astronomy & Astrophysics* **398**, 423 (2003)
[24] A Tumino *et al.*, *Prog. Theor. Phys. Suppl.* **154** 341 (2004)
[25] A Rinollo *et al.*, *Nucl. Phys.* A **758** 146 (2005)
[26] Qun-Gang Wen *et al.*, *Phys. Rev.* C **78** 035805 (2008)
[27] Qun-Gang Wen *et al.*, *J.Phys. G: Nucl. Part. Phys.*, **38**, 085103 (2011)
[28] A Tumino *et al.*, *Phys. Lett.* B **700**, 111 (2011)
[29] A Tumino *et al.*, *Phys. Lett.* B **705**, 546 (2011)
[30] L Lamia *et al.*, *Astrophysical J.* **768**, 65 (2013)

[31] A Tumino *et al.*, *Astrophysical J.* **785**, 96 (2014)
[32] R.G. Pizzone *et al. Astrophysical Journal*, **786** ,112 (2014)
[33] C Spitaleri *et al.*, *Phys. Rev.* C **90** 035801 (2014)
[34] Li Chengbo *et al.*, *Phys. Rev.* C **92**, 025805 (2015)
[35] C. Spitaleri *et al.*, *Phys. Rev.* C 91, 024612 (2015)
[36] L. Lamia *Astrophysical J.*, **850**, 175 (2017)
[37] C. Spitaleri *et al.*, *Phys. Rev.* C 95, 035801 (2017)
[38] Li Chengbo *et al.*, *Phys. Rev.* C 95, 035804 (2017)
[39] G.G. Rapisarda *et al.*, *Eur. Phys. J.* A, 54 189 (2018)
[40] L. Lamia *et al. Astrophysical Journal*, **879**, 23 (2019)
[41] M La Cognata *et al.*, *Phys. Rev. Lett.* **101** 152501 (2008)
[42] ML Sergi *et al.*, *Phys.Rev.* C **R82**, 032801 (2012)
[43] M La Cognata *et al.*, *Astrophysical J.* **708** 796 (2010)
[44] M. La Cognata *et al. Astrophysical J.* **739** L54 (2011)
[45] M La Cognata *et al.*, *Phys. Rev. Lett.* **49**, 015106 (2013)
[46] S. Cherubini *et al.*, *Phys. Rev.* C **92** 015805 (2015)
[47] R.G. Pizzone *et al. Astrophysical Journal*, **836**, 57 (2017)
[48] I. Indelicato *et al. Astrophysical Journal*, **845**, 19 2017
[49] G. D'Agata et al., *Astrophysical J.* **860**, 61 (2018)
[50] V. Burjan et al., *Eur. Phys. J.* A **55**, 114 (2019)
[51] A. Tumino et al., Eur. Phys. Journ. A, 12 327 (2001)
[52] Tumino A. et al., *Nature* **557**, 687 (2018).
[53] Garcia-Berro E. et al., *Astrophys. J.*, **286**, 765 (1997).
[54] Cooper R.L. et al., *Astrophys. J.*, **702**, 660 (2009).
[55] Becker H.W., Kettner K.U., Rolfs C. & Trautvetter H.P. *Z. Phys. A*, **303** 305 (1981).
[56] Aguilera E.F. et al., *Phys. Rev. C* **73** 064501 (2006).
[57] Spillane T. et al., *Phys. Rev. Lett.* **98**, 122501 (2007).
[58] Mazarakis M.G. & Stephens W.E. *Phys. Rev. C*, **7** 1280 (1973).
[59] M.D. High, B. Cujec, *Nucl. Phys. A* **282** 181 (1977)
[60] Kettner K.U., Lorenz-Wirzba H. & Rolfs C. *Z. Phys. A* **298** 65 (1980).
[61] Barrón-Palos L. et al., *Nucl. Phys. A* **779** 318 (2006).
[62] Abegg R. & Davis C.A. *Phys. Rev. C* **43** 6 (1991).
[63] Caciolli A. et al., *NIM B* **266** 1932 (2008).
[64] Zickefoose J. *Phys. Rev. C* **97** 065806 (2018).
[65] Jiang C.L. et al., *Phys. Rev. C* **97** 012801(R) (2018).
[66] C. Beck, arXiv:1812.08013
[67] Caughlan G.R. & Fowler W.A. *At. Data Nucl. Data Tables* **40**, 283 (1988).
[68] Pignatari M. et al., *Astrophys. J.* **762** 31 (2013).
[69] Mori K. et al., https://arxiv.org/pdf/1810.01025.

EPJ Web of Conferences **223**, 01066 (2019) https://doi.org/10.1051/epjconf/201922301066
NSD2019

Equilibration dynamics in nuclear reactions

A.S. Umar[1,*], *C.* Simenel[2,**], and *S.* Ayik[3,***] *K.* Godbey[1,****]

[1]*Department of Physics and Astronomy, Vanderbilt University, Nashville, TN 37235, USA*
[2]*Department of Theoretical Physics and Department of Nuclear Physics, Research School of Physics and Engineering, The Australian National University, Canberra ACT 2601, Australia*
[3]*Physics Department, Tennessee Technological University, Cookeville, TN 38505, USA*

Abstract. We discuss the equilibration dynamics and time-scales for various quantities that are connected to the experimentally observable entities. These include the study of mass, isospin, and total kinetic energy (TKE) equilibration time-scales as well as the time-scale for fluctuations.

1 Introduction

Heavy-ion reactions at low energies carry a wealth of information to investigate the characteristic features of nuclear dynamics. Furthermore, these reactions help explore the intriguing interplay between quantal dynamics at the microscopic level and the collective dynamics on time-scales that are too short for full equilibration. To fully classify the systematics of these reactions both theoretical and experimental studies are required for a diverse selection of systems. With the development of radioactive ion-beam facilities this goal is rapidly evolving.

In this proceeding we provide a discussion and microscopic calculations to examine the equilibration dynamics and time-scales for various quantities that are connected to the experimentally observable entities. These include the study of mass, isospin, and total kinetic energy (TKE) equilibration time-scales as well as the time-scale for fluctuations. The phenomena discussed here is directly related to the transport dynamics between reacting systems that initially carry a quantal asymmetry. A recent example of such a study is the experiment to investigate isospin equilibration in the vicinity of the Fermi energy [1,2]. Earlier studies at lower energies involved deep-inelastic reaction of systems with large isospin asymmetry [3–5].

Below, we utilize the time-dependent Hartree-Fock (TDHF) theory [6] to investigate a selected number of systems. A more comprehensive study involving many more systems will be the subject of a future publication.

2 Equilibration from TDHF

Low-energy nuclear reactions encompass very diverse time-scales depending on the type of collision studied. At times immediately after contact ($\sim 10^{-22}$ sec.) pre-equilibrium particle emission can take place. This is followed by the deep-inelastic regime (a few times 10^{-21} sec.) and the fast quasifission regime ($\sim 10^{-21} - 10^{-20}$ sec.). For these longer sticking times the outgoing channel begins to loose the memory of the entrance channel properties. Yet at longer time-scales the system becomes partially or fully equilibrated leading to slow quasifission followed by fusion-fission, and finally compound nucleus formation or compound nucleus fission. The latter processes are relatively well understood in the context of models relying on statistical equilibrium physics. The challenging aspect is the intermediary regime, where the dynamics and shell effects still play a role and equilibration of various quantities occur at different time-scales. Among these here we study the equilibration of mass, isospin, kinetic energy, and fluctuations. For numerical calculations we employ a fully 3D TDHF code [7–9] and the SLy4d [10] Skyrme energy density functional. The plotted results are comprised of calculations involving many partial waves and therefor show large fluctuations.

2.1 Mass

Quasifission provides a unique probe to quantum many-body dynamics of out-of-equilibrium nuclear systems. For instance, quasifission studies bring information on mass equilibration time-scales [11–13], on shell effects in the exit channels [14–18], dependence on the neutron-to-proton ratio N/Z of the compound nucleus [19], as well as on the nuclear equation of state [20,21]. The TDHF theory has been employed to study quasifission dynamics with considerable success [22–29]. The outcome of the calculations strongly depend on the orientation of the deformed nuclei. Here, we discuss the ^{48}Ca+^{249}Bk reaction with the TDHF approach. Unlike previous calculations where the TDHF studies of quasifission with actinide targets were restricted to one or two orientations, we performed calcu-

*e-mail: umar@compsci.cas.vanderbilt.edu
**e-mail: cedric.simenel@anu.edu.au
***e-mail: Ayik@tntech.edu
****e-mail: kylegodbey@gmail.com

lations spanning both a range of orientations and a range of angular momenta. The orientation of the deformed ^{249}Bk was changed by 15° steps to cover the full range $(0,\pi)$ with orbital angular momentum L changing in units of $10\hbar$ from 0 to quasielastic collisions.

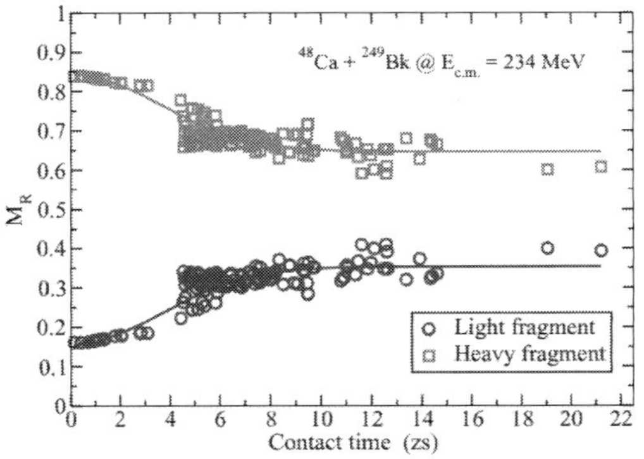

Figure 1. For the ^{48}Ca+^{249}Bk reaction; Mass ratio of fragment masses as a function of contact time at $E_{c.m.}$ = 234 MeV. The solid lines show possible fits.

Figure 1 shows the mass ratio, $M_R = M_{frag}/M_{total}$, of fragment masses as a function of contact time at $E_{c.m.}$ = 234 MeV for the ^{48}Ca+^{249}Bk reaction. We define the contact time as the time interval between the time t_1 when the two nuclear surfaces (defined as isodensities with half the saturation density $\rho_0/2 = 0.07$ fm^{-3}) first merge into a single surface and the time t_2 when the surface splits up again. The solid lines are a simple fit to the results. We observe

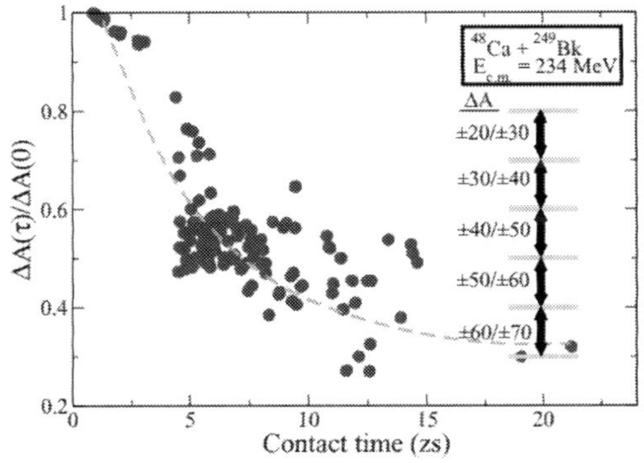

Figure 2. The ratio of the final and initial fragment masses as a function of contact time for the ^{48}Ca +^{249} Bk system at $E_{c.m.}$ = 234 MeV and for all orientations and orbital angular momenta discussed in the text. The horizontal bars on the left indicate the intervals for the net number of transferred particles. The dashed line shows one possible fit.

that mass equilibration appears to occur for times larger than about 8–10 zs, starting at about 6 zs. Another perspective of these results can be obtained by plotting the ratio of the final fragment mass difference to the initial mass difference as a function of contact time, as shown in Fig. 2 for the same system. The horizontal lines on the right indicate the intervals for the net number of transferred particles and the dashed line is a possible fit to guide the eye.

From these results and a number of others not shown here we can conclude that the mass equilibration is substantially longer as expected in comparison to the other quantities mentioned below. It is also interesting to observe that there is a bunching of results around certain mass ratios. This is shown to be related to shell effects influencing the dynamical quasifission process in Ref. [30].

2.2 Isospin and TKE

Transport properties of isospin asymmetric nuclear matter can be investigated by studying charge equilibration driven by the nuclear symmetry-energy in heavy-ion collisions. For example, collisions at Fermi energies give access to contact times which are short enough to induce only a partial charge equilibration [31] and thus can be used to determine equilibration times [1]. Due to a faster time scale for equilibration of charge and total kinetic energy (TKE), deep-inelastic collisions are the preferred venue at lower energies. We have done a systematic study of the ^{78,92}Kr + ^{208}Pb systems at 8.5 MeV/nucleon. In Fig. 3 we

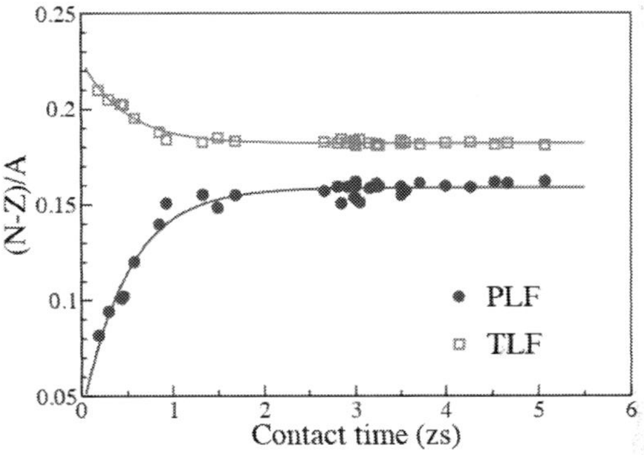

Figure 3. The $(N-Z)/A$ value of the primary light (full circles) and heavy (open squares) fragments formed in ^{78}Kr +^{208} Pb at E = 8.5 MeV/nucleon are plotted as a function of the contact time between the collision partners. The solid lines show fits to the TDHF results.

plot the ratio $(N-Z)/A$ as a function of contact time for the ^{78}Kr + ^{208}Pb system. An equilibration time $\tau \sim 0.5$ zs is obtained from the fit $(N-Z)/A = \alpha + \beta \exp(-T/\tau)$ to the TDHF results as it was done in Ref. [1]. This is in agreement with a slightly faster equilibration time obtained (~0.3 zs) for reactions near the Fermi energy [1].

In order to study the equilibration of TKE in Fig. 4 we plot the TKE of the final fragments as a function of contact time for ^{78,92}Kr + ^{208}Pb systems. The two orientations of the deformed Kr nuclei are also indicated on the plot. We observe that the equilibration of the TKE is also a relatively fast process but somewhat longer than that of isospin transfer. The TKE seems to reach its equilibrium value at around 1.5 zs.

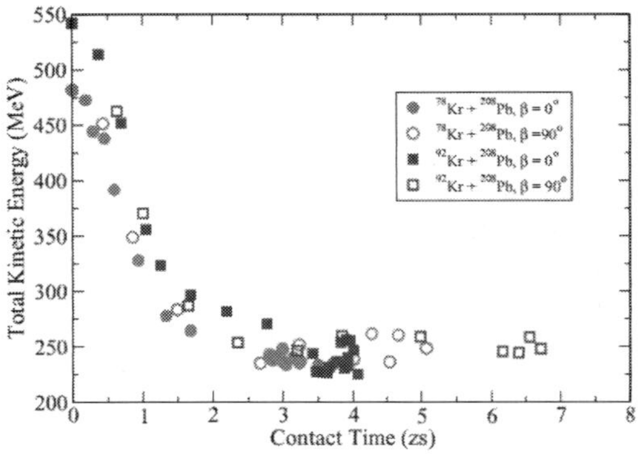

Figure 4. The exit channel TKE in ^{78}Kr +^{208} Pb reaction at E = 8.5 MeV/nucleon are plotted as a function of the contact time between the collision partners.

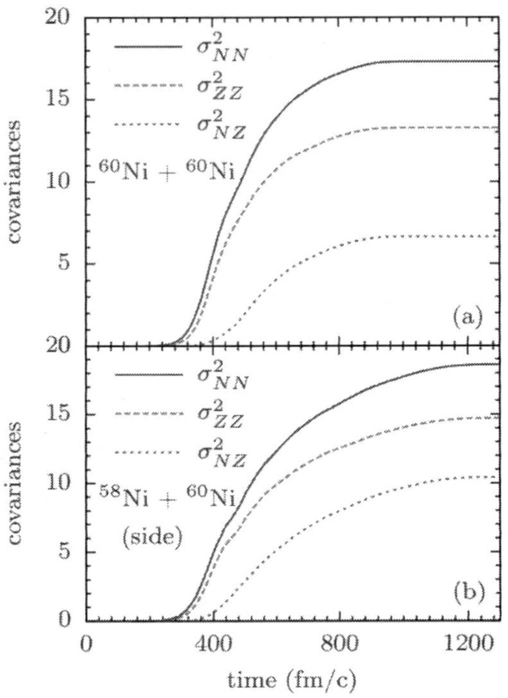

Figure 5. Co-variances in the collision of ^{60}Ni + ^{60}Ni (a) and in the collision of ^{58}Ni + ^{60}Ni in the side geometry (b) at the center of mass energy $E_{c.m.}$ = 135.6 MeV and initial orbital angular momentum $L = 73\hbar$.

2.3 Fluctuations

The mean-field description of reactions using TDHF provides the mean values of the proton and neutron drift. It is also possible to compute the probability to form a fragment with a given number of nucleons [32–35], but the resulting fragment mass and charge distributions are often underestimated in dissipative collisions [36,37]. Much effort has been done to improve the standard mean-field approximation by incorporating the fluctuation mechanism into the description. At low energies, the mean-field fluctuations make the dominant contribution to the fluctuation mechanism of the collective motion. Various extensions have been developed to study the fluctuations of one-body observables. These include the TDRPA approach of Balian and Vénéroni [38], the time-dependent generator coordinate method [39], or the stochastic mean-field (SMF) method [40,41]. In the SMF approach the collision dynamics is described in terms of an ensemble of mean-field events. The single-particle density matrix of each event, $\rho^\lambda(\vec{r}, \vec{p}, t)$, is determined by the self-consistent mean-field of the corresponding event. Here λ indicates the event label. When a di-nuclear structure is maintained during the collision, such as deep inelastic and quasi-fission reactions, we do not need to generate such an ensemble of mean-field events. In this case it is possible to describe the collision dynamics in terms of a few relevant macroscopic variables, such as mass and charge asymmetry, the relative linear momentum and the orbital angular momentum. In di-nuclear geometry, we can express these macroscopic variables in terms of the TDHF solutions with the help of window dynamics [42].

In Fig. 5a we plot the time-evolution of the calculated SMF co-variances for the collision of ^{60}Ni + ^{60}Ni, and in In Fig. 5b the same quantities for ^{58}Ni + ^{60}Ni, at center of mass energy $E_{c.m.}$ = 135.6 MeV and initial orbital angular momentum $L = 73\hbar$. As we see all of the co-variances stabilize around 800–1000 fm/c, which corresponds to approximately 2.6–3.3 zs.

This work has been supported by the U.S. Department of Energy grant Nos. DE-SC0013847 and DE-SC0015513, and by the Australian Research Council Discovery Project (project numbers DP160101254 and DP190100256) funding schemes.

References

[1] A. Jedele, A.B. McIntosh, K. Hagel, M. Huang, L. Heilborn, Z. Kohley, L.W. May, E. McCleskey, M. Youngs, A. Zarrella et al., Phys. Rev. Lett. **118**, 062501 (2017)
[2] Alan B. McIntosh, Sherry J. Yennello, Prog. Part. Nucl. Phys. **108**, 103707 (2019)
[3] R. Planeta, S.H. Zhou, K. Kwiatkowski, W.G. Wilson, V.E. Viola, H. Breuer, D. Benton, F. Khazaie, R.J. McDonald, A.C. Mignerey et al., Phys. Rev. C **38**, 195 (1988)
[4] R.T. de Souza, W.U. Schröder, J.R. Huizenga, R. Planeta, K. Kwiatkowski, V.E. Viola, H. Breuer, Phys. Rev. C **37**, 1783 (1988)
[5] R. Płaneta, K. Kwiatkowski, S.H. Zhou, V.E. Viola, H. Breuer, M.A. McMahan, W. Kehoe, A.C. Mignerey, Phys. Rev. C **41**, 942 (1990)
[6] C. Simenel, A.S. Umar, Prog. Part. Nucl. Phys. **103**, 19 (2018)
[7] A.S. Umar, M.R. Strayer, J.S. Wu, D.J. Dean, M.C. Güçlü, Phys. Rev. C **44**, 2512 (1991)
[8] A.S. Umar, V.E. Oberacker, Phys. Rev. C **73**, 054607 (2006)

[9] A.S. Umar, V.E. Oberacker, Phys. Rev. C **74**, 061601 (2006)
[10] Ka–Hae Kim, Takaharu Otsuka, Paul Bonche, J. Phys. G **23**, 1267 (1997)
[11] J. Tõke, R. Bock, G.X. Dai, A. Gobbi, S. Gralla, K.D. Hildenbrand, J. Kuzminski, W.F.J. Müller, A. Olmi, H. Stelzer et al., Nucl. Phys. A **440**, 327 (1985)
[12] W.Q. Shen, J. Albinski, A. Gobbi, S. Gralla, K.D. Hildenbrand, N. Herrmann, J. Kuzminski, W.F.J. Müller, H. Stelzer, J. Tõke et al., Phys. Rev. C **36**, 115 (1987)
[13] R. du Rietz, D.J. Hinde, M. Dasgupta, R.G. Thomas, L.R. Gasques, M. Evers, N. Lobanov, A. Wakhle, Phys. Rev. Lett. **106**, 052701 (2011)
[14] M.G. Itkis, J. Äystö, S. Beghini, A.A. Bogachev, L. Corradi, O. Dorvaux, A. Gadea, G. Giardina, F. Hanappe, I.M. Itkis et al., Nucl. Phys. A **734**, 136 (2004)
[15] K. Nishio, H. Ikezoe, S. Mitsuoka, I. Nishinaka, Y. Nagame, Y. Watanabe, T. Ohtsuki, K. Hirose, S. Hofmann, Phys. Rev. C **77**, 064607 (2008)
[16] E.M. Kozulin, G.N. Knyazheva, S.N. Dmitriev, I.M. Itkis, M.G. Itkis, T.A. Loktev, K.V. Novikov, A.N. Baranov, W.H. Trzaska, E. Vardaci et al., Phys. Rev. C **89**, 014614 (2014)
[17] A. Wakhle, C. Simenel, D.J. Hinde, M. Dasgupta, M. Evers, D.H. Luong, R. du Rietz, E. Williams, Phys. Rev. Lett. **113**, 182502 (2014)
[18] M. Morjean, D.J. Hinde, C. Simenel, D.Y. Jeung, M. Airiau, K.J. Cook, M. Dasgupta, A. Drouart, D. Jacquet, S. Kalkal et al., Phys. Rev. Lett. **119**, 222502 (2017)
[19] K. Hammerton, Z. Kohley, D.J. Hinde, M. Dasgupta, A. Wakhle, E. Williams, V.E. Oberacker, A.S. Umar, I.P. Carter, K.J. Cook et al., Phys. Rev. C **91**, 041602(R) (2015)
[20] M. Veselsky, J. Klimo, Y.G. Ma, G.A. Souliotis, Phys. Rev. C **94**, 064608 (2016)
[21] H. Zheng, S. Burrello, M. Colonna, D. Lacroix, G. Scamps, Phys. Rev. C **98**, 024622 (2018)
[22] Cédric Golabek, Cédric Simenel, Phys. Rev. Lett. **103**, 042701 (2009)
[23] David J. Kedziora, Cédric Simenel, Phys. Rev. C **81**, 044613 (2010)
[24] V.E. Oberacker, A.S. Umar, C. Simenel, Phys. Rev. C **90**, 054605 (2014)
[25] A.S. Umar, V.E. Oberacker, C. Simenel, Phys. Rev. C **92**, 024621 (2015)
[26] A.S. Umar, V.E. Oberacker, C. Simenel, Phys. Rev. C **94**, 024605 (2016)
[27] K. Sekizawa, K. Yabana, Phys. Rev. C **93**, 054616 (2016)
[28] K. Sekizawa, Phys. Rev. C **96**, 041601(R) (2017)
[29] K. Sekizawa, K. Hagino, Phys. Rev. C **99**, 051602 (2019)
[30] K. Godbey, A.S. Umar, C. Simenel, Phys. Rev. C **100**, 024610 (2019)
[31] M.B. Tsang, T.X. Liu, L. Shi, P. Danielewicz, C.K. Gelbke, X.D. Liu, W.G. Lynch, W.P. Tan, G. Verde, A. Wagner et al., Phys. Rev. Lett. **92**, 062701 (2004)
[32] S.E. Koonin, K.T.R. Davies, V. Maruhn-Rezwani, H. Feldmeier, S.J. Krieger, J.W. Negele, Phys. Rev. C **15**, 1359 (1977)
[33] C. Simenel, Phys. Rev. Lett. **105**, 192701 (2010)
[34] Kazuyuki Sekizawa, Kazuhiro Yabana, Phys. Rev. C **88**, 014614 (2013)
[35] G. Scamps, D. Lacroix, Phys. Rev. C **87**, 014605 (2013)
[36] C.H. Dasso, T. Dossing, H.C. Pauli, Z. Phys. A **289**, 395 (1979)
[37] C. Simenel, Phys. Rev. Lett. **106**, 112502 (2011)
[38] R. Balian, M. Vénéroni, Ann. Phys. **216**, 351 (1992)
[39] H. Goutte, J.F. Berger, P. Casoli, D. Gogny, Phys. Rev. C **71**, 024316 (2005)
[40] S. Ayik, Phys. Lett. B **658**, 174 (2008)
[41] S. Ayik, B. Yilmaz, O. Yilmaz, Phys. Rev. C **92**, 064615 (2015)
[42] S. Ayik, B. Yilmaz, O. Yilmaz, A.S. Umar, Phys. Rev. C **97**, 054618 (2018)

Spectroscopy of natural parity states in ^{20}Ne via the ^{19}F(p,α_0) and ^{19}F(p,α_π) reactions at low energies

Mariano Vigilante[1,*]

[1]Università degli Studi di Napoli Federico II and INFN-Sezione di Napoli, Napoli, Italy

Abstract. The ^{19}F(p,α_0)^{16}O and ^{19}F(p,α_π)^{16}O reactions can be fruitfully exploited to probe natural parity states above the proton emission threshold in the compound ^{20}Ne nucleus. However, despite the relatively large number of data sets published in the literature, it is still not possible to draw a fully unambiguous picture of the spectroscopy of ^{20}Ne, especially close to the proton emission threshold. To contribute to clarify these aspects, we have collected and reviewed all existing data sets of the ^{19}F(p,α_0)^{16}O and ^{19}F(p,α_π)^{16}O reactions to perform, for the first time, a comprehensive *R*-matrix fit including both reactions in the bombarding energy range E_p = 0.2-10 MeV. Our analysis clarifies several ambiguities present in the literature and affecting the spectroscopy of ^{20}Ne and clearly demonstrates the importance of the ^{19}F(p,α_π) reaction to better constraint the spectroscopy parameters of the fit.

1 Introduction

The investigation of low energy nuclear reactions induced by light particles (such as p or α) is a great opportunity to probe the spectroscopy (i.e. energy levels, decay widths, and J^π) of light nuclear systems formed as the compound nucleus [1–5]. Such reactions are also relevant to collateral domains of nuclear physics, such as nuclear astrophysics [4, 6, 7] and applied research [8–10].

The ^{19}F(p,α)^{16}O reaction has been shown to be a particularly powerful tool to investigate the spectroscopy of the self-conjugate ^{20}Ne nucleus at E_x > 12.884 MeV [11, 12]. Additionally, starting from even very low proton bombarding energies (few hundreds of keV) a number of reaction channels are energetically open, including the ^{19}F(p,α_π)^{16}O, where the residual ^{16}O nucleus is left in its first excited state (E_x = 6.05 MeV) and α-group followed by γ-ray emission ^{19}F(p,α_γ)^{16}O (having E_x = 6.13, 6.92, 7.12 MeV in ^{16}O). Regarding these reactions, a large body of data has been collected over the past decades, contributing to constrain the spectroscopy of ^{20}Ne in a wide energy range [13]. Despite these efforts, several ambiguities are still affecting the many states in ^{20}Ne, demanding for more detailed analyses.

In this proceeding, we propose an extensive critical analysis of all data sets published in the literature concerning the ^{19}F(p,α_0)^{16}O and ^{19}F(p,α_π)^{16}O reactions at low bombarding energies ($E_p \approx$ 0.2-10 MeV). The obtained data sets are analyzed in terms of the R-matrix theory. For the first time, we perform a systematic and simultaneous fit of such a large data set. Our results help to improve the spectroscopy of several natural-parity states in ^{20}Ne.

*e-mail: vigilant@na.infn.it

2 Data sets and preliminary R-matrix analysis

In order to build a consistent set of data we have carefully reviewed all existing measurements published in the literature. For the ^{19}F(p,α_0)^{16}O, we have complemented the data reviewed in previous analysis [14], and cov-ering the incident energy range E_p = 0.2-3.5 MeV, with the high energy data of Ref. [15] ($E_p \leq$ 12.5 MeV). The latter is mandatory to constrain the contribution of direct processes, affecting the integrated cross section and angular distributions down to very low energies [6]. Our data set includes also data of the ^{19}F(p,α_π)^{16}O*(6.05 MeV) reaction. This is particularly important to increase the reliability of the R-matrix fit. In the energy regions E_p = 0.6-1.1 MeV and E_p = 1.7-2.5 MeV, we consider the data from Refs. [1, 2, 16]. We excluded from our analysis data from Ref. [17], obtained with a pair spectrometer and potentially affected by systematic indeterminations. All data sets agree reasonably in shape. Data of Ref. [16] have been normalized by a factor 0.5 in order to match with the data published in [1]. In the energy region E_p = 1.1-1.7 MeV, we used the differential cross section measured in Ref. [18] to calculate the integrated cross section, assuming the shapes of the angular distributions as recorded in [19].

We performed a comprehensive R-matrix analysis of the above discussed data set. As initial parameters of the fit, we considered the spectroscopy of ^{20}Ne as described in [13] and taking into account the findings of Refs. [6, 11]. We also included a direct contribution to the total cross section, possibly enhanced by the cluster configuration of ^{19}F, as discussed in [6, 20, 21]. The presence of an extended data set, and the simultaneous fit of more reaction channels, allow us to refine the spectroscopy of

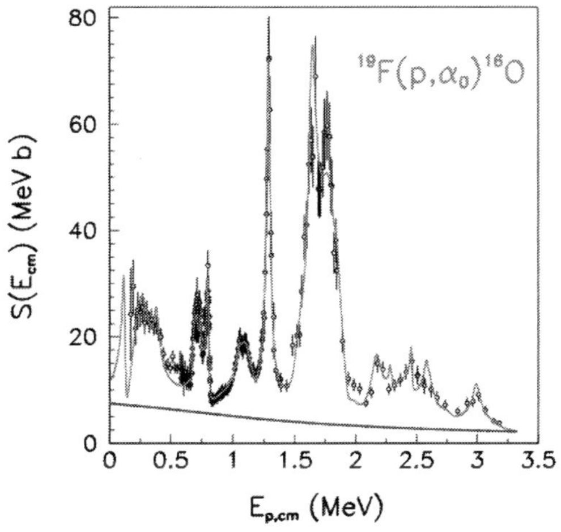

Figure 1. S-factor of the $^{19}F(p,\alpha_0)^{16}O$ reaction as a function of the energy in the center of mass reference frame. The red line is the result of the R-matrix analysis while the blue line represents the direct contribution used to reproduce the data.

natural-parity states in ^{20}Ne in the excitation energy range E_x = 13-16 MeV. Preliminary results are shown for the $^{19}F(p,\alpha_0)^{16}O$ reaction in Fig. 1, where the red line is the result of our fit procedure, while the blue line represents the expected direct contribution. The latter, has been obtained by including two broad poles in s and p waves at high energies to simulate a direct process. The overall reproduction of data is satisfactory in the whole energy range.

3 Conclusions and perspectives

In this paper, we have discussed some preliminary findings of a new comprehensive analysis of $^{19}F(p,\alpha_0)^{16}O$ and $^{19}F(p,\alpha_\pi)^{16}O^\star$(6.05 MeV) data in the bombarding energy range E_p = 0.2-10 MeV. We obtained a consistent data set by carefully reviewing all previous measurements published in the literature. We used the R-matrix theory to produce a simultaneous fit of all data. The corresponding results, that will be object of a forthcoming publication, help to improve the spectroscopy of natural-parity states in ^{20}Ne.

In the near future, we will exploit recently developed arrays for charged particles [22, 23] to carry out new experimental campaigns aimed to measure, with extremely high precision, low energy nuclear reactions to improve the spectroscopy of light nuclear systems. Additionally, information obtained by large acceptance detectors used as γ-ray arrays [24], will contribute to improve our understanding of clustering in light nuclei.

References

[1] R. Caracciolo et al., Lett. Nuovo Cim. **11**, 33 (1974)
[2] A. De Rosa, E. Perillo, P. Cuzzocrea, G. Inglima, E. Rosato, M. Sandoli, G. Spadaccini, Nuovo Cimento A Serie **44**, 433 (1978)
[3] I. Lombardo et al., Bulletin of the Russian Academy of Sciences: Physics **78**, 1093 (2014)
[4] C. Spitaleri, S.M.R. Puglia, M. La Cognata, L. Lamia, S. Cherubini, A. Cvetinović, G. D'Agata, M. Gulino, G.L. Guardo, I. Indelicato et al., Phys. Rev. C **95**, 035801 (2017)
[5] I. Lombardo et al., Phys. Rev. C **97**, 034320 (2018)
[6] I. Lombardo et al., Phys. Lett. B **748**, 178 (2015)
[7] M. La Cognata, S. Palmerini, C. Spitaleri, I. Indelicato, A.M. Mukhamedzhanov, I. Lombardo, O. Trippella, Astr. Phys. J. **805**, 128 (2015)
[8] V. Paneta et al., Nucl. Instr. Meth. Phys. Res. B **288**, 53 (2012)
[9] I. Lombardo et al., J. Phys. G. **43**, 45109 (2016)
[10] I. Lombardo et al., Nucl. Instrum. Meth. Phys. Res. **B 302**, 19 (2013)
[11] I. Lombardo et al., J. Phys. G **40**, 125102 (2013)
[12] I. Lombardo et al., J. Phys. Conf. Ser. **569**, 012068 (2014)
[13] D.R. Tilley, C.M. Cheves, J.H. Kelley, S. Raman, H.R. Weller, Nuclear Physics A **636**, 249 (1998)
[14] J.J. He, I. Lombardo, D. Dell'Aquila, Y. Xu, L.Y. Zhang, W.P. Liu, Chin. Phys. C **42**, 15001 (2018)
[15] K.L. Warsh, G.M. Temmer, H.R. Blieden, Physical Review **131**, 1690 (1963)
[16] S. Devons, G. Goldring, G.R. Lindsey, Proceedings of the Physical Society A **67**, 134 (1954)
[17] W.A. Ranken, T.W. Bonner, J.H. McCrary, Physical Review **109**, 1646 (1958)
[18] P. Cuzzocrea, A. De Rosa, G. Inglima, E. Perillo, E. Rosato, M. Sandoli, G. Spadaccini, INFN/BE-80/**5**, 1 (1980)
[19] A. Isoya, K. Goto, T. Momota, Journal of the Physical Society of Japan **11**, 899 (1956)
[20] H. Herndl, H. Abele, G. Staudt, B. Bach, K. Grün, H. Scsribany, H. Oberhummer, G. Raimann, Phys. Rev. C **44**, R952 (1991)
[21] Y. Yamashita, Y. Kudo, Progress of Theoretical Physics **90**, 1303 (1993)
[22] D. Dell'Aquila et al., Nucl. Instrum. Meth. Phys. Res. A **877**, 227 (2018)
[23] G. Pastore et al., Nucl. Instrum. Meth. Phys. Res. A **860**, 42 (2017)
[24] G. Cardella et al., Nucl. Instr. and Meth. A **799**, 64 (2015)

Identification of different reaction channels in ^{6}Li + ^{89}Y experiment by the particles-γ coincidence measurement

M. L. Wang[1], *G. X.* Zhang[1], *G. L.* Zhang[1,2,*], *S. P.* Hu[3], *W. W.* Qu[4], *H. Q.* Zhang[5], *Z.* Huang[1], *J. B.* Xiang[1], *Y. J.* Yao[1], *H. B.* Sun[3], *J. J.* Valiente-Dobón[6], *D.* Testov[7], *A.* Goasduff[7], *P. R.* John[7,8], *M.* Siciliano[6,9], *F.* Galtarossa[6], *R.* Francesco[7], *D.* Mengoni[7], and *D.* Bazzacco[7]

[1]*School of Physics and Nuclear Energy Engineering, Beihang University, Beijing 100191, China*
[2]*Beijing Advanced Innovation Center for Big Data-Based Precision Medicine, Beihang University, Beijing 100083, China*
[3]*School of Physics and Energy, Shenzhen University, Shenzhen 518060, China*
[4]*State Key Laboratory of Radiation Medicine and Protection, School of Radiation Medicine and Protection, Soochow University, Suzhou 215123, China*
[5]*China Institute of Atomic Energy, Beijing 102413, China*
[6]*INFN, Laboratori Nazionali di Legnaro, Legnaro (Padova), Italy*
[7]*Dipartimento di Fisica and INFN, Sezione di Padova, Padova, Italy*
[8]*Institut für Kernphysik der Technischen Universität Darmstadt, Germany*
[9]*Irfu/CEA, Universitéde Paris-Saclay, Gif-sur-Yvette, France*

Abstract. This short paper presents the investigation of reaction mechanism induced by ^{6}Li through a particle-γ coincidence measurement. The data have been taken from a ^{6}Li+^{89}Y experiment which is performed in INFN-LNL, Italy. In this experiment, the light charged particles are detected by a Si-ball, named EUCLIDES, and the γ rays are collected by a HPGe detector array, called GALILEO. In this contribution, scientific motivations, experimental details and some results, such as α-γ analysis, are presented.

1 Introduction

Study of fusion reaction induced by weakly bound nuclei has been one of hot research topics for nuclear physics in the past several decades [1, 2]. Compared with unstable nuclei, the stable ones, such as ^{6}Li, ^{7}Li, and ^{9}Be, have more advantages, such as higher beam intensity, smaller beam spot and so on. Therefore, plenty of works have been devoted into the reaction mechanisms induced by stable weakly bound nuclei [3, 4].

Due to the cluster structure and low breakup threshold in stable weakly bound nuclei, the total fusion (TF) processes include the contributions from complete fusion (CF) and incomplete fusion (ICF), corresponding to the ones without and following the breakup of projectile, respectively. The $\frac{ICF}{TF}$ or $\frac{1-CF}{TF}$ is called CF suppression factor. Such factor is proved to be mainly governed by the breakup thresholds of the weakly bound projectiles, and is independnent with atomic numbers of the target nuclei [5, 6]. For example, Wang et al. [7] pointed out that the CF cross sections induced by ^{6}Li on many heavy targets, have an average suppression factor around 40% at energies slightly above the Coulomb barrier. However, in a recent measurement of ^{6}Li + ^{64}Ni system, the CF suppression factor was determined as (13±7)% experimentally [8]. Therefore, more investigations on fusion processes induced by ^{6}Li should be performed, especially on the light target.

Usually, the residues survived from fusion-evaporation process can be identified and collected by measuring their characteristic γ rays. However, the residues produced from different CF and ICF processes may overlap with each other, if the total atomic number of the whole reaction system (both projectile and target) is around or less than 40. The reason is ascribed to the fact that the evaporation of charged particles is allowed when the Coulomb barrier of the fusion system becomes relatively low. Under such condition, it is impossible to clearly separate CF and ICF by simply measuring the γ rays. Furthermore, direct reation channel, such as transfer reaction, may be dominant at energies near the Coulomb barrier, which has been observed in many cases, such as ^{8}He + ^{197}Au [9] and ^{6}Li + ^{96}Zr [10].

In our previous studies, fusion processes have been investigated in ^{6}Li + ^{96}Zr [11] and ^{6}Li + ^{154}Sm [12] systems through in-beam γ-ray method at HI-13 Tandem Accelerator of China Institute of Atomic Energy (CIAE), Beijing, China. A smaller CF suppression factor around 25% was found for ^{6}Li + ^{96}Zr system, and an average suppression about 35% is found for ^{6}Li + ^{154}Sm system. Such results indicate that the systematic behaviors of the CF suppression factors for fusion processes induced by ^{6}Li are different in various situations. On the other hand, only γ rays

*e-mail: zgl@buaa.edu.cn

were measured in these experiments, thus the CF and ICF cross sections may be ambiguous to some extent for the ^{6}Li + ^{96}Zr system in which the evaporation of charged particles becomes possible. Therefore, the conclusion needs to be further investigated both theoretically and experimentally.

The first exclusive measurement, in which both γ rays and light charged particles (proton, deuteron, triton and α) are collected, has been applied in ^{7}Li + ^{198}Pt system [13] at energies near the Coulomb barrier. Through particle-γ coincidence measurement, different breakup fusion processes are identified qualitatively, such as α-capture, triton-capture, ^{5}He-capture and ^{6}He-capture processes. The absolute cross sections of different breakup fusion processes and neutron stripping reactions are measured by the collection of the γ rays in different residues. However, such method may also encounter some limitations in the identification of reaction mechanism. Taking the yields of ^{199}Au as an example: in the α-gated γ-ray spectrum (see Fig. 2(a) in Ref. [13]), the main γ rays arise from $^{198,199}Au$, which are assigned as the residues from triton-capture ICF process. However, one proton stripping reaction can also generate the target-like fragment ^{199}Au, as well as projectile-like fragment ^{6}He, which dissociates into α + 2n if the ^{6}He is excited to the continuum states through the one-proton-stripping reaction. Furthermore, the theoretical prediction of ^{199}Au cross section in which only triton-capture ICF is considered, would underestimate the experimental results, as shown in Fig. 4(a) [13]. As a conclusion, the existence of one-proton-stripping reaction cannot be excluded, and simple particle-γ coincidence measurement cannot separate it from other ICF processes.

According to the above analysis, particle-γ coincidence measurement is powerful in the separation of different ICF processes, but still has many limitations. It is deserved to develop a new particle-γ coincidence measurement in order to distinguish all the reaction mechanisms, including CF, ICF and nucleon (proton or neutron) transfer processes.

2 Experimental details

A ^{6}Li + ^{89}Y experiment was performed at the Laboratori Nazionali di Legnaro, INFN, Italy. A $^{6}Li^{3+}$ beam with an average intensity of 1.0 enA was accelerated to 34 MeV by the XTU Tandem-ALPI accelerator. The ^{89}Y target with a thickness of 550 μg/cm^2, was backed on a 340 μg/cm^2-thick ^{12}C foil in order to stop all the reaction products. The GALILEO array, which consists of 25 Compton-Suppressed Ge detectors, was employed for the collection of γ-ray. A 4π Si-ball detector array, named EUCLIDES, was used to measure the light charged particles. The EUCLIDES array was made up of 40 ΔE-E telescopes, where the thickness of ΔE and E detectors were 130 and 1000 μm, respectively. A typical particle identification plot measured by EUCLIDES was shown in Fig. 2 from Ref. [14]. The detailed information on the GALILEO and EUCLIDES arrays can be found in Refs. [12,13].

The schematic view of the current experimental setup can be found in Fig. 1 from Ref. [14]. Since the Si detectors are sensitive to radiation damage, a 200-μm-thick Al absorber was inserted between the target and EUCLIDES array to stop the elastically scattered ^{6}Li. The absorber shielded all the Si detectors excepted for those located at the angles more than 148°.

3 Results and discussions

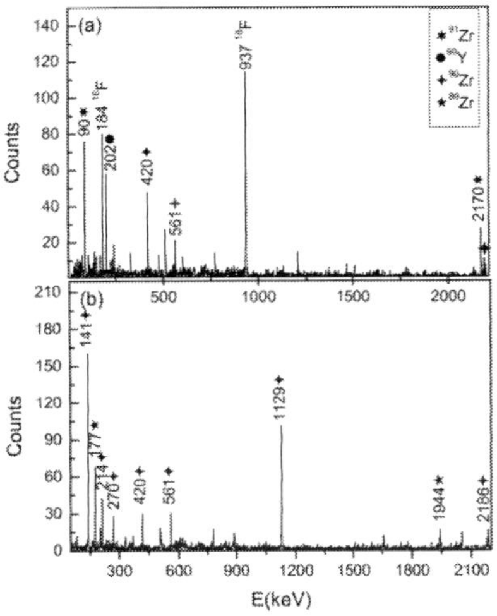

Figure 1. Gamma-ray energy spectra measured in coincidence with α detected by the shielded (a) and unshielded (b) Si detectors. Taken from Ref. [16].

The γ-ray energy spectra measured in coincidence with α which are detected in the shielded and unshielded (by Al absorber) Si-detectors are shown in Figs. 1 (a) and (b), respectively. Figure. 1 (a) shows the characteristic γ rays from ^{90}Y, ^{90}Zr and ^{91}Zr, corresponding to the products produced from one neutron, proton and deuteron stripping reactions, respectively. On the other hand, the characteristic γ rays from ^{89}Zr and ^{90}Zr, which can be ascribed evaporation residues from the deuteron-capture ICF process, are observed in Fig. 1(b).

It is noted that the γ rays in ^{90}Zr have observed simultaneously in Figs. 1(a) and (b). However, the reaction mechanisms of the productions of ^{90}Zr in Figs. 1(a) and (b) are totally different. In Fig. 1(a), the γ rays at 561 and 420 keV are the transitions which deexited the 3^- (2747 keV) and 4^- (2739 keV) states in ^{90}Zr, which is consistent with result of the (p, d) reaction in Ref. [15]. Meanwhile, the γ rays of ^{90}Zr which are shown in Fig. 1(b), are related to the transitions de-exciting the high-lying states, such as 141-, 270-, and 214-keV γ rays, being a typical feature of fusion-evaporation residues.

In addition, the energy distribution of α measured by the shielded Si-detectors is plotted in Fig. 2(a). The γ-ray spectra gated by α particles with different energies are shown in Figs. 2(b) and (c). As a result, the γ-ray spectrum

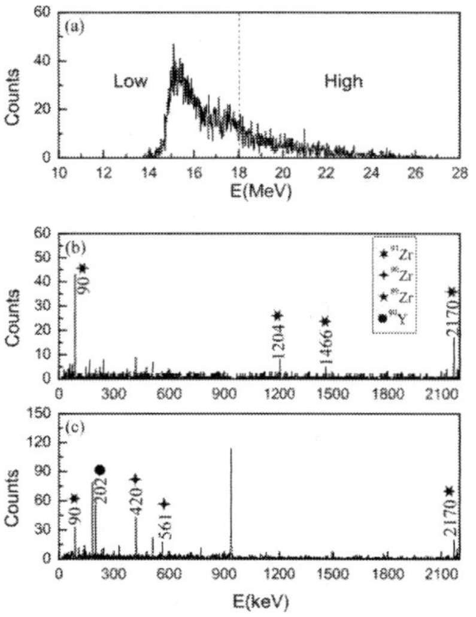

Figure 2. (a): Alpha energy spectrum measured by the shielded Si-detectors. (b) and (c): Gamma-ray energy spectra gated by α with energies higher and lower than 18 MeV, respectively. It is noted here the energies of α are the remaining ones after penetrating the absorber. Taken from Ref. [16].

with a gate on higher energy α only shows the characteristic γ rays in ^{91}Zr, being consistent with the fact that the Q value of 1d- (deuteron) stripping reaction is higher than that of 1n- (neutron) and 1p- (proton) stripping reactions. The detailed analysis of α-γ coincidence measurement can be found in earlier publication [16].

The other particle-γ analyses are still undergoing, such as proton-γ, deuteron-γ and triton-γ coincidence analyses. The detailed results will be published elsewhere soon.

4 Summary

A ^{6}Li+^{89}Y experiment has been performed at the INFN-LNL in Italy. The light charged particles and γ rays are collected by EUCLIDES and GALILEO array, respectively. The particle-γ coincidence measurement is powerful in the identification of various reaction mechanisms in this experiment, such as α-γ analysis which is discussed in this article. More information on the reaction mechanisms is expected to be obtained from other particle-γ analysis, including proton-γ, deuteron-γ and triton-γ coincidence analyses.

Acknowledgments

We are grateful to the INFN-LNL staff for providing stable ^{6}Li beam throughout the experiment. This work is supported by the National Nature Science Foundation of China under Grant Nos. 11975040, 11475013, U1832130, 11375266, 11605114, 11575118 and 11505117 as well as the National Natural Science Foundation of Guangdong, China (2016A030310042).

References

[1] L. F. Canto et al., Phys. Rep., **424**, 1 (2006)
[2] L. F. Canto et al., Phys. Rep., **596**, 1 (2015)
[3] K. J. Kook et al., Phys. Rev. Lett., **122**, 102501 (2019)
[4] L. Jin and A. M. Moro, Phys. Rev. Lett., **122**, 042503 (2019)
[5] H. Kumawat et al., Phys. Rev. C., **86**, 024607 (2012)
[6] M. K. Pradhan et al., Phys. Rev. C., **83**, 064606 (2011)
[7] B. Wang et al., Phys. Rev. C., **90**, 034612 (2014)
[8] Md. Moin Shaikh et al., Phys. Rev. C., **90**, 024615 (2014)
[9] A. Lemasson et al., Phys. Rev. Lett., **103**, 232701 (2009)
[10] S. P. Hu et al., Phys. Rev. C., **91**, 044619 (2015)
[11] S. P. Hu et al., Phys. Rev. C., **93**, 014621 (2016)
[12] C. L. Guo et al., Phys. Rev. C., **92**, 014615 (2015)
[13] A. Shrivastava et al., Phys. Lett. B., **718**, 931 (2013)
[14] J. J. Valiente-Dobòn et al., INFN-LNL Annual Report 2014.
[15] D. Testov et al., Eur. Phys. J. A., **54**, 56 (2019).
[16] S. P. Hu et al., Nucl. Instru. Methods in Phys. Res. A., **914**, 64 (2019)
[17] H. P. Block, L. Hulstman, E. J. Kaptein, J. Blok, Nucl. Phys. A., **273**, 142 (1976).

Experimental studies of neutron-rich nuclei around *N* = 126 at KEK isotope separation system

Yutaka Watanabe[1,*], *Yoshikazu* Hirayama[1], *Momo* Mukai[2], *Murad* Ahmed[2], *Yutaka* Kakiguchi[1], *Hiroari* Miyatake[1], *Michihiro* Oyaizu[1], *Peter* Schury[1], *Michiharu* Wada[1], *Sota* Kimura[3], *Hironobu* Ishiyama[3], *Sun-Chan* Jeong[4], *Jun Young* *Moon*[4], and *Jin Hyung* Park[4]

[1]Wako Nuclear Science Center, Institute of Particle and Nuclear Studies, High-Energy Accelerator Research Organization (KEK), 2-1 Hirosawa, Wako, Saitama 351-0198, Japan
[2]Graduate School of Pure and Applied Science, University of Tsukuba, 1-1-1 Tennoudai, Tsukuba, Ibaraki 305-8571, Japan
[3]Nishina Center for Accelerator-Based Science, RIKEN, 2-1 Hirosawa, Wako, Saitama 351-0198, Japan
[4]Rare Isotope Science Project, Institute for Basic Science (IBS), 70, Yuseong-daero 1689-gil, Yuseong-gu, Daejeon, Republic of Korea

Abstract. Nuclear parameters such as lifetimes and masses of the waiting point nuclei of r-process nucleosynthesis are significant to investigate its astrophysical environment. However, the difficulty in the production of extremely neutron-rich nuclei at the 126 neutron closed shell makes their experimental studies unfeasible. Therefore, the theoretical nuclear models play crucial roles in the simulation of the r-process nucleosynthesis. The systematic nuclear spectroscopy of the neutron-rich nuclei around the 126 neutron closed shell provides significant inputs to those theoretical models to improve their predictability for the waiting point nuclei. We are developing KEK Isotope Separation System (KISS) to perform the systematic nuclear spectroscopy of those neutron-rich nuclei. The nuclei of interest are produced by multi-nucleon transfer reactions between ^{136}Xe and ^{198}Pt. The experimental study demonstrated its promising potential to produce them. We have successfully performed the β-γ spectroscopy and the laser ionization spectroscopy at KISS using the nuclear production by the multi-nucleon transfer reactions.

1 Introduction

A neutron star merger was discovered by observing gravitational waves in 2017 [1]. The discovery of kilonova coincident with the gravitational waves by observing optical and near-infrared emission made neutron star mergers a promising candidate of the r-process site [2]. However the analysis of the light curves and the spectra suggests contributions of several different mass ejection mechanisms from the neutron star merger [3]. To reveal mass and electron fraction of each mass ejecta, the simulation of the r-process nucleosynthesis including properties of relevant r-process nuclei is necessary.

Nuclear properties at the 126 neutron closed shell are essential for understanding the r-process in astrophysical nucleosynthesis [4]. The r-process goes through very neutron-rich unstable region, and the nuclei with neutron magic number 126 are considered to be waiting points on the r-process path, which can be progenitors of the peak around the mass number 195 on the observed solar r-abundance distribution. The lifetime and mass of those waiting point nuclei are important parameters to identify the astrophysical environments of r-process. However the area of those nuclei remains unexplored, because it is difficult to access by fusion, fission and fragmentation with stable projectiles and targets.

Thus the theoretical nuclear models play crucial roles in the simulation of the r-process nucleosynthesis. However, predicted β-decay half-lives of the waiting point nuclei show discrepancies by about one order of magnitude among different theoretical models. Evaluation of isotopic abundance pattern variances from uncertainties of theoretical half-lives indicates significant variations [5]. The reduction of the uncertainties of theoretical predictions of nuclear properties such as half-life and mass is an important issue for comprehensive understanding of r-process nucleosynthesis.

We are now promoting KISS project to produce neutron-rich nuclei around neutron number 126 and perform their systematic nuclear spectroscopy such as measurements of lifetime and mass, β-γ spectroscopy and laser spectroscopy to investigate astrophysical environments of r-process by providing experimental inputs to the theoretical models.

* Corresponding author: yutaka.watanabe@kek.jp

2 Nuclear-production around N = 126 by MNT reactions

Recently promising expectations have been focused on multi-nucleon transfer (MNT) reactions to produce neutron-rich nuclei around the 126 neutron closed shell both theoretically and experimentally. Using a macroscopic approach based on Langevin-type dynamical equation of motion, Zagrebaev and Greiner demonstrated that the MNT reaction between the Xe projectile and the Pb target gave large cross sections enough to produce those neutron-rich nuclei [6]. The experimental measurements of MNT reactions between the Xe projectile and the Pb target performed at ANL indicated a general agreement with the macroscopic calculations [7]. However, experimental data are scarce to evaluate the production of neutron-rich nuclei with neutron number 126, which are located at more neutron-rich side of those experimental data for the elements with the atomic number less than 80.

The GRAZING model [8], which is a semi-classical approach based on the single-particle transfer probabilities, also expects large cross sections of MNT reactions between the ^{136}Xe projectile and the ^{198}Pt target enough to access neutron-rich nuclei around the neutron number 126. We adopted this reaction system to efficiently produce those neutron-rich nuclei via proton pickup and neutron stripping channels. To investigate the feasibility of the reaction system, we performed an experimental study using the large acceptance magnetic spectrometer VAMOS++ [9] coupled to the EXOGAM γ-ray array [10] at GANIL [11, 12]. The projectile-like fragments (PLFs) produced in the MNT reactions were detected by the spectrometer and the isotopic distributions of the target-like fragments (TLFs) were deduced from those measurements. The detected γ-rays confirmed the production of corresponding TLFs. It was observed that the neutron-rich side around neutron number 126 in the isotopic distribution is mainly contributed by the components with total kinetic energy loss (TKEL) less than 75 MeV and the high-TKEL components contribute to the production of neutron-deficient isotopes. Recently, Karpov and Saiko demonstrated good agreements between our experimental data and the macroscopic calculations for the total isotopic distributions and decomposed ones in different TKEL values [13]. Figure 1 shows the cross sections to produce neutron-rich isotones with neutron number 126 by the MNT reactions between ^{136}Xe and ^{198}Pt (solid circles) and by the fragmentation reactions between ^{208}Pb (1GeV/nucleon) and Be [14]. Relatively large cross sections as compared with the fragmentation reactions suggests the advantage of the MNT reactions to access neutron-rich nuclei around neutron number 126 [12].

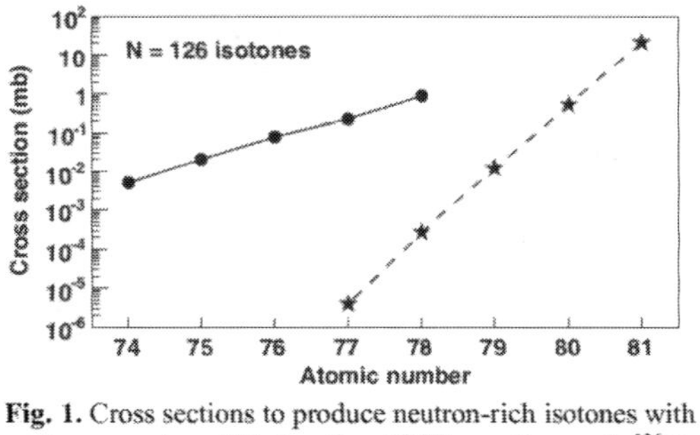

Fig. 1. Cross sections to produce neutron-rich isotones with neutron number 126 by the MNT reactions of ^{136}Xe (8 MeV/nucleon) + ^{198}Pt (solid circles) and by the fragmentation reactions between ^{208}Pb (1 GeV/nucleon) and Be [14]. (From [12])

3 KISS facility

We have been developing KEK Isotope Separation System (KISS) at RIKEN RIBF facility to separate a single nuclear species among products in the MNT reactions for the nuclear spectroscopy of the neutron-rich nuclei around neutron number 126 [15]. Figure 2 shows a schematic view of KISS. It consists of a gas cell system, a laser system, an extraction chamber, an Isotope Separator On-Line (ISOL) and a detection system for the

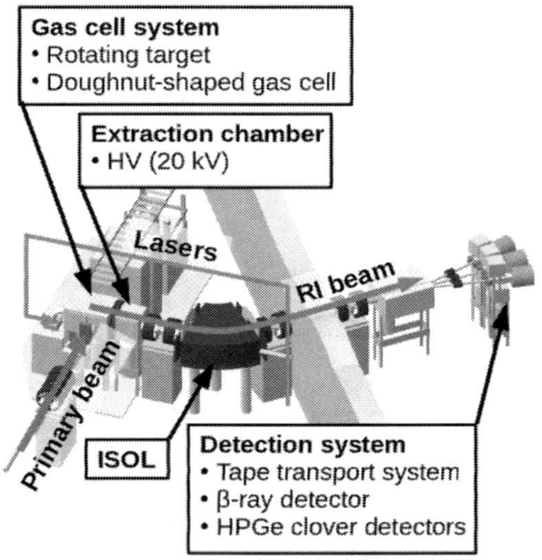

Fig. 2. Schematic view of KEK Isotope Separation System (KISS).

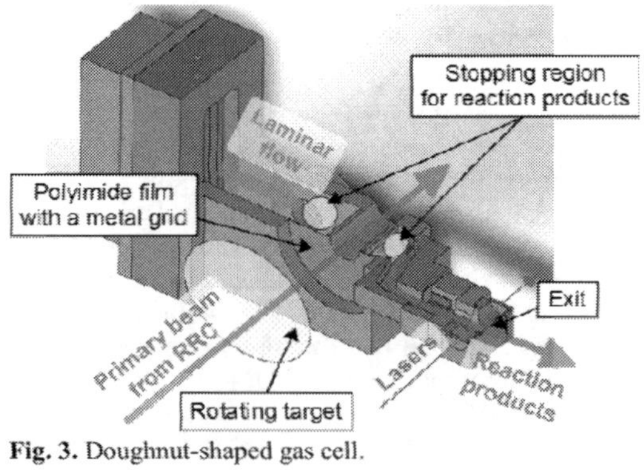

Fig. 3. Doughnut-shaped gas cell.

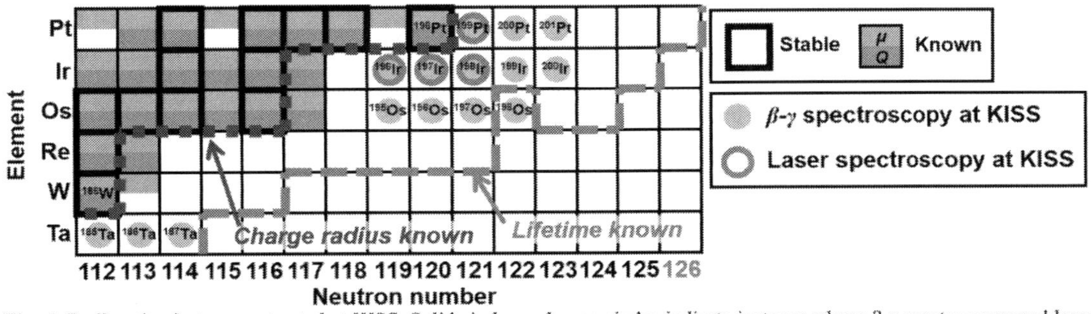

Fig. 4. Radioactive isotopes extracted at KISS. Solid circles and open circles indicate isotopes whose β-γ spectroscopy and laser ionization spectroscopy was performed, respectively. Isotopes with known magnetic dipole moments (μ) and electric quadrupole moments (Q) are indicated by boxes filled in upper and lower parts, respectively.

β–γ spectroscopy. The gas cell system encloses a doughnut-shaped gas cell [16] shown in Fig. 3. The rotating target of ^{198}Pt foils (12.5 mg/cm^2) is bombarded by the ^{136}Xe beam (10.75 MeV/nucleon) from RIKEN Ring Cyclotron after energy-degradation by three layers of titanium foils (3 μm × 3) to optimize the reaction energies in the target. The TLFs produced by MNT reactions are ejected from the target and go into the gas cell through a window made of a polyimide film (5 μm). They are thermalized and neutralized by argon gas filled in the gas cell at the pressure of 88 kPa. Those neutral atoms are transported by laminar gas flow, and are element-selectively ionized by two-color two-step laser resonance ionization technique just before the exit of the gas cell. The primary beam passes through the beam pipe with the diameter of 15 mm through the doughnut-shaped gas cell, and is stopped in the water-cooled beam dump. Thus the beam pipe prevents the primary beam from entering the gas cell to suppress the plasma effect induced in the gas by the energy deposit of the primary beam, which would neutralize the laser-ionized atoms of interest resulting in the reduction of their extraction yields. The extracted singly charged ions are transported by multipole RF ion guides and are accelerated by application of a high voltage of 20 kV. They are mass-separated with the ISOL method, and finally they are transported to the detection system for the β–γ spectroscopy.

The detection system consists of a tape transport system, a low-background gas counter called Multi-Segmented Proportional Gas Counter (MSPGC) [17] and four High-Purity Germanium (HPGe) clover detectors. The element-selected and mass-separated ions are implanted in the aluminized Mylar tape stretched vertically. The KISS beam is pulsed with certain beam-on and off periods specific for the expected half-lives for β-decay measurements. The tape is moved in synchronization of the beam pulsing to remove the background from the activity of the accumulated daughter nuclei. The tape is surrounded by the MSPGC to detect β-rays, which is a two-layered proportional gas counter. Each layer is placed concentrically with the vertical length of 200 mm, and has 16 angularly divided segments. To measure the lifetimes of rare reaction products, high efficiency for the β-ray detection and the low background rate are essential. The MSPGC achieves 40% detection efficiency for the β-rays with Q_β of 1 MeV including the geometrical acceptable solid angle of 80% of 4π. The achieved background rate is as low as 0.1 cps. The MSPGC is surrounded by four HPGe clover detectors for the γ-ray detection with compact geometry of around 5 cm spacing from the tape. The efficiency for the photo peak detection is 10% for γ-rays of 1 MeV.

4 Experimental results

Figure 4 summarizes experimental results so far at KISS. Solid circles indicate radioactive isotopes which were extracted so far. Twelve kinds of neutron-rich isotopes, $^{199-201}$Pt, $^{196-200}$Ir and $^{195-199}$Os were extracted using the ^{198}Pt target, and three kinds of neutron-rich isotopes, $^{185-187}$Ta, were extracted using a natural W target [18]. The extracted isotopes were identified by measuring half-lives of their β-decays. The β–γ spectroscopy of those extracted isotopes was performed.

The dashed lines in Fig. 4 indicate the boundary of known lifetimes before KISS experiments. We have measured the half-life of ^{198}Os for the first time.

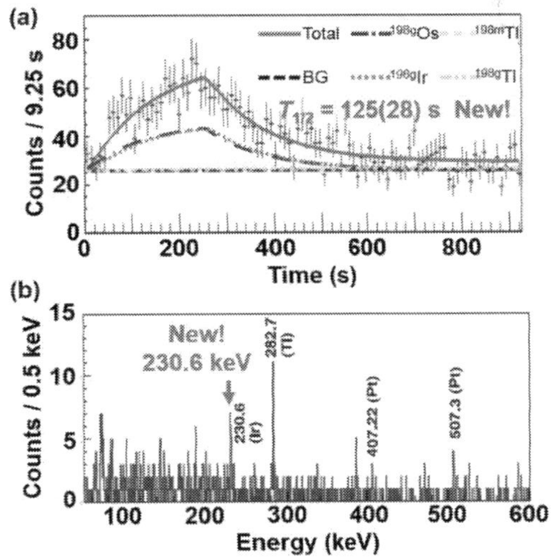

Fig. 5. (a) Growth and decay curves of the β-ray detection for ^{198}Os. (b) γ-ray energy spectrum coincident with β-ray detection for ^{198}Os. (From [19])

Figure 5(a) indicates a time spectrum of detected counts of the MSPGC for the ^{198}Os extraction with the beam-on and off periods of 252 s and 672 s, respectively. The fitting result with growth and decay curves determined the half-life of the ^{198}Os ground state 125(28) s [19]. Figure 5(b) indicates a γ-ray energy spectrum coincident with β-ray detection. A γ-ray peak at the energy of 230.6 keV was newly identified as a β-delayed γ-transition of ^{198}Os [19]. The branching ratios and log-*ft* values for β-decay branches of ^{198}Os and ^{198}Ir decay chains were evaluated, indicating intense β-decays among their ground states. Their small log-*ft* values suggested spin-parity of 0^- or 1^- for the ground state of ^{198}Ir [19].

Figure 6(a) shows the systematics of half-lives of Os isotopes. Solid circles indicate half-lives of $^{196-198}$Os measured at KISS. Open symbols indicate predictions by different theoretical models and solid squares indicate earlier experimental values. It is obvious that any single theoretical models cannot reproduce all experimental half-life values for osmium isotopes. It would be due to the difficulty of theoretical prediction of half-lives for the neutron-rich nuclei around the 126 neutron closed shell. Figure 6(b) shows the theoretical calculations of single-particle levels of ^{200}W [20]. It indicates the competition between the first-forbidden transitions and the Gamow-Teller transitions. Such a competitive condition makes it difficult to theoretically predict half-lives in this nuclear region. More experimental and theoretical studies are required for comprehensive understanding of nuclear structure of those nuclei and for more reliable prediction of nuclear properties for r-process nuclei.

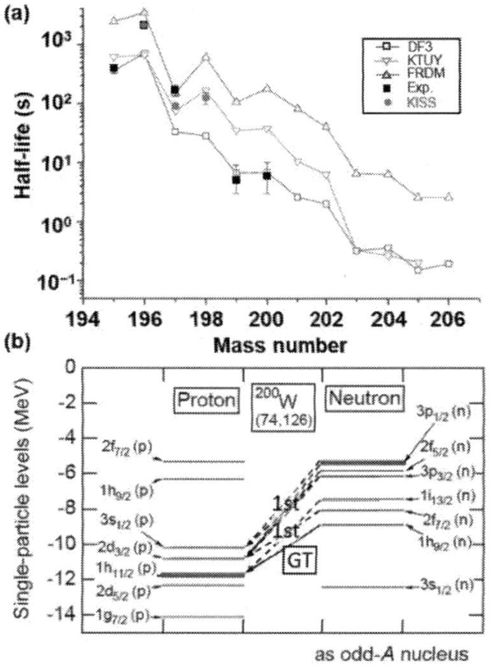

Fig. 6. (a) Systematics of half-lives of Os isotopes. (b) Theoretical calculations of single-particle levels of ^{200}W [20].

Open circles in Fig. 4 indicate isotopes, ^{199}Pt and $^{196-198}$Ir, for which the laser ionization spectroscopy was performed at KISS to measure their hyperfine structures (HFS's). HFS measurements provide electro-magnetic moments, which have information about nuclear wave-functions. Isotope shift, which has information of charge radius, is also obtained, providing hints for the nuclear deformation. The platinum isotopes are discussed to show characteristic shape transition among prolate, triaxial, oblate and finally spherical by approaching to the 126 neutron closed shell [21]. We are performing in-gas-cell laser ionization spectroscopy at KISS for study of nuclear structure of neutron-rich nuclei around neutron number 126. Figure 7 shows HFS spectrum by detecting β-rays for ^{199}Pt [22]. The spectrum is contributed by the ground state and the isomeric state of ^{199}Pt since both states are produce in the MNT reactions and extracted at KISS. Using the detected γ-rays of 392 keV, which are associated with the isomeric state decay of ^{199}Pt, the spectrum is decomposed to contributions from the ground state ^{199g}Pt and the isomeric state ^{199m}Pt of ^{199}Pt by considering the response function of the resonance curve measured for the stable

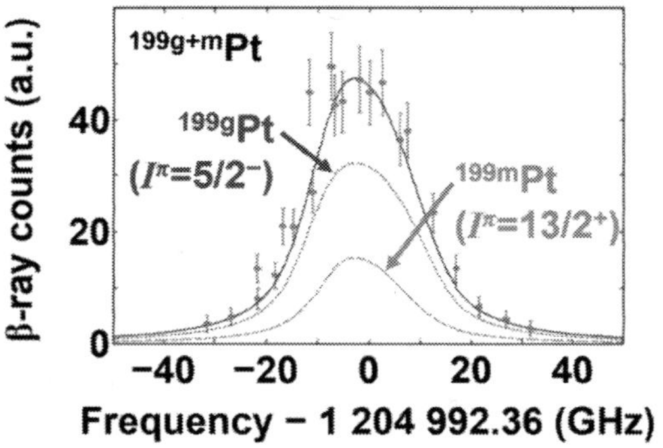

Fig. 7. HFS spectrum by detecting β-rays for ^{199}Pt [22].

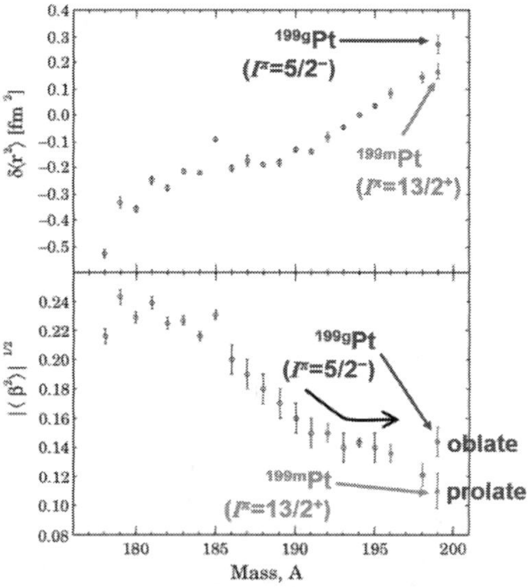

Fig. 8. Systematics of mean square charge radius variation $\delta\langle r^2\rangle$ (upper panel) and quadrupole deformation parameter $|\langle\beta^2\rangle|^{1/2}$ (lower panel) as functions of mass number of Pt isotopes [22].

even-even nucleus ^{198}Pt.

Figure 8 shows the systematics of mean square charge radius variation (upper panel) and quadrupole deformation parameter (lower panel) as functions of mass number of Pt isotopes. The solid circles indicate the ground state values measured at KISS, open circles indicate earlier experimental data. The solid triangle indicate the isomeric state value measured at KISS. The value of the deformation parameter for ^{199g}Pt is consistent with the constant value in the systematic trend for Pt isotopes with mass number lager than 192, indicating oblate deformation. The smaller deformation parameter value for ^{199m}Pt is consistent with the theoretical prediction, which indicates the prolate deformation [22].

5 R&D and future plan of KISS

For more precise studies of wave-function and deformation, high-precision in-gas-jet laser spectroscopy [23] is required. A new laser system, which consists of a narrow-band laser diode laser, a dye amplifier and intense YAG laser was installed at KISS. A de Laval nozzle was placed at the gas cell exit to form the gas jet. An S-shaped multipole RF ion guide was installed at the downstream of the gas cell to irradiate the excitation laser collinearly along the gas jet with the ionization laser running perpendicularly many times reflected by mirrors in both sides to make the ionization region wide. The off-line test of the in-gas-jet laser ionization spectroscopy using the evaporated atoms from the platinum filament resulted in the Full Width at the Half-Maximum (FWHM) of 0.6(1) GHz in the HFS spectrum for ^{194}Pt. The FWHM is obviously reduced from the present value of 12 GHz for in-gas-cell laser ionization spectroscopy. The in-gas-jet laser ionization system is ready for on-line experiments of high-precision laser spectroscopy with refractory elements.

Because the extraction yields become smaller as approaching more and more neutron-rich nuclei in the vicinity of the neutron magic number of 126, still lower background rate for the β-ray detection is necessary in order to perform their lifetime measurements. The present MSPGC achieved the background rate of 0.1 cps by enabling two-dimensional tracking on the horizontal plane. We are now adapting the MSPGC to enable three-dimensional tracking using resistive (3 kΩ/cm) carbon wires for anode wires stretched vertically. The resolution of 2 mm in FWHM was achieved for the vertical position determination by measuring the charges at both ends of the anode wires. It is good enough to reduce the background rate to the level of 0.01 cps.

In addition to the β–γ spectroscopy, a new branch was constructed in the KISS beam line for the mass measurements. We will use a multi-reflection time-of-flight mass spectrograph (MRTOF-MS) for the mass measurements, which achieved the mass resolving power of 150 000 in the atomic mass region of 200 [24]. An MRTOF-MS with a helium gas cell ion cooler, a multipole RF ion guide and a mult-trap ion buncher, which accumulate and thermalize the ions extracted from the KISS and bunch them for the MRTOF-MS, were installed already. They are now under commissioning towards the on-line experiment in December, 2019.

6 Summary

Systematic nuclear spectroscopy around the 126 neutron closed shell are significant to improve the theoretical predictions of nuclear properties such as lifetimes and masses relevant to r-process nucleosynthesis for identification of its astrophysical environments. The production of those neutron-rich nuclei by MNT reactions has been recently focused on both theoretically and experimentally. However the experimental data had been scarce. We have performed the MNT measurements of ^{136}Xe and ^{198}Pt. The results indicated the advantage of the MNT reactions over fragmentation reactions for production of the neutron-rich nuclei around neutron number 126. Encouraged by this results, we have been promoting KISS project, where the MNT reactions of ^{136}Xe and ^{198}Pt are used to produce those neutron-rich nuclei for their systematic nuclear spectroscopy. The gas cell, the laser resonant ionization and the ISOL are used for efficient collection and separation of those MNT reaction products. We have extracted so far fifteen kinds of neutron-rich radioactive isotopes, $^{199-201}$Pt, $^{196-200}$Ir, $^{195-208}$Os and $^{185-187}$Ta, and performed lifetime measurements and β–γ spectroscopy of those nuclei. The laser ionization spectroscopy was performed for ^{199}Pt and $^{196-198}$Ir. For the precise laser ionization spectroscopy, new in-gas-jet laser ionization system was installed with successful reduction of resonance width in the off-line test. The MSPGC was adapted to enable three-dimensional tracking to reduce the background rate for β-ray detection down to the level of 0.01 cps. It would make possible to perform the β–γ spectroscopy for neutron-rich nuclei produced in one order of magnitude smaller cross sections than the present setup. The preparation of the MRTOF-MS and the relevant devices for mass measurements of neutron-rich nuclei around the 126 neutron closed shell is promoting at KISS. It would be ready soon for the on-line measurements.

The experiments at the KISS were performed at the RI Beam Factory operated by RIKEN Nishina Center and CNS, University of Tokyo. The authors gratefully acknowledge the staff of accelerators for their support. This work was supported by JSPS KAKENHI Grants No. 232440660, No. 24740180, No. 26247044, No. 15H02096, No. 17H01132 and No. 18H03711.

References

1. B.P. Abbott *et al.*, Phys. Rev. Lett. **119**, 161101 (2017)
2. I. Arcavi *et al.*, Nature **551**, 64 (2017)
3. M. Tanaka *et al.*, Publ. Astron. Soc. Japan **69**, 102 (2017)

4. E.M. Burbidge *et al.*, Rev. Mod. Phys. **29**, 547 (1957)
5. M.R. Mumpower *et al.*, Prog. Part. Nucl. Phys. **86**, 86 (2016)
6. V. Zagrebaev and W. Greiner, Phys. Rev. Lett. **101**, 122701 (2008)
7. J.S. Barrett *et al.*, Phys. Rev. C **91**, 064615 (2015)
8. A. Winther, Nucl. Physics A **572**, 191 (1994); **594**, 203 (1995)
9. M. Rejmund *et al.*, Nucl. Instrum. and Methods A **646**, 184 (2011)
10. J. Simson *et al.*, Acta Phys. Hung. new Ser.: Heavy Ion Phys. **11**, 159 (2000)
11. Y.H. Kim *et al.*, EPJ Web of Conferences **66**, 03044 (2014)
12. Y.X. Watanabe *et al.*, Phys. Rev. Lett. **115**, 172503 (2015)
13. A.V. Karpov and V.V. Saiko, Phys. Rev. C **96**, 024618 (2017)
14. T. Kurtukian-Nieto *et al.*, Phys. Rev. C **89**, 024616 (2014)
15. Y. Hirayama *et al.*, Nucl. Instrum. and Methods B **353**, 4 (2015)
16. Y. Hirayama *et al.*, Nucl. Instrum. and Methods B **412**, 11 (2017)
17. M. Mukai *et al.*, Nucl. Instrum. and Methods A **884**, 1 (2018)
18. P.M. Walker *et al.*, RIKEN Accel. Prog. Rep. **51**, 72 (2018)
19. Y. Hirayama *et al.*, Phys. Rev. C **98**, 014321 (2018)
20. H. Koura, JAEA-Review **2010-056**, 83 (2010)
21. P. Sarriguren *et al.*, Phys. Rev. C **77**, 064322 (2008)
22. Y. Hirayama *et al.*, Phys. Rev. C **96**, 014307 (2017)
23. Yu. Kudryavtsev *et al.*, Nucl. Instrum. and Methods B **297**, 7 (2013)
24. P. Schury *et al.*, Phys. Rev. C **95**, 011305(R) (2017)

EPJ Web of Conferences **223**, 01070 (2019) https://doi.org/10.1051/epjconf/201922301070
NSD2019

Shape transitions between and within Zr isotopes

V. Werner[1,*], *W.* Witt[1], *T.* Beck[1], *N.* Pietralla[1], *M.* Boromiza[2], *C.* Clisu[2], *C.* Costache[2], *I. E.* Dinescu[2], *A.* Ionescu[1], *P. R.* John[1], *P.* Koseoglou[1], *N.* Margineăn[2], *R.* Margineăn[2], *C.* Mihai[2], *R. E.* Mihai[2], *A.* Mitu[2], *A.* Negret[2], *C. R.* Nita[2], *A.* Olacel[2], *S.* Pascu[2], *A.* Serban[2], *C.* Sotty[2], *L.* Stan[2], *R.* Suvaila[2], *S.* Toma[2], *A.* Turturica[2], *S.* Ujeniuc[2], and *J.* Wiederhold[1]

[1]Institut für Kernphysik, Technische Universität Darmstadt, Schlossgartenstr. 9, 64289 Darmstadt, Germany
[2]"Horia Hulubei" National Institute for Physics and Nuclear Engineering, 077125, Bucharest-Magurele, Romania

Abstract. The Zirconium isotopes across the N=56,58 neutron sub-shell closures have been of special interest since years, sparked by the near doubly-magic features of ^{96}Zr and the subsequent rapid onset of collectivity with a deformed ground-state structure already in ^{100}Zr. Recent state-of-the-art shell model approaches did not only correctly describe this shape-phase transition in the Zr isotopic chain, but also the coexistence of non-collective structures and pronounced collectivity especially in ^{96,98}Zr. The isotope ^{98}Zr is located on the transition from spherical to deformed ground state structures. We summarize recent experimental work to obtain the B(E2) excitation strengths of the first 2^+ state of ^{98}Zr, including a new experiment employing the recoil-distance Doppler-shift method following a two-neutron transfer reaction.

1 Introduction

Shape coexistence in nuclei is a well-known phenomenon since decades, but in the last years it has been observed to manifest itself in various forms all across the nuclear chart. Since nuclei are not bound to one specific configuration of nucleons, coupling among themselves in various orbitals (the valence space), they are also not prone to one specific deformation, which is typically given in terms of the classic shape parameters β and γ. Certain microscopic configurations may correspond to different shape parameters. In addition, nuclei in excited states do not necessarily remain within the configuration that they find themselves in their ground states. In turn, already in the ground state, more than one configuration may have significant contribution to the states' total wave function.

With these initial considerations, and restricting to two coexisting shape configurations, two cases of shape coexistence can be classified: first, the case where two differently deformed shapes mix into a given state or sets of states; secondly, the case where certain states or a set of states have distinct shapes. Considering the nuclear potential as a function of deformation, e.g., β deformation, the amount of mixing between different shapes can be ascribed to the interplay of the height of the potential well (barrier) between two potential minima corresponding to different shapes, and of the interaction between both shape configurations (mixing). The two cases are shematically illustrated in Fig. 1, with a spherical potential minimum at $\beta = 0$, and a deformed one at $\beta = \beta_0$. In case of a high barrier in between, sets of eigenstates should exist purely in either configuration. Only if the mixing interaction is strong enough to overcome the barrier, states from both configuration can mix with each other into eigenstates of the system. In the other extreme case, the potential well could be so low, that even with little mixing interaction all eigenstates of the system will be mixtures of both deformations, which become indistinguishable. This is the case, for example, in models such as X(5) [1], where a square well potential is used to approximate the low-barrier situation in Fig. 1, neglecting the small potential barrier between the two minima.

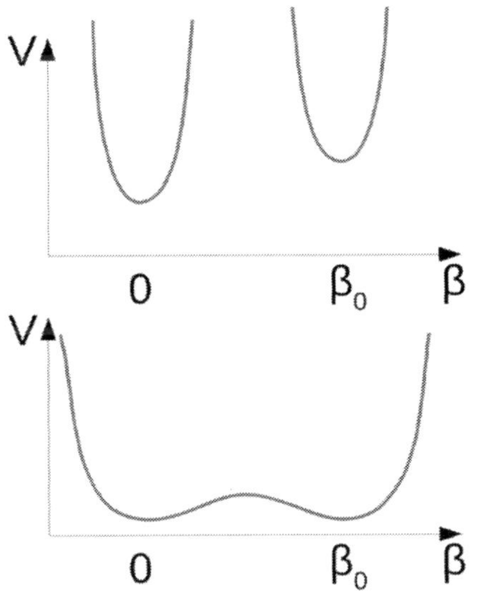

Figure 1. Schematics of nuclear potentials with local minima at β=0 and $\beta = \beta_0$, (top) in case of a high potential barrier and (bottom) in case of a low potential barrier.

*e-mail: vw@ikp.tu-darmstadt.de

https://doi.org/10.1051/epjconf/201922301070

The extreme and intermediate cases of the above-mentioned extremes have recently been worked out in Ref. [2]. In the following, we will concentrate on the high-barrier case, where eigenstates are located almost purely in either potential minimum, spherical or deformed. Such a case has recently been identified in ^{96}Zr [3]. In this case a seniority-scheme dominated, non-collective spherical ground state structure exists, coexisting with an excited-state structure which is much more collective and eventually somewhat deformed - the presently available spectroscopic data has recently been summarized in Ref. [4]. This situation has successfully been reproduced within a Monte Carlo shell model (MCSM) approach [5]. The first excited state of ^{96}Zr is located at 1750 keV, with a B(E2) value to the ground state of only 2.3 W.u. [6]. On the contrary, in ^{100}Zr a deformed ground-state structure is known, evident from a low-lying first-2^+ state at 212 keV connected by a strong B(E2) value of 75 W.u. to the ground state [7]. For ^{98}Zr, until recently [8, 9] only the rather high excitation energy of the 2_1^+ state of 1223 keV has been known, but it was not clear whether the assumingly coexisting more collective structure already mixes into its wave function. The latter can be determined from the measurement of its B(E2) transition probability to the ground state. We will summarize the recent experiments to obtain this B(E2) value and present the first estimate from another, complimentary experiment.

2 Recent data on the 2_1^+ state of ^{98}Zr

2.1 Previous works

The B(E2) value of the decay of the 2_1^+ state of ^{98}Zr has recently been constrained by two experiments. In a Coulomb excitation experiment using a radioactive ^{98}Zr beam from Cf fission at Argonne National Laboratory an upper limit for the B(E2) value has been obtained [8], which, combined with a lower limit from a fast-timing experiment [10], constrained the value to 1.83 W.u. $< B(E2; 2_1^+ \rightarrow 0_1^+) <$ 11 W.u., clearly constraining the state to a rather spherical shape, in contrast to the 2_1^+ state of ^{100}Zr with a B(E2) value of 75 W.u.

Another radioactive-beam experiment at GANIL used in-beam U fission reactions to obtain ^{98}Zr and employed the recoil-distance Doppler-shift (RDDS) technique for a direct lifetime measurement [9]. The resulting lifetime further constrained the $B(E2)$ value of interest to 2.9(6)(2) W.u., where the error bars from the measured lifetime and known branching ratio of the state are given separately.

2.2 RDDS after 2n-transfer

We employed another technique to produce ^{98}Zr in its first-excited 2^+ state, namely, a 2-neutron-transfer reaction. A stable ^{18}O beam was delivered at an energy of 49 MeV by the IFIN-HH 9-MV tandem accelerator. The energy was chosen following an excitation function and considering the reaction kinematics, such that recoiling Zr nuclei had maximum momentum in beam direction. The target was a 0.8 mg/cm^2 thick Zr foil, enriched to ^{96}Zr by 57.4 %. The recoiling Zr nuclei were subsequently stopped in an about 10 mg/cm^2 thick Au foil. The distance between both foils was varied using the IFIN plunger device. In this stable-beam experiment the production cross section of ^{98}Zr in the ^{96}Zr(^{18}O,^{16}O)^{98}Zr reaction amounted to less than 10 % of the total cross section, which, even at this sub-Coulomb-barrier energy was dominated by fusion-evaporation reactions. This includes reactions of the beam with the Zr material, but also with oxygen inclusions in the target. Therefore, the measurement suffered from significant background, which forced us to modify the usual RDDS method of measuring the ratio of Doppler-shifted and -unshifted contributions in the spectra. However, in the full spectrum of all distances, the Doppler-shifted components were visible and from their Doppler-shifts an average recoil velocity of v/c = 0.9(1) % was determined, consistent with kinematics calculations.

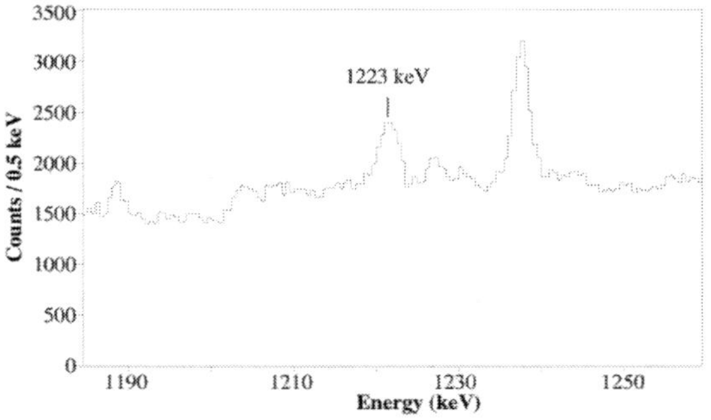

Figure 2. Sum spectrum of the forward-ring detectors, taken at a distance of 74 μm. The stopped peak component from the $2_1^+ \rightarrow 0_1^+$ transition of ^{98}Zr is marked. The Doppler-shifted component is not distinguishable from the background. (From [12].)

The Doppler-shifted peaks from in-flight γ-ray decays of the 2_1^+ state, detected by the ROSPHERE HPGe detector array [11], were too weak and too Doppler-broadened to be distinguished from the background in the individual particle-gated γ-ray spectra for each plunger distance setting and each detector ring (rings at 37°, 90°, and 143° with respect to the beam axis were available). However, the peaks from the stopped component of the decay were visible and, after normalization to other reaction channels, allowed us to obtain a decay curve. A sample spectrum is shown in Fig. 2. Data was taken at an initial three target-to-stopper distances, namely at 25 μm, 74 μm, and 148 μm, where an offset (difference between first electrical contact and actual zero-distance), which has been obtained from extrapolation of the capacitance calibration, is taken into account. After the first three distances the target needed to be cleaned in order to achieve short distances, hence, the distance calibration changed. After re-alignment another distance setting around 25 μm setting was measured. The offset between the two 25 μm settings was determined from the strong fusion-evaporation reaction producing the 2_1^+ state of ^{110}Cd with a well-known lifetime, resulting in distances of 25 μm and 30 μm, respectively. The use of the known lifetime in ^{110}Cd, hence, allowed to shift the two parts of the experiment to each other and to obtain a total decay curve. For illustration, the decay curve of the 2_1^+ state of ^{110}Cd is shown in Fig. 3. In addition, the total peak areas from transitions from fusion-evaporation and Coulomb-excitation reactions were used to normalize spectra with respect to the total amount of beam impinging on the target per distance setting.

In addition to the RDDS experiment, data was taken with a 9.2 mg/cm^2 thick ^{96}Zr target from the same material as the RDDS target, in which the recoiling ^{98}Zr nuclei were stopped, hence, minimizing Doppler shift effects. This data was used for a γ-singles and $\gamma\gamma$-coincidence analysis in order to investigate feeding of the 2_1^+ state of ^{98}Zr by other states populated in the 2n-transfer reaction. Only low-spin states (up to spin $J^\pi = 4^+$) were found to be populated. All states which were observed to feed into the 2_1^+ state had either negligible influence on the lifetime of interest from structure-based assumptions made on their lifetimes and feeding intensities, or had known lifetimes which were taken into account in the decay-curve analysis.

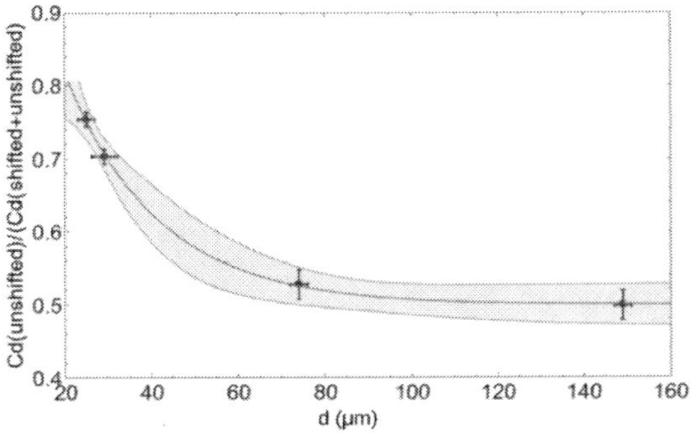

Figure 3. Determination of the distances (especially the 29-μm distance), making use of the known lifetime of the 2_1^+ state of ^{110}Cd. The values for the settings at 25, 74, and 148 μm agree well with the calibrated values, and for the point after realignment of the target 29 μm are obtained. (From [12].)

The full analysis procedure is found in [12] and final results will be given in a forthcoming publication. A preliminary value for the $B(E2; 2_1^+ \rightarrow 0_1^+)$ reduced transition probability, obtained in this RDDS experiment, is 2.8(10) W.u.

3 Discussion

The $B(E2)$ value obtained in the RDDS experiment at IFIN agrees well with the previous limits of Refs. [8, 10], as well as the $B(E2)$ value obtained in Ref. [9] of 2.9(6)(2) W.u., having a similar combined uncertainty. Therefore, in the following we will use the average value of the present work and [9] of 2.8(6) W.u. This value is incorporated into the systematics shown in Fig. 4, showing the energy differences of first and second 0$^+$ and 2$^+$ states, and the $B(E2; 2_1^+ \rightarrow 0_1^+)$ values over the relevant Zr isotopes. Firstly, the new $B(E2)$ value for ^{98}Zr is at the lower edge of the pervious limiting values, and clearly shows a non-collective excitation of the 2_1^+ state, similar to the lighter Zr isotopes, and in stark contrast to ^{100,102}Zr, where the transition is strongly collective (see Fig. 4). Secondly, the systematics of energy differences reveal that, considering a two-level crossing scenario, the 2$^+$ states are closest already in ^{98}Zr, whereas the 0$^+$ states are closest in ^{100}Zr, as shown in the bottom panel of Fig. 4. This means that the crossing of weakly and strongly collective structures occurs at smaller N for the 2$^+$ states, and the 2_1^+ state of ^{98}Zr could, in fact, be the state that belongs to the collective structure. The latter is corroborated by the strong $B(E2; 2_1^+ \rightarrow 0_2^+)$ transition of 28(6) W.u. (using the average of this work and Ref. [9]). Hence, the ground state of ^{98}Zr is the non-collective counterpart of the ground states of the lighter Zr isotopes, and the first excited 0_2^+ state has significant collectivity, with the 2_1^+ state built on it as a collective excitation.

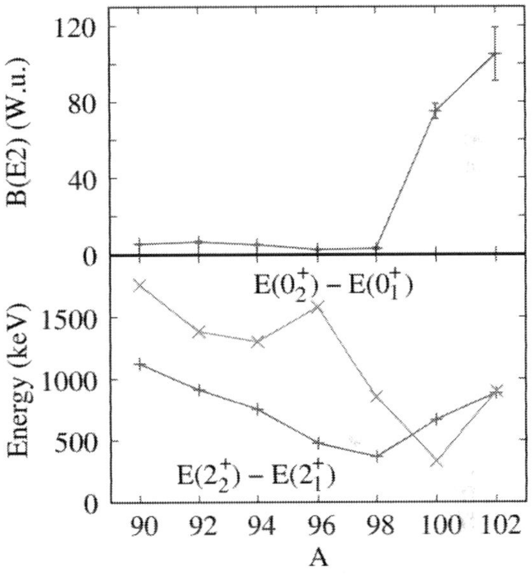

Figure 4. Top panel: Evolution of the $B(E2)$ values from the first 2$^+$ state over the Zr isotopic chain over mass number A. Bottom panel: Energy differences between the $0_{1,2}^+$ states and between the $2_{1,2}^+$ states in the Zr isotopes.

However, the $B(E2)$ value of 28 W.u. within the excited, collective structure of ^{98}Zr does not quite match the collectivity of the ground-state structure of ^{100}Zr with a corresponding $B(E2)$ value of 75 W.u. This observation triggers the idea that the collective structure itself may undergo an evolution across ^{98}Zr, while it becomes the ground-state structure, presumably changing place in energy with the non-collective structure. It is, however, an open question whether the non-collective structure actually survives in ^{100}Zr, or whether it starts mixing with other configuration and therefore looses the distinct signatures that it has up to ^{98}Zr.

This scenario has recently been theoretically fleshed out by Gavrielov, Leviatan and Iachello [13] within an algebraic approach. The finding of their work was the occurrence of an intertwined phase transition in the Zr isotopes, meaning the simultaneous occurrence of two quantum phase transitions. Firstly, two distinct and weakly mixing configurations exist - a non-collective one based on the closures of proton ($N = 40$) and neutron ($N = 50, 56$) (sub)-shells, and the other a collective configuration which in the framework of Ref. [13] starts as a spherical vibrator [U(5) in terms of the algebraic model]. A type-II phase transition occurs as the non-collective and the collective structures exchange place in energy, without substantial mixing, which corresponds to the high-barrier case sketched in Fig. 1. Moreover, due to the different energy spacings between levels within the respective configurations, the first excited 2^+ states already crossed in ^{98}Zr, whereas the lowest, i.e., the 0^+ states of the respective configurations exchanged position in ^{100}Zr. This is evident from the collective $B(E2; 2_1^+ \rightarrow 0_2^+)$ value in ^{98}Zr, whereas there is little overlap between the two configuration, witnessed by the weak observed $B(E2; 2_1^+ \rightarrow 0_1^+)$ value in this isotope.

However, a structural change within the collective structure is observed across $N = 60$. In the framework of Ref. [13] this change is described as a type-I phase transition, i.e., a quantum phase transition within the valence space given for this configuration. From the observed $B(E2)$ values the ground-state configurations of ^{100,102}Zr with substantial $B(E2)$ values should be well-deformed, hence, the transition is from the vibrational U(5) limit to the well-deformed rotor limit SU(3). This particular transition corresponds to the low-barrier case depicted in Fig. 1, which is a soft potential on the transition from a spherical-centered vibrational case and a potential with a distinct minimum at a given deformation value β_0, corresponding to the approximation of the X(5) model.

4 Conclusions

We performed a RDDS lifetime measurement on the first excited 2^+ state of ^{98}Zr. Our results are in good agreement with recent experiments. The data strongly supports the recently established picture of a intertwined quantum phase transition in the Zr isotopes across $N = 60$. A non-collective spherical structure exchanges with a collective ("intruder") structure at ^{98,100}Zr through a type-II phase transition, while simultaneously the intruder structure undergoes a type-I phase transition from presumably a spherical vibrator to a axially-deformed rotor. To further disentangle collective and non-collective structures in ^{98}Zr larger sets of electro-magnetic matrix elements among its low-lying states, are needed, especially for the lowest 2^+ states, e.g., from high-statistics multi-step Coulomb-excitation data.

Acknowledgments

We thank A. Leviatan, N. Gavrielov and F. Iachello for many discussions on the topic of quantum phase transitions in the Zr isotopes. Further, we thank the crew of the IFIN tandem accelerator for excellent beam conditions. This work was supported by German BMBF und grant no. 05P19RDFN1 and DFG grant SFB 1245, and by the Romanian MCI Project PN 19 06 01 02.

References

[1] F. Iachello, Phys. Rev. Lett. **87**, 052502 (2001).

[2] R. Budaca and A. I. Budaca, Eur. Phys. Lett. **123**, 42001 (2018).

[3] C. Kremer, S. Aslanidou, S. Bassauer, M. Hilcker, A. Krugmann *et al.*, Phys. Rev. Lett. **117**, 172503 (2016).

[4] W. Witt, N. Pietralla, V. Werner, and T. Beck, Eur. Phys. J. A **55**, 79 (2019).

[5] T. Togashi, Y. Tsunoda, T. Otsuka, and N. Shimizu, Phys. Rev. Lett. **117**, 172502 (2016).

[6] G. Kumbartzki, N. Benczer-Koller, J. Holden, G. Jakob, T.J. Mertzimekis *et al.*, Phys. Lett. B **562**, 193 (2003).

[7] B. Singh, Nucl. Data Sheets **109**, 297 (2008).

[8] W. Witt, V. Werner, N. Pietralla, M. Albers, A. D. Ayangeakaa *et al.*, Phys. Rev. C **98**, 041302(R) (2018).

[9] P. Singh, W. Korten, T. W. Hagen, A. Görgen, L. Grente *et al.*, Phys. Rev. Lett. **121**, 192501 (2018).

[10] S. Ansari, J.-M. Régis, J. Jolie, N. Saed-Samii, N. Warr *et al.*, Phys. Rev. C **96**, 054323 (2017).

[11] D. Bucurescu, I. Căta-Danil, G. Ciocan, C. Costache, D. Deleanu *et al.*, Nucl. Instrum. Methods A **837**, 1 (2016).

[12] W. Witt, Dissertation, TU Darmstadt, 2019.

[13] N. Gavrielov, A. Leviatan, and F. Iachello, Phys. Rev. C **99**, 064324 (2019).

Reactions with Exotic Nuclei at Near and Sub-barrier Energies

Lei Yang[1], *Chengjian* Lin[1,2,*], *Huiming* Jia[1], *Nanru* Ma[1**], *Dongxi* Wang[1], *Feng* Yang[1], *Lijie* Sun[1***], *Xinxing* Xu[1], *Peiwei* Wen[1****], and *Fupeng* Zhong[1,2]

[1]China Institute of Atomic Energy, P. O. Box 275(10), Beijing 102413, China
[2]Department of Physics, Guangxi Normal University, Guilin 541004, China

Abstract. We will introduce some new results derived recently by the nuclear reaction group at China Institute of Atomic Energy, on the properties of the optical model potentials of neutron-halo ^{6}He system and reaction mechanisms induced by proton-drip line nucleus ^{17}F at energies around the Coulomb barrier. For the study of optical model potentials of exotic nuclear system, we proposed a novel method, i.e., the transfer reaction method. This method has been applied to extract the optical potentials of neutron-halo ^{6}He+^{209}Bi system by measuring the one-proton transfer reactions induced by ^{7}Li on a ^{208}Pb target. A complete picture of threshold anomaly behavior was obtained in the ^{6}He+^{209}Bi system for the first time, where a decreasing trend of the depth of the imaginary potential is observed in the deep sub-barrier region, and the reaction threshold energy is extracted. Moreover, results show that the dispersion relation is not applicable for this exotic nuclear system. Reaction mechanisms of proton-rich nuclear systems ^{17}F+^{89}Y and ^{58}Ni were also studied at energies around the Coulomb barrier. Continuum discretized coupled-channels calculations indicate that the coupling effects of the continuum states of ^{17}F+^{89}Y is not significant. Thanks to the employment of a powerful ionization-chamber based detector array, the reaction products over a large Z in the systems of ^{17}F+^{58}Ni can be identified clearly. The data analysis is undergoing.

1 Introduction

The availability of relatively intense Radioactive Ion Beams (RIBs) has caused a revival of interest in the nuclear physics. For nuclear reactions at near- and sub-barrer energies, some surprising results in the elastic scattering, direct and fusion reactions induced by RIBs have been reported [1, 2]. These nuclei are often labelled as "exotic" because they have unusual properties, e.g., small separation energies for the valence particles, extensive matter distribution, exhibiting the halo/skin or cluster phenomena and so on. The nuclear interaction is a fundamental ingredient in the study of mechanisms of nuclear reactions. The optical model potential (OMP) is generally adopted to phenomenologically describe the interaction of nuclear collisions. It is well known that the OMPs are closely related to the internal structures of the colliding nuclei, thus the OMPs of exotic nuclear systems are expected to present unusual behaviors. In this paper, we will introduce some new results obtained by the nuclear reaction group at CIAE (China Institute of Atomic Energy) recently, on the properties of the OMPs of neutron-halo ^{6}He systems and the reaction mechanisms induced by the proton-drip line nucleus ^{17}F at energies around the Coulomb barrier.

*Corresponding author: e-mail: cjlin@ciae.ac.cn
**Present address: Center for Nuclear Study, University of Tokyo
***Present address: Michigan State University
****Present address: Joint Institute for Nuclear Research, Dubna

2 OMPs of neutron-halo ^{6}He systems

The OMP can be expressed as $U(r) = V(r) + iW(r)$, where the real part describes the elastic scattering process while the imaginary part represents the absorption of incident flux, i.e., all the nonelastic reaction processes. The properties of the OMPs of tightly bound systems have been recognized with decades of research, e.g., when the interaction energy gets close to the Coulomb barrier, a strong energy dependence is present in both the real and the imaginary parts, which is known as the threshold anomaly [3–5]. This behavior is characterized by a sharp decrease in the imaginary part as the bombarding energy decreases towards the Coulomb barrier, associated with a localized bell-shaped structure around the barrier in the real part. The dispersion relation can be used to connect the real and imaginary parts [6]. The dispersion relation is a natural consequence of the causality principle, describing the effect of dispersion in a medium on the properties of a wave traveling within that medium.

It is well known that the OMPs are closely related to the internal structures of the colliding nuclei, thus particular properties of the potential arising from the exotic nuclear structure can be expected [7, 8]. Conventionally, OMPs are extracted by fitting angular distributions of elastic scattering. At energies close to and below the barrier, however, the angular distributions become insensitive to the nuclear interaction potential. It becomes much more worse for the reactions induced by unstable nuclei, because of limitations of the intensity and/or the phase-space qualities of RIBs. In view of this fact, we proposed a novel method, i.e., the transfer reaction method [9], to study the OMPs of exotic systems by the utilization of a stable beam. The detail of the transfer reaction method can be found in Refs. [10, 11], where the sensitivity test and the comparison with the traditional elastic scattering method were also introduced.

We have applied the transfer reaction method to study the OMPs of neutron-halo ^{6}He systems, by measuring the one-proton transfer reactions induced by ^{7}Li beam accelerated by the HI-13 tandem accelerator at CIAE. Three systems have been measured at energies around the Coulomb barrier, i.e., ^{11}B(^{7}Li, ^{6}He)^{12}C, ^{63}Cu(^{7}Li, ^{6}He)^{64}Zn [12] and ^{208}Pb(^{7}Li, ^{6}He)^{209}Bi [11, 13, 14], to study the OMPs of the light system ^{6}He+^{12}C, the medium-mass system ^{6}He+^{64}Zn and the heavy system ^{6}He+^{209}Bi. Because the lowest terminal high voltage of the tandem accelerator is limited to 3 MV, corresponding to about 12 MeV of ^{7}Li beam energy. Therefore, the deep sub-barrier region can be achieved only for the ^{6}He+^{209}Bi system. The angular distributions of the transfer reaction ^{208}Pb(^{7}Li, ^{6}He)^{209}Bi at the lowest two incident energies (21.20 and 24.30 MeV) are shown in Fig. 1. The corresponding elastic scattering data of ^{7}Li+^{208}Pb at $E_{\rm lab}$(^{7}Li)= 24.30 MeV and ^{6}He+^{209}Bi at $E_{\rm lab}$(^{6}He)= 17.84 MeV [15] are also shown in Fig. 1 (b). The transfer reaction data were fitted by the coupled-reaction-channels (CRC) approach, to extract the OMPs of ^{6}He+^{209}He. The fitting results and that of the optical model calculations with the OMPs extracted for ^{6}He+^{209}Bi are also present in Fig. 1.

The energy dependence of the real and imaginary potentials of ^{6}He+^{209}Bi at the sensitivity radius of 13.5 fm is shown in Fig. 2. As shown in Fig. 2(b), the depth of the imaginary potential decreased as the energy reduced in the deep sub-barrier region, which vanishes at 13.73 ± 1.63 MeV (corresponding to $0.68V_{\rm B}$) according to the extrapolation, where the threshold energy emerges. It is the first time that the reaction threshold is observed for a halo nuclear system. Compared with the threshold values of $0.93V_{\rm B}$ in the tightly bound ^{16}O + ^{208}Pb system [4] and $0.81V_{\rm B}$ in the weakly bound ^{6}Li + ^{144}Sm system [16], such a low threshold energy indicates the feature of the extended matter distribution of the halo nucleus ^{6}He. Furthermore, according to the linear segment fitting of the imaginary potential as shown in Fig. 2(b), the dispersion relation calculation result with the linear schematic model [6] for the real part is shown in Fig. 2(a). It can be seen clearly the prediction of the dispersion relation cannot reproduce the experimental results. The possible reasons for the abnormal behavior are listed as follows [11, 14]: (a) the dispersion relation is deduced from the Kramers-Kronig relation, which is derived from Cauchy residue theorem. One prerequisite of the theorem is that there should be a finite list of isolated singularities, corresponding to the discrete states of the interacting nuclear system [6]. For the weakly bound system, however, the continuum states may play a more important role. Therefore, it is not mathematically rigorous to apply the dispersion relation directly to the exotic nuclear system; (b) the OMPs extracted from experimental data is a equivalent local potential, which will introduce the so-called "spurious" energy dependence. Such a "spurious" dependency is not required to follow the dispersion relation; (c) the derived OMPs are only sensitive to the region located in the external part of the potential. However, the dispersion relation is derived only for a potential that can yield the wave function over all spatial regions. Thus the phenomenological potential, which can only generate the wave function in the external region, does not necessarily need to abide by the causality property [17]. Further investigation is strongly required to discover the underlying physics.

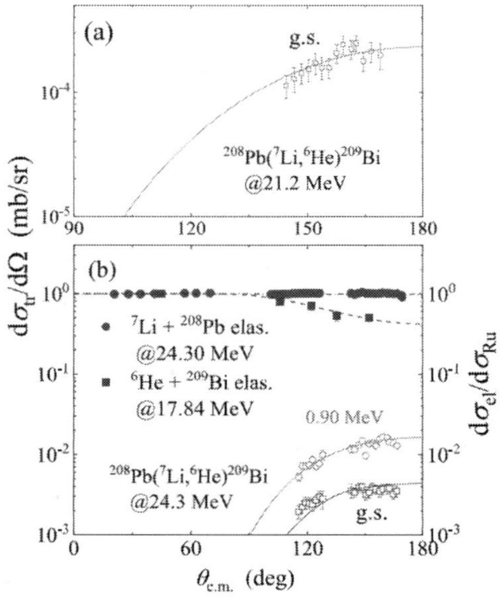

Figure 1. (Color online) Angular distributions of ^{208}Pb(^{7}Li,^{6}He) reactions for proton transferred to different states of ^{209}Bi at $E_{\rm lab}$ = 21.20 (a) and 24.30 (b) MeV, respectively. The experimental results are labelled according to the the excitation energy of ^{209}Bi. The solid curves represent the CRC fitting results. The elastic scattering of the entrance channel ^{7}Li + ^{208}Pb $E_{\rm lab}$(^{7}Li)= 24.30 MeV as well as that of the corresponding exit channel ^{6}He + ^{209}Bi at $E_{\rm lab}$ = 17.84 MeV [15] are also shown by the solid circles and squares, respectively. The dash-dotted curve is the CRC fitting result and the dashed curve represents the calculation result with the OMP extracted from the transfer reaction.

3 Reaction mechanisms of ^{17}F systems

Based on the understanding of the OMPs of weakly bound nuclear systems, we can move a further step to study the reaction mechanisms induced by exotic nuclei. There is already systematic knowledge of the reaction mechanisms of neutron-halo nuclear systems [1], for example, the neutron transfer reaction is dominant for the direct reactions; and for the fusion reaction, the fusion cross section is suppressed in the above barrier region, but enhanced below the barrier. For the proton-rich side, however, the experimental is still scarce. Therefore, we performed a series of experiments to study the reactions induced by the proton-drip line nucleus ^{17}F at energies around the Coulomb bar-

 https://doi.org/10.1051/epjconf/201922301071

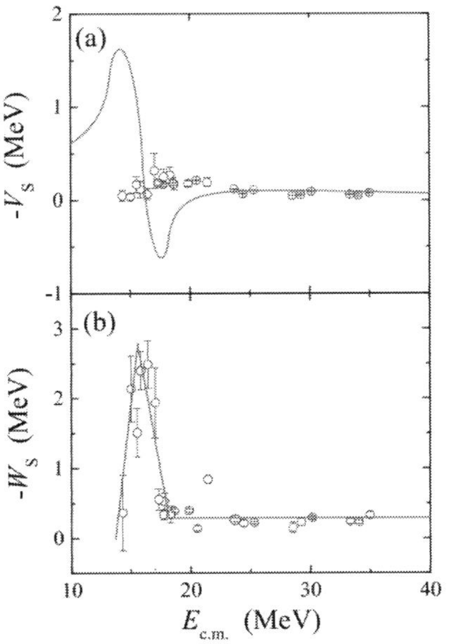

Figure 2. Energy dependence of the depths of real (a) and imaginary (b) potentials at the sensitivity radius of 13.5 fm for the ^{6}He + ^{209}Bi system [11, 13, 14]. Curves in (b) represent the linear segment fitting for the imaginary potential. The predictions of the dispersion relation according to the variations of the imaginary potentials are present in (a).

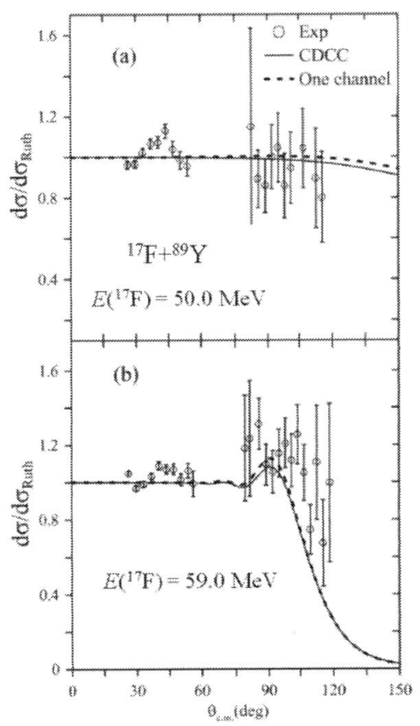

Figure 3. Angular distributions of elastic scattering of ^{17}F on ^{89}Y at $E_{lab}(^{17}F)$ = 50.0 (a) and 59.0 (b) MeV [18]. The solid and dashed curves represent the CDCC and one-channel calculations.

rier, such as ^{17}F+^{89}Y [18], ^{17}F+^{208}Pb [19] at the Radioactive Ion Beam Line at the Heavy Ion Research Facility in Lanzhou (HIRFL-RIBLL1) [20] and ^{17}F+^{58}Ni [21] at the low energy radioactive isotope beam separator CRIB (Center for Nuclear Study Radioactive Ion Beam separator) of University of Tokyo [22]. The valence proton of ^{17}F is bound only by 0.6 MeV. The root mean square (rms) radius of the ground state is 3.7 fm [23], which is significantly larger than that of the ^{16}O core. The first excited state of ^{17}F, E_x = 0.495 MeV and $J^{\pi} = 1/2^{+}$, is the only bound state below the breakup threshold, and reported to have an extended rms radius, 5.3 fm [23], exhibiting a proton-halo structure [24].

The elastic scattering of ^{17}F+^{89}Y was measured at $E_{lab}(^{17}F)$ = 50.0 and 59.0 MeV [18]. Continuum discretized coupled-channels (CDCC) and one-channel calculations were performed to understand breakup coupling effects on the elastic scattering. The experimental and theoretical results are shown in Fig.3. According to the comparison, both the CDCC and one-channel calculations can reproduce the elastic scattering data reasonably well, indicating that the effect of the coupling to continuum states (both the bound-continuum states and the continuum-continuum state couplings) is not significant.

For the measurements of ^{17}F+^{58}Ni and ^{208}Pb, an innovatively designed Multilayer Ionization-chamber (IC) Telescope Array (MITA) [21] was employed to detect the reaction products over a large range of Z. Thanks to the powerful array, both the light reaction products like p and α, and the heavy ions like ^{16}O and ^{17}F, can be clearly distinguished. The angular distributions of elastic scattering of ^{17}F+^{58}Ni are shown in Fig. 4, where the CDCC calculation results are also presented. The data analysis of these two systems is still in progress.

4 Summary

With the application of the transfer reaction method, the OMPs of neutron-halo ^{6}He+^{209}Bi system were extracted exactly. In the sub-barrier energy region, the depth of the imaginary potential increases first with energy decreasing below the barrier and then falls quickly down to 0. According to the extrapolation, the reaction threshold energy is determined for the first time for a halo nuclear system. Moreover, the dispersion relation cannot describe the connection between the real and imaginary potentials. Some possible reasons are discussed for such a peculiar behavior, but further study is still desired for the underlying physics.

We also performed experiments to study the reactions induced by the proton-rich nucleus ^{17}F. Three systems, i.e., ^{17}F+^{58}Ni, ^{89}Y and ^{208}Pb, were measured at energies around the Coulomb barrier. For the ^{17}F+^{89}Y system, only the elastic scattering data were derived due to the limited particle-identification capability of the silicon detector telescope. According to the CDCC and one-channel calculations, the breakup coupling effects on the elastic scattering of ^{17}F+^{89}Y is insignificant. For the measurement of

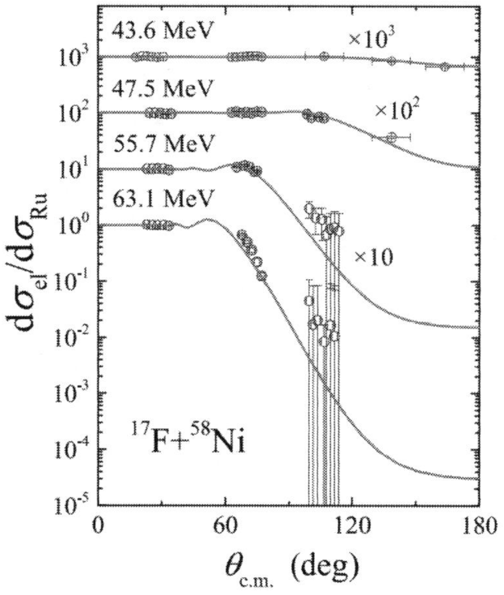

Figure 4. Angular distributions of elastic scattering of ^{17}F on ^{58}Ni at $E_{lab}(^{17}F)$ = 43.6, 47.5, 55.7 and 63.1 MeV. The solid curves represent the CDCC calculations.

^{17}F+^{89}Y and ^{208}Pb, a ionization chamber based detector array was employed, to identify the reaction products over a large *Z* region. The data analysis is still undergoing.

Acknowledgements

We acknowledge the related physicists and operations staff of the HI-13 tandem accelerator, the HIRFL-RIBLL1, and the CNS-CIRB for providing good quality beams and performing the experiments. This work is supported by the National Key R&D Program of China (Contract No. 2018YFA0404404), the National Natural Science Foundation of China (Grants Nos.11635015, U1732145, 11705285, U1867212, 11805280, and 11811530071), and the Project funded by China Postdoctoral Science Foundation (Grant No. 2017M621035).

References

[1] L. F. Canto, P. R. S. Gomes, R. Donangelo *et al.*, Phys. Rep. **596**, 1 (2015).

[2] J. J. Kolata, V. Guimarães, and E. F. Aguilera, Eur. Phys. J. A **52**, 123 (2016).

[3] B. R. Fulton, D. W. Banes, J. S. Lilley *et al.*, Phys. Lett. B **162**, 55 (1985).

[4] M. A. Nagarajan, C. C. Mahaux, G. R. Satchler, Phys. Rev. Lett. **54**, 1136 (1985).

[5] C. J. Lin, J. C. Xu, H. Q. Zhang *et al.*, Phys. Rev. C **63**, 064606 (2001).

[6] C. Mahaux, H. Ngô, G. R. Satchler, Nucl. Phys. A **449**, 354 (1986).

[7] A. M. Sánchez-Benítez, D. Escrig, M. A. G. Álvarez *et al.*, Nucl. Phys. A **803**, 30 (2008).

[8] M. S. Hussein, P. R. S. Gomes, J. Lubian and L. C. Chamon, Phys. Rev. C **73**, 044610 (2006).

[9] C. J. Lin, F. Yang, P. Zhou *et al.*, AIP Conf. Proc. **853**, 81 (2006).

[10] Z. D. Wu, L. Yang, C. J. Lin *et al.*, Chin. Phys. Lett. **31**, 092401 (2014).

[11] L. Yang, C. J. Lin, H. M. Jia *et al.*, Phys. Rev. C **96**, 044615 (2017).

[12] L. Yang, C. J. Lin, H. M. Jia *et al.*, Phys. Rev. C **95**, 034616 (2017).

[13] L. Yang, C. J. Lin, H. M. Jia *et al.*, Phys. Rev. C **89**, 044615 (2014).

[14] L. Yang, C. J. Lin, H. M. Jia *et al.*, Phys. Rev. Lett. **119**, 042503 (2017).

[15] E. F. Aguilera, J. J. Kolata, F. D. Becchetti *et al.*, Phys. Rev. C **63**, 061603(R) (2001).

[16] J. M. Figueira, J. O. Fernández Niello, A. Arazi *et al.*, Phys. Rev. C **81**, 024613 (2010).

[17] R. Lipperheide and A. K. Schmidt, Nucl. Phys. A **112**, 65 (1968).

[18] G. L. Zhang, G. X. Zhang, C. J. Lin *et al.*, Phys. Rev. C **97**, 044618 (2018).

[19] Nan-Ru Ma, Cheng-Jian Lin, Jian-Song Wang *et al.*, Chin. Phys. C **40**, 116004 (2016).

[20] Z. Sun *et al.*, Nucl. Instrum. Methods Phys. Res. A **503**, 496 (2003).

[21] N. R. Ma, L. Yang, C. J. Lin *et al.*, Eur. Phys. J. A **55**, 87 (2019).

[22] S. Kubono *et al.*, Eur. Phys. J. A **13**, 217 (2002).

[23] R. Morlock, R. Kunz, A. Mayer *et al.*, Phys. Rev. Lett. **79**, 3837 (1997).

[24] R. Lewis and A. C. Hayes, Phys. Rev. Lett. **59**, 1211 (1999).

Study of neutron-deficient mercury isotopes

Preliminary results on ^{189}Hg

Irene Zanon[1,2,*], *Marco* Siciliano[1,3], *Alain* Goasduff[4,5], and *Philipp R.* John[6]

[1]INFN, Laboratori Nazionali di Legnaro, Legnaro, Italy.
[2]Dipartimento di Fisica e Scienze della Terra, Ferrara, Italy.
[3]Irfu/CEA, Université de Paris-Saclay, Gif-sur-Yvette, France.
[4]Dipartimento di Fisica e Astronomia, Università di Padova, Padova, Italy.
[5]INFN, Sezione di Padova, Padova, Italy.
[6]Institut für Kernphysik der Technischen Universität Darmstadt, Darmstadt, Germany.

Abstract. The mercury isotopic chain is depicted as one of the best fields to observe many phenomena related to collectivity evolution as, for instance, shape transition, shape coexistence or shape staggering. In this context, the ^{189}Hg presents several interesting aspects and is still relatively unexplored. The nucleus has been studied at the Laboratori Nazionali di Legnaro using a fusion-evaporation reaction and the γ rays emitted have been detected by the GALILEO array, coupled with Neutron Wall and the GALILEO plunger. The presorting and the preliminary results are presented.

1 Introduction

The mercury isotopic chain has been and is still an important region of the nuclear chart for the understanding of the evolution of nuclear structure through spectroscopic studies. Being just two protons away of the Z=82 shell closure, it is a perfect playground for testing the effective nuclear force near the magic numbers and it presents many phenomena such as shape transition, shape coexistence and shape staggering [1]. The latter one was observed for the first time via optical pumping experiments [2], by polarizing the spins of the nuclei and then observing its polarization via the angular asymmetry of the β decay: an impressive isotope shift between $A = 185$ and $A = 187$ was observed, later understood thanks to the evidences of a rotational band built on an excited 0^+ state given by in-beam γ-ray spectroscopy [3]. Recently, a study on the characterization of the shape-staggering effect in the mercury nuclei has confirmed the measurement of Bonn et al. and further extended the systematic [4].

In addition, the neutron-deficient mercury isotopes are known to present two structures close in energy, one oblate and one prolate deformed, that compete to form the ground-state of the nucleus when approaching to the neutron mid-shell [1]. In this region, theoretical calculations predict the ground states of the Hg isotopes to be weakly oblate deformed, while the intruder band, that reaches its minimum for the ^{182}Hg nucleus, is expected to be prolate deformed. Then, for lighter isotopes, the excitation energy of the minimum increases and the shape mixing becomes less significant. At last, the ground state is expected to reach a more spherical shape in ^{176}Hg [5]. The presence of a region where these two shapes coexist is established and it is predicted to go from ^{180}Hg to ^{188}Hg [6].

While for the even mercury isotopes a wealth of studies have been published, both theoretical and experimental, for the odd isotopes the literature is scarcer. From Strutinsky shell correction calculations [7], the shape of the ^{181,183,185}Hg nuclei is predicted to be prolate deformed, while for the heavier isotopes ^{187,189}Hg is expected to become oblate deformed. These results seems to be confirmed also by more recent calculations with Hartree–Fock–Bogoliubov method employing the Skyrme SIII, SkI3 and SLy4 forces [5].

Among the mercury isotopes, the ^{189}Hg nucleus presents some interesting aspects. A $13/2^-$ isomeric states at 140 keV with a lifetime of 8.7(2) min [8] is presented as the head of a high spin band. From spectroscopic and γ-ray directional correlations (DCO) studies, the level scheme of the ^{189}Hg was established and its structure was interpreted as a ^{190}Hg core coupled to a $\nu i_{13/2}$ quasi-particle hole [9]. In addition, since the work of [11], the presence of a superdeformed band in ^{189}Hg has been demonstrated. This is in agreement with the systematics of the Hg region and the A=190 and with the theoretical predictions [10]. Several calculations [12, 13] have been performed to identify the spin of the state that forms the band head of the superdeformed band, leading to mismatching results. A new experiment was performed in order to obtain direct measurements of the lifetimes of the excited states in ^{189}Hg.

*e-mail: irene.zanon@lnl.infn.it

2 Experimental details

The experiment was performed at the Laboratori Nazionali di Legnaro and the beam was provided by the Tandem-ALPI accelerator complex. The γ rays produced in the reactions were studied using GALILEO [14], an array of 25 high-purity germanium detectors (HPGe), of which 15 were placed at backward angles with respect to the beam direction, arranged in 3 different rings (152°, 129° and 119°), and 10 in a ring at 90°. Every HPGe detector was surrounded by a Bismuth Germanate (BGO) crystal that acted as an anti-Compton shield. A thorough study of the effect of the Compton suppression and on the efficiency of GALILEO can be found in Ref. [15]. The channel of interest, expected in coincidence with neutrons, was selected by using Neutron Wall [16], an array of 45 liquid scintillators used for the detection of neutrons. In fact, thanks to the characteristic of the detectors, it was possible to perform a neutron-gamma discrimination to reject the events due to the Coulomb excitation and obtain a cleaner spectrum. More details about the selection of neutrons can be found in Ref. [17, 18]. Finally, the lifetime measurement were performed via Recoil Distance Doppler-Shift (RDDS) technique using the plunger device [19]. The GALILEO plunger is designed for lifetime measurement of the range of 1 ps to 1 ns, allowing the variation of target-stopper distances from few micrometers to millimeters with a sub-micrometer precision.

The main goal of the experiment was the study of the ^{188}Hg nucleus and the channel of interest was populated via two different fusion-evaporation reactions:

- ^{34}S beam at the energy of 185 MeV impinged onto 600 $\mu g/cm^2$ of ^{160}Gd target (2.5 mg/cm^2 thick ^{181}Ta fronting);
- ^{34}S beam at the energy of 165 MeV impinged onto 600 $\mu g/cm^2$ of ^{158}Gd target (2.5 mg/cm^2 thick ^{181}Ta fronting).

Such reactions were chosen in order to maximise the population probability of the ^{188}Hg, according to previous measurements [11, 20, 21] and to PACE4 estimations [22]. Furthermore, with the first reaction, the ^{189}Hg channel was also populated with a significant intensity, making a study of this nucleus feasible.

In order to study ^{189}Hg, the average velocity of the evaporated residue (ER) was extracted from the Doppler shift of the $17/2^+ \rightarrow 13/2^+$ transition at the different angles of the GALILEO rings, as shown in Figure 1. For a more precise estimation of the fragment velocity, a kinematic reconstruction would be necessary [23]. However, considering the large mass of the compound nucleus with respect to the evaporated particles, the difference between the direction of the beam and of the flying ER is negligible, so the average velocity can be considered a good approximation. The ratio between the shifted (in-flight) and unshifted (stopped) component E_γ/E_0 is plotted as a function of the ring angle θ and the data are fitted via the following equation:

$$\frac{E_\gamma}{E_0} = \frac{\sqrt{1-\beta^2}}{1-\beta\ \cos(\theta)} \tag{1}$$

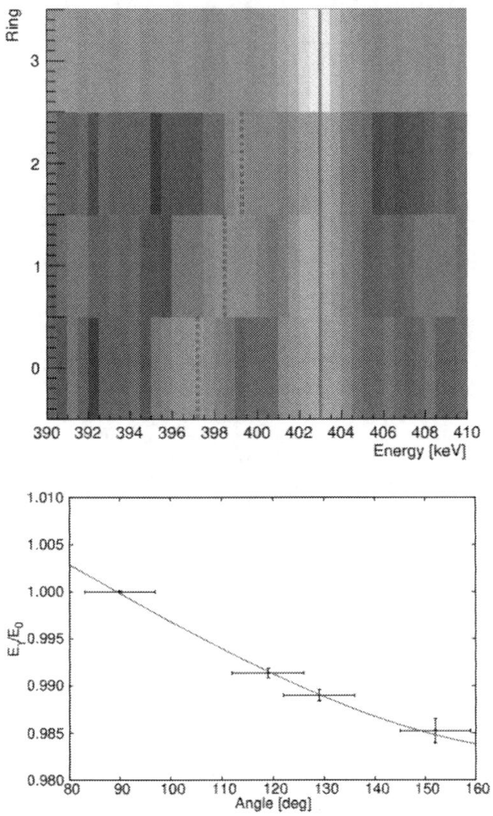

Figure 1. (top) In-flight (dashed) and stopped (solid) components of the $17/2^+ \rightarrow 13/2^+$ transition at 403 keV for the four different rings of the GALILEO array (152°, 129°, 119° and 90° respectively) at the plunger distance of 400 μm. No Doppler shift is observed at 90°. (bottom) The ration E_γ/E_0 as a function of the ring angle, fitted with Eq. 1 with β as a free parameter. The velocity resulted to be 1.7(6)%.

where β, the velocity, is a free parameter and resulted to be 1.7(6) %.

3 Preliminary results

A γ-ray energy spectrum of the ^{189}Hg nucleus was obtained by requiring the coincidence with at least one neutron, gating on the $17/2^+ \rightarrow 13/2^+$ transition at 403 keV and subtracting the background. To get rid of the in-flight and stopped components of the peaks due to the RDDS technique, only the detectors at 90° were used. The spectrum is shown in Figure 2 (left) and the identified transitions are highlighted. A partial level scheme, reconstructed with the identified γ-ray transitions, is also shown in Figure 2 (right).

A lifetime measurement of the $17/2^+$ was also attempted via RDDS technique. Six target-stopper distances were investigated. A gate on the in-flight component of the feeding transition $21/2^+ \rightarrow 17/2^+$ at 626 keV and a background subtraction was performed in order to remove any

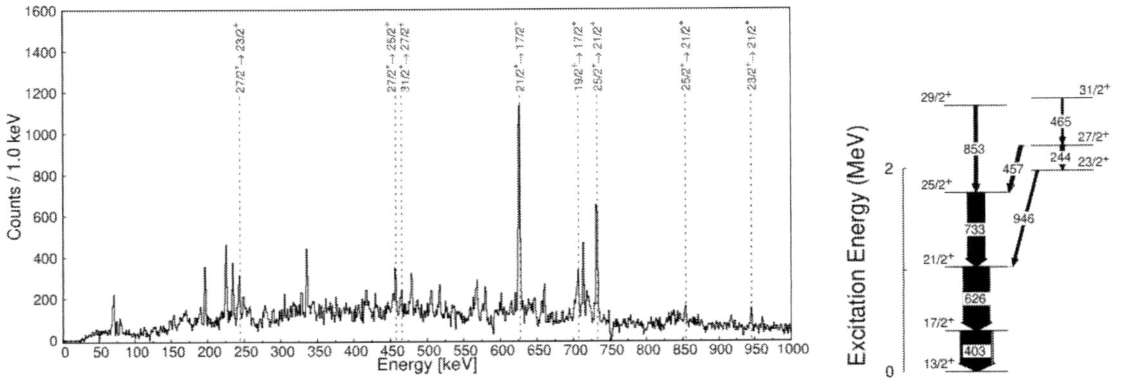

Figure 2. (left) Background-subtracted γ-ray energy spectrum of the ^{189}Hg obtained by requiring the coincidence with at least one neutron and by gating on the $17/2^+ \rightarrow 13/2^+$ transition at 403 keV. The identified transitions are highlighted in blue. (right)Partial level scheme reconstructed with the identified transitions. The efficiency-corrected intensity of the transitions is represented by the thickness of the arrows.

possible contamination from side feeders. The results for four target-stopper distances are shown in Figure 3 for detectors at 152°. The in-flight and stopped components are clearly separated and marked with a dashed and solid line respectively. The integral of the two component depends on the velocity of the nucleus, on the distance between the target and the plunger stopper and on the lifetime, that can be extracted by using the integration over time of the Bateman equation. A decay curve can be obtained and the lifetime of the state extracted.

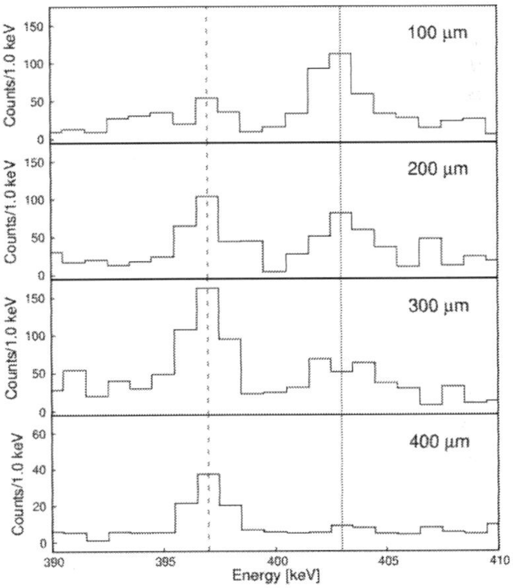

Figure 3. Background-subtracted γ-ray energy spectra of ^{189}Hg, obtained by requiring the detection of at least one neutron in Neutron Wall and the coincidence with the in-flight component of the $21/2^+ \rightarrow 17/2^+$ transition at 626 keV. The spectra are shown for 4 different target-to-stopper distances and for the GALILEO detectors at 152°. The stopped (solid line) and in-flight (dashed line) of the $17/2^+ \rightarrow 13/2^+$ transition are marked.

4 Conclusions

The ^{189}Hg was produced via a fusion-evaporation reaction at the Laboratori Nazionali di Legnaro, employing the γ spectrometer GALILEO coupled with the Neutron Wall array and the GALILEO plunger. Many transitions have been identified and a preliminary level scheme has been reconstructed. Moreover, a first attempt to measure the previously lifetime of the $17/2^+$ has been made: the in-flight and stopped component are clearly distinguishable and a decay curve can be obtained. The higher lying states are still under investigation.

References

[1] K. Heyde and J. L. Wood, Rev. Mod. Phys. **83**, 1467 (2011).

[2] J. Bonn, G. Hubert, H.J. Kluge, L. Kluger and E.W. Otten, Phys. Lett. **28B** 5, 308 (1972).

[3] N. Rud, D. Ward, H.R. Andrews, R.L. Graham and J.S. Geiger, Phys. Rev. Lett. **31** 23, 1421 (1973).

[4] B.A. Marsh, T. Day Goodacre, S. Sels, Y. Tsunoda et al., Nature **14**, 1163 (2018).

[5] S. Sakakihara and Y. Tanaka, Nuc. Phys. A **729**, 37 (2003).

[6] T. Nikšić, D. Vretenar, P. Ring and G.A. Lalazissis, Phys. Rev. C **65**, 054320 (2002).

[7] S. Fraunedorf and V.V. Pashkevic, Phys. Lett. **55B** 4, 365 (1974).

[8] V. Berg, R. Foucher and A. Höglund, Nucl. Phys. A **244**, 462 (1975).

[9] C. Bourgeois, A. Zerrouki, M.G. Desthuilliers-Porquet, F.A. Beck and T. Byrski, Z. Phys. A **299**, 85

(1981).
[10] R.V.F. Janssens et al., Nucl. Phys. A **520**, 75c (1990).
[11] I.G. Bearden, R.V.F. Jannsens et al., Nucl. Phys. A **576**, 441 (1994).
[12] R. Piepenbring and K.V. Protasov, Z. Phys. A **345**, 7 (1993).
[13] Shan-Gui Zhou and Chunkai Zheng, Phys. Rev. C **55** 5, 2324 (1997).
[14] J. J. Valiente-Dobón, D. Mengoni, F. Recchia et al., LNL-INFN Annual Report 2014, 95.
[15] I. Zanon, M. Siciliano, P.R. John and A. Goasduff, Nuovo Cimento C (in press).
[16] Ö. Skeppstedt et al., Nucl. Inst. Meth. A **421**, 3, 531 (1999).
[17] I. Zanon, M. Siciliano, P.R. John, A. Goasduff et al., LNL-INFN Annual Report 2017, 62.
[18] I. Zanon, Master Thesis, Università di Padova, 16/07/2018.
[19] C. Müller-Gatermann, F. von Spee, A. Goasduff et al., Nucl. Instr. Meth. A **920**, 95 (2019).
[20] A.N. Wilson, J. Timar et al., Phys. Lett. B **505**, 6-14 (2001)
[21] H. Hübel, A.P. Byrne, S. Ogaza, A.E. Stuchbery and G.D. Dracoulis, Nucl. Phys. A **453**, 316-348 (1986).
[22] O.B. Tarasov and D. Bazin, Nucl. Inst. Meth. B, **204**, 174 (2003).
[23] P.R. John, A. Goasduff, G. Jaworski, D. Mengoni, LNL-Annual Report 2015, 54.

AUTHOR INDEX

AUTHOR INDEX

AUTHOR INDEX

AUTHOR INDEX

AUTHOR INDEX

AUTHOR INDEX

AUTHOR INDEX